Thermoelectric Skutterudites

Thermoelectric Skutterudites

Ctirad Uher

CRC Press
Taylor & Francis Group
Boca Raton London New York

CRC Press is an imprint of the
Taylor & Francis Group, an **informa** business

First edition published 2021
by CRC Press
6000 Broken Sound Parkway NW, Suite 300, Boca Raton, FL 33487-2742

and by CRC Press
2 Park Square, Milton Park, Abingdon, Oxon, OX14 4RN

Library of Congress Cataloging-in-Publication Data

Names: Uher, Ctirad, author.
Title: Thermoelectric Skutterudite / Ctirad Uher.
Description: Boca Raton : CRC Press, 2021. | Includes bibliographical references and index.
Identifiers: LCCN 2020050057 | ISBN 9780367610791 (hardback) | ISBN 9781003105411 (ebook)
Subjects: LCSH: Skutterudite. | Thermoelectric materials. | Materials--Thermal properties.
Classification: LCC TK2950 .U34 2021 | DDC 620.1/88--dc23
LC record available at https://lccn.loc.gov/2020050057

ISBN: 978-0-367-61079-1 (hbk)
ISBN: 978-0-367-61537-6 (pbk)
ISBN: 978-1-003-10541-1 (ebk)

Typeset in Times
by Deanta Global Publishing Services, Chennai, India

Access the Support Material]: www.routledge.com/9780367610791

Dedicated to the memory of my wife Libby.

Contents

Preface

It is now 20 years since I wrote a major review article (Uher 2001) on the thermoelectric properties of skutterudites, which assessed the understanding and development of these fascinating materials when they started to attract the attention of the thermoelectric community as a class of novel, promising thermoelectric materials. Much has happened since then in the field of thermoelectricity in particular, and a considerably deeper insight has been gained into all aspects of skutterudites in general. Benefiting from new sensitive experimental tools supported by powerful theoretical calculations, modeling and simulations, the properties of skutterudites have been dramatically enhanced to a degree that efficient skutterudite-based thermoelectric modules became feasible and are being developed and tested in the laboratories worldwide. While several competing thermoelectric materials have also been identified and developed in the past dozen or so years, skutterudites have maintained their premier position and appeal as environmentally friendly, relatively inexpensive and operationally stable materials suitable for mid-temperature power generation applications, either as stand-alone couples or as segmented legs in combination with lower or higher temperature thermoelectric materials to maximize the performance. Equally exciting advances have been made regarding the superconducting and magnetic properties of skutterudites, highlighted by the discovery of $PrOs_4Sb_{12}$ as the preeminent heavy fermion superconductor. In addition, many other skutterudites show surprisingly high superconducting transition temperatures. Moreover, skutterudites have been a platform for explorations of a variety of magnetic phases that develop in the structure depending on the type of filler and its interaction with the neighboring ions of the framework. The literature on skutterudites that accumulated over the past two decades is enormous, and it is time to assess the findings and impact in a comprehensive way that would serve a broad community of readers interested in the properties of skutterudites.

Two of the first questions asked regarding skutterudites are where the strange word comes from and what it stands for. The word was first used by Wilhelm Karl von Haidinger in 1845 to classify a mineral of composition $(Co,Ni,Fe)As_3$ mined at a small town Skutterud, near Basterud in Norway. Previously, the mineral was referred to by Georgius Agricola (1529) as cobaltum eineraceum, and in 1852, Gustav Rose attempted to call it quite specifically as Arsenikkobalt. The first note on some physical properties of the mineral (hardness and density) was provided by August Breithaupt (1827), who described the mineral as 'New Kies-species from Skutterud'. Other synonyms of skutterudite minerals are modumite and smaltite. The $CoSb_3$ form of skutterudite is also known as kieftite, while the Ni-rich skutterudite mineral has previously been called chloanthite.

Skutterudites are accessory minerals of hydrothermal origin and are found in all corners of the world in moderate to high temperature veins together with other Ni–Co minerals. They are mined as an ore of Co and Ni with a byproduct of As. Mineral skutterudites have a silver-gray metallic luster, the fracture is classified as conchoidal to uneven, and the hardness is in the range of 5.5–6. The International Mineralogical Association (IMA) currently classifies mineral skutterudites as consisting of four members: skutterudites $(CoAs_{3-x})$, nickelskutterudites $(NiAs_{3-x})$, ferroskutterudites $((Fe,Co)As_3)$, and kieftite $(CoSb_3)$. However, as no pure $NiAs_3$ (mineral or synthetic) has ever been reported, and all Ni-containing mineral skutterudites invariably contain substantial amounts of Co and Fe, nickelskutterudites should properly be classified as $(Ni,Co,Fe)As_3$, as recently noted by Schumer et al. (2017). For the benefit of mineral collectors, I note that some of the best formed crystal specimens of skutterudite come from the Bousman Mine in the Bou Azer mining district of central Morocco.

The key milestone in the history of skutterudites was made in 1928 by Ivar Oftedal (1928), who identified the crystallographic structure of synthetic skutterudites as a body-centered cubic lattice in the space group $Im\bar{3}$. The early interest in transport properties of skutterudites dates back to the mid-1950s, when researchers in the then Soviet Union screened a vast number of compounds and alloys as they searched for the most promising thermoelectrics. It was L. D. Dudkin and N. K. Abrikosov (1956, 1957, 1958, and 1959) who observed that $CoSb_3$ possessed exceptionally high room temperature carrier mobilities and modest Seebeck coefficients. Unfortunately, they also noted a very high thermal conductivity of their skutterudite samples, a definitive impediment for a good thermoelectric, and soon dropped their interest in the material.

The next important step in the development of skutterudites was the work of Wolfgang Jeitschko (1977), who realized that the open structure of skutterudites, typified by the presence of two large voids in the unit cell, might be able to accept foreign ions that would effectively 'fill' the structure. Experiments with the lanthanoid group of elements verified the idea, and his group was able to synthesize a large number of filled skutterudites.

The early boom in thermoelectric research during the 1960s was precipitated by the discovery of Bi_2Te_3 in 1954 as a premier room temperature material by Julian Goldsmid (1954), the realization of the importance of forming solid solutions to lower the thermal conductivity by Abram Ioffe (1956), and the development of PbTe as an excellent mid-temperature power generating material. After these early successes, little progress was made in the field of thermoelectricity during the 1970s and 1980s. A welcomed rejuvenation of interest in thermoelectric materials started in the mid-1990s. There were two distinct reasons for this rebirth: (1) societal pressures to provide a cleaner environment and to address the looming energy crisis, and (2) scientific advances. In particular, Millie Dresselhaus (1993a,b) made a realization that lower dimensional structures offer distinct advantages for thermoelectricity. Furthermore, Glenn Slack (1995) promulgated a new paradigm that a good thermoelectric material ought to conduct heat as poorly as an amorphous material while its electronic properties should reflect that of a high-quality crystal, the so-called phonon-glass-electron-crystal (PGEC) concept.

The findings by Slack and Tsoukala (1994) that filled skutterudites, via a highly disruptive influence of the loosely bonded filler species on the thermal conductivity yet without much detriment to the electronic properties, could be promising thermoelectrics, stimulated much interest in their study. Within a few months, Morelli and Meisner (1995) demonstrated a dramatic suppression of the lattice thermal conductivity in $CeFe_4Sb_{12}$, one such filled skutterudite. From then on, the race to develop skutterudites as an outstanding thermoelectric material had commenced, and intensive worldwide efforts continue unabated. Since the early unimpressive values of the dimensionless thermoelectric figure of merit ZT measured on pure $CoSb_3$ to the current values approaching $ZT \sim 2$ of finely tuned filled skutterudites, see Figure 0.1, the progress has been spectacular and attests

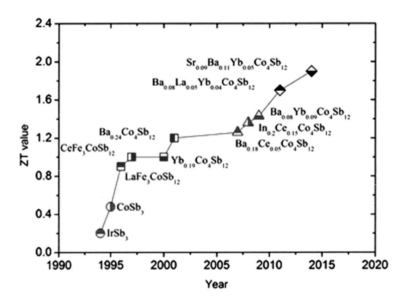

FIGURE 0.1 Chronology of the development of thermoelectric skutterudites. ZT values have progressed from binary structures (circles) to single-filled skutterudites (squares) to double-filled skutterudites (triangles) and to triple-filled skutterudites (diamonds).

to the interest and skills of scientists able to modify and engineer the structure to yield superior thermoelectric performance. Although there are many different skutterudites, from the perspective of thermoelectricity, the preponderance of bulk of studies has focused on the properties of $CoSb_3$-based structures.

This particular skutterudite compound consists of readily available chemical constituents at a moderate cost, the chemical species do not raise environmental concerns and, most importantly, $CoSb_3$-based skutterudites possess, by far, the best thermoelectric properties of all skutterudites. Thus, most of our discussion regarding thermoelectric properties of skutterudites will center on $CoSb_3$-based compounds, although other skutterudites will be mentioned when they offer some special advantages, such as the higher operational temperatures of Rh- and Ir-based skutterudites, or the exceptionally low thermal conductivities achieved with ternary skutterudites. Various types of the skutterudite structure are described in Chapter 1. Different synthesis routes of skutterudites, including their thin film forms, are presented in Chapter 2. Band structures of skutterudites are discussed in Chapter 3. Electronic transport properties of various families of skutterudites are covered in Chapter 4, followed with a description of phonon transport properties in Chapter 5. Chapter 6 is dedicated to the thermoelectric performance of skutterudites.

Skutterudites encompass a vast range of interesting physical properties. While binary skutterudites are diamagnetic semiconductors, filling their structural voids dramatically alters the band structure and gives rise to a plethora of fascinating ground states, most notably superconductivity. Since the discovery of superconductivity in $LaFe_4P_{12}$ in the early 1980s, Meisner (1981), skutterudites have been thoroughly studied for their unique superconducting properties. Focusing initially on La-filled phosphides, Shirotani et al. (1997) also observed superconductivity in some arsenide and antimonide skutterudites. Among them, a particularly surprising discovery was the superconducting state in $PrRu_4As_{12}$, where Pr is a magnetic rare-earth element. The intriguing role of Pr ions was subsequently vividly documented in the superconducting state of $PrOs_4Sb_{12}$, the first heavy fermion superconductor that is not a Ce- or U-based compound, Bauer et al. (2002), Maple et al. (2002). As more studies of $PrOs_4Sb_{12}$ followed, particularly those aided by a magnetic field, the unique crossover from the superconducting state to the high field-induced ordered phase (HFOP) emerged, Aoki et al. (2002), Ho et al. (2002), Koghi et al. (2003). The discovery of the HFOP opened the floodgate of experimental and theoretical studies hoping to shed more light on this mysterious HFOP. The outcome of the efforts is a realization that $PrOs_4Sb_{12}$ is a distinct chiral heavy fermion superconductor. Subsequent explorations of the superconducting properties of skutterudites with the $[Pt_4Ge_{12}]$ framework, Bauer et al. (2007), and particularly that of $PrPt_4Ge_{12}$, Gumeniuk et al. (2008a), revealed similarities with $PrOs_4Sb_{12}$ as far as breaking the time-reversal symmetry is concerned, but also major differences in that the $[Pt_4Ge_{12}]$-based skutterudites have charge carrier masses only mildly enhanced and nowhere near the status of heavy fermions..

The present monograph is aimed at providing a reference source on the properties of thermoelectric skutterudites, giving in-depth accounts of the relevant physical, chemical and materials issues that have captured the interest of scientists in many different branches of solid state physics. I trust it will serve well to anyone interested in the further development and use of skutterudites across the broad field of condensed matter.

It is my pleasure and honor to acknowledge my thesis advisor, Prof. Julian Goldsmid, who, many years ago, introduced me to the fascinating world of solid-state physics, and thermoelectricity in particular. Throughout my academic career, I have been blessed with many smart and talented graduate students, who are now making important scientific discoveries and have attained leadership positions in the field, e.g., Prof. Donald Morelli and Prof. Jihui Yang. I have much enjoyed wonderful collaborations with many scientists worldwide, among them with Prof. Mercouri Kanatzidis, Prof. Pierre Ferdinand Poudeu, Prof. Xinfeng Tang, Prof. Lidong Chen, and Prof. Petr Lošťák. I have learned much from them and I am grateful for their friendship. I also wish to acknowledge funding support over the past 40 years received from the US Federal Agencies, and notably from the US Department of Energy. Finally, I want to thank my most recent graduate student, now Dr. Trevor Bailey, for his careful proofreading of the manuscript and many useful suggestions to improve it.

REFERENCES

Aoki, Y., T. Namiki, S. Ohsaki, S. R. Saha, H. Sugawara, and H. Sato, *J. Phys. Soc. Jpn.* **71**, 2098 (2002).

Bauer, E., N. A. Frederick, P.-C. Ho, V. S. Zapf, and M. B. Maple, *Phys. Rev. B* **65**, 100506(R) (2002).

Bauer, E., A. Grytsiv, X.-Q. Chen, N. Melnychenko-Koblyuk, G. Hilscher, H. Kaldarar, H. Michor, E. Royanian, G. Giester, M. Rotter, R. Podloucky, and P. Rogl, *Phys. Rev. B* **99**, 217001 (2007).

Breithaupt, A., *Annalen der Physik* **85**, 115 (1827).

Dudkin, L. D. and N. Kh. Abrikosov, *Zh. Neorg. Khim.* **1**, 2096 (1956).

Dudkin, L. D. and N. Kh. Abrikosov, *Zh. Neorg. Khim.* **2**, 212 (1957).

Dudkin, L. D., *Sov. Phys.-Tech. Phys.* **3**, 216 (1958).

Dudkin, L. D. and N. Kh. Abrikosov, *Sov. Phys.-Solid State* **1**, 126 (1959).

Goldsmid, H. J., and R. W. Douglas, *Br. J. Appl. Phys.* **5**, 386 (1954).

Gumeniuk, R., W. Schnelle, H. Rosner, M. Nicklas, A. Leithe-Jasper, and Y. Grin, *Phys. Rev. Lett.* **100**, 017002 (2008a).

Hicks, L. D. and M. S. Dresselhaus, *Phys. Rev. B* **47**, 12727 (1993a).

Hicks, L. D. and M. S. Dresselhaus, *Phys. Rev. B* **47**, 16631 (1993b).

Ho, P.-C., V. S. Zapf, E. D. Bauer, N. A. Frederick, M. B. Maple, G. Giester, P. Rogl, S. T. Berger, C. H. Paul, and E. Bauer, *Int. J. Mod. Phys. B* **16**, 3008 (2002).

Ioffe, A. F., A. V. Airapetyants, A. V. Ioffe, N. V. Kolomoets, and L. S. Stilbans, *Dokl. Akad. Nauk SSSR* **106**, 981 (1956).

Jeitschko, W. and D. J. Brown, *Acta Crystallog.* **B33**, 3401 (1977).

Kohgi, M., K. Iwasa, M. Nakajima, N. Metoki, S. Araki, N. Bernhoeft, J.M. Mignot, A. Gukasov, H. Sato, Y. Aoki, and H. Sugawara, *J. Phys. Soc. Jpn.* **72**, 1002 (2003).

Meisner, G. P., *Physica B&C* **108**, 763 (1981).

Morelli, D. T. and G. P. Meisner, *J. Appl. Phys.* **77**, 3777 (1995).

Oftedal, I., *Z. Kristallogr.* **A66**, 517 (1928).

Schumer, B. N., M. B. Andrade, S. H. Evans, and R. T. Downs, *Am. Mineralogist* **102**, 205 (2017).

Shirotani, I., T. Uchiumi, K. Ohno, C. Sekine, Y. Nakazawa, K. Kanoda, S. Todo, and T. Yagi, *Phys. Rev. B* **56**, 7866 (1997).

Slack, G. A., in *CRC Handbook of Thermoelectrics*, ed. D. M. Rowe, CRC Press, Boca Raton, FL, pp. 407–440 (1995).

Slack, G. A. and V. Tsoukala, *J. Appl. Phys.* **76**, 1665 (1994).

Uher, C., Skutterudites: Prospective Novel Thermoelectrics, in *Semiconductors and Semimetals*, vol. 69, ed. T. M. Tritt, Academic Press, San Diego, pp. 139–253 (2001).

About the Author

Ctirad Uher is a C. Wilbur Peters Professor of Physics at the University of Michigan in Ann Arbor. He earned his BSc in physics with the University Medal from the University of New South Wales in Sydney, Australia. He carried out his graduate studies at the same institution under Professor H. J. Goldsmid on the topic of 'Thermomagnetic effects in bismuth and its dilute alloys', and received his PhD in 1975. Subsequently, Professor Uher was awarded the prestigious Queen Elizabeth II Research Fellowship, which he spent at Commonwealth Scientific and Industrial Research Organization (CSIRO), National Measurement Laboratory (NML), in Sydney. He then accepted a postdoctoral position at Michigan State University, where he worked with Profs. W. P. Pratt, P. A. Schroeder, and J. Bass on transport properties at ultra-low temperatures.

Professor Uher started his academic career in 1980 as an assistant professor of Physics at the University of Michigan. He progressed through the ranks and became full professor in 1989. That same year the University of New South Wales awarded him the title of DSc for his work on transport properties of semimetals. At the University of Michigan, he served as an associate chair of the Department of Physics and subsequently as an associate dean for research at the College of Literature, Sciences and Arts. In 1994, he was appointed as chair of physics, the post he held for the next 10 years.

Professor Uher has had more than 45 years of research, described in more than 520 refereed publications in the areas of transport properties of solids, superconductivity, diluted magnetic semiconductors, and thermoelectricity. In the field of thermoelectricity, to which he returned during the past 25 years, he worked on the development of skutterudites, half-Heusler alloys, modified lead telluride materials, magnesium silicide solid solutions, tetrahedrites, and Molecular Beam Epitaxy (MBE)–grown thin films forms of Bi_2Te_3-based materials. He has written a number of authoritative review articles and has presented his research at numerous national and international conferences as invited and plenary talks. In 1996, he was elected fellow of the American Physical Society. Professor Uher was honored with the title of Doctor Honoris Causa from the University of Pardubice in the Czech Republic in 2002, and in 2010 was awarded a named professorship at the University of Michigan. He received the prestigious China Friendship Award in 2011.

Professor Uher supervised 16 PhD thesis projects and mentored numerous postdoctoral researchers, many of whom are leading scientists in academia and research institutions all over the world.

Professor Uher served on the Board of Directors of the International Thermoelectric Society. In 2004–2005, he was elected vice president of the International Thermoelectric Society and during 2006–2008 served as its president.

1 Structural Aspects of Skutterudites

To describe skutterudites, it is convenient to divide them into three categories: binary skutterudites, ternary skutterudites, and filled skutterudites. Binary skutterudites are the simplest form of the skutterudite structure. A characteristic transport feature of binary skutterudites is their exceptionally high carrier mobility. However, as thermoelectric materials, binary skutterudites are not the most effective, primarily for their rather high thermal conductivity. Ternary skutterudites are man-made isoelectronic modifications of binary skutterudites intended to suppress the excessive thermal conductivity of the latter. Filled skutterudites are chemically modified binary or ternary skutterudites whereby structural voids in the skutterudite lattice are filled with foreign, loosely bonded ions. As already noted, they were first synthesized by Jeitschko and Braun (1977), and the filling feature is the primary reason why skutterudites have attracted so much interest as prospective novel thermoelectric materials. Moreover, the ability to fill the structural void with a wide variety of ions having different valence states, including magnetic $4f$-derived ions of rare earth elements, has resulted in a plethora of fascinating and often unconventional magnetic and superconducting properties observed in filled skutterudites, see, e.g., Sato et al. (2009). In this chapter, I will describe the three structural forms of skutterudites in turn.

1.1 BINARY SKUTTERUDITES

1.1.1 STRUCTURAL ASPECTS OF BINARY SKUTTERUDITES

Binary skutterudites are structures with the general formula MX_3, where M is one of the Column 9 transition metals Co, Rh, or Ir, and X stands for P, As, or Sb (elements, which together with nitrogen, are often called pnicogens, pnictogens or pnictides). I will use the name pnicogen throughout. Binary skutterudites form with all nine possible combinations of the M and X elements and crystallize with the body-centered cubic structure in the space group $Im\bar{3}$ (#204), first identified by Oftedal (1928). The unit cell contains 32 atoms arranged in eight groups of MX_3 blocks. In terms of the crystallographic designation (Wyckoff notation) applicable to the $Im\bar{3}$ space group, the metal ions occupy the 8c (0.25, 0.25, 0.25) site, and the pnicogen ions are located at the 24g (0, y, z) site, with the positional parameters $y = 0.335$ and $z = 0.160$ applicable to $CoSb_3$, for example. The skutterudite structure is often depicted as in Figure 1.1, where the unit cell is shifted by one-quarter distance along the body diagonally. In this perspective, the metal atoms M (light blue circles) form a simple cubic lattice while the pnicogen atoms X (black circles) are arranged in planar rectangular four-membered rings $[X_4]^{4-}$ that form linear arrays along the (100), (010), and (001) crystallographic directions with adjacent rings being orthogonal to each other. The unit cell can then be viewed as consisting of eight small cubes occupied by six pnicogen rings. In this rather open crystalline environment, two of the small cubes are missing pnicogen rings (the front upper left one and the back bottom right one), giving rise to two structural voids (cages) per unit cell. As we shall see, these voids can be filled by foreign species, resulting in a filled skutterudite. As can be seen in Figure 1.1, every pnicogen ring $[X_4]^{4-}$ is surrounded by eight trivalent metal cations M^{3+}; consequently, the compositional ratio M^{3+} : $[X_4]^{4-}$ = 4 : 3. Designating each void as □, the structure can be viewed equivalently as $\square_2 M_8 [X_4]_6 = 2(\square M_4 [X_4]_3) = 2(\square M_4 X_{12})$. It is customary to take just half of the unit cell, in which case the skutterudite structure becomes $\square M_4 X_{12}$, isoelectronic with the valence electron count (VEC) of 72.

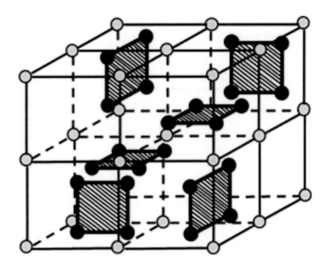

FIGURE 1.1 The unit cell of a binary skutterudite MX_3 shifted by one-quarter distance along the body diagonal. The metal atoms M (light blue circles) form a simple cubic sublattice, while the pnicogen atoms X (black circles) are arranged in planar rectangular four-membered rings. Note that there are six such pnicogen rings in the unit cell. Two of the eight small cubes are empty (the front upper left and the back bottom right), giving rise to two structural voids (often called cages) per unit cell.

From the structural and bonding perspective, a more informative picture of the skutterudite phase consists of an infinite three-dimensional array of trigonally distorted and tilted MX_6 octahedrons that share corners with six neighboring octahedrons (Figure 1.2). This depiction of the unit cell, now centered over a void at (0,0,0), the site 2a in the Wyckoff notation, nicely highlights the octahedral coordination of the metal atom M by the pnicogen atoms. It should be noted that the metal atoms are far apart from each other and no atom M has another metal atom as its nearest neighbor. It is the tilting of the octahedra that gives rise to the formation of the planar rectangular four-membered rings of pnicogen atoms X_4 that are the characteristic feature of the skutterudite structure. To grasp

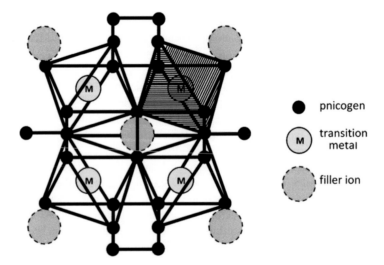

- ● pnicogen
- Ⓜ transition metal
- ◌ filler ion

FIGURE 1.2 The unit cell of the skutterudite structure centered over a void (0,0,0). The octahedral coordination of the M atom (Co) by the pnicogen atoms (black circles) is highlighted by stripes, and the tilt of the MX_6 octahedrons gives rise to the formation of the pnicogen rings depicted in Figure 1.1.

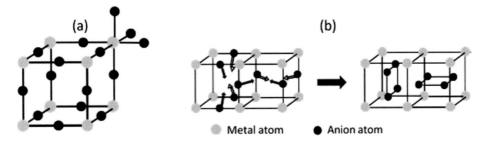

Metal atom   ● Anion atom

FIGURE 1.3   (a) Cubic structure of ReO_3, depicting the octahedral coordination of one of its metal (Re) atoms. (b) Shifts of anion atoms in the ReO_3 structure indicating the evolution of the anion rings of skutterudites.

how the tilt of the octahedrons is related to the four-membered rectangular pnicogen rings, it is convenient to consider the skutterudite structure as a severe distortion of the ReO_3 crystal structure (the tilt system $a^+a^+a^+$ discussed by Mitchell 2002) where, as illustrated in Figure 1.3a, the corner-sharing octahedrons have axes along the (001), (010), and (001) directions. As the octahedrons are tilted, while maintaining corner contacts with other octahedrons, the four anions originally located at the ReO_3 unit cell edges are displaced toward the center, giving rise to planar rectangular anion rings, as shown in Figure 1.3b.

The skutterudite structure is completely determined by giving the lattice parameter and the two positional parameters y and z specifying the exact position of the pnicogen atom (see Figure 1.4). In terms of the pnicogen positional parameters and the lattice constant a, the important interatomic distances immediately follow from Figure 1.4, as originally given by Rosenqvist (1953) and by Rundqvist and Ersson (1968):

$$D(M - X) = a\left[\left(\frac{1}{4}\right)^2 + \left(y - \frac{1}{4}\right)^2 + \left(z - \frac{1}{4}\right)^2\right]^{1/2}, \tag{1.1}$$

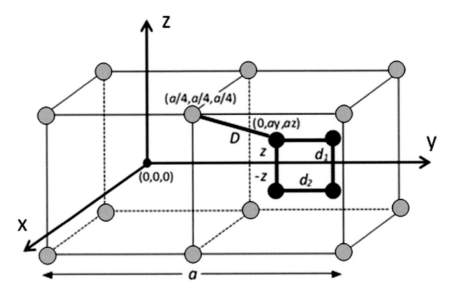

FIGURE 1.4   The definition of the positional parameters y and z of the pnicogen atoms. The lattice constant is designated as a. The important interatomic distance D between the metal atom M and the pnicogen atom X, as well as the distances d_1 and d_2 between the pnicogen atoms on the ring, are indicated.

$$d_1(X - X) = 2az, \tag{1.2}$$

$$d_2(X - X) = a(1 - 2y). \tag{1.3}$$

In his original identification of the skutterudite structure, Oftedal predicted a *square* planar configuration of the pnicogen rings, i.e., the two pnicogen distances d_1 and d_2 being equal. From Eqs. 1.2 and 1.3, this leads to what is often called the Oftedal relation:

$$2(y + z) = 1. \tag{1.4}$$

Assuming regular octahedral coordination for the metal atom M, the positional parameters y and z must further satisfy the condition

$$y(2z - 1) = z - \frac{3}{8}. \tag{1.5}$$

The simultaneous solution of Eqs. 1.4 and 1.5 yields $y = z = ¼$, the condition that signifies a structural change from the skutterudite structure to that of the more symmetric ReO_3-type structure with only one MX_3 group per unit cell and the space group *Pm3m*. Hence, to stay within the skutterudite phase, the constraints imposed by Eqs. 1.4 and 1.5 must not be viewed as rigid. As discussed by Rundqvist and Ersson (1968), violating either one of Eqs. 1.4 and 1.5 costs energy: a departure from the Oftedal relation is unfavorable to the X–X bonds, while a distortion of the ideal MX_6 octahedron environment is clearly detrimental to the M–X bonds. In fact, insisting on the Oftedal relation would make the square ring structure so large that virtually no bonding interaction would remain between the pnicogen atoms unless the M–X distance became unreasonably large. Consequently, rather than assessing a very large energy penalty for violation of only one of the constraints, it is less taxing to the structure to violate both constraints, each with only a modest energy penalty. This is exactly what happens – the octahedral MX_6 complex undergoes a slight trigonal antiprismatic distortion while the pnicogen rings assume a rectangular rather than square coordination. In binary skutterudites, the ratio of d_2/d_1 varies between 1.03 and 1.05, as follows from the data in Table 1.1.

TABLE 1.1
Structural Parameters of Binary Skutterudites

Skutterudite	Lattice Constant (Å)	y (Å)	z (Å)	D(M-X) (Å)	d_1(X-X) (Å)	d_2(X-X) (Å)	R(void) (Å)
CoP_3	7.7073	0.3482	0.1453	2.222	2.240	2.340	1.763
RhP_3	7.9951	0.3547	0.1393	2.341	2.227	2.323	1.909
IrP_3	8.0151	0.3540	0.1393	2.345	2.233	2.340	1.906
NiP_3	7.8192	0.3540	0.1417	2.280	2.216	2.283	–
$CoAs_3$	8.2055	0.3442	0.1514	2.337	2.478	2.560	1.825
$RhAs_3$	8.4507	0.3482	0.1459	2.434	2.468	2.569	1.934
$InAs_3$	8.4673	0.3477	0.1454	2.441	2.456	2.574	1.931
$CoSb_3$	9.0385	0.3351	0.1602	2.520	2.891	2.982	1.892
$RhSb_3$	9.2322	0.3420	0.1517	2.621	2.807	2.917	2.024
$InSb_3$	9.2503	0.3407	0.1538	2.617	2.850	2.943	2.040

Source: Reproduced from C. Uher, *Semiconductors and Semimetals*, Vol. 69, ed. T. M. Tritt, pp. 139–253, Academic Press, New York (2001). With permission from Elsevier.

Table 1.1 collects the important structural parameters of all bulk binary skutterudites, including the void radius R. It is interesting to point out a subtle trend in the positional parameters. Progressing from phosphides to arsenides to antimonides, in general, one positional parameter (y) decreases while the other one (z) increases. This small gradual change leads to a higher coordination number of the pnicogen atoms in the sequence MP_3, MAs_3, MSb_3 and, in turn, results in a shift from more localized bonding in phosphides to more delocalized bonding in antimonides. I should note that $FeSb_3$ can also be prepared but only as a thin film by using a nanoalloying synthesis (Hornbostel et al. 1997), or by MBE deposition (Möchel et al. 2011, Daniel et al. 2015). It is not possible to synthesize bulk forms of $FeSb_3$ because it is a metastable structure that decomposes.

The crystal structure determination of binary skutterudites of both synthetic and mineral forms was made by a number of research groups. Some of the studies assumed the validity of the Oftedal relation (Kjekshus and Rakke 1974, Zhuravlev and Zhdanov 1956), while the first hint of a distorted octahedron and the rectangular coordination of the pnicogen atoms was provided by Ventriglia (1957). Subsequent detailed studies by Rosenqvist (1953), Rundqvist and Hede (1960), Kjekshus and Pedersen (1961), Mandel and Donohue (1971), Kjekshus et al. (1973), Kaiser and Jeitschko (1999), and Schmidt et al. (1987) have firmly established small deviations from the Oftedal relation for all binary skutterudites. Kjekshus and Rakke (1974) attributed the rectangular distortions ($d_1 \neq d_2$) of the X_4 rings to their anisotropic environment – the presence of two cavities per MX_3 unit cell symmetrically arranged on opposite sides of the X_4 groups. Figure 1.5 provides graphic illustration of the Oftedal relation, the condition for ideal octahedral coordination, and the actual positional parameters of all known binary skutterudites and La-filled skutterudites.

The tilt angle ϕ for the ideal skutterudite structure is related to the M–X–M angle, see Figure 1.6, and given by O'Keefe and Hyde (1977) and Vaqueiro et al. (2008) as

$$\cos\left(M - X - M\right) = 1 - \frac{2x^2}{9}, \quad \text{where} \quad x = 2\cos\phi + 1. \tag{1.6}$$

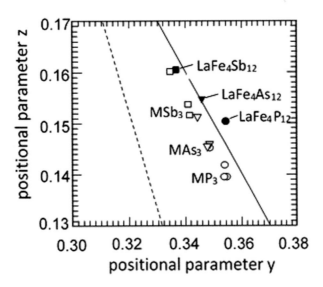

FIGURE 1.5   Positional parameters for binary skutterudites (open circles for phosphides, open triangles for arsenides, and open squares for antimonides) and for La-based filled skutterudites). The solid line is the Oftedal relation, Eq. 1.4, and the dashed line represents the ideal octahedral coordination, Eq. 1.5. The data are from Table 1.1 and from Jeitschko and Braun (1977), Braun and Jeitschko (1980a), and Evers et al. (1994). Reproduced from C. Uher, *Semiconductors and Semimetals*, Vol. 69, Ed. T. M. Tritt, Academic Press, pp. 139–253 (2001). With permission from Elsevier.

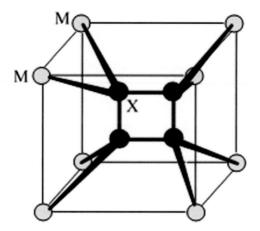

FIGURE 1.6 Coordination of pnicogen atoms (black circles) in the skutterudite structure. Pnicogen atoms form a planar rectangular cluster. Each pnicogen atom has four near neighbors: two other pnicogen atoms and two metal atoms (light blue circles). Reproduced from C. Uher, *Semiconductors and Semimetals*, Vol. 69, Ed. T. M. Tritt, Academic Press, pp. 139–253 (2001). With permission from Elsevier.

The tilt angle can also be expressed in terms of the unit cell dimension a and the M–X distance D, depicted in Figure 1.4 (Navrátil et al. 2010), as

$$\cos\phi = \frac{3a}{8D} - 0.5. \tag{1.7}$$

The dependence of the tilt angle ϕ on the shorter X–X distance d_l for binary skutterudite is shown in Figure 1.7.

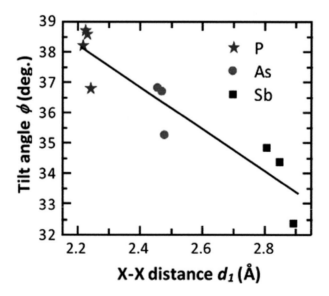

FIGURE 1.7 The tilt angle ϕ as a function of the shorter of the pnicogen-pnicogen distances d_l for all binary skutterudites. An approximately inverse relation between the tilt angle ϕ and the distance d_l is to be noted. Phosphide skutterudites are designated by blue stars, arsenide skutterudites by red circles, and antimonide skutterudites by black squares. Drawn from the data of B. N. Schumer et al., *American Mineralogist* 102, 205 (2017). With permission from the Mineralogical Society of America.

I also note a surprising report in which Raza et al. (2014), based on their DFT calculations, described the first nitrogen-based skutterudite structure PN_3, predicted to exist at pressures over 10 GPa. The phase is thought to be metallic due to nitrogen p-states being delocalized over the N_4-rings, and is estimated to become a good electron-phonon superconductor with a high T_c ~18 K. The N–N bonds at 10 GPa are extremely short at 1.30 Å. At lower pressures, the structure decomposes into N_2 pairs. Because of the very small structural voids, no filler species would fit in and no stable compounds of the type RP_4N_{12} would be stable.

1.1.2 BONDING IN BINARY SKUTTERUDITES

Addressing the issue of bonding, the key experimental input that has to be taken into account is the fact that the binary skutterudites are diamagnetic semiconductors, i.e., the relevant bonding scheme must have no unpaired spins. It has already been noted that the distance between the metal atoms is large and that there are no nearest metal atoms to any given M atom. This implies that there is no significant M-M bonding in the skutterudite structure. Referring to Figure 1.6, each pnicogen atom X has two other pnicogens as its nearest neighbors and it also bonds with the two nearest metal atoms. The pnicogen ring structure X_4 holds together *via* σ bonds, i.e., each pentavalent pnicogen atom (ns^2np^3) contributes two valence electrons, one each to bond with its two nearest neighboring pnicogen atoms. The remaining three valence electrons participate in bonding with the two nearest metal atoms. Since there are six pnicogen atoms in octahedral coordination around each metal atom M, the pnicogens contribute a total of $(5 – 2) \times \frac{1}{2} \times 6 = 9$ electrons toward the MX_6 octahedral complex. This is just enough to engage nine valence electrons of the Co-like metal (d^7s^2) to form the 18-electron rare-gas configuration that favors diamagnetism and semiconducting behavior. From the perspective of the metal atom M, it contributes $6 \times \frac{1}{2} = 3$ electrons for bonding with the six neighboring pnicogen atoms. These electrons occupy the octahedral d^2sp^3 hybrid orbitals that are the essence of the M–X bonding. The Co-like metal is thus left in the 3+ state with six nonbonding electrons that adopt a maximum spin-pairing configuration and therefore the zero-spin d^6 state. It is also interesting to note that the electronegativity of pnicogens (2.19 for P, 2.18 for As, and 2.05 for Sb) is very close to that of the Co-like metals (1.88 for Co, 2.28 for Rh, and 2.20 for Ir); thus, the M–X bond has only a small degree of ionic character. Even for CoP_3, which among the binary skutterudites has the largest electronegativity difference of 0.31 between its constituent elements, there is only a very small degree of ionicity compared to typical ionic compounds, where the electronegativity difference is ~ 1.5.

The above bonding description is a modified version of one of the bonding models developed by Dudkin and Abrikosov (1956). Although various aspects of this model have been criticized (Kuzmin 1967), particularly its apparent omission of the importance of the X_4 ring structure that some researchers view as pivotal for skutterudites, its basic premise remains valid and its predictive power served well during the past 40 years of research on these fascinating materials. Schematics of Dudkin's model are shown in Figure 1.8. On the basis of this model, one can predict the character of the skutterudite compounds. As depicted in Figure 1.8, there are no unpaired electrons in the cobalt family of skutterudites [$Co(Rh,Ir)P_3$, $Co(Rh,Ir)As_3$, and $Co(Rh,Ir)Sb_3$] and thus these structures are diamagnetic semiconductors. The hypothetical iron family of binary skutterudites [$Fe(Ru,Os)P_3$, $Fe(Ru,Os)As_3$, and $Fe(Ru,Os)Sb_3$] with one less electron in their inner d-shells should be paramagnetic semiconductors. Unfortunately, this prediction is impossible to verify because such skutterudites have not yet been synthesized in bulk form nor do they exist in nature. In the case of the equally elusive nickel family of skutterudites [$Ni(Pd,Pt)P_3$, $Ni(Pd,Pt)As_3$, and $Ni(Pd,Pt)Sb_3$], the additional nonbonding electron has no choice but to be promoted into higher energy levels (possibly into the conduction band), leading to a likely scenario that such compounds would be paramagnetic metals. Again, with the exception of NiP_3 and possibly PdP_3, no binary skutterudites with the transition metals from the Ni column are known to exist. Although Rosenqvist (1953) doubted that PdP_3 is a stable compound, the existence of NiP_3 is firmly established (Jolibois 1910, Biltz and Heimbrecht

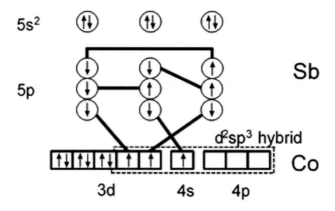

FIGURE 1.8 Schematic illustration of Dudkin's bonding model. Adapted from Dudkin and Abrikosov (1956). Reproduced from C. Uher, *Semiconductors and Semimetals*, Vol. 69, Ed. T. M. Tritt, Academic Press, pp. 139–253 (2001). With permission from Elsevier.

1938, and Rundqvist and Larsson 1959), and the VEC count of 73 with the delocalized seventh electron, indeed, leads to metallic conductivity and paramagnetism, as shown by Hulliger (1961). It is interesting to note that Ni is somewhat "tolerated" in binary phosphide skutterudites while it is distinctly less so in arsenides and even less in antimonides.

1.1.3 Solid Solutions of Binary Skutterudites

The inability to prepare pure binary skutterudites with transition metals other than those of the Co group does not mean that cobalt cannot be *partially* replaced by its immediate neighbors in the periodic table – iron and nickel. The skutterudite structure can accommodate such a partial replacement, and several studies have attempted to establish substitutional solid solution limits. It is important to realize that an elemental impurity will substitute for a component of the solid solution only if it is capable of forming the same bond as the component it replaces, and if the covalent radii of the two species are not too different. Apart from the number of d-electrons, iron and nickel have similar valence electron structure as cobalt. They can readily form octahedral bonds, and when replacing trivalent cobalt in the skutterudite structure, the valence of Fe and Ni will be such as to closely match the size of the covalent radius of Co^{3+} (1.22 Å). In the case of Ni, it is clearly its tetravalent state Ni^{4+} (1.21 Å) that preserves the d^6 configuration and promotes a single electron into a conduction band, making nickel an electron donor. When iron substitutes for cobalt, there are two valence states of iron that have comparable radii as the trivalent cobalt: the divalent state Fe^{2+} (1.23 Å) and the trivalent state Fe^{3+} (1.22 Å). Quite apart from the fact that the divalent iron does not provide enough electrons for bonding, the trivalent ion exactly matches the size of the trivalent cobalt it replaces, and thus Fe^{3+} is the most likely configuration in the solid solution. This, of course, leaves only five electrons in the d-shell and it is impossible to pair all spins. Substitution of Fe for Co is therefore likely to give rise to paramagnetism.

The actual solid solution limit, specifically in antimonide skutterudites, was first established by Dudkin and Abrikosov (1957), who found that up to 10% of cobalt atoms can be replaced by nickel. In the case of iron, this figure increases to nearly 25%. A little wider boundary is possible if one allows for a slight deviation in pnicogen stoichiometry. In arsenide skutterudites, the substitution limits are higher and the arsenide structure prefers nickel over iron. Pleass and Heyding (1962) reported 16% and 65% as the respective limits of Fe and Ni substitutions for Co. These authors confirmed the electron donor character of nickel and detected paramagnetic signals in the Fe-substituted $CoAs_3$ compounds with the effective moment per iron atom of about $2\mu_B$, where μ_B

is the Bohr magneton. This is in agreement with the presumed Fe^{3+} state and the presence of an unpaired electron in the nonbonding d-orbital. What is surprising, and also contrary to Dudkin's model that assumes "inactiveness" of iron atoms in $CoSb_3$, is the profound influence of iron on the conduction process. With as little as 0.5% of Fe substituted for Co, the structure attains metallic character (Yang et al. 1999).

Based on the solid solution field for arsenide skutterudites, Rosenboom (1962), it is possible to make a coupled replacement of two cobalt atoms by one iron and one nickel atom:

$$2Co^{3+}(d^6) \rightleftarrows Fe^{2+}(d^6) + Ni^{4+}(d^6). \tag{1.8}$$

Structures with symmetrically substituted cobalt thus preserve the total number of electrons, and all ions are in the low-spin d^6 configuration. Magnetic measurements by Nickel (1969) confirmed that $(Fe_{0.45}Ni_{0.55})As_3$ is, indeed, a diamagnetic solid, and Pleass and Heyding (1962) observed semiconducting behavior in $(Fe_{0.46}Ni_{0.54})As_3$. Symmetrically substituted Co by Ni and Fe has also been realized in phosphide skutterudites by Jeitschko et al. (2000), with interatomic distances in $Fe_{0.5}Ni_{0.5}P_3$ similar to those of CoP_3. These findings, again, do not conform to Dudkin's model that would predict a strong paramagnetism due to the unpaired spin of Fe and metallic conduction on account of the conduction electron of Ni. It does appear as if nickel and iron form pairs (Fe^-Ni^+) equivalent to Co with the conduction electron of Ni occupying the nonbonding d-orbital of Fe. In fact, the above structures are better classified as ternary skutterudites, as I discuss in Section 1.2.

Forming solid solutions is one of the most time-tested approaches to lower the lattice thermal conductivity of a material and achieve superior thermoelectric performance. The strategy works well-provided phonon scattering is enhanced more than the inevitable degradation of the carrier mobility on account of stronger carrier-point defect scattering. Exploring the limits of solid solubility among various families of skutterudite compounds has, therefore, been of considerable interest. Regarding alloying on the pnicogen site, Lutz and Kliche (1981) found that phosphide and arsenide skutterudites form a complete series of solid solutions obeying Vegard's law, e.g., $CoP_{3-x}As_x$. A solid solution with equal content of phosphorus and arsenic, $CoP_{1.5}As_{1.5}$ has been synthesized and its transport properties measured by Watcharapasorn et al. (2000). The practically more interesting solid solutions between arsenides and antimonides are, unfortunately, more restricted and, in the case of $CoAs_{3-x}Sb_x$, Lutz and Kliche (1981) determined a miscibility gap for $0.4 < x < 2.8$. As mentioned previously, solid solutions can also form on the cation site. The most explored is $Co_xIr_{1-x}Sb_3$, where a rather large 2.3% difference in the lattice parameters of $CoSb_3$ and $IrSb_3$ gives rise to immiscibility for the range of values of $0.2 < x < 0.65$, as determined by Borshchevsky et al. (1995). Somewhat uncertain is the situation in $Rh_xIr_{1-x}Sb_3$ and $Rh_xCo_{1-x}Sb_3$. In the former case, an equal part solid solution $Rh_{0.5}Ir_{0.5}Sb_3$ was prepared by Slack and Tsoukala (1994) but no attempt was made to synthesize a range of compositions to gauge how broad the solubility is. In the case of $Rh_xCo_{1-x}Sb_3$, Wojciechowski (2007) observed a linear dependence of the cell parameter for all contents x as well as linearly dependent positional parameters of Sb, suggesting that solid solutions exist across the full range of compositions. On the other hand, he also noted a rather dramatic anomaly in the transport properties for $0.2 < x < 0.5$, which casts a doubt on the complete solubility of $RhSb_3$ with $CoSb_3$.

1.1.4 STRUCTURAL STABILITY OF BINARY SKUTTERUDITES

Given an open crystalline environment of binary skutterudites containing large structural voids, it is of interest to ask how stable the structure under compression is. In fact, it is far more than a matter of curiosity as the issue is of vital importance not only during the compaction of powders of synthesized skutterudites but also in the high-pressure/high-temperature synthesis of skutterudites (see Section 2.2.6) that cannot be prepared under conditions of ambient pressure. It is thus no surprise that attempts to ascertain the stability of the $Im\overline{3}$ skutterudite structure were undertaken soon after the skutterudites have emerged as promising thermoelectric materials. In the first report by Snider et al. (2000), the authors compressed a powder of $IrSb_3$ in a diamond anvil cell and monitored its diffraction pattern to pressures of 42 GPa at ambient temperature. The structure survived this brutal pressure with no signs of collapse, amorphization, or phase transitions. Employing the Clapeyron equation:

$$\frac{dP}{dT} = \frac{\Delta S}{\Delta V},$$ (1.9)

with S and V being the entropy and volume, respectively, and the thermodynamic parameters for the most likely phase transformation of $IrSb_3$ to $IrSb_2 + Sb$, the authors also predicted that the typical powder consolidation conditions during hot pressing or spark plasma sintering (pressures up to 0.2 GPa at 1000°C) are perfectly safe with regard to the stability of the skutterudite. Subsequent measurements of the bulk modulus of CoP_3, $CoSb_3$, and several filled skutterudites by Shirotani et al. (2004) to about 12 GPa, and $CoSb_3$ by Kraemer et al. (2005) to about 20 GPa further confirmed the stability of various skutterudites to fairly high pressures. The bulk moduli B_0 and their first derivative B_0' obtained from fits of the pressure changes of the cell volume to the Birch equation of state (Birch 1947):

$$P = \frac{3}{2} B_0 \left[\left(\frac{V}{V_0} \right)^{-\frac{7}{3}} - \left(\frac{V}{V_0} \right)^{-\frac{5}{3}} \right] \left\{ 1 - \frac{3}{4} \left(4 - B_0' \right) \left[\left(\frac{V}{V_0} \right)^{-\frac{2}{3}} - 1 \right] \right\},$$ (1.10)

are collected in Table 1.2. Where available, Table 1.2 also includes the Grüneisen parameter γ_G obtained from

$$\gamma_G = \frac{2}{3} - \frac{1}{2} V \left(\frac{\partial^2 P}{\partial V^2} \right)_T \left(\frac{\partial P}{\partial V} \right)_T^{-1}.$$ (1.11)

Extending their pressure studies on $CoSb_3$ beyond 20 GPa, Kraemer et al. (2007) noted that, while the skutterudite lattice is preserved, releasing the pressure after the powder of $CoSb_3$ was compressed

---

TABLE 1.2

Bulk Modulus B_0, Its Pressure Derivative B_0', the Grüneisen Parameter γ_G, and the Critical Pressure P_c at which the Self-Insertion Reaction Starts to Take Place

Skutterudite	a (Å)	B_0 (GPa)	B_0'	γ_G	P_c (GPa)	Reference
CoP_3	7.7064	152	1	0.974		Shirotani et al. (2004)
$CoSb_3$	9.0345	–	–	1.42	–	Caillat et al. (1996)
	9.0451	81	6	1.111		Shirotani et al. (2004)
	9.0365	95	4	–		Kraemer et al. (2005)
	9.0365	93	5	–		Kraemer et al. (2007)
	9.0347	84	–	–	28	Matsui et al. (2010)
$RhSb_3$	9.2255	95	–	–	20	Matsui et al. (2010)
$IrSb_3$	9.2503	–	–	1.42	–	Slack and Tsoukala (1994)
	9.2512	136	4.8	–		Snider et al. (2000)
	9.2564	105	–	–	30	Matsui et al. (2010)
$CoAs_3$	–	101	–	–	50	Matsui et al. (2012)
$RhAs_3$	8.4446	130	–	–	35	Matsui et al. (2011b)
$IrAs_3$	–	150	–	–	45	Matsui et al. (2012)

Source: Data collected from the literature.

---

above 20 GPa, the lattice constant turned out to be somewhat larger, 9.0636 Å compared to 9.0305 Å for the pristine never compressed $CoSb_3$. Moreover, once the pressure exceeded 20 GPa at room temperature, certain Bragg reflection peaks (notably 110, 200, and 211) became dramatically suppressed and never recovered following the pressure release. Guessing correctly that they might be observing a pressure-induced redistribution of Sb, in which a fraction of Sb from the framework starts to fill the vacant voids, they carried out simulations with a structure of the form $Sb_xCo_4Sb_{4-x}$ to ascertain that, indeed, this leads to changes in the intensity of certain Bragg peaks that match the observed intensity pattern. They coined the name pressure-induced self-insertion reaction for this irreversible isosymmetric transition. The results were promptly confirmed by a sleuth of studies on $CoSb_3$ and other binary skutterudites by Matsui et al. (2010, 2011a,b, 2012) and by Miotto et al. (2011). The authors have specifically identified the critical pressure P_c (given in Table 1.2) at which the self-insertion process starts to take place. Variations of the unit cell volume as a function of applied pressure for As- and Sb-based binary skutterudites are depicted in Figure 1.9.

Using the global minimization of free energy surfaces implemented in the CALYPSO code, Ma et al. (2019) recently predicted that IrP_3 will not suffer a pressure-induced self-insertion reaction but will undergo the transformation from the skutterudite $Im\bar{3}$ structure to a more energetically favorable orthorhombic $Pmma$ structure at 47.6 GPa. Interestingly, the new high-pressure structure was calculated to be metallic with rather flat electronic bands crossing the Fermi level, and the phonon calculations *via* the PHONOPY code indicated dynamical stability of the $Pmma$ structure

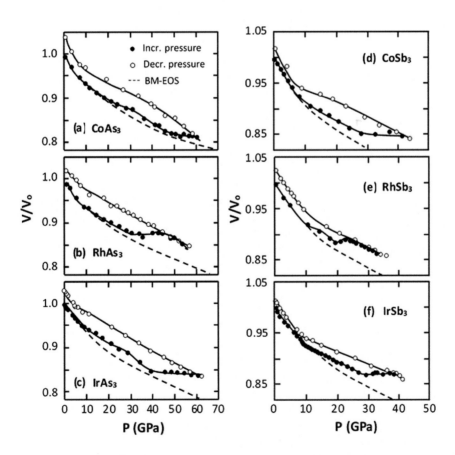

FIGURE 1.9   Pressure dependence of the relative unit cell volume V/V_0 as a function of applied pressure at room temperature for As- and Sb-based binary skutterudites. Adapted from K. Matsui et al., *Journal of the Physical Society of Japan* 81, 104604 (2012). With permission from the Physical Society of Japan.

to 100 GPa. Strong electron-phonon coupling resulted in an estimated superconducting transition temperature of 2.4 K.

1.1.5 NATIVE DEFECTS IN BINARY SKUTTERUDITES

While nominally stoichiometric samples of $CoSb_3$ with carrier densities ~10^{17} cm^{-3} have been reported as either a p-type (Morelli et al. 1995, Sharp et al. 1995, Caillat et al. 1996, Arushanov et al. 1997) or n-type (Morelli et al. 1997, Dilley et al. 2000, Kawaharada et al. 2001, Kuznetsov et al. 2003, Liu et al. 2007a,b) semiconductor, an intentionally introduced excess of either Co or Sb invariably resulted in p-type conduction. This suggested that intrinsic defects may play an important role in the nature of transport. Since binary skutterudites are composed of a transition metal and a rather volatile pnicogen element, it was always assumed that some pnicogen was lost during the synthesis process. From this perspective, it is understandable why there was an early interest to find out what is the effect of pnicogen vacancies on the electronic structure and how they affect transport properties. The early DFT study by Akai et al. (1997) suggested that Sb atoms supply electrons in the $CoSb_3$ structure, and thus their partial loss by evaporation during the synthesis results in p-type conduction. Detailed calculations based on the Korringa, Kohn, Rostoker (KKR) method within the LDA approximation were carried out by Wojciechowski et al. (2003), and the results indicated that Sb vacancies have a profound effect on the band structure, far beyond the realm of the rigid band model. Even a minute amount x of Sb vacancies rapidly closed the band gap in $CoSb_{3-x}$. In contrast, the excess of Sb, which goes into the void sites, modified the electronic structure because p-states of Sb (in the voids) formed a sharp peak inside the band gap. Unfortunately, no consideration was given to the formation energy required to create an Sb vacancy or interstitial.

In view of the fact that forming a pnicogen vacancy means severing the strong covalent bond holding together the X_4 ring, it seems rather unlikely that pnicogen vacancies would be the most important native defect states in skutterudites. The study by Park and Kim (2010) has amply documented the point. In their detailed work within the DFT formalism using the Vienna *ab initio* package in LDA approximation, they used a large 256-atom supercell ($Co_{64}Sb_{196}$), in which a single native point defect represents a defect concentration of 1.56% for vacancy at the Co site, V_{Co}, and for an interstitial Co atom, Co_i, while for the corresponding defects on the Sb site, V_{Sb} and Sb_i, this accounts for the 0.52% defect concentration. The calculations identified seven possible interstitial sites where Co can be accommodated, with the likelihood of occupancy increasing with the coordination number of the site. As the most stable site was identified seven-fold coordinated Co_i depicted in Figure 1.10a, where Co_i is bonded to five neighboring Sb atoms and to two neighboring Co atoms. This holds regardless of whether the interstitial site is neutral or charged. In the case of interstitial Sb, there are only two sites, the stable void center site and the center of the Sb_4 ring that requires 2.38 eV higher energy. Since all Co and Sb sites are equivalent, in each case, there is only one possible nonequivalent atomic site for the Co vacancy, V_{Co}, and for the Sb vacancy, V_{Sb}. The former case is shown in Figure 1.10b and the defect leads to a slight adjustment of the nearby Sb_4 ring, where the long Sb–Sb bond increases by about 0.5% and the short Sb–Sb bond increases by about 1%. In the case of V_{Sb}, the site shown in Figure 1.10c, the ring is not only broken but its remnants are rotated by as much as 6° from the original position. By calculating formation energies of the above defects, the authors were able to compare their relative stability. The results are displayed as a function of the Fermi energy in Figure 1.11 and indicate the following: in Co-rich $CoSb_3$, the defect with the exceptionally low formation energy, and therefore the dominant defect, is Co interstitial, Co_i. In Sb-rich $CoSb_3$, the defect that seems to dominate is the Co vacancy, V_{Co}. Accommodating Sb in interstitial positions requires distinctly more energy, particularly in the case of Co-rich $CoSb_3$. The data also attest to the fact that in both Co-rich and Sb-rich $CoSb_3$, Sb vacancy is a highly unlikely defect as it requires high energy to disrupt the strongly bonded Sb_4 rings. Of course, one ultimately wishes to know what role the native defects play in the transport process, i.e., are they a source of electrons or holes and how many of them. This depends on the preferred charge state of the native

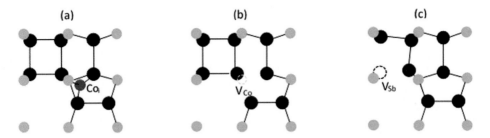

FIGURE 1.10   (a) The most stable location of the interstitial Co atom, Co_i, in the seven-fold coordination consisting of five Co_i–Sb bonds and two Co_i–Co bonds; (b) Co vacancy indicated by a dashed light blue circle; (c) antimony vacancy indicated by a dashed black circle. Such a vacancy causes the destruction of the Sb_4 ring and its remnants are rotated by about $6°$. Antimony atoms are shown as black circles, cobalt atoms as light blue circles, and the interstitial cobalt atom as a red circle. Adapted from C.-H. Park and Y.-S. Kim, *Physical Review B* 81, 085206 (2010). With permission from the American Physical Society.

defects, which can be elucidated by a direct comparison of formation energies of different charge states that show as small kinks on the formation energy curves. In Figure 1.11, they are highlighted by colored dots. From the plot of the density-of-states of Co_i and V_{Co} (see Figure 1.12), one concludes that in Co-rich $CoSb_3$, the dominant Co_i defect has the acceptor-like character. Given that Co has nine valence electrons, the calculations indicate (Figure 1.12), that there are four $3d$-like states of

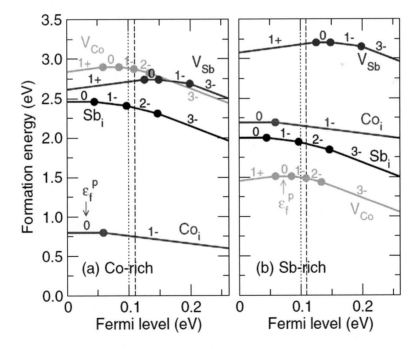

FIGURE 1.11   Calculated formation energies for various defect states as a function of the Fermi level in (a) Co-rich $CoSb_3$ and (b) Sb-rich $CoSb_3$. The dashed and dotted lines indicate the intrinsic Fermi levels at room temperature and at 600 K, respectively. The colored circles indicate breaks in the formation energy at different charge states of the defects. The pinning level by the dominant defects is indicated by ε_f^p. Reprinted from C.-H. Park and Y.-S. Kim, *Physical Review B* 81, 085206 (2010). With permission from the American Physical Society.

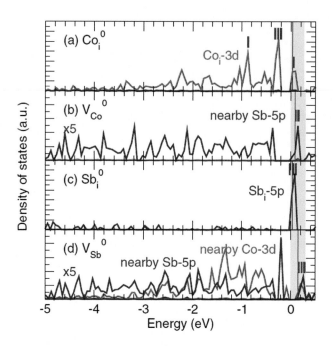

FIGURE 1.12 Projected local electronic densities of states of (a) Co_i; (b) V_{Co}; (c) Sb_i; and (d) V_{Sb} in the neutral charge states. The energy gap is shown by a gray-shaded area. The defect levels are indicated by the short thick vertical bars near the peaks. The thin vertical lines in the gap region indicate the highest singly occupied level. Reprinted from C.-H. Park and Y.-S. Kim, *Physical Review B* 81, 085206 (2010). With permission from the American Physical Society.

Co_i located inside the valence bands, and a singly occupied 3d-like state is just above the VBM and acts as an acceptor, making Co_i a shallow single acceptor in $CoSb_3$. By bonding Co_i with the five surrounding Co and two Sb atoms, the 3d-like states are stabilized by lowering their energy, which, in turn, leads to intra-atomic transfer of the two 4s electrons into the 3d-like states. The single 4s-like state of Co_i is delocalized inside the conduction band. In defect-free $CoSb_3$, the Fermi level $E_F(0)$ is located in the lower part of the band gap, calculations show it at $E_v + 0.11$ eV at 300 K. The Co_i (0/1–) transition level and the corresponding pinning energy ε_F^p, on account of the dominant presence of Co_i, is found to be lower than $E_F(0)$, giving rise to a higher concentration of holes and the p-type nature of transport in Co-rich $CoSb_3$. Co vacancy, V_{Co}, also behaves as a shallow acceptor. It can be charged up to (3–) state with the transition levels (1–/2–) and (2–/3–) at positions $E_v + 109$ meV and $E_v = 134$ meV, respectively. The presence of V_{Co} means the loss of three electrons (taking the formal oxidation state of Co as 3+) and thus three holes are generated in the Sb_4-related valence band. Hence, V_{Co} is a shallow triple acceptor. Thus, in Sb-rich $CoSb_3$, V_{Co} pins the Fermi level above VBM but below $E_F(0)$, again giving rise to a higher concentration of holes and the p-type nature of transport in Sb-rich $CoSb_3$.

Returning to the case of Co interstitials, their exceptionally low formation energy facilitates a very high density of interstitial defects, as easily judged from their population $\left[Co_i\right] = N_0 \exp\left(\Omega_f / kT_g\right)$, where N_0 is the number of available sites for cobalt interstitials, Ω_f is the formation energy of Co_i, and T_g is the growth temperature. Based on the lattice constant and 24 Co_i available sites in the conventional cell, Park and Kim (2010) quote $N_0 = 3.385 \times 10^{22}$ cm^{-3} in $CoSb_3$. With the growth temperature taken as 1400 K, the population of Co interstitials is as high as 4.5×10^{19} cm^{-3}. With such high density of interstitials, it is rather likely that pairs of Co interstitials can form. The most probable pair structure identified in calculations by Park and Kim is depicted in Figure 1.13 and it involves

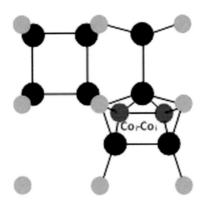

FIGURE 1.13   The most stable configuration of a pair of Co interstitials in an eight-coordinated complex. Adapted from C.-H. Park and Y.-S. Kim, *Physical Review B* 81, 085206 (2010). With permission from the American Physical Society.

two seven-coordinated Co interstitials. With an additional Co_i–Co_i bond, this gives an eight-coordinated complex, the most stable Co_i–Co_i pair. By pairing, the singly occupied highest $3d$-like state of Co_i is driven deeper into the band gap and maintains its full occupancy. Thus, the Co_i-pair has the donor character in $(Co_i$–$Co_i)^{1+}$ and $(Co_i$–$Co_i)^{2+}$ states and the acceptor character in $(Co_i$–$Co_i)^{1-}$ and $(Co_i$–$Co_i)^{2-}$ states. When the Co_i pairs are dominant in Co-rich $CoSb_3$, the Fermi level is expected to be pinned between the (1+/0) and (0/1−) transition levels at energies of E_v + 123 meV and E_v + 163 meV, respectively; see Figure 1.14a. The pinning level is higher than the Fermi energy $E_F(0)$, resulting in n-type doping in Co-rich $CoSb_3$. The pairs of Co interstitials are rather weakly bonded and at elevated temperatures they decompose into individual Co_is. The pinning energy reverts to that of a single Co_i, falls below $E_F(0)$, and results in the change of the dominant carrier species, from n-type to p-type. The relative fraction of Co interstitials and pairs of Co interstitials as a function of temperature is shown in Figure 1.14b. It is plotted for two different total concentrations of interstitials, 4.5×10^{19} cm^{-3} (solid line) and 1.5×10^{19} cm^{-3} (dashed line). At low temperatures, below about 530 K, the interstitial pairs dominate, while at higher temperatures the isolated Co interstitials become more prevalent. The often-seen transition in Co-rich $CoSb_3$ from n-type at ambient temperatures to p-type at elevated temperatures might be associated with this crossover in the population of pairs and isolated Co interstitials. In contrast, Sb-rich $CoSb_3$ compounds maintain their p-type character at all temperatures (Sharp et al. 1995, Kawaharada et al. 2001, Liu et al. 2007a,b).

The issue of intrinsic defects in $CoSb_3$ was revisited more recently by Li et al. (2016), who used DFT calculations performed with the same Vienna *ab initio* Simulation Package but, in contrast with the work of Park and Kim, used the Perdew-Burke-Ernzerhof (PBE) exchange-correlation functionals, a variant of the General Gradient Approximation (GGA). For more details concerning differences between LDA and GGA approaches used in DFT; see the section on band structure. Both DFT calculations agree that a cobalt interstitial, Co_i, is the dominant defect in the Co-rich $CoSb_3$ structure, while a cobalt vacancy, Co_v, fulfills this role in Sb-rich $CoSb_3$. Moreover, the calculations concur that both defects act as acceptors, i.e., they increase the density of holes and drive the system p-type. However, the authors provide very different explanations for the role of interstitial defects and do not agree on the charge state a pair of cobalt interstitials acquires. Li et al. argue that Co_i at the seven-fold coordinated position shown in Figure 1.10a stretches two Sb–Sb bonds on two Sb_4 rings to such a degree that it de facto breaks the two bonds, a situation depicted more clearly in Figure 1.15. The breakage of the two covalent Sb–Sb bonds requires the system to supply four electrons so that the four affected Sb atoms can form four new lone pairs. Assuming that Co_i itself can supply three electrons (an interstitial Co_i behaving chemically as the regular Co atom in the lattice), the remaining single electron must come from the surrounding environment. Co_i thus

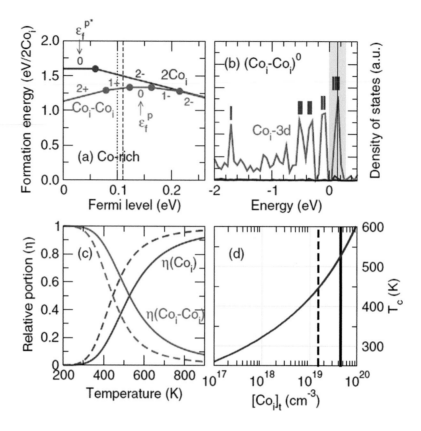

FIGURE 1.14 (a) Calculated formation energies of the most stable interstitial Co pairs, Co_i-Co_i, and two isolated Co interstitials, $2Co_i$, plotted as a function of the Fermi level in the Co-rich limit of $CoSb_3$. (b) Calculated projected local density of states of Co_i-pairs in the neutral charge state. (c) Relative fraction of isolated Co interstitials, $\eta(Co_i)$, and of the pairs of Co interstitials, $\eta(Co_i$-$Co_i)$, as a function of temperature for two populations of Co_i at the growth temperature: 4.5×10^{19} cm^{-3} (solid curves) and 1.5×10^{19} cm^{-3} (dashed curves). At low temperatures, below about 530 K, the fraction of pairs of Co interstitials is higher than the fraction of isolated Co interstitials. Above 530 K, the pairs of Co interstitials decompose into isolated Co interstitials that become dominant in numbers and drive the change over from n-type to p-type nature of transport in Co-rich $CoSb_3$. (d) Calculated decomposition temperature T_c of Co_i pairs as a function of the concentration of Co_i. Reproduced from C.-H. Park and Y.-S. Kim, *Physical Review B* 81, 085206 (2010). With permission from the American Physical Society.

acts as a single acceptor with the charge state 1−, and the system has more holes than electrons. The formation energy of various defects in both Co-rich and Sb-rich $CoSb_3$ is depicted in Figure 1.16.

While the DFT calculations of both Park and Kim and Li et al. agree on the fact that in Co-rich $CoSb_3$, the lowest formation energy is actually not that of Co_i but that of a pair of cobalt interstitials, designated as Co_i-Co_i in Park and Kim and as $Co_{i\text{-}p}$ in the paper by Li et al., they disagree on the character of such defects. Li et al. conclude that $Co_{i\text{-}p}$ behaves as an acceptor impurity with 1− charge state per Co_i, while Park and Kim obtain a donor-like nature for the interstitials, designated as Co_i–Co_i in Park and Kim and as $Co_{i\text{-}p}$ in the paper by Li et al., the same defect. Park and Kim therefore suggested that the decay of the pairs of cobalt interstitials at elevated temperatures to single interstitials that behave as acceptor defects is the reason for the often-observed crossover from the n-type to p-type conduction in $CoSb_3$ at elevated temperatures. Of course, such crossover can also arise as a consequence of a very small extrinsic carrier concentration that is taken over by intrinsic excitations as the temperature increases, with the hole mobility greatly exceeding the electron mobility, as documented in the section on transport properties. It is somewhat unsettling that

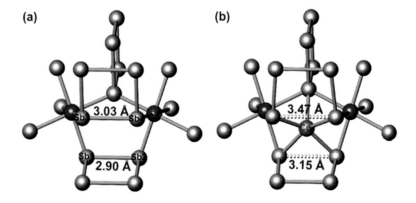

FIGURE 1.15 Coordination of the Co and Sb atoms in (a) the ideal $CoSb_3$ supercell, and (b) in a supercell containing a neutral interstitial cobalt defect Co_i. The Sb–Sb bonds stretched by incorporation of the interstitial cobalt in the lattice are indicated. Reproduced from Li et al., *Chemistry of Materials* 28, 2172 (2016). With permission from the American Chemical Society.

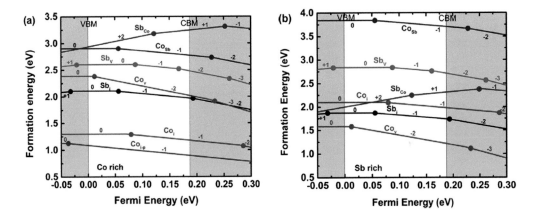

FIGURE 1.16 Calculated defect formation energies with the charge state indicated as a function of E_F in (a) Co-rich $CoSb_3$, and (b) Sb-rich $CoSb_3$. The white area represents the band gap region. Reproduced from Li et al., *Chemistry of Materials* 28, 2172 (2016). With permission from the American Chemical Society.

the outcome of the physical situation depends so critically on what approximation is used in DFT calculations, despite the works using highly refined state-of-the-art computing packages.

1.2   TERNARY SKUTTERUDITES

Ternary skutterudites are a modified version of the binary skutterudites whereby one makes isoelectronic substitutions (keeping the valence electron count at 72) either at the cation site M by a pair of elements from Column 8 and 10 forming, e.g., $Fe_{0.5}Ni_{0.5}Sb_3$ by replacing Co^{3+} with Fe^{2+} and Ni^{4+} (Kjekshus and Rakke 1974), or at the anion site X with a pair of elements from Column 14 (Ge,Sn) and Column 16 (S, Se, Te), e.g., forming $CoGe_{1.5}Se_{1.5}$ by substituting Sb^{5+} with Ge^{4+} and Se^{6+} (Korenstein et al. 1977). One can also substitute simultaneously on both the cation and anion sites and form $Fe_4Sb_8Te_4 \equiv FeSb_2Te$ by replacing divalent Fe for trivalent Co on the cation site and compensating for it on the anion site by replacing one pentavalent Sb atom with a hexavalent Te (Fleurial et al. 1997). In the extreme case, one can also consider partial substitutions on both the

TABLE 1.3

Lattice Parameters and Band Gaps for Ternary Skutterudites

Compound	Lattice Parameter (Å)	Band Gap (eV)	Reference
$Fe_{0.5}Ni_{0.5}Sb_3$	9.0904	~0.16	Kjekhus and Rakke (1974)
$Fe_{0.5}Pd_{0.5}Sb_3$	9.2048	–	Navratil et al. (2010)
$Fe_{0.5}Pt_{0.5}Sb_3$	9.1950	–	Fleurial et al. (1997)
$Ru_{0.5}Ni_{0.5}Sb_3$	9.1780	–	Fleurial et al. (1997)
$Ru_{0.5}Pd_{0.5}Sb_3$	9.2960	~ 0.60	Caillat et al. (1996)
$Ru_{0.5}Pt_{0.5}Sb_3$	-	–	Fleurial et al. (1997)
$Fe_{0.5}Ni_{0.5}As_3$	8.2560	–	Pleass and Heyding (1962)
$Fe_{0.5}Ni_{0.5}P_3$	7.7529	–	Jeitschko et al. (2000)
$CoGe_{1.5}S_{1.5}$	8.0170	–	Korenstein et al. (1977)
$CoGe_{1.5}Se_{1.5}$	8.3076	1.50	Korenstein et al. (1977)
$CoGe_{1.5}Te_{1.5}$	8.7270	–	Fleurial et al. (1997)
$CoSn_{1.5}Se_{1.5}$	8.7259	–	Fleurial et al. (1997)
$CoSn_{1.5}Te_{1.5}$	9.1284	> 2.0	Fleurial et al. (1997)
$RhGe_{1.5}S_{1.5}$	8.2746	–	Lyons et al. (1978)
$RhSn_{1.5}Te_{1.5}$	9.3064	–	Vaqueiro and Sobany (2008)
$IrGe_{1.5}S_{1.5}$	8.2970	–	Lyons et al. (1978)
$IrGe_{1.5}Te_{1.5}$	8.9632	–	Vaqueiro and Sobany (2008)
$IrGe_{1.5}Se_{1.5}$	8.5591	1.38	Lyons et al. (1978)
$IrSn_{1.5}S_{1.5}$	8.7059	–	Lyons et al. (1978)
$IrSn_{1.5}Se_{1.5}$	8.9674	1.24	Fleurial et al. (1997)
$IrSn_{1.5}Te_{1.5}$	9.3320	2.56	Fleurial et al. (1997)
$FeSb_2Se$	–	–	Fleurial et al. (1997)
$FeSb_2Te$	9.1120	0.27	Fleurial et al. (1997)
$RuSb_2Se$	9.2570	–	Fleurial et al. (1997)
$RuSb_2Te$	9.2680	1.20	Fleurial et al. (1997)
$OsSb_2Te$	9.2980	–	Fleurial et al. (1997)
$PtSn_{1.2}Sb_{1.8}$	9.3900	–	Bahn et al. (1969)
$NiGeP_2$	7.9040	–	Fleurial et al. (1997)
$NiGeBi_2$	9.4400	–	Fleurial et al. (1997)

Source: Data collected from the literature.

cation and anion sites, such as replacing binary $IrSb_3$ with ternary $PtSn_{1.2}Sb_{1.8}$ (Caillat et al. 1996). It follows that by synthesizing ternary skutterudites, one greatly expands the pallet of skutterudite compounds. A collection of ternary skutterudites with their lattice parameters and, if known, the band gap is presented in Table 1.3.

Ternary skutterudites formed by isoelectronic substitutions on the cation site fully preserve the cubic structure (space group $I m \bar{3}$) of binary skutterudites and, structurally, the only notable difference is their somewhat larger lattice parameter. Since the substitution takes place on the metal site M, which is more tolerant to disorder than the pnicogen rings, the band structure, as we discuss later, is similar to that of the corresponding binary skutterudite. While the disorder on the cation sublattice enhances phonon scattering and thus decreases the thermal conductivity, it also, unfortunately, negatively impacts the carrier mobility and strongly diminishes electrical conductivity. Most of the cation-substituted ternaries reported on in the literature are antimonides, with one representative each of the arsenide-based and the phosphide-based structures, $Fe_{0.5}Ni_{0.5}As_3$ (Pleass and Heyding 1962), and $Fe_{0.5}Ni_{0.5}P_3$ (Jeitschko et al. 2000).

In contrast, ternary skutterudites synthesized by isoelectronic substitutions on the anion site undergo a more impacting structural change. The previously homogeneous environment of the pnicogen rings of binary skutterudites now turns into a heterogeneous ring structure that must accommodate two distinct elements, one from the Column 14 (element A) and one from the Column 16 (element B). The tilt of the corner-sharing octahedrons formed by Columns 14 and 16 ions, which is comparable to that in binary skutterudites, now gives rise to two crystallographically distinct and distorted four-membered rings with stoichiometry $[(A14)_2(B16)_2]^{4-}$. The rings are ordered in alternating layers perpendicular to the [111] direction of the skutterudite unit cell and with the elements A14 and B16 in trans configuration to each other Bos and Cava (2007). The anion ordering lowers the symmetry of the structure from cubic ($Im\bar{3}$) to centrosymmetric rhombohedral with the space group $R\bar{3}$ and appears to be a common feature of all ternary skutterudites formed by isoelectronic substitutions on the pnicogen site X. The hint that such structures possess a reduced symmetry was initially found in the XRD studies by Lyons et al. (1978), Lutz and Kliche (1981), Partik et al. (1996), and Fleurial et al. (1997), where weak superstructure reflections on top of what otherwise looked like an ordinary binary skutterudite spectrum with strong reflections satisfying the body-centered condition $h + k + l = 2n$, where n is an integer, suggested specific ordering of the A14 and B16 species. Rigorous confirmation was made by powder neutron diffraction studies by Vaqueiro et al. (2006, 2008), and further supported by detailed DFT calculations by Volja et al. (2012), all indicating that the symmetry is lower than cubic. The consequence of the presence of two distinct pnicogen rings is that each metal atom M finds itself in a distorted octahedral environment with three atoms A14 and three atoms B16 forming the face of the octahedron. The X–M–X angles deviate significantly from 90°.

In an early work of Korenstein et al. (1977), it was suggested that in some cases a small fraction of atoms B16 can occupy the sites of atom A14 and vice versa, i.e., a partial disorder may occur on the pnicogen rings while preserving the overall stoichiometry. This was subsequently carefully examined in high resolution powder neutron diffraction studies of $MGe_{1.5}S_{1.5}$ by Vaqueiro et al. (2008), which indicated that the disorder model fits the experimental data marginally better than the model assuming a perfect ring order, yielding 2.5%, 0.6%, and 5.6% of sulfur atoms residing on the sites of germanium (and vice versa) for M = Co, Rh, and Ir, respectively. Of course, the above percentages represent a very small fraction of atoms, and it is not clear whether such partial disorder also applies to other ternary skutterudites formed by isoelectronic substitutions on the pnicogen sites. What is undisputable, however, is that many single-phase ternary skutterudites reported in the literature actually contain secondary phases when viewed through a sharper eye than the ordinary laboratory X-ray examinations. As an example, high-resolution powder neutron diffraction work by Vaqueiro et al. (2008) revealed that $IrGe_{1.5}S_{1.5}$ and $RhSn_{1.5}Te_{1.5}$, often described as single or near-single phase structures (Lyons et al. 1978, Partik et al. 1996, Vaqueiro et al. 2006, and Zevalkink et al. 2015), actually may contain some 10% of IrGe and $RhTe_2$, respectively.

1.3 FILLED SKUTTERUDITES

Beyond giving rise to planar rectangular rings of pnicogen atoms, the tilt of the MX_6 octahedrons also creates large icosahedral voids in the skutterudite structure. In Figure 1.1, the X_4 rings are present in only six out of eight small cubes constituting the unit cell; the two remaining cubes (the front upper left and the back bottom right) are empty. This is necessary to keep the ratio Co^{3+}:$[As_4]^{4-}$ equal to 4:3 and thus assure the overall charge neutrality of the structure. These structural voids occupy a body-centered position of the cubic lattice, called the 2a-site using Wyckoff notation. As already noted, including the void and considering only one-half of the unit cell, the skutterudite structure can be written as $\square M_4[X_4]_3$, where $\square$ designates the void. The electron count of this complex is 72, reflecting the semiconducting behavior of binary skutterudites. The voids in the skutterudite structure are large enough to accommodate foreign species. The radius of the voids of all binary

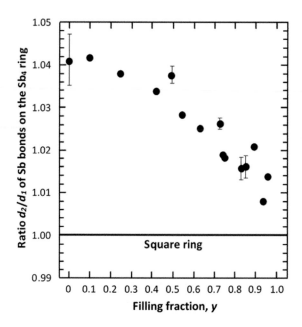

FIGURE 1.17 Ratio of the longer to the shorter bond lengths $d_2 / d_1 = (1/2 - y)/z$ as a function of the filling fraction for several filled antimony skutterudites. With the increasing filling fraction, the ratio approaches the unity, i.e., the rectangle becomes more square, closer to satisfying the Oftedal relation, $2(y + z) = 1$. The data points are taken from Chakoumakos and Sales, *Journal of Alloys and Compounds* 407, 87 (2006). The reader should note that the original Figure 6b in the paper erroneously indicates the ratio as (Pn – Pn)/(Pn – Pn)′ instead of (Pn – Pn)′/(Pn – Pn). With permission from the Elsevier.

skutterudites is given in the last column of Table 1.1, and increases from phosphide skutterudites to arsenide skutterudites to antimony skutterudites. When the voids are occupied, such skutterudites are called 'filled' (occasionally 'stuffed') skutterudites. In this form, the filler ion has 12 pnicogen nearest neighbors and 8 transition metal next-nearest neighbors. The distance between the fillers is rather large at $a \sqrt{3/2}$, where a is the lattice constant.

Reviewing existing databases, Chakoumakos and Sales (2006) summarized common structural changes observed in reported studies describing alteration of structural parameters upon a filler entering the skutterudite void. With the increasing filling fraction, the lattice parameter a increases and so do the positional parameters y and z. However, the rate of increase of the positional parameters is not the same, with z increasing faster than y. Consequently, the bond lengths d_1 and d_2 on the pnicogen ring are altered so that the shorter length d_1 approaches the longer length d_2, and the ring becomes more square. The trend toward satisfying the Oftedal relation in Eq. 1.4 was already noted in Figure 1.5. The ratio of the bond lengths d_2/d_1 as a function of filling for several antimony skutterudites is shown in Figure 1.17, replotted from the data of Chakoumakos and Sales.

More recently, using a combination of temperature-dependent synchrotron powder X-ray diffraction with Density Functional Theory (DFT), Hanus et al. (2017) explored the chemical nature of bonding in $CoSb_3$ and $Yb_yCo_4Sb_{12}$ with an emphasis on structural changes arising from Yb filling and the effect of the rising temperature. The thermal expansion causes linearly dependent increases in the lattice parameters of both $CoSb_3$ and $Yb_yCo_4Sb_{12}$, and also in the bond lengths d_1 and d_2, shown in Figure 1.18. The linear thermal expansion coefficient of $Yb_yCo_4Sb_{12}$ (10.5×10^{-6} K^{-1}) is marginally larger than that of $CoSb_3$ (9.9×10^{-6} K^{-1}). While the bond lengths d_1 and d_2 in $Yb_yCo_4Sb_{12}$ expand at not too different rates (d_1 with 12.6×10^{-6} K^{-1} and d_2 with 15.9×10^{-6} K^{-1}), the longer bond length d_2 in $CoSb_3$ expands at nearly twice the rate of the shorted bond length d_1 (15.6×10^{-6} K^{-1} compared to 8.6×10^{-6} K^{-1}). This causes the Sb_4 ring in $CoSb_3$ to depart more and

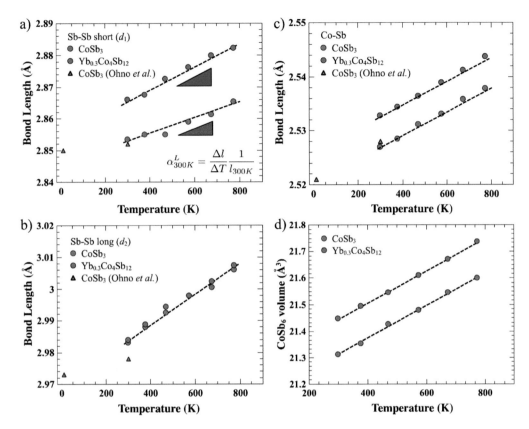

FIGURE 1.18 Temperature dependence of (a) the shorter bond length d_1, (b) the longer bond length d_2, and (c) the distance Co–Sb in CoSb$_3$ and Yb$_{0.3}$Co$_4$Sb$_{12}$ based on temperature-dependent synchrotron powder X-ray diffraction measurements. Lines correspond to linear fits used to calculate the local thermal expansion coefficient α_{300K}^{L}. For comparison, triangles represent values of Ohno et al. (2007) obtained at 10 K and 300 K based on the maximum entropy method. In CoSb$_3$, the bond lengths d_1 and d_2 expand at very different rates, 8.6×10^{-6} K^{-1} and 15.6×10^{-6} K^{-1}, respectively, while in Yb$_{0.3}$Co$_4$Sb$_{12}$ the two bond lengths expand with more similar rates, 15.9×10^{-6} K^{-1} and 12.6×10^{-6} K^{-1}, respectively. (d) Temperature dependence of the octahedral volume, showing the linear thermal expansion of the CoSb$_6$ octahedra. Reproduced from R. Hanus et al., *Chemistry of Materials*, 29, 1156 (2017). With permission from the American Chemical Society.

more from the Oftedal relation as the temperature increases, while the ring in Yb$_y$Co$_4$Sb$_{12}$ substantially maintains its shape. DFT calculations indicated that, upon filling with Yb, the significantly enhanced density of electrons populates antibonding states. The altered electronic band structure gives rise to band convergence. Details concerning the issue of band convergence in CoSb$_3$ are presented in Section 3.1 and Section 6.4.3.2.

The first filled skutterudite, LaFeP$_{12}$, was synthesized by Jeitschko and Braun in 1977. Subsequently, filled skutterudites were also prepared by Braun and Jeitschko (1980a,b), as antimonides and arsenides, respectively. Filled skutterudites were initially viewed as rather esoteric but intriguing cage compounds displaying interesting and surprising physical properties, the most notable among them being superconductivity, discovered by Meisner (1981), with an unexpectedly high transition temperature exceeding 10 K in the case of LaRu$_4$As$_{12}$ (Torikachvili et al. 1987).

The assessment of the relevance and value of filled skutterudites changed dramatically during the first half of the 1990s. It was already known from the work of Dudkin and his collaborators dating back to the late 1950s (Dudkin (1958), Dudkin and Abrikosov (1956, 1957, 1959)), that binary skutterudites have excellent electronic properties. Unfortunately, their thermal conductivity was too high to make use of them as thermoelectric materials. In their seminal work, Slack and Tsoukala (1994) pointed out a possibility of reducing the large thermal conductivity of binary $IrSb_3$ by inserting foreign species into the structural voids of the skutterudite lattice. These were expected to act as independent Einstein-like oscillators and interfere with the normal phonon modes of the framework, a strategy that nicely fit the concept of Phonon-Glass-Electron-Crystal (PGEC) promulgated by Slack (1995). Shortly thereafter, Morelli and Meisner (1995) demonstrated the effectiveness of this approach by comparing thermal conductivities of binary $CoSb_3$ and $IrSb_3$ with that of the filled $CeFe_4Sb_{12}$ skutterudite. The difference in the thermal conductivities was dramatic, with the filled structure displaying nearly an order of magnitude reduction compared to the binary skutterudites. The work of Morelli and Meisner served as an impetus for the scientific community and, since that time, filled skutterudites have become one of the most closely pursued thermoelectric materials with excellent prospects for mid-temperature power generation applications.

Depending on the constituency of the skutterudite framework, filled skutterudites can be divided into four general classes: (1) filled skutterudites with the $[M_4X_{12}]$ framework; (2) filled skutterudites of the form RT_4X_{12}, where the framework is a polyanionic complex $[T_4X_{12}]^{4-}$ with T being a Group 8 transition metal, X a pnicogen atom, and R the filler ion; (3) filled skutterudites with the $[Pt_4Ge_{12}]$ framework; and (4) the recently discovered skutterudites with anion fillers. I will discuss the different kinds of filled skutterudites in turn.

1.3.1 Filled Skutterudites with the $[M_4X_{12}]$ Framework

1.3.1.1 Void Occupancy

Filled skutterudites with the $[M_4X_{12}]$ framework are typified by $R_yCo_4Sb_{12}$, with the framework isoelectronic with that of the binary skutterudites, and the icosahedral voids are typically only partially filled ($y < 1$) with element R that can be a rare earth (up to Gd*), alkaline earth, alkali metal ion or even Tl, Sn and other species. Examples are $Ce_{0.1}Co_4Sb_{12}$ (Uher et al. 1997, Morelli et al. 1997); $Yb_{0.2}Co_4Sb_{12}$ (Nolas et al. 2000); $Tl_{0.22}Co_4Sb_{12}$ (Sales et al. 2000); and $Ba_{0.3}Co_4Sb_{12}$ (L. D. Chen et al. 2001a,b). Such skutterudites filled, or rather, partially filled with electropositive elements are invariably n-type conductors as the filler ion R, depending on its valence state, donates electrons to the structure. Moreover, they have a highly restricted void occupancy because the structure becomes rapidly saturated with the negative charge. Attempts to exceed the void occupancy inevitably result in the formation of secondary phases, unless one charge compensates by substituting some Fe for Co or Sn and Ge on the site of Sb. Due to its largest icosahedral cage, inexpensive chemical elements, and the best thermoelectric performance, the $[Co_4Sb_{12}]$ framework has been, by far, the most favored structure of all possible binary skutterudite combinations. The early effort (in the second half of the 1990s), focused on exploring the most effective filler ions and their void occupancy, called filling fraction limits (FFL), which would maximize the power factor while suppressing the lattice thermal conductivity to the greatest extent so as to achieve the highest figure of merit. It was soon realized that the trivalent rare earths, such as Ce and La, have a very limited void occupancy, $y < 0.1$ (Chen et al. 1997), and $y < 0.23$ (Nolas et al. 1998), respectively, in the $[Co_4Sb_{12}]$ framework. On account of its intermittent valence between 2+ and 3+, Yb has a somewhat higher void occupancy of $y = 0.25$ when prepared by melting and annealing, and the occupancy was later enhanced to $y = 0.29$ by Wang et al. (2016), and similar occupancy was achieved with the aid of the

---

* Rare earths are well known for their lanthanide contraction, i.e., the heavier ones have a smaller ionic radius. Rare-earth elements past Gd have too small radii compared to a large cage of $CoSb_3$, and cannot establish strong enough bonds with the framework atoms and, hence, be trapped in the cage. Gd is an interesting boundary element. While Jeitschko et al. (2000) managed to synthesize a phosphide form of the skutterudite $GdFe_4P_{12}$ from Sn flux, Gd cannot be trapped in a much larger cage of $CoSb_3$ (Mei et al. (2006)). However, it can fill a framework where Co is partly replaced by Fe, i.e., in a skutterudite $Gd_yFe_xCo_{4-x}Sb_{12}$ (Liu et al. (2011)). In fact, the Gd filling fraction y rises with the content of Fe roughly linearly as $y = x/4$ up to $x = 1.7$, at which point the structure has expanded by the presence of Fe to the extent that it can no longer hold Gd, and the structure decomposes into more stable phases.

TABLE 1.4

Void Occupancy in the CoSb$_3$ Skutterudite

Filler Ion	Maximum Occupancy (%)	Reference
Na	65	Pei et al. (2006)
K	60	Mei et al. (2008)
Rb	25	Mei et al. (2008)
Ca	20	Puyet et al. (2004)
Sr	40	Zhao et al. (2006a)
Ba	45	Chen et al. (2001b)
La	23	Nolas et al. (1998)
Ce	9	Morelli and Meisner (1995)
Nd	13	Kuznetsov et al. (2003)
Eu	44	Berger et al. (2001)
Yb	29	Wang et al. (2016)
Ga	2	Harnwunggmoung et al. (2011)
In	22	T. He et al. (2006)
Tl	22	Sales et al. (2000)
Sn	100	Takizawa et al. (2002)

Source: Data collected from the literature.

---

high-pressure synthesis (2 GPa at 590°C for 120 min) by Chen et al. (2015). With divalent Eu, the occupancy reached $y = 0.44$ (Berger et al. 2001). The divalent alkaline earth Ba can fill up to about 45% of available void sites. The sister alkaline earths, Ca, studied by Puyet et al. (2004), and Sr, explored by Zhao et al. (2006a), can also partially fill voids of CoSb$_3$, but they are less effective in improving the thermoelectric performance because of the damaging impact on the carrier mobility. Making use of high-pressure synthesis, all heavy rare earths were able to enter the voids, Kihou et al. (2004), and 100% void occupancy was also reported for Sn, Takizawa et al. (2002). Maximum void occupancies in the [Co$_4$Sb$_{12}$] framework are summarized in Table 1.4. In general, the larger the ionic radius or the lower the charge state of the filler, the larger is its occupancy. Constraints on the void occupancy also apply to arsenide and phosphide-based frameworks (Zemni et al. 1986).

The void occupancy in Table 1.4 should not be taken dogmatically but rather as a guide only. In most cases, the filling fraction limit (FFL) was determined from the flattening of the dependence of the lattice parameter on the content of the filler, or from the filler content at which secondary phases were first detected in the X-ray diffraction pattern. However, the filling fraction limit depends on the temperature, the deficiency or excess of Co and Sb in the structure, and the synthesis conditions (more specifically, the annealing treatment). Consequently, the above noted criteria for the filling fraction limit may not be sufficient. I illustrate this for the case of Yb, one of the most explored and effective filler species in CoSb$_3$ due to its heavy mass and small ionic radius.

Numerous reports (Nolas et al. 2000, Dilley et al. 2000, Anno et al. 2000, Sales et al. 2000, Nagao et al. 2002, Zhao et al. 2006b, Li et al. 2008) on filling voids of CoSb$_3$ with Yb in uncompensated samples prepared by melting and annealing near 600°C or by melt-spinning locate the FFL in the range $0.2 < y < 0.29$. This is in accord with the *ab initio* DFT calculations by Mei et al. (2006) that place the FFL of Yb at $y = 0.3$. However, annealing at higher temperatures (800 °C) for seven days increased the FFL to $y = 0.4$, as shown by Xia et al. (2012). Ball milling followed by hot pressing used by Yang et al. (2009a) was also reported to lead to a high FFL of Yb of nearly 50%, and the same level of filling was claimed by Dahal et al. (2014). The more recent ball milling synthesis by Ryll et al. (2018) placed the limit at $y = 0.44$. However, such large Yb filling fractions in uncompensated $Yb_yCo_4Sb_{12}$ are not universally accepted. By detailed X-ray, structural and quantitative compositional analysis, augmented by the composite theory of Bergman and Levy (1991), Wang et al. (2016) have shown that $Yb_yCo_4Sb_{12}$ with $y > 0.3$ is not a single-phase structure but, rather, a composite skutterudite with the $Yb_{0.3}Co_4Sb_{12}$ matrix and $YbSb_2$ inclusions. Because the $YbSb_2$ phase has a rather high room temperature thermal conductivity of 15 $Wm^{-1}K^{-1}$, it would account for increased lattice thermal conductivities seen in skutterudites filled with large nominal Yb filling contents. Of course, by charge compensating on the site of Sb with either Sn (Yang et al. 2001), or Ge (Lamberton et al. 2005), or by forming the $[Co_{4-x}Fe_xSb_{12}]$ framework where divalent Fe substitutes for trivalent Co, e.g., Chen et al. (2016), the void occupancy by Yb can be increased significantly to near $y = 0.5$. An important insight into the variability of the filling fraction limit of Yb in $CoSb_3$ was provided recently by Tang et al. (2015a) in their detailed evaluation of the equilibrium isothermal sections of the ternary phase diagram of Yb–Co–Sb at various temperatures. The results indicate a strong dependence of the FFL of Yb on the initial stoichiometry of Co and Sb and on the annealing temperature. The authors identify two stable skutterudite compositions, one for Co-rich structures where the equilibrium Yb content is $y = 0.44$, and one for Sb-rich frameworks with a considerably smaller Yb content of $y = 0.26$. This follows from the phase diagram in Figure 1.19. Starting with a slight excess of Sb and increasing the content of Yb, the system will proceed from the Sb-rich three phase region of $Yb_xCo_4Sb_{12}$ + $YbSb_2$ + liquid Sb to a two-phase region of $Yb_xCo_4Sb_{12}$ + $YbSb_2$ and then to a three-phase Co-rich region $Yb_xCo_4Sb_{12}$ + $YbSb_2$ + $CoSb_2$. If the actual Yb content is judged by the behavior of the lattice constant of such partially filled skutterudites, the lattice constant will stop increasing while the nominal composition is in the three-phase region (the phase rule here gives the degree of freedom as $F = C – P + 0 = 3 – 3 = 0$, where C is the number of components, P is the number of phases, and 0 stands for the fixed temperature and pressure) because such skutterudite phases have stable composition, designated by a blue dot in Figure 1.19a. This corresponds to the first plateau in Figure 1.19b, which could easily be mistaken for the solubility limit of Yb. As the content of Yb increases and the nominal skutterudite composition enters the two-phase region (here the phase rule yields $F = C – P + 0 = 3 – 2 = 1$), the composition of phases is not fixed, and the Yb content in the skutterudite phase increases, resulting in an

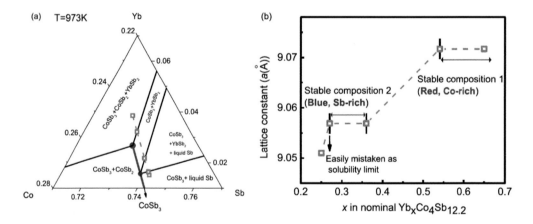

FIGURE 1.19   (a) A section of the ternary phase diagram at 973 K. Different nominal Yb contents in slightly Sb-rich $Yb_yCo_4Sb_{12.2}$ (designated by empty orange squares) result in a nonlinear dependence of the lattice constant shown in (b) because the partially filled skutterudite traverses (along the orange dashed line) different two and three phase regions of the phase diagram in (a). The initial plateau in the blue, Sb-rich region at about $y = 0.26$ (blue dot in Figure 1.19a) could be mistaken as the filling fraction limit of Yb. The actual limiting solubility of Yb in $[Co_4Sb_{12}]$ at 973 K is much higher at $y = 0.44$, designated by a red dot in Figure 1.19a. Redrawn from Y. Tang et al., *J. Materiomics* 1, 75 (2015). With permission from Elsevier.

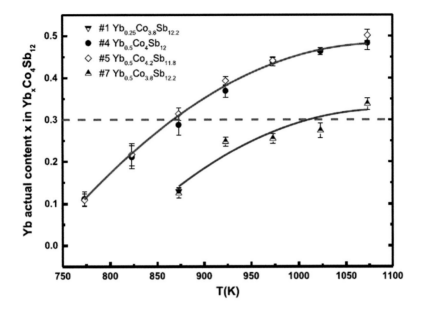

FIGURE 1.20 Temperature dependence of the filling fraction of Yb in $Yb_yCo_4Sb_{12}$ on the annealing temperature. The red curve indicates the temperature dependence of the actual Yb content in the stable skutterudite composition represented by a red dot (maximum filling fraction) in Figure 1.19a. The blue curve is the temperature dependence of the actual Yb content in the stable skutterudite composition represented by a blue dot in Figure 1.19a. The dashed line is the targeted Yb content $y = 0.3$ for optimized thermoelectric performance, which crosses the solubility curves at 873 K and 1023 K, respectively. Redrawn from Y. Tang et al., *J. Materiomics* 1, 75 (2015). With permission from Elsevier.

increased lattice constant. This continues until the Co-rich three-phase region is reached where, again, the composition is fixed and all skutterudites have the same actual Yb content (red dot), hence, the second plateau in the lattice constant in Figure 1.19b is reached. The lesson to take home is that observing a plateau in the lattice constant may not be sufficient in determining the solubility limit when one deals with a ternary system. Furthermore, both limiting filling fractions of Yb (one referring to Co-rich and the other one to Sb-rich $CoSb_3$ frameworks) are very strongly dependent on the annealing temperature, as depicted in Figure 1.20. The results thus encompass and explain the variability range of filling fraction limits of Yb reported in literature. One would expect that similar behavior pertains to the filling fraction limit of other filler species, except that no one, as yet, has looked closely into it.

I should also note that, most recently, a proposal to use Selected Area Electron Diffraction (SAED) was put forward by Wang et al. (2017) as a useful tool to determine the filling fraction limit of Yb and other filler species in the skutterudite lattice. The technique relies on the finding that, at the filling fraction limit, the crystal lattice is stressed, and the resulting lattice distortions can be detected in the SAED pattern.

A comment is in order regarding the filling fraction limit of Ce in $CoSb_3$. From the early studies by Chen et al. (1997), it was taken for granted that the FFL of Ce is very low, certainly less than $y = 0.1$. With such low FFL, it was not possible to obtain high enough electron concentration to optimize the thermoelectric power factor. This was very unfortunate because Ce, although a rare earth element, is actually the most highly abundant rare earth in the Earth's crust, with an estimated 68 ppm abundance, according to Emsley (2011). In comparison to Yb (3 ppm), generally regarded as the most effective single filler in $CoSb_3$, this represents a more than 20 times higher abundance, which is directly reflected in more than an order of magnitude cost advantage for Ce. Thus, a very limited FFL of Ce has been a major impediment to the economically viable use of single-filled $CoSb_3$ in large-scale applications. Recently, using the concept of solubility design that relies on a detailed knowledge of the Ce–Co–Sb phase diagram, Tang et al. (2015b) have shown that it is possible to double the FFL of Ce in $Ce_yCo_4Sb_{12}$ to values of $y = 0.20$. With such enhanced Ce content, the structure attains the desired 0.4–0.6 electrons per $[Co_4Sb_{12}]$ unit required to maximize the power factor (Yang et al. 2009). Because of the much-enhanced density of electrons, single-filled $Ce_{0.14}Co_4Sb_{12}$ (actual composition by EPMA) reached a ZT value of 1.3 at 850 K, matching the best values reported for Yb-filled $CoSb_3$.

1.3.1.2 Criteria for Filling

Table 1.4 lists several filler species able to occupy voids of the $CoSb_3$ framework. Are these the only lucky few or can skutterudite voids accept any foreign ion? If not, what criteria govern the choice of the fillers? An immediate constraint that comes to mind is the size of the filler ion with respect to the size of the icosahedral void. Obviously, if the radius of the ion is larger than the "radius" of the void, such an ion will not fit. From Table 1.1, it follows that antimonides have the most accommodating framework as they have the largest void radius. Based purely on the geometrical arguments, antimonide skutterudites can be filled with the largest of the rare earth and alkaline earth ions. However, a void too large may be an impediment when attempting to fill the structure with smaller size ions. The reason is that such smaller size ions cannot establish reliable bonds with the now "far-distant" pnicogen atoms of the framework, and the small ions "fall out" of the void during the synthesis. This is the reason why the desirable heavier but smaller rare earth ions (you may remember that rare earths are subject to the so-called lanthanide contraction, meaning that as the atomic number and therefore the mass of the ion increases from lanthanum (La) to lutetium (Lu), the ionic radius progressively decreases) beyond gadolinium cannot fill $CoSb_3$ voids under the normal synthesis conditions and require assistance of several GPa of high pressure (Sekine et al. 1998). But, beyond the obvious issue of relative sizes of ions and voids, is there some other fundamental reason why certain species can fill the voids while others cannot?

The first attempt to provide some guidance regarding maximum possible filling y_{max} in the voids of $CoSb_3$ was made by L. D. Chen (2002), who provided an empirical relation based on the ratio of the ionic radius of the filler ion r_{ion} and the radius of the skutterudite cage r_{cage} (the r_{ion}/r_{cage} ratio of all stable MCo_4Sb_{12} skutterudites falls between 0.6 and 0.9), and the valence of the filler n,

$$y_{max} = \frac{r_{ion}}{r_{cage}} - 0.086n - 0.24. \tag{1.12}$$

Early theoretical rationalization of void filling in the skutterudite structure was made by Løvvik and Prytz (2004) with DFT calculations exploring the stability and the filling limit of La, Y, and Sc in CoP_3 at absolute zero temperature, and by Bertini and Gatti (2004) using *ab initio* calculations to document that Sn in $CoSb_3$ can occupy both 2a and 24g positions, i.e., it can act as a filler and also substitute for Sb. Although these works presented interesting and useful assessments of particular filler species, neither work addressed the key question concerning a criterion for successful filling. This had to wait until Shi et al. (2005, 2007) carried out their detailed DFT studies of the filling fraction limit in the skutterudite structure.

An important starting point here was a realization that the formation of a stable filled skutterudite is a highly competitive process between two main phenomena: the formation of an isolated filler atom in $CoSb_3$, and the formation of secondary phases. The two processes can be viewed as chemical reactions of the form:

$$yR + Co_4Sb_{12} \rightarrow R_yCo_4Sb_{12}, \tag{1.13}$$

describing the formation of a filler atom R in $CoSb_3$, and

$$R + 2CoSb_3 \rightarrow RSb_2 + 2CoSb_2, \tag{1.14}$$

representing the formation of secondary phases, with the most stable ones being RSb_2 and $CoSb_2$. Combining Eqs. 1.13 and 1.14 with certain prefactors, the reaction describing the competition between the formation of a filled antimony skutterudite and the formation of secondary phases can be written as

$$nR + Co_4Sb_{12} \rightarrow \frac{2-n}{2-y}R_yCo_4Sb_{12} + \frac{2(n-y)}{2-y}RSb_2 + \frac{4(n-y)}{2-y}CoSb_2. \tag{1.15}$$

Reactions in Eqs. 1.13 and 1.14 are characterized by the formation enthalpies per filler; ΔH_1 for the reaction in Eq. 1.13 and ΔH_2 for the reaction in Eq. 1.14. In terms of the formation enthalpies, the Gibbs formation energy per filler is then

$$\Delta G = 2\frac{n-y}{n(2-y)}\Delta H_2 + \frac{(2-n)y}{n(2-y)} \times \left\{ \Delta H_1(y) + k_B T \left[\ln y + \frac{1-y}{y}\ln(1-y) \right] \right\}. \tag{1.16}$$

The second term in Eq. 1.16 includes $k_B T \left[\ln y + \dfrac{1-y}{y}\ln(1-y) \right]$, which stands for the configura-

tional entropy on account of the random distribution of the filler in the voids. The filling fraction limit is obtained from Eq. 1.16 by minimizing ΔG with respect to the filler fraction y. Evaluating the respective formation enthalpies ΔH_1 and ΔH_2, as done in the original publications, it can be shown that the filling fraction limit y_{max} becomes

$$y_{max} - \left(\frac{y_{max}}{2}\right)^2 = C_1\left(1 - e^{-0.25(x_{Sb}-x_R)}\right) + C_2. \tag{1.17}$$

Constants C_1 and C_2 can be specified by linearly fitting Eq. 1.17 where x_{Sb} and x_R are the Pauling electronegativities of Sb and the filler, respectively. Since the maximum filling fraction should be greater than zero, i.e., $y_{max} > 0$, Eq. 1.17 can be simplified to yield a particularly appealing criterion for the formation of filled antimony skutterudites,

$$x_{Sb} - x_R > 0.80. \tag{1.18}$$

Thus, an atom R is expected to fill the voids of $CoSb_3$ if its Pauling electronegativity with respect to the electronegativity of Sb satisfies Eq. 1.18. Atoms R for which $x_{Sb} - x_R$ is less than 0.80 are predicted to have their FFL zero and cannot fill antimonide skutterudites.

It should be understood that the DFT calculations leading to Eq. 1.18 were carried out at the absolute zero temperature and that temperature effects could somewhat modify the numerical value of the criterion. Nevertheless, at least for rare earth and alkaline earth species, the agreement with the experimental results is excellent, as shown in Figure 1.21.

Having a simple criterion for filling has stimulated a vigorous research activity to identify new filler species in $CoSb_3$, beyond rare earth and alkaline earth ions. Specifically, due to their very low mass, alkali metals have long been considered of no interest as possible fillers because it was deemed highly unlikely that they could significantly impair heat transport in skutterudites and, consequently, no one has tried to use them as fillers. However, theoretical calculations along the lines outlined above by Zhang et al. (2006) have brought to light a surprising finding; Na and K turned out to have very large FFLs of some 60% in $CoSb_3$, with the computational results promptly confirmed in experiments by Pei et al. (2006). Subsequently, all alkali metals have been screened as possible filler species by Mei et al. (2008). High FFLs of Na and K have been confirmed and Rb filling was estimated at a much smaller value of 25%, likely due to its rather large ionic radius (1.52 Å). The largest and the smallest alkali metal atoms, Cs and Li, respectively, however, could not at that

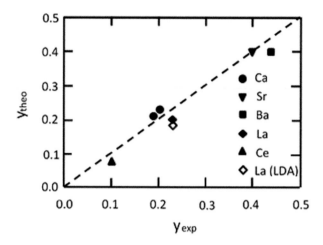

FIGURE 1.21 Calculated filling fraction limits vs. the experimentally measured ones. The dashed line indicates $y_{theo} = y_{exp}$. Redrawn from X. Shi et al., *Physical Review Letters* 95, 185503 (2005). With permission from the American Physical Society.

time fill the skutterudite voids. The ionic radius of Cs^{1+} (1.88 Å) is nearly the size of the void, and, as such, Cs ions had little chance of entering. On the other hand, the ionic radius of Li^{1+} (0.92 Å) is too small relative to the void radius of $CoSb_3$, and Li was unable to establish stable bonds with Sb atoms at ambient pressure. However, as shown by Zhang et al. (2011), applying high pressures of 3–4 GPa for a couple of hours at temperatures of 1073 K, referred to as High Pressure High Temperature (HPHT) synthesis, resulted in single phase $Li_xCo_4Sb_{12}$ with the Li content up to $x = 0.4$. The point is that in DFT calculations of FFL, where the formation enthalpy determines whether filling will succeed or secondary phases will form, the enthalpy is taken at ambient pressure. However, since it varies with pressure, and the pressure of 3–4 GPa used in the HPHT synthesis is rather high, species that are predicted as not being able to fill the skutterudite voids at ambient conditions may do so under a large applied pressure.

1.3.1.3 Column 13 Elements (Ga, In, and Tl) as Fillers

The filling criterion expressed in Eq. 1.18 serves reasonably well for rare earths, alkaline earths, and alkali fillers, but fails rather miserably for some other fillers. The case in point are Column 13 elements, Ga, In, and Tl. Having Pauling electronegativities of 1.81, 1.78, and 2.04, respectively, they should have no chance to enter voids of $CoSb_3$ according to the criterion in Eq. 1.18, because their electronegativities are too close to that of Sb (2.05), especially for Tl. Yet, Tl was one of the first fillers positively identified by Sales et al. (2000), and its FFL determined as $y = 0.22$. Subsequent studies by Harnwunggmoung et al. (2010) confirmed the FFL in the range of 20%–25% filling. For Ga and In, their role as fillers in $CoSb_3$ has been amply demonstrated experimentally, but with an interesting twist. Both Ga and In act as amphoteric impurities, meaning that in a given semiconductor ($CoSb_3$ in this case), they can behave as both donors and acceptors, depending on which lattice site they enter. If they are void fillers, they donate electrons, while if they substitute on the site of Sb, they become acceptors, as their valence state is less than that of Sb. Moreover, there is also a theoretical possibility that they might substitute for cobalt because the radii of Ga^{3+} and In^{3+} ions are very similar to that of the Co^{3+} ion. However, the formation energy of this defect makes it energetically far less likely. Experiments by Harnwunggmoung et al. (2011), attempting to fill $CoSb_3$ with Ga, i.e., forming $Ga_yCo_4Sb_{12}$ with up to $y \sim 0.3$, have invariably resulted in a filled skutterudite $Ga_{0.02}Co_4Sb_{12}$ plus a GaSb secondary phase, suggesting that the FFL of Ga in $CoSb_3$ is exceedingly small at $y = 0.02$. In a detailed study of the prospect of filling voids with Ga, Qiu et al.

(2013) came to a remarkable conclusion that Ga actually plays a dual role in the skutterudite lattice; it acts as a filler and simultaneously substitutes on the site of Sb, forming a Ga_{Sb} defect. Here, in the case of Ga, the simultaneous occupancy of voids and Sb lattice sites is unique in the sense that the two processes are coupled. The electrons donated by Ga that fill the voids saturate the dangling bonds created by Ga substituting for Sb, giving rise to a dual site occupied skutterudite described as $Ga_yCo_4Sb_{12-y/2}(Ga_{Sb})_{y/2}$. First principles calculations supported by experimental work set the maximum value of y at 0.1 when Ga acted as a dual site occupant. This value of y is much larger than if Ga acted exclusively as a filler or only substituted for Sb. The situation is nicely visualized with the use of an expanded region of the ternary phase diagram near the $CoSb_3$ composition, where the range of solubility of Ga is extended when its content y aims at the formation of dual site occupied $Ga_yCo_4Sb_{12-y/2}(Ga_{Sb})_{y/2}$, shown by a red line in Figure 1.22. When the nominal content of Ga exceeds $y = 0.1$, the dual site occupied skutterudite $Ga_yCo_4Sb_{12-y/2}(Ga_{Sb})_{y/2}$ is accompanied by secondary phases of GaSb and CoGa. Deviating from the red line either to the left or right, i.e., attempting to preferentially form either filled skutterudite or preferentially place Ga on the site of Sb, leads to a majority skutterudite phase but with considerably less Ga in the system than the nominal composition would suggest.

As far as indium is concerned, its filling limit in $CoSb_3$ was initially set by T. He et al. (2006) at $y = 0.22$, and the value confirmed in subsequent studies by Mallik et al. (2008, 2009). Later measurements by Leszczynski et al. (2013) set the limit at $y = 0.26$. However, it was also reported that the ingots of $Ce_xIn_yCo_4Sb_{12}$ with the FFL of indium less than $y = 0.15$ (Li et al. 2009), and ingots of $Yb_xIn_yCo_4Sb_{12}$ with the indium FFL limit less than $y = 0.20$ (Graff et al. 2011), contained an InSb minority phase, suggesting that the FFL of indium might be lower than $y = 0.22$. In a study by Sesselmann et al. (2011), it was found that indium was present at grain boundaries when its nominal content was just 0.20. Thus, there is plenty of experimental evidence to suggest that In, just like Ga, enters the voids of the structure as well as substitutes for Sb. Moreover, Grytsiv et al. (2013), studying phase relations in the In–Co–Sb system, reported that there are actually two solubility

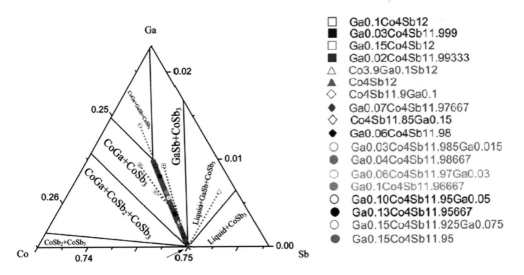

FIGURE 1.22   A section of the ternary phase diagram near $CoSb_3$ at 923 K. On account of the stability of Ga in dual site occupancy, the solubility range of Ga is extended in the direction of $Ga_yCo_4Sb_{12-y/2}(Ga_{Sb})_{y/2}$ to $y = 0.1$, following the heavy red line in the figure. The nominal sample compositions are shown as open symbols. Closed symbols represent the skutterudite composition $Ga_yCo_4Sb_{12-y/2}(Ga_{Sb})_{y/2}$ determined by EPMA value for Ga and are connected to the corresponding nominal skutterudite composition with a dotted line. Adopted and redrawn from Y. Qiu et al., *Advanced Functional Materials* 23, 3194 (2013). With permission from John Wiley & Sons.

limits of In in $In_yCo_4Sb_{12}$, depending on the exact Co/Sb ratio in the starting composition. When the skutterudite phase is in equilibrium with $CoSb_2$ and InSb, the FFL is $y = 0.22$, while when in equilibrium with InSb and Sb, the FFL is much lower at $y = 0.09$. These seemingly conflicting results were placed into proper perspective by the work of Tang et al. (2014) who carefully explored the ternary In–Co–Sb phase diagram in the proximity of the $CoSb_3$ composition by DFT calculations and experimental studies. The authors concluded that indium displays a dual character; it fills the voids and it substitutes on the sites of Sb. This time, and unlike the case of Ga, indium enters the void and donates only one electron, i.e., its ion is in the valence state In^{1+}. However, when In substitutes on the site of Sb, the skutterudite is deficient of two electrons. Consequently, a perfectly charge balanced dual site indium occupied skutterudite can be designated as $In_{2y/3}Co_4Sb_{12-y/3}(In_{Sb})_{y/3}$. Similar to the case of Ga, the authors constructed a ternary In–Co–Sb phase diagram in the proximity of $CoSb_3$ at 873 K, illustrated in Figure 1.23. The FFL of indium in skutterudite compositions following the $In_{2y/3}Co_4Sb_{12-y/3}(In_{Sb})_{y/3}$ stoichiometry is enhanced to a value $y = 0.27$, as shown by the thick red line in Figure 1.23. Nominal compositions that are slightly Sb-rich, such as $In_{0.3}Co_{4-\delta}Sb_{11.9+\delta}$, $\delta > 0$, are in a two-phase region of liquid + $CoSb_3$, and here the phase rule gives the degree of freedom of 1. Therefore, this additional degree of freedom can be used to describe the tie line between phases with variable indium content.

As Figure 1.23 indicates, the tie line connecting the nominal composition and the skutterudite phase is rather steep, resulting in a considerably lower content of indium in the skutterudite and an In-rich liquid In-Sb impurity phase. Nominal compositions that are somewhat Co-rich, e.g., $In_{0.3}Co_{4-\delta}$

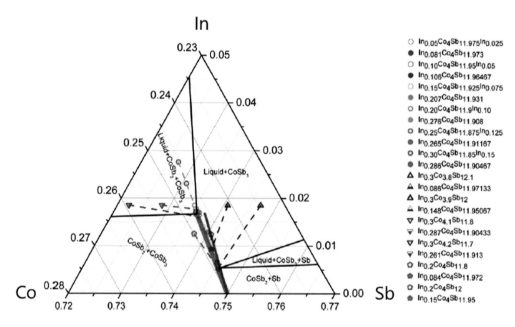

FIGURE 1.23 Enlarged phase region of the In-containing $CoSb_3$ at 873 K. The diagram depicts several phases with In either filling the voids or substituting at sites of Sb or both. Note that because of the stability of In in dual sites (voids and substituting for Sb), the region of In solubility is extended along the red line, i.e., in the direction of $In_{(v)2y/3}Co_4Sb_{12-y/3}In_{(Sb)y/3}$ up to $y = 0.27$. Here $In_{(v)}$ and $In_{(Sb)}$ mean In at the void and In substituting for Sb, respectively. The nominal sample compositions are shown as open symbols. Solid symbols refer to compositions of the skutterudite majority phase determined by the experimental EPMA value for In content, which is below 0.27 for all samples. The nominal and experimental compositions are connected with a dotted line. Reproduced from Y. Tang et al., *Energy & Environmental Science* 7, 812 (2014). With permission from the Royal Society of Chemistry.

$Sb_{11.9+\delta}$, $\delta < 0$, are in a three-phase region with fixed compositions of liquid InSb, $In_{0.27}Co_4Sb_{11.9}$ and $CoSb_2$. Therefore, all nominal compositions in this three-phase region will produce the same skutterudite with the actual indium content of $y = 0.27$. Thus, to maximize the content of indium in $CoSb_3$, it is more advantageous to start with a slightly Co-rich stoichiometry.

When comparing skutterudites with the same contents of Ga and In, the one containing indium always has much higher density of electrons, implying that the exactly charge balanced chemical formula above cannot be a fully accurate representation of the structure containing indium. Rather, there must be more than twice the number of In atoms entering the voids than those substituting on Sb sites. In other words, the dual site indium entry into the skutterudite lattice must coexist with simple indium filling. When this is the case, the single-phase region (red line in Figure 1.23) will slightly shift toward the blue line ($In_yCo_4Sb_{12}$).

It should be recognized that filling the voids with electropositive ions is an exceptionally effective doping mechanism whereby the density of electrons can be enhanced by two to four orders of magnitude to densities as high as $10^{20} - 10^{22}$ cm^{-3}, depending on the degree of filling and the valence state of the filler ion. The carrier concentration of electrons as a function of the filling fraction of several fillers is depicted in Figure 1.24.

As we discuss in more details in a section describing the influence of filler species on the thermal conductivity, there are two distinct visions of the role fillers play in the skutterudite structure. The first one, vastly more popular, is based on the PGEC concept of Slack (1995), where the filler acts as an independent localized (Einstein-like) phonon mode that resonantly interacts with the skutterudite's normal phonon modes of comparable frequencies, and this interaction leads to a degradation of the heat conduction. The alternative viewpoint, based primarily on the documented coupling of vibrations of the filler and framework atoms (Koza et al. 2008), relies on a reduction in the phonon group velocity near the frequency of the filler that is caused by a downshift of the phonon modes responding to the presence of the filler in the voids. Within the PGEC concept, because the localized vibrations of fillers are substantially harmonic (Feldman et al. 2000), one can model them as a filler of mass m attached at the end of a spring with the spring constant k that makes displacements x from the center position of the void. The reader may recall from introductory physics that the vibration energy of the spring is $E = \frac{1}{2} kx^2$. Consequently, each respective filler has its own resonant frequency $\omega_0 = (k/m)^{1/2}$. Spring constants and the resonant

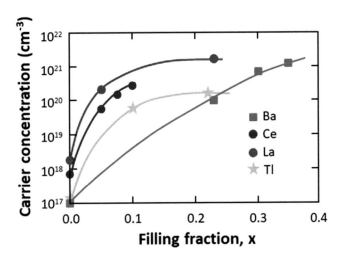

FIGURE 1.24 Carrier concentration of electrons as a function of filling fraction for some partially filled skutterudites of the form $R_xCo_4Sb_{12}$, R = Ba, Ce, La, and Tl.

frequencies of notable fillers from among rare earths, alkaline earths and alkalis were calculated by Yang et al. (2007a) and range from a low frequency of 42 cm^{-1} for heavy Yb to a high frequency of 141 cm^{-1} for light K. Since skutterudites are cubic systems, the parameter values are isotropic. Within the same family of fillers, e.g., rare earths, or alkaline earths, there are small differences in the resonant frequency. However, the differences are substantial between fillers from different families.

While proposals to fill the voids with more than a single type of filler (Nolas et al. 1998), and successful attempts to realize them in experiments have been made earlier (Chen et al. 2001a, Bérardan et al. 2003, 2005a, and Lu et al. 2005), the impact of seeing the actual frequency range one can cover with different fillers has stimulated many new efforts to double and even triple fill the skutterudite structure. With carefully selected filler species to cover a wide range of resonant frequencies, numerous studies have shown that such multifilling is superior to single filling and is able to reduce the thermal conductivity to a much greater extent. Moreover, the power factor is often better optimized with multifilling rather than with a single filler. Thus, with single-filled skutterudites, the highest ZT values hover around unity with occasional reports of $ZT \sim$ 1.2, such as in the case of $Ba_{0.3}Co_{3.95}Ni_{0.05}Sb_{12}$ (Dyck et al. 2002), and $In_{0.25}Co_4Sb_{12}$ (T. He et al. 2006). Judiciously chosen double-filler species can raise the ZT to values in the range 1.3–1.4. Particularly effective double-filling combinations are Ba+Yb (Shi et al. 2008), Ba+Ce (Bai et al. 2009), Ba+In (Zhao et al. 2009), and In+Ce (Li et al. 2009). To maximize the degradation of the lattice thermal conductivity, even several triple-filling combinations of elements, aiming to cover a broad range of localized vibrational frequencies, have been explored (Graff et al. 2011), Zhang et al. 2009), Rogl et al. 2011), and Ballikaya et al. 2012). Among these, $Ba_{0.08}La_{0.05}Yb_{0.04}$ Co_4Sb_{12} achieved the record high $ZT = 1.7$ at 850 K (Shi et al. 2011). I also mention the filling of skutterudites with the misch-metal by Yang et al. (2007b), an alloy of rare-earth elements in their naturally occurring proportions, dominated by Ce, La, Pr, and Nd with trace impurities of Fe, Si, Al, and O. Misch-metal is an intermediate product in the preparation of all high-purity rare-earth elements, and the content of rare-earths is some 95.5%. Finally, impressive values approaching $ZT \sim 2$ were obtained by filling the skutterudite cages with didymium, an alloy consisting of 4.76 mass% Pr and 95.24 mass% Nd, and applying severe plastic deformation using high-pressure torsion (Rogl et al. 2010, 2014). Full discussion on these is given in the thermal transport section of the book.

It had always been assumed that partial filling of voids in $CoSb_3$ and other binary skutterudites is a random process. This seems reasonable at very low filling fractions. However, as the filling fraction increases, interaction between filler species may overcome entropy and partial filler ordering may take place. This was first documented by Kim et al. (2010) in the case of Ba, which has a rather high filling fraction in $CoSb_3$. Combining *ab initio* calculations with the cluster expansion method, which captures all possible ways of distributing Ba and vacancies over the void sites, one constructs an onsite Hamiltonian that is expressed exactly as a series expansion of configurational basis functions. DFT is then used to parameterize the effective cluster interactions and determine the truncation of the series. Upon application of Monte Carlo simulations to the cluster-expanded Hamiltonian, one can construct a temperature-composition phase diagram, such as shown in Figure 1.25a. Atomic structures of the two respective ordered phases γ and α are depicted in Figures 1.25b and 1.25c. The γ phase has composition $x = 0.25$ and is highly anisotropic with Ba–Ba nearest neighbors along [111] and Ba-vacancy pairs along the other <111> family axes. The phase disorders at $T \approx 350$ K. In contrast, the α phase has $x = 0.5$ and a high symmetry with the filled Ba sites forming a diamond network. This phase disorders at $T \approx 750$ K. Because the phase transitions are all first order, the phases coexist at intermediate content x and low temperatures. At higher temperatures, the phases coexist with a solid solution. The presence of order-disorder phase transitions and the two-phase mixture have major implications on the thermal conductivity as I discuss in Section on phonon transport.

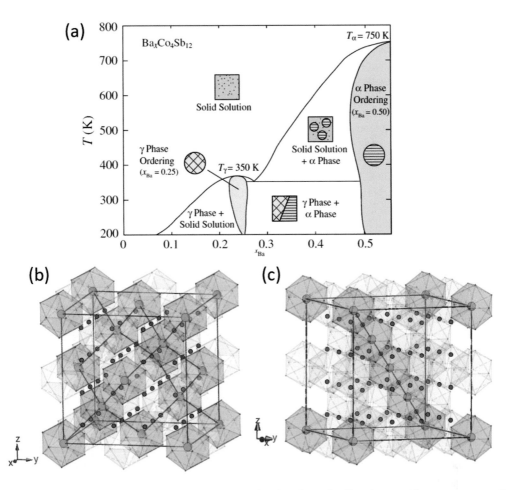

FIGURE 1.25 (a) The phase diagram of $Ba_xCo_4Sb_{12}$ showing the order-disorder transition temperatures T and T between phase α with the Ba content $x = 0.5$ and phase γ with $x = 0.25$. Atomic structure of (b) phase γ and (c) phase α. Ba atoms are represented by green circles and Co atoms by small blue circles. Atoms of the polyhedra (Sb) are not shown to avoid crowding. Adapted from H. Kim et al., *Physical Review Letters* 105, 265901 (2010). With permission from the American Physical Society.

The formation of the ordered γ and α phases depicted in Figure 1.25 and, specifically, their mixture for Ba contents $0.25 < x < 0.5$, have been explored by positron annihilation measurements by Zhang et al. (2015). The measurements rely on the positron lifetime being determined by trapping and annihilation within open volumes of defects, such as voids in the skutterudite structure. As such, positron spectroscopy provides useful information on the size of the defects, their concentration, and on their neighboring chemical environment. The reader interested in the topic is referred to a monograph by Krause-Rehberg and Leipner (1999) and a review by Tuomisto and Makkonen (2013). In the measurements of Zhang et al. (2015), the major component of the positron lifetime τ_I = 263 ± 2 ps (96% − 98.7% of the intensity) was associated with positron annihilation within the voids while a minor component τ_2 (less than 4% of intensity) was due to annihilation in the pores between the grains of spark plasma sintered powders. The positron lifetime increased with the increasing content of Ba, and by comparing the experimental trend with their calculation results for various Ba contents, the authors were able to confirm the formation of ordered phases γ near $x = 0.25$ and α near $x = 0.5$, and the mixture they form for $0.25 < x < 0.5$.

In multifilling skutterudite voids to achieve as low a thermal conductivity as possible, one usually neglects interactions between different filler species. Hu et al. (2018), based on their studies of local structure and local dynamics, pointed out that the interactions are actually significant. Using a combination of Yb and In fillers in $CoSb_3$ where the filling fraction of Yb was high, the authors observed Yb pushing out In fillers from their 2a filler sites into 24g sites of Sb, where they replaced Sb.

In a recent theoretical study of the electronic structure and phase stability of Yb-filled $CoSb_3$, one of the most prospective filled skutterudites for thermoelectric applications, Isaacs and Wolverton (2019) noted a mild tendency of $Yb_xCo_4Sb_{12}$ to phase separate into Yb-rich and Yb-poor regions. Such a phase separation has never been observed in experiments because a very small energy lowering of only about 10 meV per Yb/void site is likely trumped by configurational entropy that stabilizes the single-phase structure. Considerably more favorable conditions for phase separation were found in $La_{0.8}Ti_{0.1}Ga_{0.1}Fe_4Sb_{12}$ by Ren et al. (2017), where the computed energy reduction of 0.6 eV actually resulted in the formation of La-rich and La-poor skutterudite phases. The ensuing multiscale strain field fluctuations very effectively scattered phonons and led to an exceptionally low lattice thermal conductivity approaching the theoretical minimum.

1.3.1.4 Skutterudites as Zintl Phases

Skutterudites are often viewed as Zintl phases. What do I mean by that statement? Zintl phases (after a German chemist Eduard Zintl) are electronic structures where an electropositive metallic element reacts with a nonmetallic element or metalloid by transferring its valence electrons, and the electronegative element accepts as many of them as fully fill its valence shell. The process is similar to ionic salts like NaCl, except that rather than achieving an electronic octet as isolated species, the anions bond together to form a polyanion to fulfill the octet rule. Precise electron counting is at the heart of the Zintl concept and is useful for identifying new thermodynamically stable semiconducting structures. In the case of skutterudites, the polyanion is $[X_4]^{4-}$ and the binary skutterudite can be equivalently written as $M^{3+}_4[X_4]^{4-}$, reflecting the Zintl concept. Such a binary skutterudite has 24 electrons per formula unit (or, as I noted in Section 1.1, a VEC of 72 per one-half of the unit cell).

The full power of the Zintl concept is realized when considering filled skutterudites and aiming to achieve the desired semiconducting structure. A fully filled skutterudite RM_4X_{12} with the transition metal M in the low spin d^6 configuration, is expected to be a stable semiconducting structure when its valence electron count is $4 \times 24 = 96$ electrons per formula unit. Therefore, if one wishes to fill the structure fully with a trivalent rare earth element, such as La, one must remove three electrons from the pnicogen site X in order to keep the electron count at 96 and thus maintain a semiconducting structure. Among the first reports on this balancing act is the work of Tritt et al. (1996) with $LaIr_4Sb_9Ge_3$, where three Ge atoms were substituted for Sb to compensate for three electrons donated by La. In the case of divalent alkaline earth Ba, one must remove two electrons from the pnicogen site X to achieve a semiconducting skutterudite, for instance by synthesizing $BaCo_4Sb_{10}Sn_2$, where two tetravalent Sn atoms replace two pentavalent Sb atoms. Such precise valence electron counting enables the expansion of the family of semiconducting skutterudites to numerous new compositions, as demonstrated by Lou et al. (2015). The approach was supported by DFT calculations and verified by synthesizing some 63 new single-phase skutterudites, with a potential to expand the synthesis to hundreds of new structures when including phosphide- and arsenide-based skutterudites and not just antimonides. Although I doubt that many of these new precise count-synthesized skutterudites will turn out to be outstanding thermoelectric materials (the disruption of the pnicogen rings might lower lattice thermal conductivity but it will also dramatically degrade the carrier mobility, just as was the case of ternary skutterudites), expanding the spectrum of skutterudite compounds should open exciting possibilities for the study of new physical phenomena that will shed more light on these remarkable structures.

1.3.1.5 Atomic Displacement Parameter

A convenient measure of the strength of bonding of a filler in the skutterudite cage is the so-called atomic displacement parameter (ADP), in years past also known as the thermal parameter. It specifies the mean-square displacement amplitude of an atom about its equilibrium lattice site, in other words, it reflects on how "vigorously" the atom vibrates and possibly also on any present static disorder at that site. In crystalline solids, the ADPs acquire the underlying structural anisotropy, and to describe the displacement amplitude by just a single parameter, the ADPs are converted into an isotropic quantity U_{iso} (usually given in units of Å^2) that provides the mean-square displacement amplitude of the atom averaged over all directions. The ADPs are collected by X-rays or neutron scattering. From the time the first filled skutterudites were synthesized, it was clear that the ADPs of the filler ions are unusually large in comparison to the other atoms of the structure. In fact, the term "rattling" was first used in the original paper of Braun and Jeitschko (1980a). ADPs are useful as an estimate of physical parameters, such as the Debye temperature, lattice specific heat, and sound velocity. If the rattling ion is treated as a localized harmonic oscillator (Einstein oscillator), the mean-square displacement amplitude $< u^2 >$ is given by

$$U_{iso} \equiv \left\langle u^2 \right\rangle = \frac{h}{8\pi^2 mf} \coth \frac{hf}{2k_B T}, \qquad (1.19)$$

where m is the mass of the rattler and f is its frequency. At high T, Eq. 1.19 reduces to the classical formula $U_{iso} \cong k_B T/K$, where $K = m (2\pi f)^2$ is the force constant of the oscillator. Thus, from the slope of U_{iso} vs. T, one can estimate the Einstein temperature $\theta_E = hf/k_B$. An example of the ADPs in Figure 1.26 shows the U_{iso} for La$_{0.75}$Fe$_3$CoSb$_{12}$ that clearly indicates an unusually large ADP for the filler La. In fact, Sales et al. (1999) proposed that the U_{iso} can be used to estimate thermal conductivity of skutterudites. The key point here is the use of the kinetic formula for thermal conductivity, $\kappa = \frac{1}{3} C_v v_s l$, with C_v being the specific heat at constant volume, v_s the speed of sound, and the assumption that the mean free path of phonons l is primarily limited by the separation of the local oscillators (rattlers). The authors obtained a very good agreement between room temperature thermal conductivities estimated from the U_{iso} and the experimental values and, as such, the ADPs are useful tools for screening prospective novel thermoelectrics.

Some dozen years later, Mi et al. (2011) used synchrotron radiation X-ray diffraction measurements between 90 K and 700 K on several skutterudites filled with rare-earth elements, R$_y$Co$_4$Sb$_{12}$ with R = La, Ce, Nd, Sm, Yb, and Eu, all filled with the same fixed content $y = 0.1$. They compared

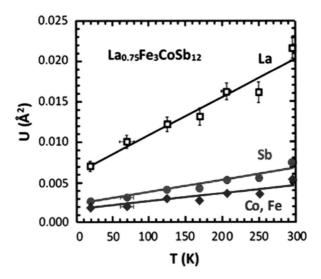

FIGURE 1.26 Atomic displacement parameters of La, Fe, Co, and Sb in the filled skutterudite La$_{0.75}$Fe$_3$CoSb$_{12}$, showing a significantly larger ADP of the La filler compared to other species of the structure. Reproduced from B. C. Sales et al., *Physical Review B* 56, 15081 (1997). With permission of the American Physical Society.

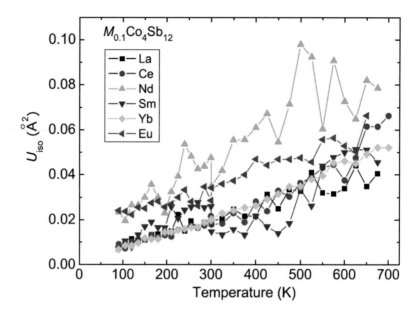

FIGURE 1.27 Temperature dependence of the isotropic ADP, i.e., U_{iso}, for rare-earth fillers in partially filled $R_yCo_4Sb_{12}$. Reproduced from J. L. Mi et al., *Physical Review B* 84, 064114 (2011). With permission from the American Physical Society.

the ADPs and obtained transport parameters for their samples. Across the entire temperature range covered, the largest ADP, by far, was observed for Nd, reflecting its weak bonding in the void. Isotropic ADPs, in terms of U_{iso}, of all samples studied are shown in Figure 1.27. Later on, in Section 5.13.3.2, we shall see that Nd also attains the lowest lattice thermal conductivity, for the same filling fraction, of all single-filled skutterudites, further attesting to the validity of the proposal by Sales et al. that the ADPs of the fillers and the lattice thermal conductivities of filled skutterudites are closely related.

1.3.2 Filled Skutterudites with the $[T_4X_{12}]^{4-}$ Framework

1.3.2.1 $[T_4X_{12}]^{4-}$ Polyanion and Valency of the Fillers

What is generally referred to as filled skutterudites are structures with the framework that is not iso-electronic with the binary skutterudites but rather has a general form RT_4X_{12}, where R is an electropositive filler (lanthanide, actinide, alkaline-earth ion, alkali ion), T stands for the transition metal of Group 8 elements (T = Fe, Ru, Os), and X is the pnicogen atom (X = P, As, Sb). It is important to recognize that it is not the neutral $[M_4X_{12}]$ complex of binary skutterudites we have dealt with in Section 1.3.1, but the charged $[T_4X_{12}]^{4-}$ complex that forms the basis of the filled skutterudites. This fundamental polyanionic building block of the structure, $[T_4X_{12}]^{4-} \sim [TX_3]^{1-}$, contains an iron-like Group 8 transition metal in its low-spin d^6 configuration (in the case of Fe it is exclusively Fe^{2+} state), just as was the Group 9 cobalt-like element. While binary skutterudites with the Group 8 transition metals do not form in bulk ($FeSb_3$ was, however, synthesized in thin film form, as discussed later) because of the obvious charge imbalance, the filler ion R is supposed to supply the missing electrons, saturate the bond, and electrically neutralize the structure. To do so, the polyanion $[T_4X_{12}]^{4-}$ should couple with a tetravalent R^{4+} ion, resulting in a diamagnetic and semiconducting filled skutterudite $R^{4+}[T_4X_{12}]^{4-}$ with the valence electron count of 72. This happens very rarely and is plausible only with Ce (assuming Ce is in the 4+ state) among the rare-earth elements, and possibly Th (exclusively tetravalent) and U among actinides. As a result of the semiconducting nature of conduction, a small magnetic susceptibility, and a pronounced dip in the cell volume of $CeFe_4P_{12}$, Jeitschko and Braun

(1977) took it as a sign of Ce being tetravalent in $CeFe_4P_{12}$. Later, following the discovery of filled arsenide skutterudites, the same reasoning was applied to $CeFe_4As_{12}$. Lacking the dip in the cell volume and displaying metallic conduction, the corresponding antimonide $CeFe_4Sb_{12}$ was assumed filled with trivalent Ce (Grandjean et al. 1984). Subsequent X-ray absorption near-edge spectroscopy (XANES) by Xue et al. (1994) and Grandjean et al. (2000), as well as X-ray photoemission spectroscopy by Grosvenor et al. (2006) revealed that even in $CeFe_4P_{12}$, the filler Ce enters the voids as trivalent ($4f^1$) rather than tetravalent ($4f^0$) ion. The gap resulting in the semiconducting behavior likely arises due to a contribution of Ce $4f^1$ orbital to bonding states near the Fermi level. In other words, $4f$-orbital hybridization appears an important aspect of bonding in these filled skutterudites.

Most of the rare earths are trivalent, some of them (notably Yb and to some extent Sm) display an intermediate valence, and Eu usually favors a divalent state. The departure of a rare earth element from the trivalent state is vividly documented by a graph of the lattice constant of $LnFe_4P_{12}$ skutterudites as a function of the rare-earth filler shown in Figure 1.28. Ce, Eu, and Yb deviate substantially from the otherwise straight-line dependence characterizing the decreasing lattice parameter with the increasing atomic number (smaller and heavier) of the rare earth filler.

Other filler species, such as alkaline earths, are divalent, and alkali metal fillers are monovalent. Consequently, the valence count for skutterudites fully filled with ions of valence less than 4+ will be smaller than 72, and such filled skutterudites will tend to be paramagnetic metals, unless magnetic interactions set in at lower temperature or the structure becomes a superconductor. From the perspective of thermoelectricity, the metallic state is not favorable since metals invariably have small Seebeck coefficients. To force the compound back into its semiconducting regime (and the valence electron count of 72), the structure must be charge compensated. Such charge compensation can be attempted either on the pnicogen rings (replacing some pnicogen atoms in the 24g positions with the Group IV elements, such as Ge or Sn), or on the 8c sites occupied by the iron-group metal. Both approaches not only greatly expand the scope of filled skutterudites but also result in a plethora of fascinating physical properties well beyond thermoelectric applications. A variety of possible ground states arising from a strongly correlated nature of electrons in filled skutterudites (Maple et al. 2003, Sato et al. 2009), presents an opportunity to observe phenomena such as superconductivity (both BCS-type as

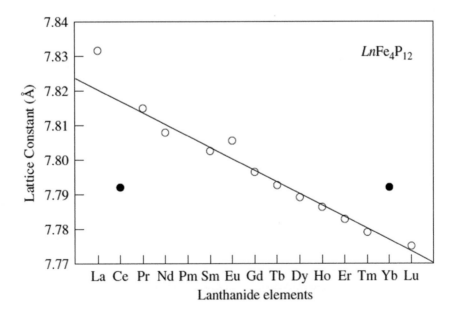

FIGURE 1.28 Lattice constant of iron-phosphide skutterudites as a function of the rare-earth filler. $CeFe_4P_{12}$ and $YbFe_4P_{12}$ and are ashown by black circles. Reproduced from I. Shirotani et al., *Physica B* 382, 8 (2006). With permission from Elsevier.

well as highly exotic), various long-range magnetic orders, heavy fermion behavior, non-Fermi liquid states, mixed valence, and others, a truly fertile ground for modern solid-state physics.

1.3.2.1.1 Charge Compensation

Although a disruption of the pnicogen ring environment leads to a dramatic reduction in the lattice thermal conductivity, it also has a highly adverse effect on the carrier mobility. Consequently, one has a better chance to obtain a good thermoelectric material if charge compensation is done on the site of the iron-group metal. By far the most explored possibility is a substitution of Co for Fe used in the original studies by Fleurial et al. (1996) and Chen et al. (1997) with $Ce_yFe_{4-x}Co_xSb_{12}$, by Sales et al. (1996, 1997) in their work with $La_yFe_{4-x}Co_xSb_{12}$, by Grytsiv et al. (2002) with $Eu_yFe_{4-x}Co_xSb_{12}$, and by Bérardan et al. (2005b) with (Ce/Yb) double-filled frameworks of $[Fe_{4-x}Co_xSb_{12}]$ and $[Fe_{4-x}Ni_xSb_{12}]$. As the amount of Co substituting for Fe (or Ni substituting for Fe) increases, the nature of conduction changes from that of a metal to a semiconductor. For instance, given a trivalent filler, simple crystal chemistry anticipates attaining the semiconducting state when $y = (4 - x)/3$ for Co replacing Fe atoms and when $y = (4 - 2x)/3$ for Ni replacing Fe atoms. Thus, under the full void occupancy, this would mean Co substituting for one-quarter of Fe atoms, while Ni would have to substitute for one-eighth of Fe atoms. Unfortunately, here is where the simplistic approach breaks down, as the allowed void occupancy enters the picture. While the full occupancy of voids is possible when $x = 0$, as the content of Co increases, the void occupancy rapidly diminishes. As we have already seen, in pure $CoSb_3$ it is severely limited to $y \leq 0.1$ of Ce and $y \leq 0.23$ of La. By a careful structural analysis, phase diagrams depicting the interdependence of the filling fraction y of any given filler ion on the degree of charge compensation can be established. Figure 1.29 shows an example relevant to $Ce_yFe_{4-x}Co_xSb_{12}$.

The $Ce_yFe_{4-x}Co_xSb_{12}$ system has drawn much attention as it harbors several compositions with promising thermoelectric performance that benefit not just from filling but also from the presence of Fe/Co, which further degrades the thermal conductivity while good electronic properties are maintained by the covalently-bonded framework. It was therefore deemed important to know in detail how the structure responds to filling and to replacement of some of Fe atoms by Co (or by Ni). Kitagawa et al. (2000), using X-ray diffraction accompanied by energy dispersive X-ray (EDX)

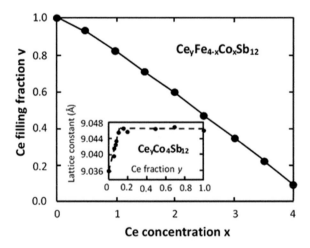

FIGURE 1.29 Phase diagram showing the fractional void occupancy by Ce as a function of Co content in $Ce_yFe_{4-x}Co_xSb_{12}$. The inset shows the variation of the lattice constant of $Ce_yCo_4Sb_{12}$ with increasing cerium filling. A sharp break near $y = 0.1$ indicates the maximum content of Ce that can be accommodated in the Co_4Sb_{12} cage. Reproduced from C. Uher, *Semiconductors and Semimetals*, Vol. 69, ed. by T. M. Tritt, Academic Press, San Diego, pp. 139–253 (2001). With permission from Elsevier.

measurements, established the variation of the lattice parameter and the positional parameters of Sb as a function of Ce content and the fraction of Co replacing Fe. The variation of the lattice parameter and the positional parameters is chiefly given by the filling fraction y with a minor contribution of the fraction x of Fe, according to the fitted equations:

$$a = \left(-0.00102 \pm 0.00009\right)x + \left(0.00094 \pm 0.00036\right)y + 0.9127 \pm 0.00069 \text{ in nm,} \qquad (1.20)$$

$$z_1 = 10^{-4}\left[\left(4.4 \pm 3.7\right)x + \left(33.9 \pm 14.4\right)y + 3307 \pm 25\right], \qquad (1.21)$$

$$z_2 = 10^{-4}\left[\left(-2.0 \pm 1.8\right)x + \left(9.5 \pm 6.9\right)y + 1580 \pm 13\right]. \qquad (1.22)$$

In the above equations, I have designated the positional parameters of Sb by z_1 and z_2 instead of the usual y and z to avoid confusion with the filling fraction y. The lattice parameter increases as a function of y and decreases with x. Both positional parameters increase with the increasing filling and their sum approaches the Oftedal relation, $z_1 + z_2 \longrightarrow 1/2$. Kitagawa et al. also noted that upon increasing filling, the intensity ratio of the (211) diffraction peak to the (310) diffraction peak rather sharply decreases, following the curve in Figure 1.30. The graph is a useful tool to quickly extablish the content of Ce in the voids of the $[Fe_{4-x}Co_xSb_{12}]$ framework in the absence of EDX measurements.

Based on the measurements of the thermal conductivity of $Ce_yFe_{4-x}Co_xSb_{12}$ skutterudites and monitoring the void occupancy for each respective Fe/Co composition, Meisner et al. (1998) ratio-nalized the behavior of the thermal resistivity as a function of the Ce content in terms of a solid solu-tion of fully filled skutterudite $CeFe_4Sb_{12}$ and the unfilled skutterudite $\square Co_4Sb_{12}$, where $\square$ stands for a vacancy. Writing the solid solution in a form of $(CeFe_4Sb_{12})_\alpha(\square Co_4Sb_{12})_{1-\alpha}$ and plotting the experimental lattice parameter as a function of α, excellent Vegard's law behavior was obtained. This simple model was useful in determining the optimum composition to achieve high figure of merit via looking for the maximum thermal resistivity in the structure. Bérardan et al. (2003) synthesized a series of $Ce_yFe_{4-x}Co_xSb_{12}$ skutterudites with $x = 0, 1, 2, 3,$ and 4 by arc melting and long-term annealing and noticed phase separation (see Figure 1.31), in structures annealed at 550°C and 650°C (but not at 700°C), which prompted them to question whether the structures are true solid solutions as proposed by Meisner et al.

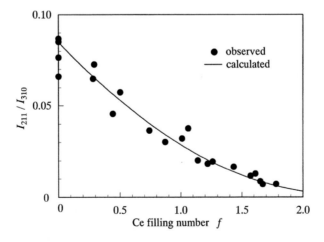

FIGURE 1.30   Ratio of X-ray intensity of (211) and (310) diffraction peaks as a function of Ce filling frac-tion. Reproduced from H. Kitagawa et al., *Materials Research Bulletin* 35, 185 (2000). With permission from Elsevier.

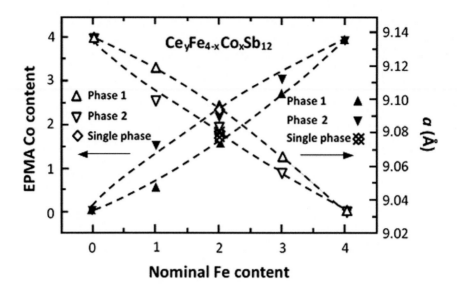

FIGURE 1.31 EPMA cobalt content (left scale) and the lattice parameter (right scale) as a function of the Fe content in $Ce_yFe_{4-x}Co_xSb_{12}$. Redrawn from D. Bérardan et al., *Journal of Alloys and Compounds* 350, 30 (2003). With permission from Elsevier.

Stimulated by impressive values of the figure of merit $ZT \geq 1$ reported by Nolas et al. (2000) in n-type $Yb_{0.19}Co_4Sb_{12}$ and by Tang et al. (2001) in p-type $Ce_{0.28}Co_{2.48}Fe_{1.52}Sb_{12}$, Bérardan et al. (2005a) double-filled the $[Fe_{4-x}Co_xSb_{12}]$ and $[Fe_{4-x}Ni_xSb_{12}]$ frameworks with a combination of Ce and Yb in equal amounts, i.e., forming $Ce_{y/2}Yb_{y/2}Fe_{4-x}Co_xSb_{12}$ and $Ce_{y/2}Yb_{y/2}Fe_{4-x}Ni_xSb_{12}$ skutterudites. X-ray absorption spectroscopy indicated a strictly trivalent nature of Ce while Yb had a mixed valence that rapidly decreased with the increasing content of Yb in the structure. Refined positional parameters of Sb remained at $y = 0.335$ and $z = 0.159$ in both series, indicating that double-filling by Ce and Yb has no effect on the shape of the Sb octahedron, while its size follows the variation of the lattice parameter depicted in Figures 1.32a and 1.32b.

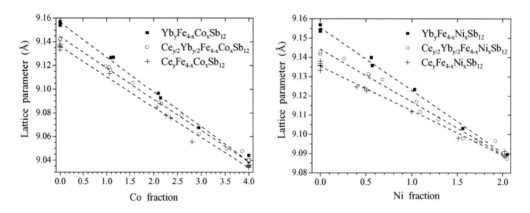

FIGURE 1.32 Lattice parameter as a function of (a) Co fraction in $(Ce/Yb)_yFe_{4-x}Co_xSb_{12}$, and (b) as a fraction of Ni in $(Ce/Yb)_yFe_{4-x}Ni_xSb_{12}$. Note, the lattice parameter of the Ce–Yb double-filled skutterudite falls between the lattice parameters of single-filled skutterudites. Adapted and redrawn from D. Bérardan et al., *Materials Research Bulletin* 40, 537 (2005b). With permission from Elsevier.

Regardless of whether Fe is partly replaced by Co or Ni, the lattice parameter of the double-filled structure falls between the lattice parameters of the single-filled skutterudites, in accord with the notion that the double-filled structures are solid solutions of the respective single-filled skutterudites.

The assessment of the overall structural response to filling the skutterudite lattice was carried out by Chakoumakos and Sales (2006), who used the available Crystal Structure Database (ICSD). Selecting reliable entries collected on both single crystals and powders by X-ray or neutron diffraction, they carried out Rietveld refinement of the data using the General Structure Analysis System (GSAS) developed at the Los Alamos National Laboratory (Report LAUR 86-748 (2000)). The outcome of this effort can be summarized in the following key points: (1) because the position of the transition metal as well as of the filler species is determined by the $Im\bar{3}$ symmetry of the skuttrudite structure, the only variable atomic parameter in the structure are the positional parameters y and z of the pnicogen atom X; (2) the four-membered pnicogen rings are a unique identifier of the skutterudite structure and pnicogen ring-orbitals play an important role in the form of a band structure, which, in turn, determines the physical properties of each skutterudite. In addition to Eqs. 1.1–1.3 relating the distance $D(M–X)$ between the transition metal and pnicogen atoms in terms of the positional parameters of the pnicogen, and the two sides d_1 and d_2 of a X_4 rectangular ring, in filled skutterudite one must add one more equation describing the separation between a filler A and the nearest neighbor pnicogens X on the icosahedron,

$$A - X = a\left(y^2 + z^2\right)^{1/2}, \tag{1.23}$$

where a is the lattice parameter. As I have already discussed, both the octahedra and icosahedra are slightly distorted from the regular shape, leading to small deviations from the Oftedal's relation ($y + z = \frac{1}{2}$), and giving rise to rectangular rather than square pnicogen rings. While the positional parameters of pnicogen atoms in binary skutterudites lie below the Oftedal line, as shown in Figure 1.5, filling of the structure shifts their position in such a way that the deviation from the Oftedal relation become smaller, and the positional parmaters of filled skutterudites cluster on both sides of the Oftedal line, depicted in Figure 1.33. The figure also includes positional parameters of skutterudites with the [Pt$_4$Ge$_{12}$] framework introduced in Section 1.3.3, and positionl parameters of anion-filled Fe-antimonide skutterudites discussed in Section 1.3.4. While the positional parameters of the former are situated above the Oftedal line, the latter fall distinctly below the Oftedal line. As Figure 1.33 indicates, the pnicogen positional parameters also tend to group, such as in the case of antimonides, where (Co,Fe,Ni) compositions are distinctly separated from (Ru,Rh,Os, and Ir) compositions; (3) for any given pnicogen X, the positional paramaters y and z increase with the increasing level of filling. The results of Kitagawa et al. (2000) for the Ce-filled [Fe$_{4-x}$Co$_x$Sb$_{12}$] framework in Eqs. 1.21 and 1.22, were generalized by Chakoumakos and Sales (2006) for any filler in antimonide skutterudites having the [Fe$_{4-x}$Co$_x$Sb$_{12}$] framework by equations

$$y = 0.3353 + 0.001348x + 0.00121x^2, \text{ with } r^2 = 0.86 \text{ and} \tag{1.24}$$

$$z = 0.1579 + 0.0007772x + 0.00213x^2, \text{ with } r^2 = 0.90, \tag{1.25}$$

obtained by fitting the data for the positional parameters as a function of the filling fraction. The lattice parameter increases with the content of Fe and with the filling fraction. A multiple linear regression fit made by Chakoumakos and Sales to the data of Bérardan et al. (2005b) for

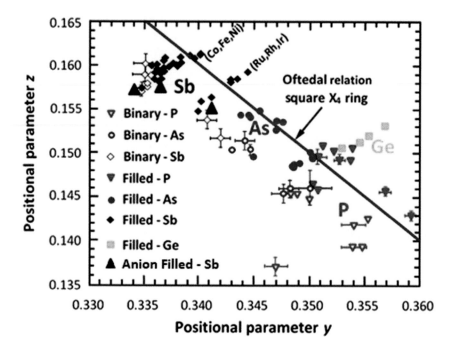

FIGURE 1.33 Positional parameters y and z of both binary and filled skutterudites. The blue line indicates the Oftedal relation. The data from Chakoumakos and Sales, *Journal of Alloys and Compounds* 407, 87 (2006) are supplemented by more recent entries measured on several high-pressure synthesized arsenides and on monovalent fillers in antimonides. All binary skutterudites fall below the Oftedal line, while the filled skutterudites are positioned both below and above the Oftedal line and closer to it. Included also are positional parameters of Ge in the $[Pt_4Ge_{12}]$ framework skutterudites, which all fall above the Oftedal line; and positional parameters of a few anion-filled antimony skutterudites, which generally fall in the range of other filled skutterudites but below the Oftedal line. Adapted from B. C. Chakoumakos and B. C. Sales, *Journal of Alloys and Compounds* 407, 87 (2006). With permission from Elsevier.

$Ce_{y/2}Yb_{y/2}Fe_{4-x}Co_xSb_{12}$ and $Ce_{y/2}Yb_{y/2}Fe_{4-x}Ni_xSb_{12}$ skutterudites yielded the lattice parameter of the form

$$a = 9.0372(9) + 0.0254(7)Fe - 0.020(2)Ce + 0.016(2)y. \qquad (1.26)$$

The numbers in the parentheses are the standard errors of the coefficients. Here Fe and Ce stand for the content of Fe and Ce per formula unit, and y represents the filling fraction. Given a particular framework, the cell volume increases with the level of filling. In antimony skutterudites, the cell volume typically expands by as much as 3.6%. Monovalent fillers expand the cell volume more than divalent fillers and, in turn, divalent fillers swell the unit cell more than trivalent fillers; (4) The ratio of the sides of the pnicogen ring d_2/d_1 is independent of the lattice parameter. The largest distortion of the ring from a regular square is observed in binary skutterudites. Upon filling, the ring tends to be more square, can become perfectly square (the positional parameters y and z falling exactly on the Oftedal line), and may also distort in opossite sense, i.e., instead of $d_2 > d_1$ and the positional parameters laying below the Oftedal line, d_1 becomes larger than d_2 with a shift of the positional parameters above the Oftedal line. As already noted, with the increasing filling fraction the intensity of lower angle reflections, such as (110), (200), and (211), decreases with respect to the strongest (310) reflection and, when properly calibrated, the ratio can be used as a measure of the filling fraction.

1.3.3 SKUTTERUDITES WITH THE $[PT_4GE_{12}]$ FRAMEWORK

Two research groups, one a consortium of Vienna University of Technology and Vienna University, Bauer et al. (2007), and the other one from the Max Planck Institute of Chemical Physics of Solids in Dresden, Gumeniuk et al. (2008), submitted their discovery of an entirely new skutterudite framework based on $[Pt_4Ge_{12}]$ polyanions for publication in 2007 within a week of each other. One of the motivations for searching for a new Ge-based skutterudite framework were exciting results achieved with the Ge-based clathrate compounds, which feature cage-like structures reminiscent of skutterudites (many scientists working in thermoelectricity study both skutterudites and clathrates). Moreover, the fact that Ge has an excellent compatibility with electronic devices was an additional enticement to explore whether Ge could play a more prominent role than merely an element that may partly replace Sb on the skutterudite rings.

Although the new framework is entirely devoid of pnicogen atoms, the fundamental structural aspects of filled $[Pt_4Ge_{12}]$-based skutterudites are essentially identical with the pnicogen-based filled skutterudites. We still deal with the $Im\bar{3}$ structure formed by the corner-sharing octahedrons, the tilt of which gives rise to near square planar $[Ge_4]$ rings and two large structural cavities per unit cell that must be filled to stabilize the framework. There are no structural transitions between the liquid helium temperature range and room temperature, the rigid-band picture holds with the charge being transferred from the filler ion to the $[Pt_4Ge_{12}]$ polyanion, and the position of the Fermi level can be adjusted using fillers with different valence, or replacing some of Pt with aliovalent species, such as Au. There is also, of course, a possibility of substituting a small amount of Group III or Group V elements for Ge, provided the structure maintains its integrity. Important structural data for several filled $[Pt_4Ge_{12}]$-based skutterudites are given in Table 1.5.

Perhaps the greatest structural difference between the two types of skutterudites is the size of the cage that in RPt_4Ge_{12} skutterudites is somewhat smaller than in RT_4Sb_{12} skutterudites. This will have implications on the vibration motion of the larger fillers, such as Ba, that might lose their rattling character. From the perspective of physical properties, such as transport, magnetism, and superconductivity, the difference between the two types of skutterudites is in the nuances of their band structures, reflecting which species contribute most to the density-of-states (DOS) at the Fermi energy. In general, the DOS at the Fermi level of RT_4X_{12} skutterudites is dominated by $3d$ states of the transition metal that often hybridize with p-states of pnicogen atoms. In contrast, in RPt_4Ge_{12} skutterudites, the DOS at the Fermi energy is essentially all due to Ge $4p$ states, with Pt $5d$ states lying at a somewhat lower energy. As I discuss in Chapter 3, the 72-electron rule, so important in pnicogen-based filled skutterudites does not seem to apply to RPt_4Ge_{12} skutterudites, and the Zintl principle is violated. Assuming four electrons from each Ge atom and one s-electron of each Pt, gives $4 \times 1 + 12 \times 4 = 52$ electrons, well below 72 characterizing the $[T_4X_{12}]$ framework of pnicogen-based skutterudites. To get to the valence count of 72, each Pt atom would be expected to contribute five of its nine d-electrons, neglecting electrons from the filler ions. In reality, bonding analysis based on the concept of electron localizability indicator performed by Rosner et al. (2009) indicated that about 0.5 platinum d-electrons contribute an electron.

TABLE 1.5

Lattice Constant a, Interatomic Distances, Cage Radius, All in Å, and the Ratio of the Cage Radius to the Filler Ion Radius (using Shannon (1976) Compilation) for Several APt_4Ge_{12} Skutterudites

	$SrPt_4Ge_{12}$	$BaPt_4Ge_{12}$	$LaPt_4Ge_{12}$	$CePt_4Ge_{12}$	$PrPt_4Ge_{12}$	$NdPt_4Ge_{12}$	$EuPt_4Ge_{12}$	$ThPt_4Ge_{12}$	UPt_4Ge_{12}
a	8.6509	8.6838	8.6235	8.610	8.6111	8.602	8.6435	8.5931	8.5887
A-Ge	3.3468	3.3706	3.3290	3.3207	3.3175	3.3142	3.3289	3.321	3.2927
Pt-Ge	2.4983	2.5050	2.4870	2.4843	2.4828	2.4814	2.4899	2.481	2.4729
d_1	2.5052	2.4906	2.5014	2.5077	2.5090	2.5112	2.5140	2.482	2.528
d_2	2.6309	2.6623	2.6169	2.6081	2.6052	2.6022	2.6150	2.604	2.584
r_{cage}	2.06	2.08	2.05	2.04	2.04	2.04	2.05	2.05	2.01
r_{cage}/r_{ion}	1.64	1.51	1.77	1.79	1.81	1.84	1.71	1.95	2.01

Source: Entries for $BaPt4Ge_{12}$, $LaPt_4Ge_{12}$, $CePt_4Ge_{12}$, $PrPt_4Ge_{12}$, $NdPt_4Ge_{12}$, and $EuPt_4Ge_{12}$ are the data of Gumeniuk et al. (2010). Entries for $SrPt_4Ge_{12}$ are from Bauer et al. (2007), and entries for $ThPt_4Ge_{12}$ and UPt_4Ge_{12} are from Bauer et al. (2008). Lattice constants of $CePt_4Ge_{12}$ and $NdPt_4Ge_{12}$ from Toda et al. (2008) and that of $EuPt_4Ge_{12}$ from Grytsiv et al. (2008).

1.3.4 SKUTTERUDITES FILLED WITH ELECTRONEGATIVE FILLERS

So far, all filler species R in RT_4X_{12} and RPt_4Ge_{12} skutterudites were electropositive species (cations) supplying electrons to the respective frameworks and, in turn, stabilizing the structure. It was thus rather surprising when Fukuoka and Yamanaka (2010) reported that they were able to fill $RhSb_3$ with iodine, and the filler attained the valence 1-, i.e., it acted as an electronegative ion (acceptor) in IRh_4Sb_{12}. An appealing feature of the possible electronegative fillers is the natural p-type conduction that otherwise requires chemical manipulation on the framework (e.g., replacing some Co with Fe) to introduce enough positive carriers (holes) to counteract electrons generated by the usual filling with electropositive fillers. The advantage here is the much-reduced charge carrier scattering on the structural disorder as no adjustments on the framework need to be made. To prepare such acceptor-filled skutterudites, high-pressure synthesis (12 GPa at 850°C) is necessary to trap the volatile iodine in the voids. Remarkably, very high filling fraction was achieved with up to 95% of voids filled with iodine in this mostly single-phase skutterudite structure. Because of a fairly large I^{1-} ion, the lattice parameter $a = 9.2997$ Å of IRh_4Sb_{12} was larger than that of the parent $RhSb_3$, $a = 9.2322$ Å. Forcing iodine into $RhSb_3$ under 12 GPa of pressure at 850°C, rather than reacting the elements, was also successful and the resulting skutterudite had a comparable lattice parameter $a = 9.2906$ Å. This approach has worked even at a much-reduced temperature of 300°C, yielding a structure with $a = 9.301$ Å, albeit of lower crystallinity. However, relying purely on the vapor pressure of iodine to fill the voids of $RhSb_3$ by heating a quartz tube filled with $RhSb_3$ and iodine up to 600°C has failed. Clearly, high external pressure seemed to be essential to synthesize IRh_4Sb_{12}. Moreover, because of iodine's volatility and rather loose bonding, heating IRh_4Sb_{12} in a quartz tube at 500°C decomposed the structure into a binary $RhSb_3$ skutterudite.

The authors carried out a detailed structural analysis of IRh_4Sb_{12} showing that while the Rh–Sb bond length is essentially unchanged, 2.63 Å in IRh_4Sb_{12} vs. 2.621 Å in $RhSb_3$, the $[Sb_4]$ ring distances $d_1 = 2.88$ Å and $d_2 = 2.96$ Å are somewhat longer than in $RhSb_3$ ($d_1 = 2.80$ Å, $d_2 = 2.92$ Å). Likewise, the I–Sb bond length of 3.48 Å is slightly longer than in a typical rare earth-filled $LaFe_4Sb_{12}$ (3.41 Å) or alkaline earth-filled $BaFe_4Sb_{12}$ (3.46 Å). Positional parameters of Sb in the acceptor-filled antimonide skutterudites fall within the band of other antimonide skutterudites but stay below the Oftedal curve, see Figure 1.33. X-ray photoelectron spectroscopy confirmed that the oxidation state of iodine is 1−. Exploratory measurements of the electrical resistivity and magnetic properties indicated a metallic nature of transport with a rather high resistivity, and no magnetic ordering down to 2 K. Fukuoka and Yamanaka (2010) also performed DFT calculations using the WIEN2k package, Blaha et al. (1990), to obtain the band structure of this new anion-filled skutterudite. Details are presented in a section on band structure.

It is no surprise that the exciting discovery of an anion-filled skutterudite led to more explorations of similar structures and their detailed physical properties. The first in line was Li et al. (2014) who showed that the high-pressure synthesis works equally well for trapping iodine in the $CoSb_3$ framework. Attempting to find out how much of iodine is possible to fill in, the authors started with the nominal iodine contents $y = 0.4$, 0.8, and 1.2, which returned the iodine filling fractions 0.13, 0.72, and 0.79, respectively. It thus appeared that the maximum filling is limited to about $y = 0.8$, a remarkably high content of iodine in the voids of $CoSb_3$. In fact, by far, the largest level of filling achieved in $CoSb_3$. Such a large amount of the filler expanded the lattice constant of $CoSb_3$ (9.036 Å) to $a = 9.1217$ Å. The summary of the structural parameters of $I_yCo_4Sb_{12}$ are given in Table 1.6. Transport properties, and specifically a very low thermal conductivity of iodine-filled $CoSb_3$, are discussed in the section on heat conduction. The calculated band structure of fully filled ICo_4Sb_{12} is discussed in the section on band structure.

TABLE 1.6

Nominal and Actual Content of Iodine in the Voids of $I_yCo_4Sb_{12}$, the Lattice Parameter a, and the Positional Parameters y and z

Nominal Content	Actual Filling	a (Å)	y	z
$I_{0.4}Co_4Sb_{12}$	$I_{0.13}Co_4Sb_{12}$	9.0374	0.3360	0.1580
$I_{0.8}Co_4Sb_{12}$	$I_{0.72}Co_4Sb_{12}$	9.1211	0.3364	0.1577
$I_{1.2}Co_4Sb_{12}$	$I_{0.79}Co_4Sb_{12}$	9.1217	0.3372	0.1571

Source: The table is constructed from the data of X.-G. Li et al., *Journal of Alloys and Compounds* **615**, 177 (2014). With permission from Elsevier.

TABLE 1.7

Nominal Composition, the Actual Filling Fraction of Iodine, and the Lattice Parameter of $I_yFe_xCo_{4-x}Sb_{12}$ High-Pressure Synthesized Iodine-Filled Skutterudites

Nominal Composition	Actual Filling y	a (Å)
$IFe_{0.3}Co_{3.7}Sb_{12}$	0.69	9.1307
$IFe_{0.5}Co_{3.5}Sb_{12}$	0.52	9.1194
$IFe_{0.7}Co_{3.3}Sb_{12}$	0.5	–

Source: The table prepared from the data of L. Zhang et al., *Materials Letters* **139**, 249 (2015). With permission from Elsevier.

In their subsequent work, filling iodine into the $Fe_xCo_{4-x}Sb_{12}$ framework using the high-pressure synthesis, the same team, Zhang et al. (2015), explored an interplay between the filling fraction of iodine and the content of Fe. As the data in Table 1.7 indicates, aiming for $y = 1$ while the fraction of Fe was $x = 0.3, 0.5,$ and 0.7 impacted the actual content of iodine in an inverse order to what we have seen with electropositive fillers. Namely, the greater the presence of Fe, the smaller the content of iodine in the voids. This should not be surprising, as both Fe and iodine drive the skutterudite p-type, and there is only so much of the positive charge the structure can tolerate before it collapses.

Although filling the skutterudite void with an electronegative iodine filler is an exciting discovery, two strikes might go against their practical applications: an expensive synthesis relying on a complex high-pressure apparatus, and the fact that the structure is unstable at elevated temperatures where such p-type filled skutterudite would be expected to operate. But perhaps other halide fillers might fare better. In this regard, it is important to note the work of Ortiz et al. (2016) who succeeded in synthesizing bromine-filled $Br_yCo_4Sb_{12}$ skutterudite with the filling fraction as high as $y = 0.4$ via reactive hot-pressing at 600°C under an easily manageable pressure of 80 MPa. Rietveld refinement of a series of structures with different Br content y revealed a surprisingly very small and substantially nonlinear lattice expansion. A structure with the highest Br filling $y = 0.4$ had the lattice parameter merely 9.0446 Å. Just as with iodine, Br contributes holes and at $y = 0.2$ filling the concentration of holes reaches 8×10^{19} cm^{-3}. An important positive outcome is a high-carrier mobility that is substantially independent of the content of Br up to at least $y = 0.2$. Details concerning transport properties are provided in section on transport. Chlorine with $y = 0.2$ was successfully filled into the voids of CoSb$_3$ under ambient pressure by Duan et al. (2016).

The existence of halide-filled skutterudites, be they prepared with the aid of the high-pressure synthesis or not, offers enticing possibilities to explore electronegative filler species that might show novel physical properties, and perhaps even point out the way how to achieve filled skutterudites with superior thermoelectric performance.

In Section 1.3.1, we have seen that in order for an ion to enter the void in the [Co$_4$Sb$_{12}$] framework and stabilize the structure, rather than forming a secondary phase, there should be a sufficiently large difference between the electronegativities of Sb and the filler given by Eq. 1.18. I also pointed out that certain species, namely Ga and In, do not satisfy Eq. 1.18, yet they are undoubtedly found as occupants of the voids. The dilemma was subsequently explained by Xi et al. (2015) as a consequence of self-compensation, whereby Ga and In form charged dual-site defects consisting of donor-like void filling and acceptor-like substitution on the Sb site, which lowers the formation energy. Recently, Duan et al. (2016) inquired about potential filler species from Column 14 (previously Group VIA) comprising elements S and Se. While on their own S and Se are unstable as fillers in CoSb$_3$, the formation energy calculations indicated that when the structure is charge compensated by substituting some Te for Sb, the formation energy becomes negative and both S and Se may enter the void. S as well as Se act as electronegative, i.e., acceptor-like filler species, similar to iodine, bromine, and chlorine. Calculations also indicated that, unlike S, Se has a dual-site nature, it can fill the voids and it can substitute at the Sb site. Both elements form very strong covalent bonds with the neighboring Sb. The difference between the halide fillers and S and Se rests in the nature of bonding with Sb. Because the difference in electronegativities Δx of iodine or bromine and Sb is large ($\Delta x \sim 0.9$), the bond between the filler and Sb is substantially ionic. In contrast, a much smaller electronegativity difference ($\Delta x \sim 0.5$) between Sb and S (or Se) results in a bond with the strong polar covalent character. This is nicely illustrated by calculated maps of the charge density difference in a plane of the Sb$_4$ ring shown in Figure 1.34, where the strong bond between S and Sb results in Sb being drawn toward S with a bond length of merely 2.57 Å, much shorter than the usual filler-Sb distance of 3.40 Å. In the process, S gains 1.04 electrons from the nearest neighbor Sb.

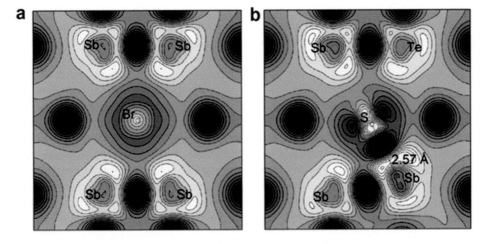

FIGURE 1.34 Calculated charge density maps at the Sb$_4$-ring plane of (a) Br$_{0.063}$Co$_4$Sb$_{12}$ and (b) S$_{0.063}$Co$_4$Sb$_{11.5}$Te$_{0.5}$. Reproduced from B. Duan et al., *Energy & Environmental Science* 9, 2090 (2016). With permission from the Royal Society of Chemistry.

In contrast, the ionic bond between Br and Sb leaves the ring structure little perturbed. The strong bond between S and Sb greatly influences the lattice dynamics as it causes S and Sb to vibrate as a unit ("cluster vibration"), resulting in low frequency optical mode of 35–50 cm^{-1}. Such low frequency optical modes interfere with the normal phonon modes of the structure, leading to a very low thermal conductivity of 0.8 $Wm^{-1}K^{-1}$ at 750 K for $S_{0.26}Co_4Sb_{11.11}Te_{0.73}$, some 60% reduction vis-à-vis an unfilled skutterudite. In comparison, the halide-filled $[Co_4Sb_{12}]$ framework, because of the low mass of the filler and its ionic bonding, shows a more limited reduction in the thermal conductivity with respect to the unfilled structure. The highest figure of merit attained with the electronegative sulfur filler was about 1.5 at 850 K, the value rivaling the performance of the best triple-filled skutterudites. A similar outcome, namely much enhanced solubility of Te in the presence of sulfur in the skutterudite voids, leading to very low lattice thermal conductivity and the figure of merit of about 1.5 at 850 K, was reported by Wang et al. (2018). Moreover, because S does not require any nonequilibrium synthesis, the resulting filled skutterudite has an excellent thermal stability, as documented by no significant changes in the transport properties of structures annealed at 875 K for 15 days. Recently, instead of substituting Te on the Sb rings, to enable sulfur to fill the voids and stabilize the structure, Li et al. (2019) replaced a fraction of Co with Ni and synthesized $S_yCo_{4-x}Ni_xSb_{12}$ in the range $0 \leq y \leq 0.35$ and $0.6 \leq x \leq 1.0$. Very low vibration frequency of sulfur of 54 cm^{-1} (determined by Raman spectroscopy) resulted in low values of the lattice thermal conductivity (see Section 5.7.2), although the figure of merit turned out to be only about one half of the value achieved with Te doping on the pnicogen rings. The authors also noted that the presence of S in the voids greatly enhanced the solubility of Ni on the Co sites. The study was extended by Wan et al. (2018) to substitute Pd on the site of Co, and by Tu et al. (2019) who substituted Pd as well as Pt for Co.

1.4 COMPOSITE SKUTTERUDITES

Composite structures based on skutterudites include both binary and filled forms of the material. I like to divide them into two distinct categories: (a) intrinsic composites that form in situ as a result of segregation of some particular second phase in skutterudites, typically caused by a large deviation in the stoichiometry, or, as a result of rapid quenching; and (b) extrinsic composites prepared by adding unrelated phases into the skutterudite matrix. I describe them in turn.

1.4.1 INTRINSICALLY FORMED COMPOSITE SKUTTERUDITES

In Section 1.1.3, we have seen that the solubility limit of Fe in $CoSb_3$ is between 16 at% and 25 at%. In one of the first reports documenting an improvement in the thermoelectric properties upon the presence of a secondary phase in the skutterudite matrix, Katsuyama et al. (1998a) synthesized $Co_{1-x}Fe_xSb_3$ with $x = 0 - 0.4$. Samples with $x \leq 0.06$ turned out to be single-phase skutterudites. Structures with the Fe content $0.06 \leq x < 0.25$ indicated the presence of Sb, which disappeared upon hot pressing at 923 K with 10 MPa for 1 h, presumably by Sb evaporating. However, samples with $x \geq 0.25$ contained $FeSb_2$ as a second phase, which reduced the electrical resistivity and the thermal conductivity, and resulted in a much-enhanced thermoelectric figure of merit reaching 0.37 at 773 K for $x = 0.25$, compared to a maximum figure of merit of pure $CoSb_3$ at about 480 K of merely 0.154. The $Co_{1-x}Fe_xSb_3$ system was revisited a decade later by Zhou et al. (2011), aiming to promote the formation of the $FeSb_2$ nanophase by synthesizing $Co_{0.9}Fe_{0.1+x}Sb_{3+2x}$ with $x = 0, 0.05$, 0.1, and 0.2. While X-ray diffraction detected the $FeSb_2$ phase only in the highest $x = 0.2$ sample, high resolution field emission scanning electron microscopy revealed the $FeSb_2$ nanophase in all $x > 0$ structures. The presence of the $FeSb_2$ nanophase not only strongly scatters phonons and thus lowers the thermal conductivity, but it also alters the scattering parameter of holes, which favors the Seebeck coefficient. Consequently, an improved $ZT = 0.59$ was obtained at 788 K in a sample with $x = 0.05$, the value twice as large as measured for stoichiometric $Co_{0.9}Fe_{0.1}Sb_3$.

Demonstrating an improvement in the thermoelectric properties when the $FeSb_2$ phase precipitated in the over-stoichiometric skutterudite matrix, the same team (Katsuyama et al. 2003), looked into the effect of overstoichiometric Ni in $Co_{1-x}Ni_xSb_3$ with x up to 0.4, well in excess of the solubility limit of about 10 at%. Consequently, a second phase, $NiSb_2$, should be present and influence the transport behavior. In contrast to $Co_{1-x}Fe_xSb_3$, where Fe has one less electron than Co and thus the structures are p-type materials, an extra Ni electron results in an n-type semiconducting structure. Making use of exactly the same synthesis as in the case of Fe-substituted $CoSb_3$, the resulting $Co_{1-x}Ni_xSb_3$ was a single-phase material for $x \leq 0.10$, but contained a mixture of $CoSb_3$, Sb, and $NiSb_2$ (with the CdI_2 structure) at concentrations of Ni higher than 10 at%. Upon hot pressing, the structure possessed NiSb instead of $NiSb_2$, which likely formed by decomposition of $NiSb_2$ into NiSb and Sb, the latter subsequently lost by evaporation. While samples with low Ni content $x \leq 0.06$ improved the thermoelectric performance over pure $CoSb_3$ with the figure of merit peaking at 0.49 at 663 K for $Co_{0.04}Ni_{0.06}Sb_3$, at higher Ni content the precipitated NiSb was no longer beneficial due to its positive Seebeck coefficient acting against the negative thermopower of $Co_{1-x}Ni_xSb_3$, and due to its quite high thermal conductivity.

In situ–formed composites can also be prepared utilizing the matrix of ternary skutterudites, such as $FeSb_2Te$. The choice of this particular matrix over the binary $CoSb_3$ is rationalized based on two factors: (a) given that p-type forms of skutterudites lack behind the thermoelectric performance of n-type skutterudites, it was desirable to explore possible improvements by forming a composite structure with p-type $FeSb_2Te$ as the matrix; (b) band structure calculations by Tan et al. (2013a) (see Chapter 3) indicated that, unlike a single linearly dispersing valence band in $CoSb_3$, the presence of Fe in the ternary skutterudite tends to flatten the top of the valence band and brings into play several heavy bands that make up the region near the valence band edge, a situation similar to what happens upon filling the voids in skutterudites. This should enhance the effective mass and support robust Seebeck coefficients. As a desirable nanoinclusion phase was chosen InSb for its very high electron mobility, expected to counter the mobility degradation at the grain boundaries. The in situ formation of InSb nanoinclusions in the $FeSb_2Te$ matrix was realized by Tan et al. (2015) using the traditional melting/annealing/grinding/SPS processing of elemental constituents weighed according to the nominal composition $FeSb_{2.2}Te_{0.8} + x$ mol% InSb, where $x = 0, 1,3,6$, and 10. The best performance was achieved in a composite with the InSb molar faction of 3%. The increase in the electrical conductivity balanced the decreased Seebeck coefficient, resulting in the power factor of close to 2 $mWm^{-1}K^{-2}$ at 750 K. The real gain came from a significantly reduced thermal conductivity that at 800 K decreased by 23% compared to pure $FeSb_{2.2}Te_{0.8}$, resulting in $ZT = 0.76$ at 800 K, and improvement of some 15% over the figure of merit of $FeSb_{2.2}Te_{0.8}$, measured previously by Tan et al. (2013b). In fact, this stands as the highest ZT of any p-type unfilled skutterudite.

In filled skutterudites, there is an additional route to form a secondary phase, this time via filling the voids with amounts of fillers that exceed the filling fraction limit. This has been well documented by Zhao et al. (2006b), where an excess of Yb reacted with oxygen (presumably picked up when the powder was exposed to air) to form Yb_2O_3 precipitates spanning in size from a few tens of nanometers up to about a micron and well-distributed within the grains as well as decorating grain boundaries. From the early 2000s, Yb was known as perhaps the most effective filler to reduce the thermal conductivity of $CoSb_3$, resulting in ZT values of unity at its maximum filling of around $y = 0.2$ (Nolas et al. 2000, Sales et al. 2001, and Anno et al. 2002). Thus, it was important to show that even in this premier thermoelectric material one may benefit from the presence of a secondary phase. Indeed, extra scattering of phonons on Yb_2O_3 precipitates brought down the thermal conductivity close to its theoretical minimum, enhancing in turn the figure of merit to a value of 1.3 at 850 K in a composite structure $Yb_{0.25}Co_4Sb_{12}/Yb_2O_3$. Although the presence of Yb_2O_3 was well documented in back-scattered electron and TEM images, the exact amount of Yb_2O_3 in the composite structure was not provided. Similar measurements were also performed by Ding et al. (2013) with intentionally annealed $Yb_{0.3}Co_4Sb_{12}$ in quartz tubes containing 2 kPa pressure of oxygen at a temperature of 873 K for 10, 20, and 30 days. During this process, a small fraction of Yb escapes the voids and reacts with oxygen to form Yb_2O_3, the amount of which increases with the annealing time and appears to be quite homogeneously distributed following the heat treatment for 30 days. In their subsequent study with a much greater Yb content of $y = 0.6$, Ding et al. (2014) observed nanometer and micrometer size particles of Yb_2O_3 distributed chiefly on the grain boundaries.

An interesting study of the effect of Yb_2O_3 oxide on the thermoelectric performance of triple-filled $(SrBaYb)_yCo_4Sb_{12}$ was made by Rogl et al. (2014). The authors intentionally increased the presence of Yb to a composition $Sr_{0.125}Ba_{0.125}Yb_{0.25}Co_4Sb_{12.5}$ (also a small excess of Sb), which resulted in a structure with several skutterudite phases. The material was homogenized by additional annealing (750°C for 7 days) followed by ball milling and hot pressing, and the composition settled around $Sr_{0.10}Ba_{0.12}Yb_{0.04}Co_4Sb_{12}$ with micron-size domains of Yb_2O_3 (see Figure 1.35a). Such a composite skutterudite with micron-size Yb_2O_3 domains had a rather high resistivity, poor power factor, and an elevated thermal conductivity. Nanostructuring the material by crushing the ingot and subjecting the powder to ball milling (14 h) with the eventual compaction done by hot pressing for 2 h at 650°C under 56 MPa, yielded a material with finely dispersed Yb_2O_3 nanoparticles, shown in Figure 1.35b. The nanostructuring treatment resulted in a significant reduction of the electrical resistivity, an essentially unchanged Seebeck coefficient, and decreased thermal conductivity, yielding an excellent figure of merit of 1.6 at 835 K. With subsequently applied high-pressure torsion, the figure of merit reached 1.9 at 835 K.

Apart from Yb-filled antimonide skutterudites, another promising thermoelectric system is $In_xCo_4Sb_{12}$, reported by T. He et al. (2006) to reach $ZT = 1.2$ at 575 K for $x = 0.25$. Interestingly, according to Shi et al. (2005), indium should not have been able to enter the voids of $CoSb_3$ as the electronegativity difference between In and Sb is merely 0.27 and thus much smaller than 0.80 required by Eq. 1.18. Nevertheless, there are several reports on In-filled antimonide skutterudite cages. Among them is an attempt by Li et al. (2009) to double-fill the structure with a combination of In and Ce. In a series of samples $In_xCe_yCo_4Sb_{12}$ with $x = 0.15$, $y = 0.15$; $x = 0.2$, $y = 0.15$; $x = 0.2$, $y = 0.2$; and $x = 0.25$, $y = 0.05$, X-ray diffraction patterns always revealed the presence of an InSb phase. This phase appeared as evenly distributed nanograins of 30–80 nm size, the density of which increased with the increasing content x of In. Again, the InSb phase formed in situ as a consequence of an overstoichiometric amount of indium in the starting powder. Enhanced phonon scattering on such nanophase led to an improved figure of merit $ZT = 1.4$ at 800 K for a composition $In_{0.2}Ce_{0.15}Co_4Sb_{12}$. I suspect that similar nanophase would be found in a singly-filled $In_xCo_4Sb_{12}$ with a sufficiently high In content, if one looked for it.

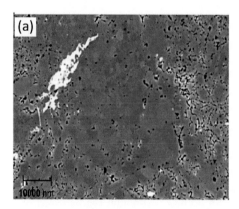 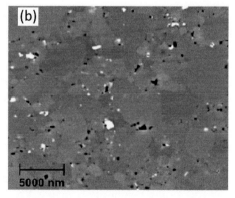

FIGURE 1.35   Scanning electron micrograph of (a) $Sr_{0.10}Ba_{0.12}Yb_{0.04}Co_4Sb_{12}$ with micron-size domains of Yb_2O_3, yielding a material with an ordinary $ZT < 1.2$ at 835 K. (b) The same skutterudite after nanostructuring consisting of crushing the material, ball milling the powder for 14 h, and hot pressing, resulting in a finely dispersed, nanometer-size particles of Yb_2O_3, and $ZT = 1.6$ at 835 K. Adapted and reproduced from G. Rogl et al., *Acta Materialia* 63, 30 (2014). With permission from Elsevier.

Promoting the formation of the InSb phase in the matrix of $In_xCo_4Sb_{12}$, Peng et al. (2012) synthesized a series of overstoichiometric compounds $In_{0.2+x}Co_4Sb_{12+x}$ with $x = 0.2, 0.4, 0.6$, and 0.8. Confirming that the filling fraction limit of indium in $CoSb_3$ is somewhere between 0.22 and 0.32, the highest content of In, the total of 0.32, was observed in the nominal $In_{0.8}Co_4Sb_{12.6}$ composition. Differential thermal analysis (DTA) indicated not only the dominant endothermic peak at 877°C, reflecting melting of the compound, but also a distinct endothermic peak near 492°C, associated with the melting of InSb/Sb eutectic, which formed during the synthesis process. The overstoichiometric starting powders were initially heated at 1030°C for 24 h and then the temperature was lowered to 650°C and held there for 4 days to anneal the structure. At this temperature, the minor InSb + Sb phase that formed as the result of the excess of In and Sb, was a liquid In–Sb phase at the grain boundaries. During slow cooling to room temperature it solidified, forming InSb/Sb eutectic, consistent with the In–Sb phase diagram. All structures, regardless of the content of In, had a similar grain size in the range 5–20 μm. Such grain size was apparently not fine enough to scatter phonons significantly, and the highest reported thermoelectric figure of merit did not exceed unity.

An interesting twist how the InSb phase forms in situ was observed by Eilertsen et al. (2012) in their preparation of nominal $In_{0.1}Co_4Sb_{12}$. An ingot synthesized by solid-state reaction of the constituents was ball-milled for various periods of time up to 40 hr, and the powder consolidated by hot pressing at 600°C for 20 min under 200 MPa. With the increasing ball-milling time, greater and greater fraction of InSb was observed in the samples. Since the starting nominal In content of $x = 0.1$ was well below the solubility limit, the formation of the increasing fraction of the InSb phase had to proceed via attrition-enhanced indium diffusion out of the voids of the skutterudite structure during ball-milling. Such metastability in the indium void occupancy is consistent with the fact that its electronegativity is too close to that of Sb and does not conform to the criterion expressed in Eq. 1.18. ZT values were relatively low (≤ 0.5), reflecting the low In filling fraction.

ZT values approaching unity at 575 K were achieved in $In_xCo_4Sb_{12}/y(InSb)$ nanocomposites by Gharleghi et al. (2016a). The structures were synthesized by combining the hydrothermal growth of $CoSb_3$ powders with the solid-liquid processing, during which the powders were reacted with pieces of In to fill the voids, and the excess of In bonded with Sb to form the InSb nanophase. The grain size of InSb was in the range of 50–100 nm. Although the initial overstoichiometry of In amounted to $x = 0.42$, the actual content of indium in the voids was surprisingly low at $x = 0.06$.

A dramatically better result with the overstoichiometry of In was achieved recently by Khovaylo et al. (2017), who prepared $In_xCo_4Sb_{12}$ with the nominal indium contents $x = 0.2, 0.6$, and 1.0 and some 10% excess of Sb by induction melting for 2 min in ordinary alumina crucibles and quenched the ingots. Following a homogenizing anneal at 973 K for 5 hr, the ingots were pulverized by ball milling and consolidated by SPS. The InSb nanophase formed in situ and segregated at grain boundaries and in the main skutterudite phase. A sample with the nominal indium content $x = 1$ reached $ZT = 1.5$ at 725 K, the highest reported value for a single-filled skutterudite.

Similar to In, the electronegativity of Ga (1.81) is also too close to that of Sb (2.05) and Ga should not be found in the voids of $CoSb_3$ at ambient temperature. However, according to temperature-dependent Gibbs free energy calculations noted by Xiong et al. (2010), Ga might enter the voids at temperatures around 1100 K. Given that GaSb forms a eutectic with Sb at 859 K when the composition ratio GaSb : Sb = 0.22 : 0.78, here is a possibility to form GaSb nanoinclusions in situ in the $CoSb_3$–Ga system by precipitating the GaSb nanophase by cooling the high-temperature solid solution. Using this approach, Xiong et al. synthesized nominal compositions $Yb_{0.26}Co_4Sb_{12}/yGaSb$ with $y = 0, 0.1, 0.2$, and 0.3 by melting the charge at 1393 K for 10 h, rapidly quenching the material, and subsequently annealing it 1073 K for 168 h. Eventual pulverization of the ingots into fine powders followed by the SPS consolidation at 873 K for 8 min under 60 MPa, gave a fully compacted material. All structures with the nominal GaSb content $y \geq 0.2$ showed the presence of uniformly dispersed GaSb nanostructure with the grain size 5–20 nm. Because the melting point of GaSb ($T_m = 985$ K) is nearly two hundred degrees higher than that of InSb ($T_m = 800$ K), the in situ–formed GaSb nanostructure is more in accord with the temperature range where $Yb_xCo_4Sb_{12}$ would be expected to operate than is the InSb

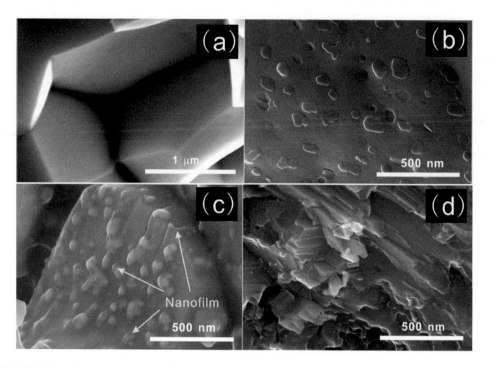

FIGURE 1.36 SEM images of $Nd_{0.6}Fe_2Co_2Sb_{12-x}Ge_x$ with (a) $x = 0$, (b) $x = 0.15$, (c) $x = 0.3$, and (d) $x = 1$. The structure with $x = 0.3$ is covered by a very thin, 10-nm-thick film precipitated in situ. Reproduced from L. Zhang et al., *Journal of Applied Physics* 114, 083715 (2013). With permission from the American Institute of Physics.

nanostructure. In their subsequent publication, Xiong et al. (2011) rationalized their synthesis strategy by drawing attention to the behavior of the Gibbs free energy as a function of temperature. Provided the Gibbs free energy becomes negative at a temperature lower than the peritectic temperature of $CoSb_3$, the filler enters the void. Upon cooling, the filler is expelled from the void as the Gibbs free energy turns positive. The synthesized $Yb_{0.26}Co_4Sb_{12}$ attained the $ZT = 1.45$ at 850 K. This synthesis strategy will not work for other interesting fillers, such as In and Tl, because the Gibbs free energy becomes negative at too high temperatures only, well above the peritectic point.

Interesting in situ precipitates of Fe_3Ge_2Sb and Sb have formed in p-type $Nd_{0.6}Fe_2Co_2Sb_{12-x}Ge_x$ with the increasing content of Ge, Zhang et al. (2013). While $Nd_{0.6}Fe_2Co_2Sb_{12}$ has clean grains with no sign of precipitates, as the Ge content approaches $x = 0.15$, nanoparticles of a couple of hundred nanometers form, and as the Ge content reaches $x = 0.3$, an unusual, first time seen thin film less than 10 nm thick forms over the grains (Figure 1.36). Remarkably, this single-filled structure attained the lowest room temperature thermal conductivity ever measured on skutterudites with the total thermal conductivity of only 1 $Wm^{-1}K^{-1}$, yielding the ZT value of 1.1 at 700 K.

1.4.2 Extrinsically Formed Composite Skutterudites

Extrinsic composite materials based on skutterudites, i.e., structures prepared by intentionally adding foreign phases into the skutterudite matrix, started to draw attention at the same time the first in situ–generated intrinsic skutterudite composites were reported on. The effort was led by Katsuyama et al. (1998b, 2000) who dispersed a powder of marcasite structure $FeSb_2$ in a matrix of $CoSb_3$ in a ratio $CoSb_3$: $FeSb_2 = (1-x)/x$ with $x = 0.08 - 0.40$ by mechanical grinding (ball milling) for up to 40 h. The resulting product was compacted by hot pressing at 923 K for 1 h with 10 MPa under vacuum. While pure $CoSb_3$ had a higher figure of merit than all samples containing $FeSb_2$ at temperatures below 500 K, a significantly decreased thermal conductivity of the composites (enhanced boundary scattering) and their only a modestly degraded electrical conductivity resulted in the figure of merit $ZT = 0.42 - 0.46$ measured at 756 K on a composite with a molar ratio $CoSb_3 : FeSb_2 = 0.8 : 0.2$. The ZT far exceeded the maximum figure of merit of 0.154 at 482 K of similarly prepared pure $CoSb_3$.

Right from the early studies of composite systems, it was of interest to learn what might be the effect of inert nanoparticles, such as oxides, on the thermoelectric performance of the $CoSb_3$ matrix. By dispersing oxides nanoparticles in the matrix of $CoSb_3$, one hopes that phonon scattering on such nanophase is more effective than scattering of the charge carriers, i.e., a reduction in the thermal conductivity weighs over a degradation of the carrier mobility. There is also an opportunity to enhance the Seebeck coefficient due to energy-dependent scattering of electrons (energy filtering) at the interface of nanoinclusions and the matrix, provided the potential barrier is optimized.

In the first attempt to form a $CoSb_3$/oxide composite structure, Katsuyama et al. (1999) used mechanical grinding to disperse several commercial oxide powders (MoO_2, WO_2, and Al_2O_3) in $CoSb_3$ in the ratio of $CoSb_3$: oxide = $(1 - x)/x$, with x varying from $x = 0.1$ to $x = 0.3$. While all oxides reduced the thermal conductivity of $CoSb_3$, the conducting MoO_2 and WO_2 oxides degraded the electrical resistivity considerably less than the insulating Al_2O_3 oxide. Nevertheless, the gain in the figure of merit over pure $CoSb_3$ was marginal even for the two conducting oxides. No beneficial effect of Al_2O_3 in either n-or p-type skutterudites was confirmed nearly two decades later by Rogl et al. (2017). The above studies suggested that forming composites of $CoSb_3$ or of its filled forms with insulating oxides might not be the most effective way of improving the thermoelectric performance. In the subsequent work, this time with the $CeFe_3CoSb_{12}$ matrix, Katsuyama and Okada (2007) again used MoO_2 and WO_2 and noted that 5 mol% of MoO_2 in the matrix improved the figure of merit to a value of 1.22 at 773 K, nearly a 20% increase over the value measured on the matrix. Other molar fractions of MoO_2, as well as all molar fractions of WO_2 led to a deterioration of the thermoelectric performance. From the above studies, it appeared that the Seebeck coefficient of composites with oxides of MoO_2, WO_2, and Al_2O_3 is degraded in comparison to the Seebeck coefficient of the pure matrix. Clearly, there was no benefit gained from any assumed energy filtering at the grain boundaries. However, subsequent studies have indicated that there are examples where energy filtering is effective in enhancing the Seebeck coefficient. Among them is the work by Chubilleau et al. (2012a, 2014) with ZnO in $CoSb_3$ and in $In_{0.2}Co_4Sb_{12}$, respectively, and the study of TiO_2 inclusions in $Co_4Sb_{11.7}Te_{0.3}$, Zhu et al. (2012), and in $Ba_{0.3}Co_4Sb_{12}$ by Zhou et al. (2014).

As pointed out by He et al. (2007a) in their study with a ZrO_2 nanophase dispersed in $CoSb_3$, making comparisons between transport properties of composite structures and the parent matrices, it is important to take into account the porosity of the structure. Meaningful comparisons can be made only when the porosities of the samples are comparable. There is also a possibility that, as one tries to incorporate an oxide nanophase in $CoSb_3$, a small amount of metallic Sb might precipitate during the processing, as reported by He et al. (2007b) in their work with $CoSb_3/ZrO_2$ nanocomposites. Moreover, as noted by Z. He et al. (2006), the electrical conductivity and the thermal conductivity seem to depend on the sintering temperature (both increase with the increasing sintering temperature) and any comparison between the conductivities of the composite structures and the matrix should be based on samples sintered at the same temperature. In any case, improvements in ZT upon dispersing oxide nanophases in $CoSb_3$ are usually not very large (rarely exceeding 15%) and, if present, are observed only when the oxide content is rather low. This has to do with the fact that light-element oxides have high thermal conductivity, and while one always benefits from enhanced phonon scattering upon the presence of a secondary oxide phase, when its content is high, it may start thermally shorting the composite. This has been seen, for instance, by Xiong et al. (2009) and Zhou et al. (2014) in their work with TiO_2 dispersed in $Ba_{0.22}Co_4Sb_{12}$ and in other studies. The latter authors used hierarchically structured TiO_2 inclusions by combining a commercially available TiO_2 nanoparticles with a diameter ~ 20 nm with larger, spherically shaped TiO_2 particles with a diameter of about 600 nm prepared in-house by a carbon sphere-templated method, all dispersed in the $Ba_{0.22}Co_4Sb_{12}$ matrix by ball milling. TEM and HRTEM images of the TiO_2 nanophase in the matrix are shown in Figure 1.37. The idea here was to scatter a broader range of phonons on a wider range of nanoparticle sizes. The resulting $ZT = 1.2$ at 813 K of a composite structure containing 0.5

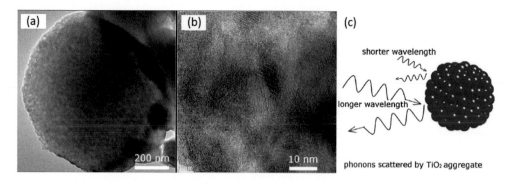

FIGURE 1.37 (a) TEM image of a larger, spherical TiO_2 nanoparticle and (b) small TiO_2 nanoparticles in the matrix of $Ba_{0.3}Co_4Sb_{12}$. (c) A schematic diagram illustrating hierarchical phonon scattering on small and large nanoparticles. Reproduced from X. Y. Zhou et al., *Journal of Materials Chemistry A* 2, 20629 (2014). With permission from the Royal Society of Chemistry.

wt% of the commercial nanoparticles and 0.5 wt% of larger in-house synthesized TiO_2 particles represented about 17% improvement over the matrix material. Another oxide tried as an additive to the $CoSb_3$ matrix was a commercial 16 nm CeO_2 nanopowder dispersed by mechanical alloying at concentrations of 1–3% by Alleno et al. (2010). An observed reduction in the thermal conductivity of about 15% has two aspects: enhanced phonon scattering on CeO_2 nanoinclusions at the grain boundaries, and about a factor of three reduction in the grain size of the $CoSb_3$ matrix upon the presence of CeO_2 nanoparticles that acted effectively as a grain growth inhibitor during the SPS compaction. Since no data on the power factor were provided, no assessment of the figure of merit of the composite could be made. In their subsequent study, Alleno et al. (2013) added comparable amounts of CeO_2 powders to $Co_{0.97}Pd_{0.03}Sb_3$, hoping to reduce further the thermal conductivity by Rayleigh-type scattering as Pd substituted for Co. In this case, the power factors were provided and the ZT reached 0.65 at 650 K when 2 vol% of CeO_2 were added to the matrix. Without CeO_2, the matrix only attained $ZT = 0.49$ at 650 K.

The case of CuO_2 nanoparticle additions is controversial. According to Battabyal et al. (2015), 1.4 wt% of CuO_2 nanoparticles dispersed by ball milling in $Ba_{0.4}Co_4Sb_{12}$ decreased the electrical resistivity, somewhat improved the Seebeck coefficient, especially above 450 K, and suppressed the lattice thermal conductivity by about 25%. The overall effect was a large figure of merit $ZT \sim 0.9$ at 575 K, a dramatic improvement over the figure of merit of their $Ba_{0.4}Co_4Sb_{12}$ matrix. However, detailed measurements by Rogl et al. (2017) of the effect of CuO_2 nanoparticles in both n-type $(Mm,Sm)_{0.60}Co_4Sb_{12}$ and p-type $DD_yFe_3CoSb_{12}$ skutterudite matrices (commercially available from Treibacher Industrie AG, Austria) cast serious doubt about the beneficial effect of CuO_2 nanoparticles. Rogl et al. prepared the series of composites with 1, 2, and 4 wt% of CuO_2 in both n- and p-type matrices by mixing the powders of skutterudites and CuO_2 by high-energy ball milling and compacting the milled product by hot pressing. The X-ray diffraction data revealed that the presence of CuO_2 in the p-type matrix leads to multiphase structures as a result of CuO_2 reacting with the skutterudite matrix and partly decomposing it. While the n-type composites were single-phase structures, some oxidation must have taken place, as the band gap of the material decreased with the increasing content of CuO_2. The unfortunate consequence was a much higher electrical resistivity, low power factor, and poor figure of merit that reached the highest value of only 0.2 at 623 K for the composite with 1 wt% of CuO_2, nowhere near the claimed value of $ZT \sim 0.9$ at 575 K reported by Battabyal et al. (2015) for their 1.4 wt% CuO_2 composite. The writer's own experience is that any form of Cu introduced into the skutterudite matrix will likely lead to decomposition of the skutterudite at elevated temperatures. Thus, in my opinion, bringing Cu into the skutterudite structure should be discouraged.

The other oxide used as an additive was WO_3 dispersed in n-type $CoSb_{2.925}Te_{0.075}$ by ball milling in the work of Zhao et al. (2015). The composite structure containing 1.5 wt% of WO_3 nanoparticles of an approximate size of 50 nm yielded $ZT = 0.71$ at 750 K, benefitting from a much decreased thermal conductivity and somewhat increased Seebeck coefficient that overcame the reduced electrical conductivity. I should also mention that CeO_2 was added to $In_{0.25}Co_4Sb_{12}$ by Benyahia et al. (2018), but for a very different reason, namely, to act purely as an inhibitor of grain growth structure. Apart from insulating oxides, borides, such as $Fe_{2.25}Co_{0.75}B$ and $Ta_{0.8}Zr_{0.2}B$, were investigated as part of various additives in the high-performing n-type $(Mm,Sm)_{0.60}Co_4Sb_{12}$ and p-type $DD_yFe_3CoSb_{12}$ commercially available skutterudite matrices in already mentioned paper by Rogl et al. (2017). Somewhat effective were the metallic nanoparticles of $Ta_{0.8}Zr_{0.2}B$ that in concentrations of 0.5 and 1 wt% in $DD_yFe_3CoSb_{12}$ raised the figure of merit to over 1.3 at 823 K from the value of 1.23 in $DD_yFe_3CoSb_{12}$ with no additives. The paper also contains data on $La_{1.85}Sr_{0.15}CuO_4$ additives in the n-type $(Mm,Sm)_{0.60}Co_4Sb_{12}$ matrix and p-type $DD_yFe_3CoSb_{12}$. The presence of Cu had no detrimental effect on the skutterudite structure as Cu was apparently strongly bonded in the additive's lattice. 2 wt% of $La_{1.85}Sr_{0.15}CuO_4$ in the n-type matrix marginally increased the figure of merit by about 8%, while no benefit was gained when the additive was introduced into the p-type matrix. I include the effect of various additives explored by Rogl et al. in Figures 1.38a and 1.38b.

Semiconducting nanoparticles have been a frequent choice of additives in $CoSb_3$. In Section 1.4.1, I discussed the effect of in situ–formed NiSb precipitates arising from over-stoichiometric amounts of Ni used in the synthesis of $Co_{1-x}Ni_xSb_3$ skutterudites by Katsuyama et al. (2003). In this publication, the authors also described the influence of NiSb particles dispersed in the $CoSb_3$ matrix by mechanical grinding. The prolonged grinding turned out to be more effective in reducing the thermal conductivity than the in situ–formed precipitates, particularly at low temperatures.

In subsequent years, numerous studies have been conducted with different additives to the skutterudite matrix, some of them semiconducting or metallic nanoinclusions and others being various forms of carbon. A special class of composites are structures where $CoSb_3$ nanoparticles are dispersed in the micron-size $CoSb_3$ or in the filled skutterudite matrix. In all cases, the emphasis has been on reducing the thermal conductivity but, with dispersed conducting additives, it is also of interest to see how the electronic properties of such composite structures are affected.

The effect of semiconducting nanoinclusions in the $CoSb_3$ matrix is described in several publications. Chubilleau et al. (2012b) used PbTe nanoparticles with an exceptionally small diameter

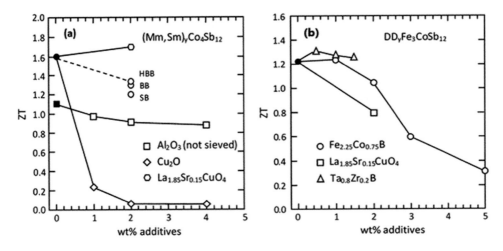

FIGURE 1.38   Figure of merit of (a) n-type composite skutterudites based on the $(Mm,Sm)_yCo_4Sb_{12}$ matrix and (b) p-type skutterudites based on the $DD_yFe_3CoSb_{12}$ matrix. HBB, BB, and SB refer to a sieved powder of $(Mm,Sm)_yCo_4Sb_{12}$ mixed with 2 wt% of an Al_2O_3 powder and treated by ball milling under different conditions: SB standing for milling with small balls of 3 mm diameter; BB indicates milling with large balls of 10 mm diameter; and HBB represent high-energy milling with large balls. Adapted and reproduced from G. Rogl et al., *Journal of Alloys and Compounds* 695, 682 (2017). With permission from Elsevier.

of 6 nm formed by laser fragmentation of micron-sized particles in water, the process described by Chubilleau et al. (2011). The composites of $CoSb_3$ + x wt% PbTe nanoparticles ($0 \leq x \leq 8$) were prepared by dispersing the PbTe phase in ethanol in an ultrasonic bath for 20 min. Ethanol was subsequently removed by freeze-drying, and the resulting powder densified by SPS at 873 K under 50 MPa for 6 min. The presence of the PbTe nanophase led to an improvement of the room temperature figure of merit from about 0.012 for $CoSb_3$ to 0.024 for the 8 wt% PbTe nanocomposite. Interestingly, the nature of the conduction changed from p-type in $CoSb_3$ to n-type whenever PbTe nanoparticles were present in the structure. This was explained by a substitution of a small amount of Sb by Te. Duan et al. (2012) dispersed commercial nanoparticles of TiN with the average size below 50 nm in $CoSb_{2.875}Te_{0.125}$, the matrix where a fraction of Sb is replaced by Te to dope the structure n-type and to reduce its lattice thermal conductivity. The powders of $CoSb_{2.875}Te_{0.125}$ and TiN were mixed by a planetary mill for 20 min under argon according to a stoichiometry $CoSb_{2.875}Te_{0.125}$ + yTiN, where y = 0, 0.3, 0.6, and 1 vol%. Subsequently, the admixed powder was consolidated by SPS at 873 K for 7 min under a pressure of 40 MPa. Although the electrical conductivity decreased slightly with the increasing content of TiN, the Seebeck coefficient remained essentially unchanged and the thermal conductivity decreased more substantially. Consequently, the figure of merit ZT improved by some 10% over TiN-free $CoSb_{2.875}Te_{0.125}$ and reached $ZT = 1.0 \pm 0.1$ at 800 K in the composite containing 1 vol% of TiN nanoparticles. Moreover, the mechanical properties, assessed via the flexural strength and fracture toughness, increased by 31% and 40%, respectively. The improvement was explained by the pinning effect of TiN nanoparticles, which formed interlocking interfaces between TiN nanoparticles and the matrix grains. A rather complex nanocomposite comprising $Yb_{0.25}Co_4Sb_{12}$ with 4 wt% of $(Ag_2Te)_x(Sb_2Te_3)_{1-x}$ for x = 0.36, 0.40, 0.42, and 0.46 was prepared by Zheng et al. (2015) by pre-synthesizing the component compounds, ball milling them to achieve thorough mixing, and hot pressing the final powder. Benefitting from the improved power factor and reduced thermal conductivity, the composite with x = 0.42 attained ZT = 1.3 at 773 K, some 40% improvement over the pristine $Yb_{0.25}Co_4Sb_{12}$. Gharleghi et al. (2016b) made a composite from a pre-synthesized micron-sized WTe_2 powder that was randomly embedded in a nano-sized $CoSb_3$ matrix prepared via hydrothermal synthesis, similar to the one described in Section 2.10. The respective powders were mixed according to $CoSb_3$ + $yWTe_2$, where y = 0, 3, 6, 9, and 12 wt%. The admixed powder was cold-pressed into disks, sealed in a quartz tube and sintered at 580°C for 5 h. Since the sintered material was not pulverized and compacted by SPS or hot pressing, the composites had a very low density of merely 58%–78% of the theoretical density. As such, the structures were not really representatives of a well-compacted material. Nevertheless, a composite with y = 6 wt% attained quite an impressive $ZT \approx 0.8$ at 575 K, primarily due to its high power factor of 1.9 $mWm^{-1}K^{-2}$ at 575 K, even though the thermal conductivity actually increased slightly over a similarly sintered $CoSb_3$.

Adding and dispersing carbon-based materials has been a favored approach to form composite skutterudite structures. Shi et al. (2004) added fullerene (C_{60}) at up to 8 mass% to stoichiometric powders of Co and Sb and heated the well-mixed powders at 943 K under Ar atmosphere for 150 h. The product was then ground into a fine powder and consolidated by SPS at 848 K for 15 min to obtain fully densified pellets. The resulting composites contained slightly less fullerene that the amount nominally added. For instance, in the 8 mass% sample, the amount of fullerene detected by EPMA was only 6.54 mass%. Interestingly, while small amounts of fullerene did not improve the thermoelectric properties, the composite containing 6.54 mass% showed a significantly enhanced ZT originating primarily from the reduced thermal conductivity as phonons were effectively scattered by clusters of C_{60} that tended to agglomerate at the grain boundaries when the amounts of fullerene was high. There may also have been a contribution arising from a crossover of the impurity scattering regime of charge carriers in composites with low fullerene content to grain-barrier scattering of charge carriers when the C_{60} content exceeded about 5 mass%. Another form of graphite – graphene – was incorporated in $CoSb_3$ by Feng et al. (2013). The preparation started with forming graphite oxide via the modified Hummers' method, Hummers and Offeman (1958), which was ultrasonically dispersed in triethylene glycol to obtain exfoliated graphene oxide sheets. To this solution were added ingredients necessary for the solvothermal synthesis of $CoSb_3$ (see Section 2.10). The resulting product, p-type composite of $CoSb_3$ with well-dispersed graphene, designated as $CoSb_3/G$ and estimated to contain 1.5 weight% of graphene, was densified by hot pressing at 873 K under 80 MPa for 2 h. The much-enhanced electrical conductivity of the $CoSb_3/G$ composite over a similarly prepared pure $CoSb_3$, little affected Seebeck coefficient, and reduced lattice thermal conductivity resulted in a $ZT \approx 0.6$ at 800 K, some 130% improvement over graphene-free $CoSb_3$ (≈ 0.26). A similar solvothermal synthesis was used by Che et al. (2017) to prepare a composite consisting of La-filled $CoSb_3$ with up to 6 wt% of finely dispersed multi-walled carbon nanotubes. In this case, the value of $ZT = 0.26$ at 780 K, improved only modestly compared to $ZT = 0.19$ for nanotube-free $La_{0.3}Co_4Sb_{12}$ at the same temperature, being limited by a large thermal conductivity of the carbon nanotubes. A composite of p-type skutterudite $La_{0.8}Ti_{0.1}Ga_{0.1}Fe_3CoSb_{12}$ with graphene was also made via a plasma enhanced chemical vapor deposition process by Qin et al. (2018), where under a flowing reaction gas mixture of $Ar:CH_4 = 90:50$, graphene grew on the surface of the powdered skutterudite grains under a radio frequency power of 200 W at 600°C for 30 min. The resulting material was consolidated by hot pressing. The presence of graphene was detected by Raman spectra and XRD, but the amount was not specified. The effect of graphene was primarily to lower the lattice thermal conductivity by as much as 25% at 675 K, compare to the graphene-free sample, while the electronic properties remained essentially unchanged. The maximum ZT of the composite reached unity, some 25% higher value than for the graphene-free skutterudite. Graphene was also used to modify grain boundary scattering in skutterudites, achieving significantly reduced thermal conductivity while having little effect on the electronic properties, Zong et al. (2017). Using the already mentioned Hummer's method, the authors prepared graphite oxide from purified natural graphite, and dispersed it in water. The solution was added dropwise to pre-prepared fine powders of $Ce_{0.85}Fe_3CoSb_{12}$ (p-type) or $Yb_{0.27}Co_4Sb_{12}$ (n-type) dispersed in water. After ultrasonic mixing, the composites were vacuum filtered and dried, and then re-grinded and compacted by SPS, resulting in fully densified n- and p-type disk-shaped composite samples with 0, 0.56, 1.4, and 2.8 vol% of reduced graphite oxide (rGO). The best n-type composite with 1.4 vol% of rGO attained $ZT = 1.5$ at 850 K, while the best p-type composite with the same fraction of rGO reached 1.06 at 700 K. In fact, the authors assembled 8 n-p couples into a module and achieved the conversion efficiency of 8.4%, significantly higher than 6.8% efficiency obtained with a similar 8-couple module but with no graphene added.

A strictly metallic nanophase in the matrix of $Ba_{0.3}Co_4Sb_{12}$ dispersed by ball milling was tried initially by Zhou et al. (2012), and the promising 30% improvement obtained in the figure of merit over the $Ba_{0.3}Co_4Sb_{12}$ matrix was subsequently explored further by Schmidt et al. (2014) and Peng et al. (2015). They used commercial (Alfa Aesar) Ag nanoparticles with a grain size in the 20–40 nm range that were added at concentrations spanning from 0 to 10 wt% to the $Ba_{0.3}Co_4Sb_{12}$ matrix. Composites with less than 2 wt% of Ag showed uniform dispersion of Ag nanoparticles at grain boundaries and on the surface grains on the matrix. However, at higher concentrations, the Ag_3Sb phase was detected. The best thermoelectric performance was obtained with 2 wt% Ag composite, which featured a much enhanced electrical conductivity, a slightly improved Seebeck coefficient, and essentially unchanged total thermal conductivity (the lattice thermal conductivity component was suppressed but was compensated by an increased electronic contribution). Overall, the figure of merit reached $ZT = 1.4$ at 800 K, an impressive 40% enhancement over the $Ba_{0.3}Co_4Sb_{12}$ matrix material. The authors also noted an enhanced densification of the structure with Ag acting as a lubricant. An interesting Ni nanophase in the skutterudite matrix was prepared by Fu et al. (2015) by blending nickel acetate into $Yb_{0.2}Co_4Sb_{12}$ by ball milling. The well-mixed powder was subsequently heated at 623 K for 30 min under flowing reducing gas (95% Ar + 5% H_2) to decompose the acetate into Ni nanoparticles, and compacted into pellets by hot pressing during which Ni dissolved into the skutterudite matrix. About 0.5 wt% of Ni could be dispersed in the matrix before the NiSb secondary phase formed. The enhanced carrier concentration upon Ni doping and a surprisingly improved carrier mobility (for Ni contents up to 0.2 wt%), with the Seebeck coefficient little affected, resulted in a near doubling of the power factor compared to pristine $Yb_{0.2}Co_4Sb_{12}$. Coupled with a slightly reduced thermal conductivity, the presence of 0.2 wt% of Ni resulted in a $ZT = 1.07$ at 723 K.

Finally, there are several reports on dispersing $CoSb_3$ nanoparticles in the micron-sized $CoSb_3$ matrix or its filled skutterudite matrices. Again, the intent of forming such "homo"-composites was to enhance phonon scattering and reduce the thermal conductivity. Ji et al. (2007) prepared a homo-composite of $CoSb_3$ by a process they called solvothermal nano-plating. The initial step is the preparation of bulk $CoSb_3$ powder by the usual heating and annealing of stoichiometric quantities of Co and Sb powders, with the product pulverized by ball milling. A part of this $CoSb_3$ powder is mixed with $CoCl_2$ and $SbCl_2$ precursors (Co : Sb molar ratio of 1:3) in absolute ethanol. The admixture is then transferred into a Teflon-lined autoclave where ethylenediamine and $NaBH_4$ redundant are added (see the solvothermal synthesis in Section 2.2.10). The sealed autoclave is maintained at 240°C for 80 h and then naturally cooled. The precipitate is filtered, washed, dried and hot-pressed at 500°C for 30 min under a pressure of about 200 MPa to attain 95% of the theoretical density. $CoSb_3$ nanoparticles growing on the surface of bulk grains of $CoSb_3$ prior to hot pressing are shown in Figure 1.39a. $CoSb_3$ nanoparticles squeezed between the large grains of $CoSb_3$ in hot-pressed samples are shown in Figure 1.39b. The thermal conductivity of the $CoSb_3$ homo-composite was, indeed, reduced from 7 $Wm^{-1}K^{-1}$ to 4 $Wm^{-1}K^{-1}$ following the "nano-plating" process. However, such homo-composites have very poor thermal stability as documented by the thermal conductivity reverting to its original room temperature value of 7 $Wm^{-1}K^{-1}$ following annealing at 850 K for a few hours. The general strategy of preparing homo-composites by employing a solvothermal synthesis to obtain nanosized particles and the traditional melting/ annealing approach to make micron-sized particles and then mixing the two together with a final compaction by SPS or hot-pressing has been repeated in several studies. The difference between the reports is the matrix material in which the nanometer-sized $CoSb_3$ particles are embedded. Mi et al. (2007) and Yang et al. (2009) used $CoSb_3$ as the matrix material, while Alboni et al. (2008) chose $La_{0.9}CoFe_3Sb_{12}$ and Mi et al. (2008) opted for $Yb_{0.15}Co_4Sb_{12}$. Regardless of the matrix material, all studies reported a significant reduction in the thermal conductivity, which played the decisive role in an enhancement of the figure of merit. In that sense, the studies documented the benefits of having nanostructural inclusions in the matrix material. The unsettled issue with these homo-composite structures is their thermal stability. Although some of the studies tried to address the issue by annealing their best performing composite and inspecting its microstructure and transport properties with encouraging results of no significant changes, the annealing temperatures were generally below 800 K and for the duration of no more than a few days. Whether the nanometer-sized particles would significantly ripen or perhaps "dissolve" in the matrix when more rigorous annealing tests were carried out, remains a question.

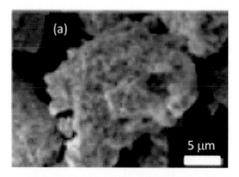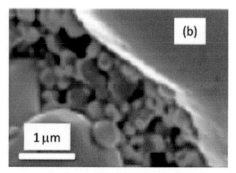

FIGURE 1.39 Scanning electron microscopy image of (a) $CoSb_3$ nanoparticles as-grown on the surface of bulk grains of CoSb3; (b) hot-pressed sample showing the $CoSb_3$ nanoparticles squeezed between the grains of $CoSb_3$. Adapted from X. Ji et al., *Physica Status Solidi* (RRL) 1, 229 (2007). With permission from John Wiley & Sons.

REFERENCES

Akai, K., H. Kurisu, T. Shimura, and M. Matsuura, *Proc. 16th Inter. Conf. on Thermoelectrics*, IEEE Catalog Number 97TH8291, Piscataway, NJ, p. 334 (1977).

Alboni, P. N., X. Ji, J. He, N. Gothard, and T. M. Tritt, *J. Appl. Phys.* **103**, 113707 (2008).

Alleno, E., L. Chen, C. Chubilleau, B. Lenoir, O. Rouleau, M. F. Trichet, and B. Villeroy, *J. Electron. Mater.* **39**, 1966 (2010).

Alleno, E., M. Gaborit V. Ohorodniichuk, B. Lenoir, and O. Rouleau, *J. Electron. Mater.* **42**, 1835 (2013).

Anno, H., Y. Nagamoto, K. Ashida, E. Taniguchi, T. Koyanagi, and K. Matsubara, *Proc. 19th Inter. Conf. on Thermoelectrics*, ed. D. M. Rowe, Babrow Press, UK, p. 90 (2000).

Anno, H., J. Nagao, and K. Matsubara, *Proc. 21st Int. Conf. on Thermoelectris*, IEEE Catalog Number 02TH8657, Piscataway, NJ, p. 56 (2002).

Arushanov, E., K. Fess, W. Kaefer, C. Kloc, and E. Bucher, *Phys. Rev. B* **56**, 1911 (1997).

Bahn, S., T. Gödecke, and K. Schubert, *J. Less Common Met.* **19**, 121 (1969).

Bai, S. Q., Y. Z. Pei, L. D. Chen, W. Q. Zhang, X. Y. Zhao, and J. Yang, *Acta Mater.* **57**, 3135 (2009).

Ballikaya, S., N. Uzar, S. Yildirim, J. R. Salvador, and C. Uher, *J. Solid State Chem.* **193**, 31 (2012).

Battabyal, M., B. Priyadarshini, D. Sivaprahasam, N. S. Karthiselva, and R. Gopalan, *J. Phys. D: Appl. Phys.* **48**, 455309 (2015).

Bauer, E., A. Grytsiv, X.-Q. Chen, N. Melnychenko-Koblyuk, G. Hilscher, H. Kaldarar, H. Michor, E. Royanian, G. Giester, M. Rotter, R. Podloucky, and P. Rogl, *Phys. Rev. Lett.* **99**, 217001 (2007).

Bauer, E., X.-Q. Chen, P. Rogl, G. Hilscher, H. Michor, E. Royanian, R. Podloucky, G. Giester, O. Sologub, and A. P. Gonçalves, *Phys. Rev. B* **78**, 064516 (2008).

Benyahia, M., V. Ohorodniichuk, E. Leroy, A. Dauscher, B. Lenoir, and E. Alleno, *J. Alloys Compd.* **735**, 1096 (2018).

Bérardan, D., C. Godart, E. Alleno, S. Berger, and E. Bauer, *J. Alloys Compd.* **351**, 18 (2003).

Bérardan, D., E. Alleno, C. Godart, M. Puyet, and B. Lenoir, *J. Appl. Phys.* **98**, 033710 (2005a).

Bérardan, D., E. Alleno, C. Godart, O. Rouleau, and J. Rodriguez-Carvajal, *Mater. Res. Bull.* **40**, 537 (2005b).

Berger, St., Ch. Paul, E. Bauer, A. Grytsiv, P. Rogl, D. Kaczorowski, A. Saccone, R. Ferro, and C. Godart, *Proc. 20th Inter. Conf. on Thermoelectrics*, IEEE Catalog 01TH8589, Piscataway, NJ, p. 77 (2001).

Bergman, D. J., and O. Levy, *J. Appl. Phys.* **70**, 6821 (1991).

Bertini, L. and C. Gatti, *J. Chem. Phys.* **121**, 8983 (2004).

Biltz, W. and M. Heimbrecht, *Z. Anorg. Allg. Chem.* **237**, 132 (1938).

Blaha, P., K. Schwarz, P. Sorantin, and S. B. Trickey, *Comp. Phys. Commun.* **59**, 399 (1990).

Borshchevsky, A., J.-P. Fleurial, E. Allevato, and T. Caillat, *Proc. 13th Inter. Conf. on Thermoelectrics*, Kansas City, American Institute of Physics, p. 3, (1995).

Bos, J. W. G. and R. J. Cava, *Solid State Commun.* **141**, 38 (2007).

Braun, D. J. and W. Jeitschko, *J. Less-Common. Met.* **72**, 147 (1980a).

Braun, D. J. and W. Jeitschko, *J. Solid State Chem.* **32**, 357 (1980b).

Caillat, T., J. Kulleck, A. Borshchevsky, and J.-P. Fleurial, *J. Appl. Phys.* **79**, 8419 (1996).

Chakoumakos, B. C. and B. C. Sales, *J. Alloys Compd.* **407**, 87 (2006).

Che, P., B. B. Wang, C. Y. Sun, Y. S. Han, and W. J. Li, *J. Alloys Compd.* **695**, 1908 (2017).

Chen, B., J. X. Xu, C. Uher, D. T. Morelli, G. P. Meisner, J.-P. Fleurial, T. Caillat, and A. Borshchevsky, *Phys. Rev. B* **55**, 1476 (1997).

Chen, L. D., X. F. Tang, T. Kawahara, J. S. Dyck, W. Chen, C. Uher, T. Goto, and T. Hirai, *Proc. the 20th Int. Conf. on Thermoelectrics*, IEEE Catalog Number 01TH8589, Piscataway, NJ, p. 57 (2001a).

Chen, L. D., T. Kawahara, X. F. Tang, T. Goto, T. Hirai, J. S. Dyck, W. Chen, and C. Uher, *J. Appl. Phys.* **90**, 1864 (2001b).

Chen, L. D., *Proc. the 21st Int. Conf. on Thermoelectrics*, Long Beach, CA, IEEE Catalog Number 02TH8657, Piscataway, NJ, p. 42 (2002).

Chen, Y., Y. Kawamura, J. Hayashi, and C. Sekine, *Jap. J. Appl. Phys.* **54**, 055501 (2015).

Chen, Y., Y. Kawamura, J. Hayashi, K. Takeda, and C. Sekine, *J. Appl. Phys.* **120**, 235105 (2016).

Chubilleau, C., B. Lenoir, S. Migot, and A. Dauscher, *J. Colloid. Interf. Sci.* **357**, 13 (2011).

Chubilleau, C., B. Lenoir, P. Masschelein, A. Dauscher, and C. Godart, *J. Electron. Mater.* **41**, 1181 (2012a).

Chubilleau, C., B. Lenoir, A. Dauscher, and C. Godart, *Intermetallics* **22**, 47 (2012b).

Chubilleau, C., B. Lenoir, C. Candolfi, P. Masschelein, A. Dauscher, E. Guilmeau, and C. Godart, *J. Alloys Compd.* **589**, 513 (2014).

Dahal, T., Q. Jie, G. Joshi, S. Chen, C. F. Guo, Y. C. Lan, and Z. F. Ren, *Acta Mater.* **75**, 316 (2014).

Daniel, M. V., L. Hammerschmidt, C. Schmidt, F. Timmermann, J. Franke, N. Jöhrmann, M. Hietschold, D. C. Johnson, B. Paulus, and M. Albrecht, *Phys. Rev. B* **91**, 085410 (2015).

Dilley, N. R., E. D. Bauer, M. B. Maple, and B. C. Sales, *J. Appl. Phys.* **88**, 1948 (2000).

Ding, J., H. Gu, P. F. Qiu, X. H. Chen, Z. Xiong, Q. Zheng, X. Shi, and L. D. Chen, *J. Electron. Mater.* **42**, 382 (2013).

Ding, J., R. H. Liu, H. Gu, and L. D. Chen, *J. Inorgan. Mater.* **29**, 209 (2014).

Duan, B., P. C. Zhai, P. F. Wen, S. Zhang, L. Liu, and Q. J. Zhang, *Scripta Mater.* **67**, 372 (2012).

Duan, B., Jiong Yang, J. R. Salvador, Y. He, B. Zhao, S. Wang, P. Wei, F. S. Ohuchi, W. Q. Zhang, R. P. Hermann, O. Gourdon, S. X. Mao, Y. W. Cheng, C. M. Wang, J. Liu, P. C. Zhai, X. F. Tang, Q. J. Zhang, and Jihui Yang, *Energy Environ. Sci.* **9**, 2090 (2016).

Dudkin, L. D., *Sov. Phys.-Tech. Physics* **3**, 216 (1958).

Dudkin, L. D. and N. Kh. Abrikosov, *Zh. Neorg. Khim.* **1**, 2096 (1956).

Dudkin, L. D. and N. Kh. Abrikosov, *Zh. Neorg. Khim.* **2**, 212 (1957).

Dudkin, L. D. and N. Kh. Abrikosov, *Sov. Phys.-Solid State* **1**, 126 (1959).

Dyck, J. S., W. Chen, C. Uher, L. D. Chen, X. F. Tang, and T. Hirai, *J. Appl. Phys.* **91**, 3698 (2002).

Eilertsen, J., S. Rouvimov, and M. A. Subramanian, *Acta Mater.* **60**, 2178 (2012).

Emsley, J., in *Nature's Building Blocks: An A-Z Guide to the Elements.* Oxford University Press (2011).

Evers, C. B. H., L. Boonk, and W. Jeitschko, *Z. Anorg. Allg. Chem.* **620**, 1028 (1994).

Feldman, J. L., D. J. Singh, I. I. Mazin, D. Mandrus, and B. C. Sales, *Phys. Rev. B* **61**, R9209 (2000).

Feng, B., J. Xie, G. S. Cao, T. J. Zhu, and X. B. Zhao, *J. Mater. Chem. A* **1**, 13111 (2013).

Fleurial, J.-P., A. Borshchevsky, T. Caillat, D. T. Morelli, and G. P. Meisner, *Proc. 15th Inter. Conf. on Thermoelectrics*, IEEE Catalog Number 96TH8169, Piscataway, NJ, p. 91 (1996).

Fleurial, J.-P., T. Caillat, and A. Borshchevsky, *Proc. 16th Inter. Conf. on Thermoelectrics*, IEEE Catalog Number 97TH8291, Piscataway, NJ, p. 1 (1997).

Fu, L. W., J. Y. Yang, J. Y. Peng, Q. H. Jiang, Y. Xiao, Y. Luo, D. Zhang, Z. W. Zhou, M. Y Zhang, Y. D. Cheng, and F. Q. Cheng, *J. Mater. Chem. A* **3**, 1010 (2015).

Fukuoka, H. and S. Yamanaka, *Chem. Mater.* **22**, 47 (2010).

Gharleghi, A., P.-C. Hung, F.-H. Lin, and C.-J. Liu, *ACS Appl. Mater. Interfaces* **8**, 35123 (2016a).

Gharleghi, A., Y. F. Liu, M. H. Zhou, J. He, T. M. Tritt, and C.-J. Liu, *J. Mater. Chem. A* **4**, 13874 (2016b).

Graff, J., S. Zhu, T. Holgate, J. Peng, J. He, and T. M. Tritt, *J. Electron. Mater.* **40**, 696 (2011).

Grandjean, F., A. Gérard, D. J. Braun, and W. Jeitschko, *J. Phys. Chem. Solids* **45**, 877 (1984).

Grandjean, F., G. J. Long, R. Cortes, D. T. Morelli, and G. P. Meisner, *Phys. Rev. B* **62**, 12569 (2000).

Grosvenor, A. P., R. G. Cavell, and A. Mar, *Chem. Mater.* **18**, 1650 (2006).

Grytsiv, A., P. Rogl, S. Berger, C. Paul, E. Bauer, C. Godart, B. Ni, M. M. Abd-Elmeguid, A. Saccone, R. Ferro, and D. Kaczorowski, *Phys. Rev. B* **66**, 094411 (2002).

Grytsiv, A., X.-Q. Chen, N. Melnychenko-Koblyuk, P. Rogl, E. Bauer, G. Hilscher, H. Kaldarar, H. Michor, E. Royanian, R. Podloucky, M. Rotter, and G. Giester, *J. Phys. Soc. Jpn.* **77**, 121 (2008).

Grytsiv, A., P. Rogl, H. Michor, E. Bauer, and G. Giester, *J. Electron. Mater.* **42**, 2940 (2013).

Gumeniuk, R., W. Schnelle, H. Rosner, M. Nicklas, A. Leithe-Jasper, and Y. Grin, *Phys. Rev. Lett.* **100**, 017002 (2008).

Gumeniuk, R., H. Borrmann, A. Ormeci, H. Rosner, W. Schnelle, M. Nicklas, Y. Grin, and A. Leithe-Jasper, *Z. Kristallogr.* **225**, 531 (2010).

Hanus, R., X. Y. Guo, Y. L. Tang, G. D. Li, G. J. Snyder, and W. G. Zeier, *Chem. Mater.* **29**, 1156 (2017).

Harnwunggmoung, A., K. Kurosaki, H. Muta, and S. Yamanaka, *Appl. Phys. Lett.* **96**, 202107 (2010).

Harnwunggmoung, A., K. Kurosaki, T. Plirdpring, T. Sugihara, Y. Ohishi, H. Muta, and S. Yamanaka, *J. Appl. Phys.* **110**, 013521 (2011).

He, T., J. Z. Chen, H. D. Rosenfeld, and M. A. Subramanian, *Chem. Mater.* **18**, 759 (2006).

He, Z., C. Stiewe, S. H. Li, D. Platzek, G. Karpinski, E. Müller, M. Toprak, and M. Muhammed, *Proc. 25th Int. Conf. on Thermoelectrics*, IEEE Catalog Number 06TH8931, Piscataway, NJ, p. 701 (2006).

He, Z., C. Stiewe, D. Platzek, G. Karpinski, and E. Müller, *J. Appl. Phys.* **191**, 043707 (2007a).

He, Z., C. Stiewe, D. Platzek, G. Karpinski, E. Müller, S. H. Li, M. Toprak, and M. Muhammed, *Nanotech.* **18**, 235602 (2007b).

Hornbostel, M. D., E. J. Hyer, J. Thiel, and D. C. Johnson, *J. Am. Chem. Soc.* **119**, 2665 (1997).

Hu, Y., J. R. Salvador, N. Chen, and Y.-J. Kim, *Phys. Rev. Mater.* **2**, 082401® (2018).

Hulliger, F., *Helv. Phys. Acta* **34**, 782 (1961).

Hummers, W. S. and R. E. Offeman, *J. Amer. Chem. Soc.* **80**, 1339 (1958).

Isaacs, E. B. and C. Wolverton, *Chem. Mater.* **31**, 6154 (2019).

Jolibois, P., *C. R. Acad. Sci.* **150**, 106 (1910).

Jeitschko, W. and D. J. Braun, *Acta Crystallog.* **B33**, 3401 (1977).

Jeitschko, W., A. J. Foecker, D. Paschke, M. V. Dewalsky, C. B. H. Evers, B. Künnen, A. Lang, G. Kotzyba, U. C. Rodewald, and M. H. Möller, *Z. Anorg. Allg. Chem.* **626**, 1112 (2000).

Ji, X., J. He, P. Alboni, Z. Su, N. Godard, B. Zhang, T. M. Tritt, and J. W. Kolis, *Phys. Stat. Solidi (RRL)* **1**, 229 (2007).

Kaiser, J. W. and W. Jeitschko, *J. Alloys Compd.* **291**, 66 (1999).

Katsuyama, S., Y. Shichijo, M. Ito, K. Majima, and H. Nagai, *J. Appl. Phys.* **84**, 6708 (1998a).

Katsuyama, S., Y. Kanayama, M. Ito, K. Majima, and H. Nagai, *Proc. 17th Int. Conf. on Thermoelectrics*, IEEE Catalog Number 98TH8365, Piscataway, NJ, p. 342 (1998b).

Katsuyama, S., H. Kusaka, M. Ito, K. Majima, and H. Nagai, *Proc. 18th Int. Conf. on Thermoelectrics*, IEEE Catalog Number 99TH8407, Piscataway, NJ, p. 348 (1999).

Katsuyama, S., Y. Kanayama, M. Ito, K. Majima, and H. Nagai, *J. Appl. Phys.* **88**, 3484 (2000).

Katsuyama, S., M. Watanabe, M. Kuroki, T. Maehata, and M. Ito, *J. Appl. Phys.* **93**, 2758 (2003).

Katsuyama, S. and H. Okada, *J. Jpn. Soc. Powder Power Metall.* **54**, 375 (2007).

Kawaharada, Y., K. Kurosaki, M. Uno, and S. Yamanaka, *J. Alloys Compd.* 315, 193 (2001).

Khovaylo, V. V., T. A. Korolkov, A. I. Voronin, M. V. Gorshenkov, and A. T. Burkov, *J. Mater. Chem. A* **5**, 3541 (2017).

Kihou, K., I. Shirotani, Y. Shimaya, C. Sekine, and T. Yagi, *Mater. Res. Bull.* **39**, 317 (2004).

Kim, H., M. Kaviany, J. C. Thomas, A. van der Ven, C. Uher, and B. Huang, *Phys. Rev. Lett.* **105**, 265901 (2010).

Kitagawa, H., M. Hasaka, T. Morimura, H. Nakashima, and S. Kondo, *Mater. Res. Bull.* **35**, 185 (2000).

Kjekshus, A., D. G. Nicholson, and T. Rakke, *Acta Chem. Scand.* **27**, 1307 (1973).

Kjekshus, A. and G. Pedersen, *Acta Cryst.* 14, 1065 (1961).

Kjekshus, A. and T. Rakke, *Acta Chem. Scand. A* **28**, 99 (1974).

Korenstein, R., S. Soled, A. Wold, and G. Collin, *Inorg. Chem.* **16**, 2344 (1977).

Koza, M. M., M. R. Johnson, R. Viennois, H. Mutka, L. Girard, and D. Ravot, *Nature Mater.* **7**, 805 (2008).

Kraemer, A. C., C. A. Perottoni, and J. A. H. da Jornada, *Solid State Commun.* **133**, 173 (2005).

Kraemer, A. C., M. R. Gallas, J. A. H. da Jornada, and C. A. Perottoni, *Phys. Rev. B* **75**, 024105 (2007).

Krause-Rehberg, R. and H. S. Leipner, in *Positron Annihilation in Semiconductors: Defect Studies*, Springer, Vol. 127 (1999).

Kuzmin, R. N., in *Chemical Bonds in Semiconductors and Solids*, ed. N. N. Sirota, Consultants Bureau, New York (1967).

Kuznetsov, V. L., L. A. Kuznetsova, and D. M. Rowe, *J. Phys.: Condens. Matter* **15**, 5035 (2003).

Lamberton, G. A., R. H. Tedstrom, T. M. Tritt, and G. S. Nolas, *J. Appl. Phys.* **97**, 113715 (2005).

Leszczynski, J., V. Da Ros, B. Lenoir, A. Dauscher, C. Candolfi, P. Masschelein, J. Hejtmánek, K. Kutorasinski, J. Tobola, R. I. Smith, C. Stiewe, and E. Müller, *J. Phys. D: Appl. Phys.* **46**, 495106 (2013).

Li, G.-D., S. Bajaj, U. Aydemir, S.-Q. Hao, H. Xiao, W. A. Goddard, III, P.-C. Zhai, Q.-J. Zhang, and G. J. Snyder, *Chem. Mater.* **28**, 2172 (2016).

Li, H., X. F. Tang, Q. J. Zhang, and C. Uher, *Appl. Phys. Lett.* **93**, 252109 (2008).

Li, H., X. F. Tang, Q. J. Zhang, and C. Uher, *Appl. Phys. Lett.* **94**, 102114 (2009).

Li, J. L., B. Duan, H. J. Yang, H. T. Wang, G. D. Li, J. Yang, G. Chen, and P. C. Zhai, *J. Mater. Chem. C* **7**, 8079 (2019).

Li, X.-D., B. Xu, L. Zhang, F.-F. Duan, X.-L. Yan, J.-Q. Yang, and Y.-J. Tian, *J. Alloys Compd.* **615**, 177 (2014).

Liu, R. H., X. H. Chen, P. F. Qiu, J. F. Liu, J. Yang, X. G. Huang, and L. D. Chen, *J. Appl. Phys.* **109**, 023719 (2011).

Liu, W.-S., B.-P. Zhang, J.-F. Li, and L.-D. Zhao, *J. Phys. D* **40**, 566 (2007a).

Liu, W.-S., B.-P. Zhang, J.-F. Li, and L.-D. Zhao, *J. Phys. D* **40**, 6784 (2007b).

Løvvik, O. M. and Ø. Prytz, *Phys. Rev. B* **70**, 195119 (2004).

Lu, Q. M., J. X. Zhang, X. Zhang, Y. Q. Liu, D. M. Liu, and M. L. Zhou, *J. Appl. Phys.* **98**, 106107 (2005).

Luo, H., J. W. Krizan, L. Muechler, N. Haldolaarachchige, T. Klimczuk, W. Xie, M. K. Fuccillo, C. Felser, and R. J. Cava, *Nature Commun.* **6**, 6489 (2015).

Lutz, H. D. and G. Kliche, *J. Solid State Chem.* **40**, 64 (1981).

Lyons, A., R. P. Gruska, C. Case, S. N. Subbarao, and A. Wold, *Mat. Res. Bul.* **13**, 125 (1978).

Ma, X. J., D. Zhou, Y. Yan, J. Xu, S. Y. Liu, Y. L. Wang, M. N. Cui, Y. H. Cheng, Y. Miao, and Y. H. Lin, *Phys. Chem. Chem. Phys.* **21**, 21262 (2019).

Mallik, R. C., J. Y. Jung, S. C. Ur, and I. H. Kim, *Metals Mater. Int.* **14**, 223 (2008).

Mallik, R. C., C. Stiewe, G. Karpinski, R. Hassdorf, and E. Müller, *J. Electron. Mater.* **38**, 1337 (2009).

Mandel, N. and J. Donohue, *Acta Cryst. B* **27**, 2288 (1971).

Maple, M. B., E. Bauer, N. A. Frederick, P.-C. Ho, W. A. Yuhasz, and V. S. Zapf, *Physica B* **328**, 29 (2003).

Matsui, K., J. Hayashi, K. Akahira, K. Ito, K. Takeda, and C. Sekine, *J. Phys.: Conf. Series* **215**, 012005 (2010).

Matsui, K., J. Hayashi, K. Akahira, K. Ito, Y. Fukushi, K. Takeda, and C. Sekine, *J. Phys.: Conf. Series* **273**, 012043 (2011a).

Matsui, K., J. Hayashi, S. Mitsuka, H. Nakamura, K. Takeda, and C. Sekine, *J. Phys. Soc. Jpn.*, **80**, SA031 (2011b).

Matsui, K., K. Yamamoto, T. Kawaai, Y. Kawamura, J. Hayashi, K. Takeda, and C. Sekine, *J. Phys. Soc. Jpn.*, **81**, 104604 (2012).

Mei, Z. G., W. Zhang, L. D. Chen, and J. Yang, *Phys. Rev. B*, **74**, 153202 (2006).

Mei, Z. G., Jiong Yang, Y. Z. Pei, W. Zhang, L. D. Chen, and Jihui Yang, *Phys. Rev. B* **77**, 045202 (2008).

Meisner, G. P., *Physica B&C* **108**, 763 (1981).

Meisner, G. P., D. T. Morelli, S. Hu, J. Yang, and C. Uher, *Phys. Rev. Lett.* **80**, 3551 (1998).

Mi, J. L., X. B. Zhao, T. J. Zhu, and J. P. Tu, *Appl. Phys. Lett.* **91**, 172116 (2007).

Mi, J. L., X. B. Zhao, T. J. Zhu, and J. P. Tu, *J. Phys. D: Appl. Phys.* **41**, 205403 (2008).

Mi, J. L., M. Christensen, E. Nishibori, and B. B. Iversen, *Phys. Rev. B* **84**, 064114 (2011).

Miotto, F., C. A. Figueiredo, G. R. Ramos, C. L. G. Amorim, M. R. Gallas, and C. A. Perottoni, *J. Appl. Phys.* **110**, 043529 (2011).

Mitchell, R. H., in *Perovskites, Modern and Ancient*, Almaz Press, Thunder Bay, Ontario, (2002).

Möchel, A., I. Sergueev, N. Nguyen, G. J. Long, F. Grandjean, D. C. Johnson, and R. P. Hermann, *Phys. Rev. B* **84**, 064302 (2011).

Morelli, D. T. and G. P. Meisner, *J. Appl. Phys.* **77**, 3777 (1995).

Morelli, D. T., G. P. Meisner, B. Chen, S. Hu, and C. Uher, *Phys. Rev. B* **56**, 7376 (1997).

Nagao, J., D. Nataraj, M. Ferhat, T. Uchida, S. Takeya, T. Ebinuma, H. Anno, K. Matsubara, E. Hatta, and K. Musaka, *J. Appl. Phys.* **92**, 4135 (2002).

Navrátil, J., F. Laufek, T. Plecháček, and J. Plášil, *J. Alloy Compd.* **493**, 50 (2010).

Nickel, E. H., *Chem. Geol.* **5**, 233 (1969).

Nolas, G. S., J. L. Cohn, and G. A. Slack, *Phys. Rev. B* **58**, 164 (1998).

Nolas, G. S., M. Kaeser, R. T. Littleton, and T. M. Tritt, *Appl. Phys. Lett.* **77**, 1855 (2000).

Oftedal, I., *Z. Kristallogr.* **A66**, 517 (1928).

Ohno, A., S. Sasaki, E. Nishibori, S. Aoyagi, M. Sakata, and B. B. Iversen, *Phys. Rev. B* **76**, 64119 (2007).

O'Keeffe, M. and B. G. Hyde, *Acta Cryst. B* **33**, 3802 (1977).

Ortiz, B. R., C. M. Crawford, R. W. McKinney, P. A. Parilla, and E. S. Toberer, *J. Mater. Chem. A* **4**, 8444 (2016).

Park, C.-H. and Y.-S. Kim, *Phys. Rev. B* **81**, 085206 (2010).

Partik, M., C. Kringe, and H. D. Lutz, *Z. Kristallogr.* **211**, 304 (1996).

Pei, Y. Z., L. D. Chen, W. Zhang, X. Shi, S. Q. Bai, X. Y. Zhao, Z. G. Mei, and X. Y. Li, *Appl. Phys. Lett.* **89**, 221107 (2006).

Peng, J. G., X. Y. Liu, L. W. Fu, W. Xu, Q. Z. Liu, and J. Y. Yang, *J. Alloys Compd.* **521**, 141 (2012).

Peng, K., L. Guo, G. Wang, X. Su, X. Y. Zhou, X. F. Tang, and C. Uher, *Sci. Adv. Mater.* **7**, 1 (2015).

Pleass, C. M. and R. D. Heyding, *Canad. J. Chem.* **40**, 590 (1962).

Puyet, M., B. Lenoir, A. Dauscher, M. Dehmes, C. Stiewe, and E. Müller, *J. Appl. Phys.* **95**, 4852 (2004).

Qin, D. D., Y. Liu, X. F. Meng, B. Cui, Y. Qi, W. Cai, and J. H. Sui, *Chin. Phys. B* **27**, 048402 (2018).

Qiu, Y., L. Xi, X. Shi, P. Qiu, W. Zhang, L. S. Chen, J. R. Salvador, J. Y. Cho, J. Yang, Y.-C. Chien, S.-W. Chen, Y. Tang, and G. J. Snyder, *Adv. Funct. Mater.* **23**, 3194 (2013).

Raza, Z., I. Errea, A. R. Oganov, and M. Saitta, *Sci. Rep.* **4**, 5889 (2014).

Ren, W., H. Y. Geng, Z. H. Zhang, and L. X. Zhang, *Phys. Rev. Lett.* **118**, 245901 (2017).

Rogl, G., A. Grytsiv, N. Melnychenko-Koblyuk, E. Bauer, S. Laumann, and P. Rogl, *J. Phys.: Condens. Matter* **23**, 275601 (2011).

Rogl, G., A. Grytsiv, E. Bauer, P. Rogl, and M. Zehetbauer, *Intermetallics* **18**, 57 (2010).

Rogl, G., A. Grytsiv, P. Rogl, N. Peranio, E. Bauer, M. Zehetbauer, and O. Eibl, *Acta Mater.* **63**, 30 (2014).

Rogl, G., A. Grytsiv, F. Failamani, M. Hochenhofer, E. Bauer, and P. Rogl, *J. Alloys Compd.* **695**, 682 (2017).

Rosenboom, E. H. Jr., *Am. Mineralogist* **47**, 310 (1962).

Rosenqvist, T., *Acta Met.* **1**, 761 (1953).

Rosner, H., J. Gegner, D. Regesch, W. Schnelle, R. Gumeniuk, A. Leithe-Jasper, H. Fujiwara, T. Hauptricht, T. C. Koethe, H.-H. Hsieh, H.-J. Lin, C. T. Che, A. Ormeci, Y. Grin, and L. H. Tjeng, *Phys. Rev. B* **80**, 075114 (2009).

Rundqvist, S. and N.-O. Ersson, *Ark. Kemi* **30**, 103 (1968).

Rundqvist, S. and A. Hede, *Acta Chem. Scand.* **14**, 893 (1960).

Rundqvist, S. and E. Larsson, *Acta Chem. Scand.* **13**, 551 (1959).

Ryll, B., A. Schmitz, J. de Boor, A. Franz, P. S. Whitfield, M. Reehuis, A. Hoser, E. Müller, K. Habicht, and K. Fritsch, *ACS Appl. Energy Mater.* **1**, 113 (2018).

Sales, B. C., D. Mandrus, and R. K. Williams, *Science* **272**, 1325 (1996).

Sales, B. C., D. Mandrus, and B. C. Chakoumakos, V. Keppens, and J. R. Thompson, *Phys. Rev. B* **56**, 15081 (1997).

Sales, B. C., B. C. Chakoumakos, D. Mandrus, and J. W. Sharp, *J. Solid State Chem.* **146**, 528 (1999).

Sales, B. C., B. C. Chakoumakos, and D. Mandrus, *Phys. Rev. B* **61**, 2475 (2000).

Sales, B. C., B. C. Chakoumakos, and D. Mandrus, *Mater. Res. Soc. Symp. Proc.* **626**, Z7.1.1 (2001).

Sato, H., H. Sugawara, Y. Aoki, and H. Harima, in *Handbook of Magnetic Materials*, Vol. 18, Ch. 1, Elsevier, p. 1 (2009).

Schmidt, R. D., E. D. Case, Z. Lobo, T. R. Thompson, J. S. Sakamoto, X. Y. Zhou, and C. Uher, *J. Mater. Sci.* **49**, 7192 (2014).

Schmidt, Th., G. Kliche, and H. D. Lutz, *Acta Cryst. C* **43**, 1678 (1987).

Schumer, B. N., M. B. Andrada, S. H. Evans, and R. T. Downs, *Amer. Mineral.* **102**, 205 (2017).

Sekine, C., H. Saito, T. Uchiumi, A. Sakai, and I. Shirotani, *Solid. State Commun.* **106**, 441 (1998).

Sesselmann, A., T. Dasgupta, K. Kelm, E. Muller, S. Perlt, and S. Zastrow, *J. Mater. Res.* **26**, 1820 (2011).

Shannon, R. D., *Acta Cryst. A* **32**, 751 (1976).

Sharp, J. W., E. C. Jones, R. K. Williams, P. M. Martin, and B. C. Sales, *J. Appl. Phys.* **78**, 1013 (1995).

Shi, X., L. D. Chen, J. Yang, and G. P. Meisner, *Appl. Phys. Lett.* **84**, 2301 (2004).

Shi, X., W. Q. Zhang, L. D. Chen, and J. Yang, *Phys. Rev. Lett.* **95**, 185503 (2005).

Shi, X., W. Q. Zhang, L. D. Chen, J. Yang, and C. Uher, *Phys. Rev. B* **75**, 235208 (2007).

Shi, X., H. Kong, C.-P. Li, C. Uher, J. Yang, J. R. Salvador, H. Wang, L. D. Chen, and W. Q. Zhang, *Appl. Phys. Lett.* **92**, 182101 (2008).

Shi, X., J. Yang, J. R. Salvador, M. F. Chi, J. Y. Cho, H. Wang, S. Q. Bai, J. H. Yang, W. Q. Zhang, and L. D. Chen, *J. Amer. Chem. Soc.* **133**, 7837 (2011).

Shirotani, I., T. Noro, J. Hayashi, C. Sekine, R. Giri, and T. Kikegawa, *J. Phys.: Condens. Matter* **16**, 7853 (2004).

Slack, G. A. and V. G. Tsoukala, *J. Appl. Phys.* **76**, 1665 (1994).

Slack, G. A., in *CRC Handbook of Thermoelectrics*, ed. D. M. Rowe, CRC Press, Boca Raton, FL, pp. 407–440 (1995).

Snider, T. S., J. V. Badding, S. B. Schujman, and G. A. Slack, *Chem. Mater.* **12**, 697 (2000).

Takizawa, H., K. Okazaki, K. Uhedu, T. Endo, and G. S. Nolas, *Mater. Res. Soc. Symp. Proc.* **691**, 37 (2002).

Tan, G., Y. Zheng, and X. F. Tang, *Appl. Phys. Lett.* **103**, 183904 (2013a).

Tan, G., W. Liu, H. Chi, X. Su, S. Wang, Y. G. Yan, X. F. Tang, W. Wong-Ng, and C. Uher, Acta Mater. 61, 7693 (2013b).

Tan, G., H. Chi, W. Liu, Y. Zheng, X. F. Tang, J. He, and C. Uher, *J. Mater. Chem. C* **3**, 8372 (2015).

Tang, X. F., L. M. Zhang, R. Z. Yuan, L. D. Chen, T. Goto, T. Hirai, J. S. Dyck, W. Chen, and C. Uher, *J. Mater. Res.* **16**, 3343 (2001).

Tang, Y., Y. Qiu, L. Xi, X. Shi, W. Zhang, L. D. Chen, S.-M. Tseng, S.-W. Chen, and G. J. Snyder, *Energy Environ. Sci.* **7**, 812 (2014).

Tang, Y., R. Hanus, S.-W. Chen, and G. J. Snyder, *Nat. Commun.* **6**, 7584 (2015b).

Tang, Y., S.-W. Chen, and G. J. Snyder, *J. Materiomics* **1**, 75 (2015a).

Toda, M., H. Sugawara, K. Magishi, T. Saito, K. Koyama, Y. Aoki, and H. Sato, *J. Phys. Soc. Jpn.* **77**, 124702 (2008).

Torikachvili, M. S., J. W. Chen, Y. Dalichaouch, R. P. Guertin, M. W. McElfresh, C. Rossel, M. B. Maple, and G. P. Meisner, *Phys. Rev. B* **36**, 8660 (1987).

Tritt, T. M., G. S. Nolas, G. A. Slack, A. C. Ehrlich, D. J. Gillespie, and J. L. Cohn, *J. Appl. Phys.* **79**, 8412 (1996).

Tu, Z., X. Sun, X. Li, R. X. Li, L. Xi, and J. Yang, *AIP Adv.* **9**, 045325 (2019).

Tuomisto, F. and I. Makkonen, *Rev. Mod. Phys.* **85**, 1583 (2013).

Uher, C., in *Semiconductors and Semimetals*, Vol. 69, Ch. 5, ed. T. M. Tritt, Academic Press, San Diego, p. 139 (2001).

Uher, C., B. Chen, S. Hu, D. T. Morelli, and G. P. Meisner, *Mater. Res. Soc. Symp. Proc.* **478**, 315 (1997).

Vaqueiro, P., G. G. Sobany, A. V. Powell, and K. S. Knight, *J. Solid State Chem.* **179**, 2047 (2006).

Vaqueiro, P., G. G. Sobany, and M. Stindl, *J. Solid State Chem.* **181**, 768 (2008).

Vaqueiro, P. and G. G. Sobany, *Mater. Res. Soc. Symp. Proc.* Vol. 1044, Materials Research Society, 1044-U05-08 (2008).

Ventriglia, U., Periodico Mineral. (Rome) **26**, 345 (1957).

Volja, D., B. Kozinsky, A. Li, D. Wee, N. Marzari, and M. Fornari, *Phys. Rev. B* **85**, 245211 (2012).

Wan, S., P. F. Qiu, X. G. Huang, Q. F. Song, S. Q. Bai, X. Shi, and L. D. Chen, *ACS Appl. Mater. Interfaces* **10**, 625 (2018).

Wang, H. T., B. Duan, G. H. Bai, J. L. Li, Y. Yu, H. J. Yang, G. Chen, and P. C. Zhai, *J. Electron. Mater.* **47**, 3061 (2018).

Wang, S., J. R. Salvador, J. Yang, P. Wei, B. Duan, and J. Yang, *NPG Asia Mater.* **8**, e285 (2016).

Wang, Y., J. Mao, Q. Jie, B. Ge, and Z. Ren, *Appl. Phys. Lett.* **110**, 163901 (2017).

Watcharapasorn, A., R. C. DeMattei, R. S. Feigelson, T. Caillat, A. Borshchevsky, G. J. Snyder, and J.-P. Fleurial, *Mat. Res. Soc. Symp. Proc.*, Vol. 626, Materials Research Society, Z1.4.1 (2000).

Wociechowski, K. T., J. Tobola, and J. Leszczynski, *J. Alloys Compd.* **361**, 19 (2003).

Wojciechowski, K. T., *J. Alloys Compd.* **439**, 18 (2007).

Xi, L., Y. Qiu, S. Zheng, X. Shi, Jiong Yang, L. D. Chen, D. J. Singh, Jihui Yang, and W. Q. Zhang, *Acta Materialia* **85**, 112 (2015).

Xia, X., P. Qui, X. Shi, X. Li, X. Huang, and L. D. Chen, *J. Electron Mater.* **41**, 2225 (2012).

Xiong, Z., X. Chen, X. Y. Zhao, S. Q. Bai, X. G. Huang, and L. D. Chen, *Solid State Sci.* **11**, 1612 (2009).

Xiong, Z., X. Chen, X. Y. Huang, S. Q. Bai, and L. D. Chen, *Acta Mater.* **58**, 3995 (2010).

Xiong, Z., L. Xi, J. Ding, X. Chen, X. Y. Huang, H. Gu, L. D. Chen, and W. Q. Zhang, *J. Mater. Res.* **26**, 1848 (2011).

Xue, J. S., M. R. Antonio, W. T. White, and L. Sonderholm, *J. Alloys Compd.* **207–208**, 161 (1994).

Yang, J., G. P. Meisner, and C. Uher, *Proc. 18th Int. Conf. on Thermoelectrics*, IEEE Catalog Number 99TH8407, Piscataway, NJ, p. 458 (1999).

Yang, J., G. P. Meisner, W. Chen, J. S. Dyck, and C. Uher, *Proc. 20th Inter. Conf. on Thermoelectrics*, IEEE Catalog Number 01TH 8589, Piscataway, NJ, p. 73 (2001).

Yang, J., W. Zhang, S. Q. Bai, Z. Mei, and L. D. Chen, *Appl. Phys. Lett.* **90**, 192111 (2007a).

Yang, J., G. P. Meisner, C. J. Rawn, H. Wang, B. C. Chakoumakos, J. Martin, G. S. Nolas, B. L. Pedersen, and J. K. Stalick, *J. Appl. Phys.* **102**, 083702 (2007b).

Yang, Jiong, Q. Hao, H. Wang, Y. Lan, Q. He, A. Minnich, D. Wang, J. Harriman, V. Varki, M. S. Dresselhaus, G. Chen, and Z. Ren, *Phys. Rev. B* **80**, 115329 (2009a).

Yang, Jiong, L. Xi, W. Zhang, L. D. Chen, and J. Yang, *J. Electron. Mater.* **38**, 1397 (2009b).

Yang, L., H. H. Hng, D. Li, Q. Y. Yan, J. Ma, T. J. Zhu, X. B. Zhao, and H. Huang, *J. Appl. Phys.* **106**, 013705 (2009).

Zemni, S., D. Tranqui, P. Chaudouet, R. Madar, and J. P. Senateur, *J. Solid State Chem.* **65**, 1 (1986).

Zevalkink, A., K. Star, U. Aydemir, G. J. Snyder, J.-P. Fleurial, S. Bux, T. Vo, and P. von Allmen, *J. Appl. Phys.* **118**, 035107 (2015).

Zhang, J., B. Xu, L.-M. Wang, D. Yu, Z. Liu, J. He, and Y. Tian, *Appl. Phys. Lett.* **98**, 072109 (2011).

Zhang, L., A. Grytsiv, P. Rogl, E. Bauer, and M. Zehetbauer, *J. Phys. D: Appl. Phys.* **42**, 225405 (2009).

Zhang, L., F. F. Duan, X. D. Li, X. L. Yan, W. T. Hu, L. M. Wang, Z. Y. Liu, Y. J. Tian, and B. Xu, *J. Appl. Phys.* **114**, 083715 (2013).

Zhang, L., B. Xu, X.-D Li, F.-F. Duan, X.-L. Yan, and Y.-J. Tian, *Mater. Lett.* **139**, 249 (2015).

Zhang, T., K. Zhou, X. F. Li, Z. Q. Chen, X. L. Su, and X. F. Tang, *J. Appl. Phys.* **117**, 055103 (2015).

Zhang, W., X. Shi, Z. G. Mei, Y. Xu, L. D. Chen, J. Yang, and G. P. Meisner, *Appl. Phys. Lett.* **89**, 112105 (2006).

Zhao, W., P. Wei, Q.-J. Zhang, C.-L. Dong, L. Liu, and X. F. Tang, *J. Amer. Chem. Soc.* **131**, 3713 (2009).

Zhao, X. Y., X. Shi, L. D. Chen, W. Zhang, W. B. Zhang, and Y. Z. Pei, *J. Appl. Phys.* **95**, 053711 (2006a).

Zhao, X. Y., X. Shi, L. D. Chen, W. Q. Zhang, S. Q. Bai, Y. Z. Pei, X. Y. Li, and T. Goto, *Appl. Phys. Lett.* **89**, 092121 (2006b).

Zhao, D., M. Zuo, Z. Q. Wang, X. N. Teng, and H. R. Geng, *J. Nanosci. Nanotechnol.* **15**, 3076 (2015).

Zheng, J., J.-Y. Peng, Z.-X. Zheng, M.-H. Zhou, E. Thompson, J.-Y. Yang, and W.-L. Xiao, *Frontiers in Chem.* **3**, article 53 (2015).

Zhou, C., J. Sakamoto, D. T. Morelli, X. Y. Zhou, G. Wang, and C. Uher, *J. Appl. Phys.* **109**, 063722 (2011).

Zhou, X. Y., G. Y. Wang, L. Zhang, H. Chi, X. Su, J. Sakamoto, and C. Uher, *J. Mater. Chem.* **22**, 2958 (2012).

Zhou, X. Y., G. W. Wang, L. J. Guo, H. Chi, G. Wang, Q. Zhang, C. Q. Chen, T. Thompson, J. Sakamoto, V. P. Dravid, G. Z. Cao, and C. Uher, *J. Mater. Chem. A* **2**, 20629 (2014).

Zhu, Y. G., H. L. Shen, and H. L. Chen, *Rare Metals* **31**, 43 (2012).

Zhuravlev, N. N. and G. S. Zhdanov, *Kristallografiya* **1**, 509 (1956).

Zong, P. A., R. Hanus, M. Dylla, Y. S. Tang, J. C. Liao, Q. H. Zhang, G. J. Snyder, and L. D. Chen, Energy & Environ. Sci. **10**, 183 (2017).

2 Fabrication of Skutterudites

To characterize thermoelectric Skutterudite, one needs sufficiently large and homogeneous samples on a scale of a few millimeters. Of course, to assemble thermoelectric modules, the amount of the material required is significantly larger. Although from the perspective of transport it is always advantageous to have available single crystals, they are not the absolute necessity. It is only when one desires to ascertain the role of grain boundaries and establish intrinsic values of the carrier mobility that single crystals are essential. Since skutterudites are cubic structures, intrinsic anisotropy is not a concern and polycrystalline forms of skutterudites are generally perfectly adequate. However, one should be aware of sample anisotropy induced extrinsically as a consequence of certain processing techniques, most notably the application of high pressure during hot pressing and during the spark plasma sintering process, which often result in a preferred texture of the samples.

Identification of the optimal processing conditions leading to a material with the desired physical/chemical properties is usually a tedious trial and error process that explores various different synthesis routes. Once the optimal material parameters are established, it is the task of a materials technologist to come up with a fabrication approach that not only faithfully reproduces the optimal material characteristics but that is also scalable and economically viable.

2.1 PHASE DIAGRAM OF SKUTTERUDITES

Skutterudites are compounds that form from elements with significantly different melting points and vapor pressures. In such cases, the successful synthesis requires a thorough knowledge of the phase diagram that describes thermodynamic equilibrium between various phases of the system. In fact, skutterudites are stable compounds with the highest pnicogen concentration that form by fusing with Co, Rh, and Ir at ambient pressure. As there are other, lower pnicogen content compounds within any one of the phosphide, arsenide and antimonide series, the phase diagram of skutterudites is rather complex, as illustrated in Figure 2.1 for the case of $CoSb_3$, Feschotte and Lorin (1989). Other binary skutterudites have substantially similar phase diagrams.

From the phase diagram, one immediately notes that there are three stable compounds in the higher pnicogen concentration range: CoSb (γ-phase), which grows congruently and is a metal, and two semiconducting phases, $CoSb_2$ (δ-phase) and $CoSb_3$ (ϵ-phase), each forming peritectically. The skutterudite structure – the ϵ-phase $CoSb_3$ – is obtained in a peritectic reaction at 873°C from the δ-phase and the liquid phase. To remind the reader, a peritectic reaction involves formation of a homogeneous compound or solid solution from another (different) solid phase and a liquid at a specific fixed temperature called the peritectic temperature, T_p. In the case of $CoSb_3$, the reaction can be described as

$$\delta\text{-phase} + \text{Liq} \leftrightarrow \epsilon\text{-phase} \quad \text{at } T_p = 873°C. \tag{2.1}$$

Peritectic temperatures for other binary skutterudites are given in Table 2.1. Because of very low kinetics, growth rarely proceeds under conditions that the peritectic reaction is fully completed and extensive annealing (several days to a couple of weeks) is required to achieve the skutterudite phase. With these facts in place, there are several strategies one can use to prepare skutterudites of either polycrystalline or single crystal nature.

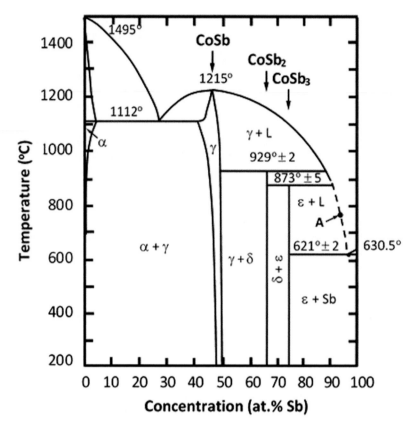

FIGURE 2.1   Phase diagram of CoSb$_3$. Point A indicates a possible starting composition to grow a single crystal from melt. Diagram data from P. Feschotte and D. Lorin, *J. Less-Common Metals* 155, 255 (1989). Reproduced from C. Uher, *Semiconductors and Semimetals*, Vol. 69, Ed. T. M. Tritt, Academic Press, San Diego, pp. 139–253 (2001). With permission from Elsevier.

TABLE 2.1

Peritectic Temperatures of the Binary Skutterudites

Skutterudite	Peritectic Temperature (°C)	Reference
CoP$_3$	> 1000	Biltz and Heimbrecht (1939)
CoAs$_3$	960	Caillat et al. (1995)
CoSb$_3$	873	Feschotte and Lorin (1989)
RhP$_3$	> 1200	Odile et al. (1978)
RhAs$_3$	> 1000	Caillat et al. (1995)
RhSb$_3$	900	Caillat et al. (1996)
IrP$_3$	> 1200	Rundqvist and Ersson (1968)
IrAs$_3$	> 1000	Kjekshus (1961)
IrSb$_3$	1141	Caillat et al. (1993)
NiP$_3$	> 850	Rundqvist and Larsson (1959)

2.2   SYNTHESIS OF SKUTTERUDITES

There are numerous synthesis processes that can be used in the preparation of skutterudites, and the one chosen depends very much on the skill and experience of researchers, the required form and size of specimens, and experimental facilities available. I outline here techniques that have been found useful.

2.2.1   Synthesis by Melting and Annealing

The early favored method of preparing polycrystalline samples, the variants of which are followed till this day, is to dissolve a stoichiometric quantity of cobalt in antimony at 1220°C - 1250°C and cool the material to room temperature, Zobrina and Dudkin (1960). To ensure completion of the peritectic reaction leading to the formation of the $CoSb_3$ phase, the samples are annealed for a week or two at a somewhat lower temperature (550–600°C) than that corresponding to the fusion of the eutectic with antimony. To increase sample density and uniformity, the compound is crushed and hot pressed at 500°C under pressure of 3000 kg/cm^2. Finally, the sample is annealed at 800°C (below the peritectic temperature) in argon for a couple of days. An alternative initial step is to use induction melting and quench the melt by pouring it into a copper mold, Nakagawa et al. (1996).

2.2.2   Solid-Liquid Sintering

Polycrystalline forms of skutterudite can also be synthesized by the solid-liquid sintering process, consisting of sealing stoichiometric amounts of the constituent elements in an evacuated quartz ampoule and heating the mixture for a prolonged time (a week or so) at temperatures just below the peritectic temperature. This yields a compound that is typically 80–85% dense and that likely contains small amounts of other phases. The phase purity is improved by repeating grinding and reannealing until X-rays show complete absence of any impurity phases. Further densification of the structure to above 97% of the theoretical density can be achieved by hot pressing or SPS. This approach works well for most of the skutterudites.

2.2.3   Mn-Reduction of Oxides

A novel synthesis method for skutterudites, inspired by the industrial pyrometallurgical process of Kroll (1940), was described recently by Le Tonquesse et al. (2019). The technique relies on Mn-reduction of Co_4O_4 and Sb_2O_4 powders that are mixed by ball-milling, cold-pressed into pellets, and stacked on top of a Mn chips bed in an argon-filled vessel heated to 810 K and held there for 3–4 days. Mn chips "suck out" oxygen, leaving behind pellets of pure $CoSb_3$, easily separated from MnO. The resulting pellets are powdered and consolidated by SPS. The ingot is a single-phase $CoSb_3$ structure with fine grains of 500 nm–800 nm. Mixing in the starting powders also an appropriate amount of $In_{0.10}Co_{2.90}O_4$, the authors synthesized $In_{0.13}Co_4Sb_{12}$ with quite respectable $ZT =$ 0.75 at 650 K.

2.2.4   Single Crystal Growth

2.2.4.1   Single Crystals from Nonstoichometric Melts

An inspection of the phase diagram in Figure 2.1 suggests that skutterudites can also be grown from melts. There is a narrow window of nonstoichiometric melt compositions (such as represented by point A in Figure 2.1) where the melt is richer in pnicogen that the concentration corresponding to the intersection of the peritectic isotherm with the liquidus, but not as rich as the eutectic point near the pnicogen's endpoint. As the melt cools and its temperature reaches the liquidus, the skutterudite phase starts to emerge. Slowly lowering furnace temperature while maintaining a large temperature gradient (to aid phase separation), and using a quartz ampoule with a pointed bottom (to promote nucleation of a single crystal), it is possible to grow large single crystals of skutterudites. A pure skutterudite phase grows at the lower part of the ampoule until the melt cools close to the temperature of the eutectic point. Afterwards, the remaining melt solidifies as a eutectic mixture of the $\mathcal{E}$-phase and the pnicogen phase at the upper part of the ampoule. The two different phases-pure skutterudite and eutectic mixture – are easily discerned upon examination of the ingot. Caillat et al. (1996) used this so-called gradient freeze technique with great success in preparing large single crystals of $CoSb_3$ and $RhSb_3$. Doping elements can be introduced directly into the melt and the crystals are robust with at least 99.5% theoretical density. The gradient freeze technique is less successful when the peritectic region extends too close to the eutectic point, as in the case of $IrSb_3$. This leaves only a very narrow range of compositions with steep temperature variation available for crystal growth, and the resulting ingot consists of a mixture of $IrSb_3$ and the eutectic with no clear delineation between the two phases.

2.2.4.2 Flux Growth of Crystals

One of the most productive techniques to grow single crystals is the flux-assisted growth. This technique often reflects the intuition and years of experience of a crystal grower in identifying a suitable flux from which the desired crystal could be grown at a considerably lower temperature, and where the flux does not form phases that could impede nucleation of the intended compound. In the optimal case, the flux would be one of the constituents. This works exceptionally well with antimonide skutterudites, as shown by Mandrus et al. (1995), where a large excess of Sb serves as a flux. The excess flux is spun off while still molten, leaving behind shiny crystals with dimensions of a few millimeters on a side. Skutterudites also grow well using tin flux, the method used first by Jolibois (1910) more than 100 years ago, and employed extensively by Biltz and Heimbrecht (1938) in their preparation of CoP_3 and NiP_3, and later by Odile et al. (1978) and Torikachvili et al. (1987) in their growth of single crystals of several filled phosphide skutterudites. Excess tin is easily removed by leaching in diluted HCl.

Unfortunately, a very high vapor pressure of arsenic of over 35 atm (3.5 MPa) at its melting point of 817°C was a serious impediment to the self-flux growth of filled arsenide skutterudites and, for a number of years, precluded the studies of their physical properties. The problem was solved by Henkie et al. (2008) who noted that the melting point (and therefore the vapor pressure) of As can be significantly lowered by diluting As with Cd, Figure 2.2. By trial and error, it was found that single crystals of filled arsenide skutterudites can be prepared from Cd–As fluxes when the Cd:As

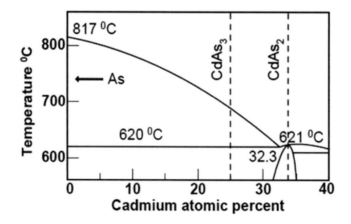

FIGURE 2.2 Arsenic-rich part of the Cd-As phase diagram showing a rapid lowering of the melting point with the increasing content of Cd. Horizontal line at 620°C denotes melting of $CdAs_2$ compound. Successful growth of single crystals was achieved for fluxes with the Cd:As ratio between one-quarter and one-third. Reprinted from Z. Henkie et al., *Journal of the Physical Society of Japan* 77, Suppl. A, 128 (2008). With permission from the Physical Society of Japan.

ratio is between one-quarter to one-third. Fluxes with the Cd content over 50% have not produced single crystals. The actual growth was carried out in a 2 mm thick quartz tube that was placed in a high-pressure cell filled with Ar, the task of which was to balance the vapor pressure of the flux inside the quartz tube at the desired growth temperature, typically between 730°C and 800°C. The synthesis process took 3–4 weeks. The solidified flux containing single crystals of typically 1–2.5 mm in size was then placed in a 20 cm–long quartz tube and the flux was removed by sublimation from a high temperature zone held at 600°C to a low temperature zone of 400°C. The techniques enabled preparation of numerous different filled arsenide skutterudites and the study of their transport, magnetic, and superconducting properties.

2.2.4.3 Growth of Crystals by Chemical Vapor Transport

The chemical vapor transport method is a classical growth technique that has a potential to yield high-quality single crystals. Several attempts have been made to apply this technique to skutterudites. The most notable effort was that of Ackermann and Wold (1977) who grew CoP_3, $CoAs_3$, and $CoSb_3$. The transport agent in this case was chlorine, and they obtained single crystals with dimensions up to $3 \times 3 \times 1$ mm^3 over a period of 3 weeks. Optical measurements indicated that while CoP_3 has a gap of 0.45 eV, $CoAs_3$ and $CoSb_3$ appeared, surprisingly, gapless and displayed a metallic character in their temperature dependent resistivity. Thus, considerable nonstoichiometry had to be present in the crystals despite having the appearance of a very high-quality crystalline structure. Although the technique is simple in principle, it requires good control of temperature and the growth takes quite a lot of time. The resulting crystals are rather small, typically not much larger than a couple of millimeters. Clearly, the vapor transport technique is hopelessly inadequate for any large-scale fabrication of skutterudites.

2.2.5 Rapid Fabrication Techniques

2.2.5.1 Melt Spinning Technique

Numerous attempts have been made to speed up the processing of skutterudites, especially to cut down on long hours and days required for the solid-liquid sintering process. Spin casting under argon on the copper drum rotating with various surface speeds, also referred to as melt-spinning, has been tried with both binary and filled skutterudites. The technique produces ribbons of typically 20 μm thickness with submicron grain sizes that decrease with the increasing surface speed of the copper drum. As shown by Morimura et al. (1997) and Kitagawa et al. (1998), the ribbons become phase pure skutterudite after annealing at 600°C–700°C for just a few hours. An alternative and even faster process consists of collecting the ribbons, crushing them, and consolidating the powder by SPS for a few minutes, during which the skutterudite phase fully forms (Tan et al. 2013).

2.2.5.2 Ball Milling

Ball milling, also known as mechanical alloying, is a versatile fabrication process used with many thermoelectric materials and beyond. The high energy generated in ball milling is usually sufficient to form the desired phase. Unfortunately, ball milling does not directly fuse elemental powders into the skutterudite phase, even after prolonged ball milling times of 50 h, as was found early on by Yang et al. (2004, 2006). However, the fine grained (submicron size) powders can be consolidated by hot pressing or SPS, and it is during this stage of the process that the skutterudite phase forms (Jie et al. 2013). Because the ball milled powders are very fine, great care must be exercised to prevent their exposure to air, otherwise significant oxidation takes place. Moreover, the powders are also often "dirtied" by the contact with the walls of the vessel and the milling balls.

2.2.5.3 Melt Atomization

Another approach that aims at cutting down on the processing time is to start with powders of small diameter. An efficient way how to obtain such powders is to atomize the melt by purified argon, Uchida et al. (1998). This fast processing method results in a fine-grained powder of less than 100 μm

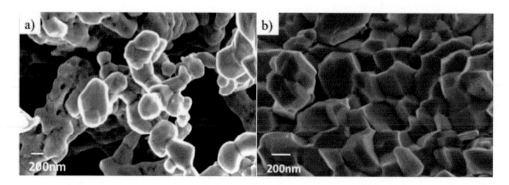

FIGURE 2.3 SEM micrographs of (a) as precipitated powder of $Fe_xCo_{1-x}Sb_3$, and (b) SPS compacted bulk ingot. Reproduced from M. Y. Tafti et al., *Materials Research Society Symposia Proceedings* 1543, 105 (2013). With permission of the Materials Research Society.

diameter that is not contaminated, as often happens during the ball milling process. The subsequent compaction can be done by hot pressing or SPS, the latter requiring considerably less time.

2.2.5.4 Chemical Alloying

Yet another synthesis route that yields very small grain powders, now in the submicron range, is chemical alloying used by Tafti et al. (2013). The process consists of two steps: (1) co-precipitation of precursor materials (oxalene compounds) with the desired metal atom composition, and (2) calcination at 350°C to obtain metal oxides that are subsequently reduced under hydrogen atmosphere. The synthesis process was used to prepare $Fe_xCo_{1-x}Sb_3$ skutterudite powders of submicron size (~200 nm) that were consolidated by SPS. Although there is a small grain growth during the SPS process, the compacted high-density $Fe_xCo_{1-x}Sb_3$ remains a nanostructured skutterudite, as shown in Figure 2.3.

2.2.6 HIGH-PRESSURE SYNTHESIS

High-pressure-assisted synthesis has served well in the preparation of filled skutterudites that otherwise are difficult or cannot be formed via more conventional methods (Takizawa et al. 1999). I have already mentioned that high-pressure-assisted synthesis allowed for inserting, among others, tin (Takizawa et al. 2002), and lithium (Zhang et al. 2011), into the voids of $CoSb_3$, and high pressure was also instrumental in increasing the filling fraction limit of Pr in the $[Fe_4Sb_{12}]$ framework to near unity (Tanaka et al. 2007). Beyond enhancing the filling fraction limit, the high-pressure synthesis was essential in the preparation of rare earth-filled skutterudites with fillers heavier and smaller than Gd (Sekine et al. 1998, 2000, Shirotani et al. 2003), and in the synthesis of skutterudites with electronegative fillers, notably iodine (Fukuoka and Yamanaka 2010). The high-pressure technique relies on the use of cubic multi-anvil wedge-type apparatus, where anvils press in a coordinated fashion on the faces of a cube made from pyrophyllite, which houses a boron nitride sleeve containing a sample under study. The BN sleeve is enclosed by a graphite heater that allows heating to the desired temperature monitored by a thermocouple. More specifics regarding high-pressure measurements can be found, for instance, in a paper by Sekine and Mori (2017). Most of high-pressure studies on skutterudites have used pressures to about 6 GPa, but nearly an order of magnitude higher pressures can be reached, depending on the design of the apparatus. The pressure is applied at the desired temperature, maintained during the synthesis, and not released until the apparatus has cooled down to room temperature. Of course, all structures thus prepared, whether polycrystalline or single crystals, are more or less metastable compounds. This is fine for studies of physical properties, but an early decomposition at elevated temperatures might be a serious problem for their operation in power generating thermoelectric modules.

2.2.7 MICROWAVE-ASSISTED SYNTHESIS

Certain inorganic substances can couple to microwave radiation, resulting in rapid dielectric heating of the material. The absorbed energy can facilitate atomic interdiffusion of reactants and yields a reacted product in a very short time. The microwave technique has been used to prepare numerous functional oxide materials, the interested reader will find useful a review article by Rao et al. (1999). It was therefore tempting to explore the microwave synthesis with skutterudites and $In_{0.2}Co_4Sb_{12}$ served as a test case (Biswas et al. 2011). Stoichiometric quantities of elemental powders were sealed in a silica tube and the tube was "buried" in CuO, a strong microwave susceptor, inside of an alumina combustion boat. Alumina, in turn, was surrounded by high temperature firebricks and the entire assembly was placed on a rotating glass table inside a microwave oven. The oven operated with 750 W for 2 min. The CuO susceptor acted as a small localized furnace with an extremely high ramping rate, much higher that what a conventional furnace can achieve. Immediately after the microwave exposure, the powder contained small amounts of $CoSb_2$ and Sb secondary phases that were completely eliminated during the subsequent sintering process carried out at $675°C$ under the flow of N_2/H_2 for 4 hr. Thermoelectric performance of microwave fabricated samples was comparable to those synthesized by conventional processing with the $ZT \sim 0.85$ at 600 K. Microwave assisted synthesis was also used by Ioannidou et al. (2014) to synthesize a series of $Co_{1-x}Fe_xSb_3$ skutterudite powders with $x = 0, 0.1, 0.2,$ and 0.3 that were subsequently sintered by SPS. The best performing composition $Co_{0.8}Fe_{0.2}Sb_3$ reached $ZT = 0.33$ near 800 K. A series of nearly single-phase $Co_4Sb_{11.9-x}Te_xSe_{0.1}$ ($x = 0.2 - 0.6$) was prepared recently by Lei et al. (2019) with the highest $ZT = 0.81$ at 773 K. The microwave-assisted synthesis significantly cuts on the processing time by essentially eliminating the calcination step, reduces energy use, and is potentially scalable.

2.2.8 SELF-PROPAGATING HIGH-TEMPERATURE SYNTHESIS (SHS)

Perhaps the most exciting novel synthesis approach of thermoelectric materials is the use of self-propagating-high-temperature synthesis (SHS) (Su et al. 2014), the technique that relies on exothermic reactions taking place among the constituting elements, which, under the right circumstance, can acquire a self-propagating character and sweep through the entire material in a few seconds. The SHS synthesis is well known in fabrication of high temperature intermetallics and ceramics (Merzhanov 1992 and Moore and Feng 1995), but has not been used with semiconductors on account of the previously held opinion (the Merzhanov criterion) (Munir and Anselmi-Tamburini 1989 and Merzhanov 1990), that the successful use of SHS requires temperatures of at least $1800°C$. Of course, the vast majority of semiconductors, including essentially all thermoelectric materials, would melt and evaporate at such high temperatures and, thus, no one has tried to use it. We have shown that the application of SHS is actually governed by a different and more intuitive criterion that requires the adiabatic temperature (the highest temperature the reaction zone reaches) to be higher than the melting point of the lower melting point constituent. The molten state of this element then dramatically speeds up the fusion process as the layers of the material in the proximity of the reaction zone are ignited, and the fusion acquires a chain reaction character as the reaction zone sweeps through. SHS can be used in the preparation of most of the existing thermoelectric materials and, in collaboration with the Wuhan University of Technology, we have demonstrated its specific relevance to skutterudites in a paper by Liang et al. (2014). The technique does not rely on any expensive equipment and it only requires that a small section of the sample (typically the bottom or the top of the cold-pressed ingot of elemental powders) be heated rapidly to initiate the reaction. Typical speed of propagation of the reaction zone through the length of the ingot are 1–3 mm/s, and the ingot is fully reacted (pure $CoSb_3$ phase) after the passage of the reaction zone. The reacted ingot often changes its shape and is not fully dense. This, however, is easily handled by crushing the ingot into powder and briefly applying SPS. Remarkably, the thermoelectric performance of such SHS–SPS-prepared $CoSb_3$ is at least as good as the best $CoSb_3$ compounds reported in literature.

The technique is not only rapid and uses minimum energy, but it is also readily scalable. There might even be an aspect of sample purification as the reaction zone sweeps through, reminiscent of the well-known zone refining process. This, however, is a distinct disadvantage when attempting to apply SHS to the synthesis of filled skutterudites. The weakly bonded filler species are simply swept through with the reaction zone, and we have not yet been able to directly synthesize filled skutterudites. On the other hand, dopants substituting on the pnicogen rings are strongly bonded and remain in the structure after the passage of the reaction zone, as was shown with Te-doped $CoSb_3$ by Liang et al. (2014).

2.2.9 SELECTIVE LASER MELTING

The traditional fabrication of TE devices and modules, starting with the synthesis of TE materials and culminating in the assembly of reliable modules is a complex process that includes, among others, slicing, electroplating, dicing, assembly and soldering. The process is not only time con-suming, but it is also wasteful of materials lost during the processing. Consequently, it incurs a rather high cost penalty, which is a series constraint on a large-scale deployment of TE technology, see, e.g., Bell (2008), Zhao et al. (2008), and Wang et al. (2012). One of the highly prospective alternatives to the traditional TE manufacturing is Selective Laser Melting (SLM) (Kruth et al. 1998, Bremen et al. 2012, and Martin et al. 2017), a relatively novel and rapid fabrication technol-ogy that has become perhaps the most developed in the field of additive manufacturing and widely applied to the printing of metallic structural materials (Campanelli et al. 2014, Cherry et al. 2015). SLM uses a laser to melt a thin layer of dispersed powder, which rapidly solidifies. 3D objects can be built by successive steps of powder deposition followed by laser-induced melting. Provided the SLM technology can be applied to a rapid processing of n- and p-type TE legs and their bond-ing to the electrode stripes, there is a tantalizing potential of a one-step rapid manufacturing of TE modules. This would not only much shorten the production cycle but it would also eliminate wasteful slicing and dicing of the material. Importing the SLM technology to the fabrication of TE materials and modules is challenging, primarily because of much lower melting points, low ther-mal conductivity, weak mechanical strength and poor thermal shock resistance of thermoelectric materials compared to metals. Nevertheless, first important steps to utilize this modern manufac-turing approach in the field of thermoelectric materials have been taken recently. El-Desouky et al. (2016) demonstrated processing of Bi_2Te_3 using SLM and a team from the Wuhan University of Technology, Mao et al. (2017), successfully applied the SLM method to the fabrication of n-type $Bi_2Sb_{2.7}Se_{0.3}$ and determined the optimal processing parameters leading to high-quality layers. Most recently, the team has extended the SLM technique to the processing of n-type $CoSb_{2.85}Te_{0.15}$, Yan et al. (2018). Current efforts, combining simultaneous SLM processing of both n- and p-type layers are the next important milestone to reach to accomplish a truly one-step fabrication of TE modules. Interestingly, SLM can also be used with rather thick powder layer beds of up to about 2 mm (scan speed of 200 mm per minute, hatching spacing 3.75 mm, 1 mm diameter laser spot with optimized power of 150 W), which are then ground into fine powders and a 10 min SPS processing under 60 MPa at 883 K yields the pure skutterudite phase. Including filler species in the starting powder bed, Chen et al. (2018) prepared $Yb_{0.4}Co_4Sb_{12}$ and $In_{0.5}Co_4Sb_{12.1}$ with $ZT = 1.23$ at 850 K and $ZT = 1.09$ at 700 K, respectively.

2.2.10 HYDROTHERMAL AND SOLVOTHERMAL GROWTH

As I discuss in Chapter 5, to minimize the thermal conductivity, one of the approaches is to intro-duce nanostructural features into a sample to enhance phonon boundary scattering. In its ultimate form, one can make samples by compacting nanometer-scale particles. In one of the first attempts to prepare nanostructured $CoSb_3$ by chemical solution process, Wang et al. (1999) used a two-step synthesis consisting of (1) co-precipitation of Sb_2O_3 (needle-like particles of about 100 nm diameter) and $CoC_2O_4.2H_2O$ (rock-like particles of about 1 μm size) from a solution of 1 M $SbCl_3$ in 3 M HCl, 1 M $CoCl_2$, 0.2 M of $H_2C_2O_4$, and 1 M NH_3, all dissolved in deionized water; and (2) calcination of the co-precipitates and subsequent reduction under flowing hydrogen at around 500°C.

A more effective way of making nanometer-size powders of skutterudites is the hydrothermal and solvothermal synthesis. The two synthesis routes are very similar, differing only in a medium in which the chemical reaction takes place. In the former case it is an aqueous solution that is heated in an autoclave under pressure to temperatures above the boiling point of water, while in the latter case it is a nonaqueous solution (e.g., ethanol or tetraethylene glycol). Hydrothermal growth of filled skutterudite $NaFe_4P_{12}$ single crystalline nanowires with the diameter of 25–80 nm and 500 – 4000 nm length (depending on the exact temperature in the range 160°C – 180°C) was achieved by Liu et al. (2002) by reacting iron chloride, sodium hydroxide and white phosphorus. Gharleghi et al. (2014) synthesized n-type $CoSb_3$ nanopowders by the hydrothermal growth using as reactants $CoCl_2.6H_2O$ (source of Co) and $SbCl_3$ (source of Sb) in the molar ratio to yield $CoSb_3$ dissolved in deionized water and sonicated for 15 min. The critical issue for a successful reaction is a reducing agent. Sodium borohydride, $NaBH_4$, has been found an excellent reducing agent, and it is a matter of trial and error to identify its optimal amount. The precursors are placed in Teflon-lined stainless steel autoclave. The autoclave is sealed and heated in a one-step or two-step fashion to temperatures typically between 190°C and 290°C (the Teflon liner cannot withstand higher temperatures) for the duration ranging from 12 h to 2–3 days. It is believed that the formation of $CoSb_3$ proceeds via the following reaction steps:

$$CoCl_2 + 2NaBH_4 \rightarrow Co + 2BH_3 + 2NaCl + H_2 \tag{2.2}$$

$$2SbCl_3 + 6NaBH_4 \rightarrow 2Sb + 6BH_3 + 6NaCl + 3H_2 \tag{2.3}$$

$$Co + 2Sb \rightarrow CoSb_2 \tag{2.4}$$

$$Co + 3Sb \rightarrow CoSb_3 \tag{2.5}$$

$$CoSb_2 + Sb \rightarrow CoSb_3 \tag{2.6}$$

Reactions of the chlorides with $NaBH_4$ in Eqs. 2.2 and 2.3 bring about a rapid reduction in the oxidation state of Co and Sb from Co^{2+} and Sb^{3+} to Co and Sb, which then react to form either a minority diantimonide phase $CoSb_2$ or the majority skutterudite phase $CoSb_3$. The diantimonide is an intermediate product that further reacts with Sb to form the skutterudite phase. The respective amounts of $CoSb_2$ and $CoSb_3$ depend critically on the amount of $NaBH_4$, on the temperature of the reaction, and the duration of the reaction. Higher amount of $NaBH_4$, higher temperatures, and longer reaction times all favor the formation of the skutterudite phase. However, too much of $NaBH_4$ leads to a degraded crystalline quality of the grains and, as already mentioned, the temperature is limited by the use of Teflon lining. The reacted product (black powder) is washed several times in distilled water and ethanol and dried for several hours. Under the optimal synthesis conditions of reaction temperature of 290°C, it consisted of $CoSb_3$ with a minute presence of $CoSb_2$. Typical grain size is in the range of tens of nanometers. Applying hot pressing or SPS to consolidate the powder, the trace of the $CoSb_2$ phase disappeared, revealing a pure $CoSb_3$ skutterudite. The conversion is also achieved by simply annealing the powder under vacuum at around 580°C for 5 h. During the consolidation of the powder, there is an unavoidable grain growth to sizes of 100–200 nm. Typical size and shape of the nanoparticles of $CoSb_3$ synthesized at 170°C and subsequently cold-pressed to a rectangle and annealed at 580°C for 5 h is shown in Figure 2.4. Gharleghi et al. (2016) also successfully used the hydrothermal synthesis in the preparation of doubly substituted $CoSb_3$, i.e., $Co_{1-x-y}Ni_xFe_ySb_3$ with a range of x and y values between 0 and 0.14.

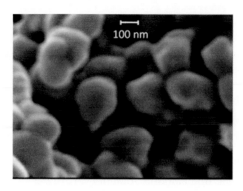

FIGURE 2.4 SEM image of a polished surface of a hydrothermally synthesized $CoSb_3$ at 170°C, subsequently cold-pressed and annealed under vacuum at 580°C for 5 h. The material is a single-phase skutterudite. Note a rather large porosity of some 71–78% due to merely cold pressing the powder. Adapted from A. Gharleghi et al., *Journal of Materials Chemistry C* 2, 4213 (2014). With permission from the Royal Society of Chemistry.

As already noted, the solvothermal synthesis is carried out in a non-aqueous solution, usually an ethanol or tetraethylene glycol, but otherwise the reactants, the reducing agent, and the instrumentation are very similar if not identical. The technique was used by, among others, Xie et al. (2004), Mi et al. (2007), Ji et al. (2008), and Bhaskar et al. (2015a) to prepare $CoSb_3$. A similar process, using appropriate metal chlorides, was used by Kadel and Li (2014) to synthesize Yb-filled $CoSb_3$, and by Bhaskar et al. (2015b) to prepare $Ce_xCo_4Sb_{12}$ with an enhanced Ce filling fraction $x = 0.15 - 0.30$.

2.2.11 GROWTH OF THIN FILMS OF SKUTTERUDITES

In the early 1990s, Hicks and Dresselhaus (1993a,b) predicted that the sharply peaked electronic density of states in lower dimensional structures might lead to much increased power factors. Combined with enhanced boundary scattering of phonons on numerous interfaces, should then result in greatly improved thermoelectric figure of merit ZT. The predictions stimulated intensive experimental efforts to verify such tantalizing prospects.

Early exploratory studies of skutterudite thin films included $RhSb_3$ prepared by Chen et al. (1997) using electron beam evaporation, $CoSb_3$ synthesized by Anno et al. (1996) with the aid of RF sputtering, and by Christen et al. (1997) using pulsed laser deposition (PLD). The latter technique was also used by Caylor et al. (1999) to grow thin films of $IrSb_3$. A variety of thin film skutterudite structures were synthesized by Hornbostel et al. (1997a) using the technique called modulated elemental reaction method (MERM).

Apparent breakthroughs came soon with Venkatasubramanian et al. (2001) reporting room temperature values of $ZT = 2.4$ for their thin film Bi_2Te_3/Sb_2Te_3 superlattice structures and Harman et al. (2002) claiming $ZT = 1.3 - 1.6$ for their p-type $PbTe/PbTe_{1-x}Se_x$ quantum dot superlattices. Although until now, neither result has been independently confirmed, the work generated much excitement and was behind the growing interest to explore properties of thin films of thermoelectric materials. Regarding skutterudites, there was an additional impetus to study thin films because of a chance to stabilize structures that in the bulk form could not be synthesized, such as $FeSb_3$. There are variety of techniques one can use to make thin films and the most frequently used ones are outline here in turn. Before I do so, I want to make a comment regarding substrates, as they are essential for whatever form of thin film deposition.

The choice of a substrate is, of course, important for many reasons, including compatibility with the depositing film, determining morphology of the structure, and applications the film is intended for. Since there is not really a very good lattice match with any of the usual substrates used in thin film deposition, it is somewhat unrealistic to expect that skutterudite films would grow as epitaxially formed single crystals. Skutterudite films condensed at room temperature from the vapor phase (atomic beams or plumes) are always of amorphous nature and require annealing at elevated temperatures to transform into a crystalline skutterudite phase. In the case of $CoSb_3$, the crystallization temperature is typically $(150 \pm 20)°C$. The annealing is usually carried out starting at $200°C$ and might extend to temperatures a couple of hundred degrees higher before the film with its substrate is cooled down to room temperature. This brings into considerations a relative thermal expansion of the skutterudite film and the substrate. If the mismatch between the two is large, there is a very good chance that the film will be full of cracks. With its linear thermal expansion coefficient in the range $(8.5 - 9.1) \times 10^{-6}$ K^{-1}, Möchel et al. (2011), and Rogl et al. (2010), one should aim for substrates with comparable thermal expansion coefficient. While there are several types of glass and some ceramics that fit the bill, the often used thermally oxidized Si, i.e., SiO_2/Si is actually a notably poor choice in this regard as its thermal expansion is merely 2.6×10^{-6} K^{-1}. $CoSb_3$ films deposited on such substrate are likely to suffer serious cracks upon annealing, as well documented by Daniel et al. (2013).

2.2.11.1 MBE Fabrication of Skutterudite Films

Being a nonequilibrium phase, it should be possible to prepare films of $FeSb_3$ by a quintessential nonequilibrium technique, namely, Molecular Beam Epitaxy (MBE). Daniel et al. (2015a), indeed, succeeded to do so by coevaporating Fe and Sb from effusion cells in appropriate flux rates on room temperature–held thermally oxidized silicon or glass substrates. The deposited film, typically 30 nm thick, was subsequently annealed in deep vacuum while the structure evolution was monitored by X-rays. Crystallization of $FeSb_3$ commenced near $151°C$, and this phase always formed when the Sb content was above 75 at%. Pushing temperatures much higher to near $300°C$, the skutterudite started to decompose and, by $400°C$, there was no sign of the skutterudite phase with just Sb and some unknown phase remaining on the substrate. Such phase evolution was also confirmed by differential scanning calorimetry. The lattice constant of $FeSb_3$ was measured as 9.154 Å, in close agreement with the value extrapolated from the lattice parameters of bulk $Fe_{1-x}Co_xSb_3$ compounds (9.126 Å). The positional parameters of Sb turned out to be $y = 0.334$ and $z = 0.158$. The conduction of these MBE-deposited $FeSb_3$ films was distinctly metallic.

Using the same codeposition MBE growth, the team explored the synthesis of $CoSb_3$, Daniel et al. (2015b, 2016). The emphasis was on establishing the appropriate flux rates and annealing conditions to obtain the skutterudite phase. Two distinct approaches were tried: co-deposition on a substrate (thermally oxidized SiO_2/Si) held at room temperature with the structure post-annealed in deep vacuum for 1 h at three different temperatures of $175°C$, $200°C$, and $230°C$ to transform the film from its amorphous state to the crystalline $CoSb_3$ phase; and co-deposition on a heated substrate. The first option is more flexible in terms on not needing very precise control of the individual fluxes as long as the Sb flux was high enough to yield a film with at least 75 at% Sb content. The point is that volatile Sb tends to evaporate easily and any excess of it above 75 at% would be taken care of. The latter option, depositing on a heated substrate, might be less time-consuming as it avoids subsequent annealing, but it requires a rather precise flux control to keep the impinging atomic beams very near the 75 at% of Sb content. In either case, there is usually a fair amount of trial and error to establish the optimal growth parameters. Rotating the substrate during the deposition is advantageous as it tends to make uniform films. $CoSb_3$ films should not be exposed to temperatures much above $300°C$ for a long time as they would start decomposing. It may seem that the time to prepare a $CoSb_3$ film might be speeded up significantly if higher flux rates of Sb were used. Unfortunately, too high flux rates, such as 1.2 Å/s tested by Daniel et al. (2015b), cause serious damage to the film in the form of rough surfaces and the presence of secondary phases and, as such, should be avoided. In general, annealed films have smoother surface than films deposited on heated substrates. On the other hand, they tend

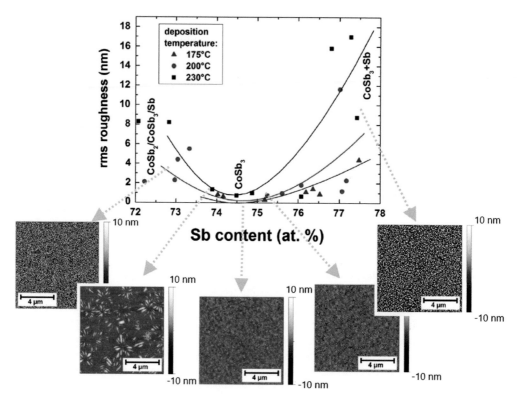

FIGURE 2.5   The root-mean-square surface roughness and AFM images of the surfaces as a function of Sb content of CoSb3 films annealed at different temperatures. Clearly, a stoichiometric CoSb₃ composition results in very smooth films and depends little on the annealing temperature. Reproduced from M. V. Daniel et al., *Journal of Alloys and Compounds* 624, 216 (2015). With permission from Elsevier Publishers.

to have larger grain size than films deposited on substrates heater to elevated temperatures. The relation between the Sb content and the film morphology is vividly revealed in Figure 2.5.

The same MBE setup was used by Lonsky et al. (2016) to prepare Yb-filled cobalt triantimonide of the composition $Yb_yCo_4Sb_{12}$ with $y = 0.05, 0.09, 0.12$, and 0.27. With the deposition parameters for CoSb₃ established, it was a matter of determining an appropriate flux rate of the Yb effusion cell. The films were deposited on SiO₂/Si at room temperature and subsequently annealed at 300°C. The elevated temperature tended to improve crystallinity but, at the same, somewhat depressed the filling fraction of Yb. The grain size was in the 3–10 μm range.

2.2.11.2   Modulated Elemental Reaction Synthesis of Skutterudite Films

The essence of the modulated elemental reaction method (MERM) described initially in the papers of Hornbostel et al. (1997a) and Sellinschegg et al. (1998) is the low-temperature (~10 K) deposition of short modulation wavelength multilayers (repeat distance typically 20–30 Å) consisting of the elemental layers. Provided the layers are thin enough, they will mix (interdiffuse) before nucleation sets in, and will form an amorphous intermediate product. The composition of this amorphous intermediate is controlled by changing the ratio of the elemental layer thickness. The compound crystallizes upon annealing at relatively low temperatures, and the process is controlled by nucleation kinetics rather than being diffusion driven. The resulting compounds are thermodynamically metastable and decompose at temperatures around 500°C into a mixture of elemental constituents or more stable binary compounds. The technique works well for both binary and filled skutterudites. As an example, it was possible for the first time to prepare FeSb₃ and even skutterudites with a filler such as Sn. Furthermore, the technique produced the full range of lanthanide-filled skutterudites, including $LuCo_4Sb_{12}$, Hornbostel et al. (1997b). Lutetium is a cousin of lanthanum (La having an empty $4f$ electron shell while Lu has the $4f$ shell fully occupied) but is much smaller and heavier and thus has more space to rattle in the skutterudite void. Referring specifically to $FeSb_3$, Williams et al. (2001) has shown that, provided the layer thickness of Fe and Sb (both evaporated using effusion cells with a rate of 0.5 Å/s) is below about 35 Å and the composition contains at least 75 at% of Sb, the skutterudite $FeSb_3$ phase will crystallize at 134°C. However, if either the Sb content is below 75 at% or the layers are thicker than 35 Å, it will be iron diantimonide, $FeSb_2$, that nucleates at a higher temperature of about 200°C. Such thickness dependence reflects the role of diffusion: in thin-layered strata the layers diffuse completely and form an amorphous intermediate that subsequently nucleates homogeneously, while in the deposits with thick layers, heterogeneous nucleation takes place prior to the diffusion process is completed. More than a decade later (Möchel et al. 2011) revisited the MERM technique, now calling it nanoalloying, and used it to synthesize thin films of $FeSb_3$ on Kapton foils. This time, the Kapton foil was held at room temperature, and multiple alternate layers of Fe and Sb with the overall molar stoichiometry corresponding to $FeSb_3$ were deposited on it. Following annealing under nitrogen atmosphere at 137°C yielded $FeSb_3$. The overall film thickness was in the range of 1–1.5 μm. In stark contrast to MBE-prepared $FeSb_3$ by Daniel et al. (2015a), these MERM-fabricated films basically showed a semiconducting transport character with the band gap of about 16 meV. The lattice constant at room temperature was 9.2383 Å, a high value in comparison to what one would expect based on the extrapolation of the lattice parameters of $Co_{1-x}Fe_xSb_3$ skutterudites (expected 9.126 Å). By systematically comparing the lattice dynamics of $FeSb_3$ with $CoSb_3$, it became evident that the bonding in $FeSb_3$ is considerably softer, and this should have a major influence on the thermal conductivity. However, that is a topic for Chapter 5. The MERM approach was also used successfully by Smalley et al. (2003) to prepare films of $CoSb_3$. Because of the higher melting point of Co, it was necessary to use e-beam source while Sb flux was generated using an effusion cell. The respective deposition rates were 0.5 Å/s and 0.8–1.0 Å/s, with the elemental atomic fluxes deposited alternately on substrates, such as SiO_2/Si, quartz, and glass. The deposited strata were annealed in steps of 100°C from 200°C to 600°C under either nitrogen atmosphere for 1 h or under vacuum from 30 min to 8 hours to evaluate the effect of annealing on the composition and transport nature of the films. Differential scanning calorimetry indicated a sharp exothermic peak around 160°C revealing nucleation of the $CoSb_3$ phase. Electrical resistivity increased upon annealing due to the decreased carrier concentration as defects were progressively annealed out.

Until recently, the MERM synthesis was very much a "chancy" process with *a priori* unpredictable outcomes regarding the final product, which therefore required considerable trial and error approach to identify the right composition and thickness of the precursor elemental amorphous layers to obtain the desired structure after crystallization. This changed with the work of Bauers et al. (2015) who tried to bring more clarity into the process by employing pair distribution function (PDF) analysis to show that precursors that nucleate metastable $FeSb_3$ skutterudite show similar local structure as does the final product. Moreover, and perhaps even more surprising was a finding that precursors that directly nucleate the thermodynamically stable marcasite $FeSb_2$ structure also contain the structural motif of the metastable $FeSb_3$ phase. This motif is a network of corner-sharing octahedra $FeSb_6$ present even in precursors with the Fe : Sb ratio 1 : 2, i.e., intended to nucleate $FeSb_2$. Previously it was though that what structure ($FeSb_2$ or $FeSb_3$) actually nucleates in the MERM synthesis is determined by the local composition, with different compositions nucleating different structures. The authors surmised that the reason why the metastable skutterudite phase nucleates before does the thermodynamically stable marcasite phase is the energy needed to reorient the octahedra of the latter structure so that each corner Sb atom is shared by three octahedra rather than by two as in the skutterudite structure. The power of the PDF analysis rests in a possibility to optimize the local structure of a precursor in compliance with the dominant motif of the desired final product and thus assures the targeted crystallization.

2.2.11.3 DC and RF Sputtering of Skutterudite Films

Sputtering is one of the most economical and frequently used physical deposition methods of thin films. It relies on accelerated ionized gas molecules bombarding a target material and sputter off its atoms into a glow discharge plasma. Such vaporized atoms then condense as a thin film on the substrate. The chief advantage of sputtering is the replicated film stoichiometry of the target material, allowing deposition of complicated structures. There are two basic forms of sputtering, DC and RF, depending on the power supply used.

In DC sputtering, schematically shown in Figure 2.6a, a negative potential difference in the range of −2 to −5 kV is applied between the target material (cathode) and a substrate (anode) located several cm to tens of cm above and parallel to the target in a vacuum chamber. The chamber is initially evacuated and then back-filled with high purity inert gas (most often argon) to a pressure in the range 0.1–10 Pa. The process starts with ionization of argon by inelastic collisions with electrons, which must be accelerated to about 15 eV in order to strip an electron from the neutral Ar atoms The ionized gas species are accelerated towards the cathode and their energy and momentum "kicks out" target atoms that are collected at the substrate. The substrate is often provided with a small negative bias to assist in controlling the quality of the depositing films. Typical deposition rates achieved in DC sputtering are in the range of 100 Å/min. DC sputtering works exceptionally well with metallic targets but fails when attempting to deposit dielectric films with resistivities above 10^6 Ωcm. The reason is a surface charge buildup on the target that creates arcing and eventually leads to a secession of operation.

RF (radio frequency) sputtering, a sketch of which is depicted in Figure 2.6b, can be used for any kind of target material, as the alternating polarity on the target prevents a buildup of charge or, if you like, cleans the target surface during one half of the period. The frequency of the power supply is standardized at 13.56 MHz, the frequency high enough so that the ions (heavy species) can no longer respond to alterations of the voltage polarity while the light electrons have no problem to do so. The plasma in RF sputtering tends to spread throughout the chamber, the deposition proceeds at

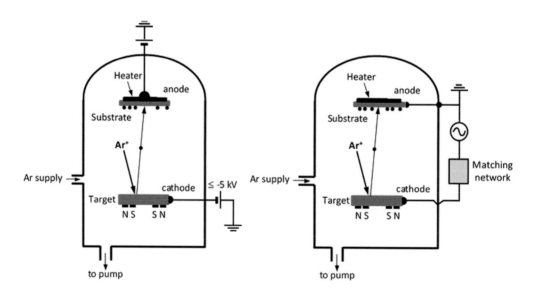

FIGURE 2.6 (a) DC magnetron sputtering system, and (b) RF magnetron sputtering system. The power source operates at a standard frequency of 13.56 MHz and has to be impedance matched using a matching network. Magnetron operation in both DC and RF systems is indicated by small but powerful permanent magnets attached at the backside of the target that provide an approximately cross-field with the electric field and impose a helical motion on electrons, which enhances their probability to ionize Ar gas. The substrate is provided with a heater for deposition at elevated temperatures or for sample annealing.

a lower pressure (perhaps by as much as an order of magnitude), and this improves the line-of-sight deposition. The flexibility of RF sputtering in being able to deposit any kind of material is balanced by generally lower deposition rates than in DC sputtering, and much higher cost of the power supply that requires a special impedance matching network between the power supply and the sputtering chamber.

Both DC and RF sputtering benefit from an introduction of magnetic field, the process known as magnetron sputtering. This is achieved by assembling small but powerful permanent magnets behind the target in a circular fashion and a configuration providing approximately cross field with the electric field generated between the anode and cathode. The idea here is to use the Lorentz force to shape the trajectory of electrons into a helix and thus increase the probability of electrons ionizing argon gas. Moreover, the magnetic field helps to confine the glow discharge plasma close to the target, increasing the density of ions, which leads to higher deposition rates and an opportunity to lower the pressure to 0.07 Pa (~ 0.5 mTorr). Confining the plasma to the vicinity of the target also reduces electron bombardment of the growing film on the substrate.

The first $CoSb_3$ films made by sputtering (RF magnetron) were reported on by Anno et al. (1996, 1997), who used GaAs (100) as substrates. The films were polycrystalline and, as deposited, had a rather low carrier concentration of 1.2×10^{19} cm^{-3}, which further decreased to 4.0×10^{18} cm^{-3} upon annealing at 600°C and to 4.1×10^{17} cm^{-3} when annealed at 750°C. No structural nor compositional analysis was made following annealing at such high temperatures and it is not clear how much Sb remained in the films. Films with the smallest thickness of 71 nm were reported to reach the power factor of 2×10^{-4} Wm^{-1}K^{-2} at 700 K. Savchuk et al. (2002, 2003) used DC magnetron sputtering to deposit $CoSb_3$ films of thickness around 190 nm from compound targets of $Co_{20}Sb_{80}$ composition on thermally oxidized Si and on sapphire held at temperatures between 20°C and 200°C. As expected, films deposited at room temperature were amorphous and transformed into polycrystalline films at 153°C, with the preferential growth orientation along the (310) direction. The authors explored the dependence on source power and the best films, as judged by the highest Seebeck coefficient of −250 μV/K near 550 K, were deposited with 150 W and deposition rate of 6.3 nm/min.

Since then, numerous reports on sputtered skutterudite films and their properties have been published, particularly in the last 5 years. In a series of papers by Zheng et al. (2015, 2017a,b,c,d, 2018) and Fan et al. (2015), the authors explored various aspects of sputter deposition of $CoSb_3$ and its partially filled forms. Among the studies was room temperature co-sputtering of In and $CoSb_3$ alternating with the sputtering of $CoSb_3$ on polyimide flexible substrates to obtain $In_xCo_4Sb_{12}$, where Zheng et al. (2015) claimed a better control of In content by depositing multilayer structures, shown schematically in Figure 2.7. The films were subsequently annealing at 245°C for 1 h. The content of In varied between 0 and 18%, and the measured room temperature ZT improved from 0.05 for pure $CoSb_3$ to a respectable 0.56 obtained with the 14% In content in $CoSb_3$.

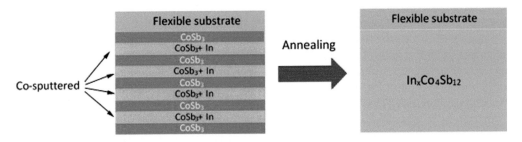

FIGURE 2.7   Schematics of sputter deposition of alternate $CoSb_3$ and cosputtered $CoSb_3$ + In layers and subsequent annealing to form $In_xCo_4Sb_{12}$ used by Zheng et al., *Journal of Alloys and Compounds* 639, 74 (2015). With permission from Elsevier.

In their subsequent work, Zheng et al. (2017c) explored the excess of Co and Sb in the films of $CoSb_3$ by co-sputtering $CoSb_3$ (via RF sputtering) with either Co or Sb deposited by DC sputtering. All films were annealed at 325°C for 1 h. While the power factor of pure $CoSb_3$ was 0.6×10^{-4} $Wm^{-1}K^{-2}$, the Co-rich and Sb-rich structures attained power factors of 2.4×10^{-4} $Wm^{-1}K^{-2}$ and 6.9×10^{-4} $Wm^{-1}K^{-2}$, respectively, suggesting that some non-stoichiometry is beneficial to the thermoelectric performance.

Depending on the deposition parameters, particularly the content of Sb in the target structure, and the annealing temperature, the same team reported the maximum power factor of 1.71×10^{-4} $Wm^{-1}K^{-2}$ at 423 K, Fan et al. (2015), and 9.8×10^{-4} $Wm^{-1}K^{-2}$ at 575 K, Zheng et al. (2017b) for their films of $CoSb_3$. In their most recent study, Zheng et al. (2018) used a multilayer sputtering and deposited a trilayer by first sputtering $CoSb_3$ from a target with the composition Co : Sb = 1 : 3.5, followed by sputtering a layer of Ag, and completing the deposition by again sputtering $CoSb_3$. Altering the deposition time, i.e., the thickness of the layers, controlled the content of Ag in $CoSb_3$. All trilayer structures were annealed at the same 325°C for 1 h under Ar. XRD analysis (lattice expansion) coupled with DFT calculations of the cohesive energy (the largest negative value when Ag occupied voids) indicated that Ag enters the voids of $CoSb_3$ rather than substitutes either Co or Sb. The largest void occupancy measured was about 2.7%. The authors observed that both the Seebeck coefficient and the electrical conductivity increased upon the presence of Ag in the structure as a result of the measured increase in the carrier concentration and the hypothesized increase in the carrier effective mass. The highest power factor of 1.1×10^{-4} $Wm^{-1}K^{-2}$ at 573 K was attained for the sample containing 2.2% of Ag. Although this value was some five times higher than the value of the power factor of $CoSb_3$ with no Ag in this particular experiment, in comparison to their previously reported power factors reported on sputtered $CoSb_3$ films, the value is not at all spectacular.

Partially Yb-filled $CoSb_3$ films of thickness between 130 nm and 860 nm were DC sputtered on SiO_2/Si substrates by Fu et al. (2015) from a commercial target of, unfortunately, unspecified Yb content. The target material was merely as-quenched and contained mostly $CoSb_2$, Sb, and some $CoSb_3$. For comparison, the authors also made a second target from this material by annealing it at 747°C for 120 h to promote the formation of $CoSb_3$. The films were deposited from both types of targets at room temperature and annealed at various temperatures between 597°C and 747°C. The measured content of Yb in the films was estimated as corresponding to a composition $Yb_{0.15}Co_4Sb_{12}$. The highest power factor of about 7.7×10^{-4} $Wm^{-1}K^{-2}$ was measured on a film of thickness 130 nm at 700 K. Since the authors also measured the thermal conductivity using a 3ω-method, the actual ZT value of 0.48 was reported at 700 K for this sample.

Because sputtering tends to faithfully replicate the composition of targets in the growing films, by far most of the sputtered films of skutterudites have been prepared from skutterudite targets of the desired composition, e.g., Jeong et al. (2015). However, just as in the MBE deposition, one may resort to co-deposition by sputtering Co and Sb simultaneously from separate targets and independently control the individual fluxes. This was an approach preferred by Ahmed and Han (2017) in their RF-cosputtering of $CoSb_3$ films on SiO_2/Si substrates held at a room temperature. Argon served as a sputtering gas and its pressure was maintained at 0.4 Pa with 12 sccm. While the power supplied to the Co target was fixed at 60 W, the power at the Sb target was varied between 55 W and 84 W to alter the Sb content in the deposited films between 70 at% and 79 at%. All films were annealed at the same 300°C for 3 h under Ar atmosphere to transform the amorphous films into crystalline structures. Similarly to what was seen in the MBE deposition, films with the substoichiometric content of Sb contained $CoSb_2$, either as a minority phase when the content of Sb was between 73 at% and 75 at%, or as the majority phase in films with the Sb content below 73 at%. In contrast, films with the Sb content between 75 at% and 77 at% were single-phase skutterudite structures with the preferential growth along the (013) plane. All films possessed excellent electronic properties as judged by the value of the power factor measured for the stoichiometric $CoSb_3$ film of 1.26×10^{-3} $Wm^{-1}K^{-2}$. However, what was even more remarkable were high values of the Seebeck coefficient observed in substoichiometric films, which, combined with very high values of the electrical conductivity due to a substantial presence of the $CoSb_2$ phase, resulted in an exceptionally high power factor of 7.92×10^{-3} $Wm^{-1}K^{-2}$ at 750 K in a highly Sb depleted 70 at% film. While the power factor of most of thin films of $CoSb_3$ is some 3–5 times smaller than that of the bulk, the power factors reported by Ahmed and Han (2017), particularly those for sub-stoichiometric $CoSb_3$ films, rival the power factors of bulk forms of $CoSb_3$.

2.2.11.4 Pulsed Laser Deposition of Skutterudite Films

Pulsed laser deposition (PLD) is one of the most widely used thin film fabrication techniques applicable to a vast range of materials, including both unfilled and partially filled skutterudites. It requires a laser, typically KrF (248 nm wavelength) or ArF (193 nm wavelength) excimer laser operated in a pulsed mode, the energy of which impinges on a target material (skutterudite pre-synthesized by one of the bulk techniques described previously) located in a chamber that can be evacuated and back-filled with typically 1–10 Pa (10–100 mTorr) of argon gas. The role of the background gas is to partially thermalize the ablated species so that they do not arrive at a substrate (held typically 4–7 cm from the target, within the plume) with too high kinetic energy to damage the growing film layers. Typical laser energy densities used vary between 2.5 and 5 J/cm^2. It is advantageous to provide for the rotation of both the target and the substrate so that the target is ablated more uniformly and the depositing film is more homogeneous. Since films deposited at room temperature are amorphous, the substrate should be provided with a heater to allow deposition at temperatures above 200°C, or facilitate post-annealing to transform the amorphous film to the crystalline skutterudite phase. As already mentioned, the transformation in $CoSb_3$ starts at temperatures around 150°C, but depends somewhat on the composition of the film. The advantage of the PLD is its simplicity and the fact that, similarly as in the case of sputtering, the deposited film tends to maintain the composition of the target. The drawbacks are rather rough film surfaces that may blister, particularly when no background gas or too little of it is used and the ablated species arrive at the substrate with too high kinetic energy, and often observed micron-size droplets on the surface. Although the droplets have the same composition as the target, they make for rough, ugly looking films.

PLD was first tried with skutterudites in the work of Christen et al. (1997) who deposited $CoSb_3$ on rotated substrates of Si, CaF_2, sapphire, $SrTiO_3$, and the near-lattice matched (001)-oriented TiO_2 between 250°C and 500°C using a KrF excimer laser. The deposition was carried out in a chamber filled with 10.5 Pa of Ar atmosphere to thermalize ablated species. Although the films grew with a distinct (310)-texture, no epitaxy was observed. The work was followed shortly by Caylor et al. (1999, 2001a) who successfully deposited films of $CoSb_3$ and $IrSb_3$ on either a thermally oxidized SiO_2/Si or on Si covered with 60 nm of low-stress Si_xN_y coating. The aim was to explore the dependence of the film quality on the laser fluence (ablation threshold for $CoSb_3$ of about 2.5 J/cm^2), deposition temperature (optimum around 270°C), and the distance between the target and the substrate (6.5 cm). While the uniformity of the film improved with the increasing target-substrate distance and films suffered less stress, too large separation led to the preferential formation of diantimonide rather than the skutterudite phase. The authors also found the range of stability of the films improved dramatically when the target material contained a slight excess (0.75 at%) of antimony. In their subsequent study, Caylor et al. (2001b) attempted to grow $CoSb_3$ epitaxially on the (001) surface of InSb. I already mentioned that there is no usual substrate that matches the lattice spacing of skutterudites. Consequently, they had to use a rather atypical substrate hoping that a 45° rotational relationship between $CoSb_3$ with its $a = 9.034$ Å would bring it close to the lattice constant (6.478 Å) of the zinc blende InSb structure. This apparently succeeded, as confirmed by detailed x-ray diffraction and transmission electron microscopy examinations, the latter depicted in Figure 2.8.

Durant et al. (2000) confirmed the ablation threshold for $CoSb_3$, and the fact that at modest annealing temperatures the stoichiometry of the target material is faithfully replicated in the growing film. At too high annealing temperatures, in their case above 380°C, the films tended to lose Sb due to its high volatility. Because the deposition was carried out with no background gas, the films contained many droplets on the surface.

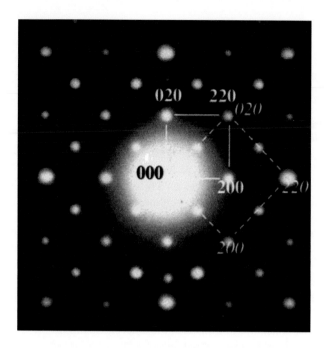

FIGURE 2.8 Electron diffraction pattern showing the lattice matching between the $CoSb_3$ film and the InSb substrate via the coincident [001] zone axis. Diffraction spots due to the $CoSb_3$ film are labeled in bold and marked with a solid line, while those due to the InSb substrate are labeled in itallics and marked with a dashed line. Reproduced from J. C. Caylor et al., *Journal of Materials Research* 16, 2467 (2001). With permission from the Materials Research Society.

Once the films of binary skutterudites were successfully deposited, effort has focused on the growth of filled or partially filled skutterudite films. The first in this regard was Suzuki (2003) who prepared fully filled $LaFe_3CoSb_{12}$ films from targets of the similar but slightly Sb-rich composition using an ArF excimer laser with a shorter wavelength of 193 nm compared to more frequently employed KrF lasers with the wavelength of 248 nm. This may have been the reason for a rather low laser fluence of 1.5 J/cm² with which the films could be deposited. In experiments of Dauscher et al. (2004) and Conceag et al. (2007) it was revealed that the laser wavelength also has a strong influence on the film formation and its quality. Using a Q-switched Nd:YAG laser working at 355 nm and 532 nm, in the latter paper with 266 nm and 532 nm, the authors documented that it is possible to use the visible range of wavelengths to deposit skutterudites, provided the fluence and the substrate temperature is carefully optimized for the growth. In fact, films prepared with the visible laser light seemed to have far fewer droplets on the surface than films grown using UV wavelengths. Documenting that both n- and p-type films of filled skutterudites can be prepared by PLD, Zeipl et al. (2010) grew $Yb_{0.14}Co_4Sb_{12}$ and $Ce_{0.09}Fe_{0.67}Co_{3.33}Sb_{12}$ on quartz from targets of composition $Yb_{0.19}Co_4Sb_{12}$ and $Ce_{0.1}Fe_{0.7}Co_{3.3}Sb_{12}$. The best structures, as measured by their thermoelectric properties, were deposited at the substrate temperature of 250°C using laser energy density of 3 J/cm². Films of double-filled $CoSb_3$ combining In and Yb fillers (since the authors were not sure whether they filled or merely doped the structure, they called them dopants) were synthesized by Kumar et al. (2011) on fused quartz substrates. The target composition $In_{0.2}Yb_{0.2}Co_4Sb_{12}$ was replicated in the deposited single-phase films grown within a narrow optimal window of 260°C in 1.3 Pa of Ar gas. Films of about 190 nm were grown with a rather large laser energy density between 4 and 4.6 J/cm² compensated for by a larger 9 cm separation between the target and the substrate. I should add that some of the PLD-prepared films of $CoSb_3$ by Kumar et al. (2011) were subsequently implanted with Ni under the fluences of 3×10^{15}, 6×10^{15}, and 1.5×10^{16} ions/cm² by Bala et al. (2018). Although the carrier concentration has increased, as one would expect by Ni knocking out some Co atoms and forming a $Co_{0.75}Ni_{0.25}Sb_3$ secondary phase, the overall effect on the transport was disappointing because the carrier mobility was damaged to a greater extent than the gain coming from the enhanced carrier concentration.

2.2.11.5 Electrodeposition of Skutterudite Films

Electrodeposition (plating) is a well understood and widely used industrial process. The technique has several notable virtues, such as low deposition temperature, high rates of deposition, control of the deposition process via electrical parameters that are easier to adjust precisely than temperature, simple apparatus, and good prospects for mass production of films. It works exceptionally well when depositing a single element, such as chromium or gold. However, attempting to electrodeposit two elements simultaneously raises problems as one must assure proper deposition rates of both species at a fixed electrochemical potential. If the two elements have comparable reduction potentials, the situation is manageable, as demonstrated with successful electrodeposition of Bi_2Te_3 films, e.g., Lim et al. (2002) and Martin-Gonzales et al. (2002). Unfortunately, this is not the case of Co and Sb, where vastly different reduction potentials (Co^{2+}/Co of −0.28 V while SbO^{1+}/Sb of +0.21 V, Vanysek 2012), make it difficult to attain comparable deposition rates unless one controls the concentration of the constituents. This is the primary reason why there are rather few reports on electrodeposition of films of $CoSb_3$. Nevertheless, several attempts have been made, some of them in the context of trying to fabricate one-dimensional wires of $CoSb_3$ in porous anodic aluminum oxide templates (Behnke et al. 1999, Chen et al. 2006, and Quach et al. 2010). Basically, there are two kinds of electrodeposition available for skutterudite films: using aqueous solutions or using non-aqueous solutions. I outline both in turn.

2.2.11.5.1 *Electrodeposition from Aqueous Solutions*

The first attempt to electrodeposit Co–Sb compounds was made by Sadana and Kumar (1980), who considered the films to be useful protective and decorative coatings. Although not specifically addressing the deposition of $CoSb_3$, they came up with a basic composition of the bath, which, with minor modifications, has been used in all subsequent studies. To partly bridge the large separation of 0.49 V between the reduction potentials of Co and Sb, they used citrate ions to complex the metals and reduce the difference down to about 0.16 V, a more manageable situation. Moreover, using citrate and citric acid considerably enhances solubility of Sb_2O_3, the precursor for Sb. The recipe was adopted in all works that followed, including Behnke et al. (1999), Cheng et al. (2008), Quach et al. (2010), and Yadav et al. (2017). Starting with cobalt and antimony precursors prepared by mixing 0.003 M Sb_2O_3 and 0.172 M $CoSO_4.7H_2O$ aqueous solutions and adding 0.125 M $C_6H_5Na_3O_7.2H_2O$ (sodium citrate monobasic) and 0.196 M $C_6H_8O_7$ (citric acid), the relevant reactions are,

$$Sb_2O_3 + 2H^+ \rightarrow 2SbO^+ + H_2O \text{ and} \tag{2.7}$$

$$SbO^+ + H_2O \rightarrow HSbO_2 + H^+. \tag{2.8}$$

The actual electrodeposition of $CoSb_3$ proceeds by the reduction of Co^{2+} and $HSbO_2$ on the electrode, i.e.,

$$Co^{2+} + 2e^- \rightarrow Co \text{ and} \tag{2.9}$$

$$HSbO_2 + 3H^+ + 3e^- \rightarrow Sb + 2H_2O. \tag{2.10}$$

Co and Sb react with each other and the overall process can be expressed as

$$Co^{2+} + 3HSbO_2 + 9H^+ + 11e^- \rightarrow CoSb_3 + 6H_2O. \tag{2.11}$$

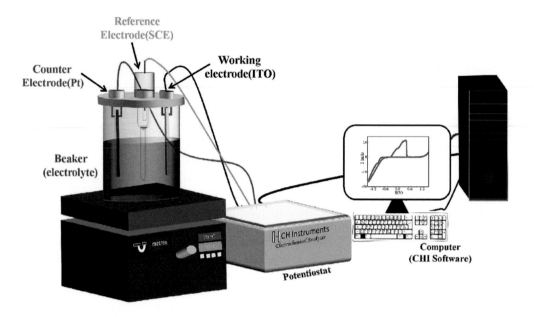

FIGURE 2.9 Schematic of the electrodeposition setup. Reproduced from S. Yadav et al., *RCS Advances* 7, 20336 (2017). With permission from the Royal Society of Chemistry.

The deposition is usually carried out with the aid of a computer controlled potentiostat comprising three electrodes: working electrode, Ag/AgCl reference electrode, and a counter electrode (Au or Pt), a schematic of a typical set up is shown in Figure 2.9. The films grow on a variety of substrates, including gold, Quach et al. (2010), and indium tin oxide, Yadav et al. (2017), that should be thoroughly cleaned electrochemically by cycling the potential between 0 and about 1.5 V several times. In cyclic voltammetry scans, one looks for characteristic features that marked the deposition process, such as potentials at which the reduction of ions commences, current plateaus indicating the diffusion-limited deposition, and potentials at which the current starts to rapidly increase. Correlating the voltammetry scans with the structural information from XRD, EDX, SEM, and perhaps the transport properties collected on the deposited films, one can establish the optimal deposition potential, composition of the electrolyte, and its temperature that yield the desired $CoSb_3$ phase. XRD patterns obtained on thin films prepared at different deposition potentials and the associated variation of Co and Sb is shown in Figures 2.10a and 2.10b. The correlation between the surface morphology of the deposited $CoSb_3$ films and the deposition potential is depicted in Figure 2.11.

2.2.11.5.2 *Electrodeposition from Nonaqueous Solutions*

Rather than using aqueous solutions to electrodeposit cobalt–pnicogen compounds, it is possible to use molten salts, as first tried by Chene (1941) with a combination of Co and P. Starting with melts containing cobalt oxide and sodium metaphosphate, Chene deposited several Co–P films, but no mention was given of specifically depositing CoP_3. Some 60 years later, DeMattei et al. (2001) revisited the issue and successfully deposited a single-phase CoP_3 by properly adjusting the composition of CoO and $(NaPO_3)_3$ melts, and using graphite electrodes. The experiments were conducted under flowing nitrogen at rather high temperatures between 650°C and 700°C.

Interestingly, the traditional electrodes, such as Pt and Ag did not work well, the former being attacked by phosphorus forming PtP_2, while the latter resulted in a meager deposition. The authors also deposited $CoAs_3$ from the $CoO + NaAsO_2$ melt and $CoSb_3$ from the $CoO + Na_2Co_3 + Sb_2O_3$ melt but did not optimized the melt composition to yield single-phase structures.

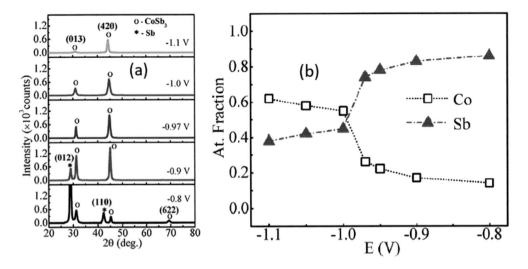

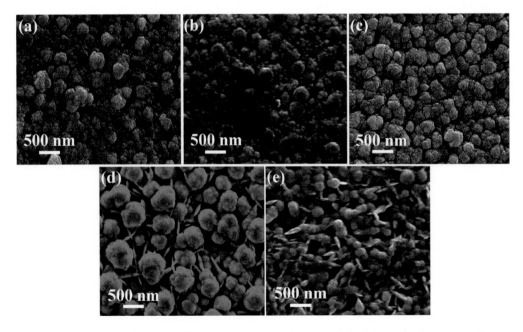

FIGURE 2.10 (a) XRD patterns of electrodeposited CoSb₃ thin films grown at different deposition potential. (b) Atomic % of Co (empty squares) and Sb (red triangles) in films prepared at various deposition potentials. Reproduced from S. Yadav et al., *RSC Advances* 7, 20336 (2017). With permission from the Royal Society of Chemistry.

FIGURE 2.11 Surface morphology of eletrodeposited films of CoSb₃ at various deposition potentials: (a) −0.8 V; (b) −0.9 V; (c) −0.97 V; (d) −1.0 V; and (e) −1.1 V. At the optimal deposition potential of −0.97 V, the grains attain the most uniform globular shape. Reproduced from S. Yadav et al., *RCS Advances* 7, 20336 (2017). With permission from the Royal Society of Chemistry.

As Eq. 2.8 indicates, electrodeposition of CoSb₃ from aqueous solutions inevitably leads to the generation of H₂ gas. To avoid this complication, which it might be in a large-scale production, Yamamoto et al. (2012) developed an electrodeposition process using ethylene glycol–CoCl₂–SbCl₃ electrolyte that, in comparison to the molten salt bath, works well at a much lower temperature of 120°C. The composition of the deposited film was controlled by the contents of the bath and the cathodic current density. The $CoSb_3$ phase was deposited when the bath consisted of the ethylene glycol : $CoCl_2$: $SbCl_3$ ratio of (90.0 : 9.3 : 0.7) mol%, while maintaining the constant potential of 0.1 V.

2.2.12 STABILITY OF THIN FILMS OF CoSb₃

For any practical applications it is essential to know the range of temperatures where a device, using particular structural elements, is intended to operate. In bulk form, $CoSb_3$ is stable to about 750°C and perhaps to a somewhat higher temperature when protected by some form of surface coating. A question is how stable are thin films of $CoSb_3$?

This important issue was looked into in a few reports. Daniel et al. (2015c) tested the stability of some 30 MBE-prepared thin films of $CoSb_3$ with a thickness of 30 nm codeposited on SiO_2/Si substrates held at room temperature and having the nominal Sb content between 75 at% and 77 at%. The deposited amorphous films were annealed at a progressively higher temperature, starting at 200°C and extending to 700°C, and were exposed for times between 15 min and 4 h. The heating rate to reach the final annealing temperature was 10°C/min. The film stability was judged by structural changes observed on XRD patterns, surface morphology changes monitored by AFM, and transport changes revealed in measurements of the electrical resistivity. Samples with the stoichiometric content of Sb (75 at%) were stable to 500°C. These nominally stoichiometric films started to lose Sb when annealed at 550°C, at which point their Sb content dropped to 72.5 at%. At 650°C, there was no $CoSb_3$ left as the phase decomposed into CoSb and Sb evaporated. The Sb-rich series of samples with the initial Sb content of 77 at% started to decompose at 650°C, where the content of Sb decreased to 73 at%. In other words, a slight Sb overstoichiometry resulted in more stable films.

A somewhat lower range of stability was reported by Kumar et al. (2011) for their PLD-prepared $CoSb_3$ films of 190 nm thickness on fused quartz substrates that were completely destroyed by heating at 477°C. On the other hand, a more extended range of stability up to 575°C was observed by Schüpp et al. (2003) in their DC-magnetron sputtered films on SiO_2/Si substrates, which, however, were thicker at 200 nm. Interestingly, their partly iron-substituted films, (Fe,Co)Sb_3, showed the range of stability extended to about 640°C. The films of $FeSb_3$, prepared by the modulated elemental reaction synthesis (MERS), Hornbostel et al. (1997a), decomposed above 500°C. Thus, in comparison to bulk forms of $CoSb_3$, the useful operation range of $CoSb_3$ films is some 100°C–200°C lower, and this must be taken into account in any of their potential applications.

REFERENCES

Ackermann, J. and A. Wold, *J. Phys. Chem. Solids* **38**, 1013 (1977).

Ahmed, A. and S. Han, *Appl. Surf. Sci.* **408**, 88, (2017).

Anno, H., K. Matsubara, Y. Notohara, T. Sakakibara, K. Kishimoto, and T. Koyanagi, *Proc. 15th Int. Conf. on Thermoelectrics*, IEEE Catalog Number 96TH8169, Piscataway, NJ, p. 435 (1996).

Anno, H., T. Sakakibara, Y. Notohara, H. Tashiro, T. Koyanagi, H. Kaneko, and K. Matsubara, *Proc. 16th Int. Conf. on Thermoelectrics*, IEEE Catalog Number 97TH8291, Piscataway, NJ, p. 338 (1997).

Bala, M., A. Masarrat, A. Bhogra, R. C. Meena, Y.-R. Lu, Y.-C. Huang, C.-L. Chen, C.-L. Dong, S. Ojha, D. K. Avasthi, S. Annapoorni, and K. Asokan, *ACS Appl. Energy Mater.* **1**, 5879 (2018).

Bauers, S. R., S. R. Wood, K. M. Ø. Jensen, A. B. Blichfeld, B. B. Iversen, S. J. L. Billinge, and D. C. Johnson, *J. Amer. Chem. Soc.* **137**, 9652 (2015).

Behnke, J. F., A. L. Prieto, A. M. Stacy, and T. Sands, *Proc. 18th Int. Conf. on Thermoelectrics*, IEEE Catalog Number 99TH8407, Piscataway, NJ, p. 451 (1999).

Bell, L. E., *Science* **321**, 1457 (2008).

Bhaskar, A., Y.-W. Yang, and C.-J. Liu, *Ceram. Int.* **41**, 6381 (2015a).

Bhaskar, A., Y.-W. Yang, Z.-R. Yang, F.-H. Lin, and C.-J. Liu, *Ceram. Int.* **41**, 7989 (2015b).

Biltz, W. and H. Heimbrecht, *Z. Anorg. Allg. Chem.* **237**, 132 (1938).

Biltz, W. and H. Heimbrecht, *Z. Anorg. Allg. Chem.* **241**, 349 (1939).

Biswas, K., S. Muir, and M. Subramanian, *Mater. Res. Bull.* **46**, 2288 (2011).

Bremen, S., W. Meiners, and A. Diatlov, *Laser Technik J.* **9**, 33 (2012).

Caillat, T., A. Borshchevsky, and J.-P. Fleurial, *J. Alloys Compd.* **199**, 207 (1993).

Caillat, T., J.-P. Fleurial, and A. Borshchevsky, *Proc. 30th* Intersoc. Energy Conv. Eng. Conf., Amer. Soc. Mech. Eng., ed. D. Y. Goswami, L. D. Kannberg, T. R. Mancini, and S. Somasundaram, Vol. 3, p. 83 (1995).

Caillat, T., J.-P. Fleurial, and A. Borshchevsky, *J. Cryst. Growth* **166**, 722 (1996).

Campanelli, S. L., N. Contuzzi, A. D. Ludovico, F. Caiazzo, F. Cardaropoli, and V. Sergi, *Materials* **7**, 4803 (2014).

Caylor, J. C., A. M. Stacy, T. Sands, and G. Gronsky, *Mat. Res. Soc. Symp. Proc.* **545**, 327 (1999).

Caylor, J. C., A. M. Stacy, G. Gronsky, and T. Sands, *J. Appl. Phys.* **89**, 3508 (2001a).

Caylor, J. C., M. S. Sander, A. M. Stacy, J. S. Harper, G. Gronsky, and T. Sands, *J. Mater. Res.* **16**, 2467 (2001b).

Chen, B., J.-H. Xu, S. Hu, and C. Uher, *Mat. Res. Soc. Symp. Proc.* **452**, 1037 (1997).

Chen, F., R. H. Liu, Z. Yao, Y. F. Xing, S. Q. Bai, and L. D. Chen, *J. Mater. Chem. A* **6**, 6772 (2018).

Chen, L.-J., H. Hu, Y.-X. Li, G.-F. Chen, S.-Y. Yu, and G.-H. Wu, *Chem. Lett.* **35**, 170 (2006).

Chene, P. M., Ann. Chim. (Paris) **15**, 187 (1941).

Cheng, H., H. H. Hng, J. Ma, and X. J. Xu, *J. Mater. Res.* **23**, 3013 (2008).

Cherry, J. A., H. M. Davies, S. Mehmood, N. P. Lavery, S. G. R. Brown, and J. Sienz, *Int. J. Adv. Manuf. Technol.* **76**, 869 (2015).

Christen, H. M., D. G. Mandrus, D. P. Norton, L. A. Boatner, and B. C. Sales, *Mater. Res. Soc. Symp. Proc.*, Vol. 478, p. 217 (1997).

Conceag, D., A. Dauscher, B. Lenoir, V. DaRos, R. Birjega, A. Moldovan, and M. Dinescu, *Appl. Sur. Sci.* **253**, 8097 (2007).

Daniel, M. V., M. Friedemann, N. Jöhrmann, A. Liebig, J. Donges, M. Hietschold, G. Beddies, and M. Albrecht, *Phys. Stat. Solidi A* **210**, 140 (2013).

Daniel, M. V., L. Hammerschmidt, C. Schmidt, F. Timmermann, J. Franke, N. Jöhrmann, M. Hietschold, D. C. Johnson, B. Paulus, and M. Albrecht, *Phys. Rev. B* **91**, 085410 (2015a).

Daniel, M. V., C. Brombacher, G. Beddies, N. Jöhrmann, M. Hietschold, D. C. Johnson, Z. Aabdin, N. Peranio, O. Eibl, and M. AlbrechtL, *J. Alloys and Comp.* **624**, 216 (2015b).

Daniel, M. V., M. Friedemann, J. Franke, and M. Albrecht, *Thin Solid Films* **589**, 203 (2015c).

Daniel, M. V., M. Lindorf, and M. Albrecht, *J. Appl. Phys.* **120**, 125306 (2016).

Dauscher, A., M. Puyet, B. Lenoir, D. Conceag, and M. Dinescu, *Appl. Phys. A* **79**, 1465 (2004).

DeMattei, R. C., A. Watcharapasorn, and R. S. Feigelson, J. Electrochem. Soc. **148**, D109 (2001).

Durant, H.-A., K. Nishimoto, K. Ito, and I. Kataoka, *Appl. Surf. Sci.* **154–155**. 387 (2000).

El-Desouky, A., M. Carter, M. A. Andre, P. M. Bardet, and S. Leblanc, *Mater. Lett.* **185**, 598 (2016).

Fan, P., Y. Zhang, Z.-H. Zheng, W.-F. Fan, J.-T. Luo, G.-X. Liang, and D.-P. Zhang, *J. Electron. Mater.* **44**, 630 (2015).

Feschotte, P. and D. Lorin, *J. Less-Common Metals* **155**, 255 (1989).

Fu, G. S., L. Zuo, J. Chen, M. Lu, and L. Y. Yu, *J. Appl. Phys.* **117**, 125304 (2015).

Fukuoka, H. and S. Yamanaka, *Chem. Mater.* **22**, 47 (2010).

Gharleghi, A., Y. H. Pai, F. H. Lin, and J.-C. Liu, *J. Mater. Chem. C* **2**, 4213 (2014).

Gharleghi, A., Y. H. Chu, F. H. Lin, Z. R. Yang, Y. H. Pai and J.-C. Liu, *ACS Appl. Mater. Interf.* **8**, 5205 (2016).

Harman, T. C., P. J. Taylor, M. P. Walsh, and B. E. LaForge, *Science* **297**, 2229 (2002).

Henkie, Z., M. B. Maple, A. Pietraszko, R. Wawryk, T. Cichorek, R. E. Baumbach, W. M. Yuhasz, and P.-C. Ho, *J. Phys. Soc. Jpn.*, **77**, 128 (2008).

Hicks, L. D. and M. S. Dresselhaus, *Phys. Rev. B* **47**, 12727 (1993a).

Hicks, L. D. and M. S. Dresselhaus, *Phys. Rev. B* **47**, 16631 (1993b).

Hornbostel, M. D., E. J. Hyer, J. Thiel, and D. C. Johnson, *J. Amer. Chem. Soc.* **119**, 2665 (1997a).

Hornbostel, M. D., E. J. Hyer, J. H. Edvalson, and D. C. Johnson, *Inorg. Chem.* **36**, 4270 (1997b).

Ioannidou, A. A., M. Rull, M. Martin-Gonzales, A. Moure, A. Jacquot, and D. Niarchos, *J. Electron. Mater.* **43**, 2637 (2014).

Jeong, M., A. Ahmed and S. Han, *Sci. Adv. Mater.* **7**, 68 (2015).

Ji, X.-H., J. He, P. N. Alboni, T. M. Tritt, and J. W. Kolis, *Mater. Res. Soc. Symp. Proc.* Vol. 1044, ed. T. P. Hogan, J. Yang, R. Funahashi, and T. M. Tritt, 3 (2008).

Jie, Q., H. Z. Wang, W. S. Liu, H. Wang, G. Chen, and Z. F. Ren, *Phys. Chem. Chem. Phys.* **15**, 6809 (2013).

Jolibois, P., *C. R. Acad. Sci.* **150**, 106 (1910).

Kadel, K. and W. Li, *Crystal Res. & Technol.* **49**, 135 (2014).

Kitagawa, H., M. Hasaka, T. Morimura, and S. Kondo, *Proc. 17th Int. Conf. Thermoelectrics*, IEEE Catalog Number 98TH8365, Piscataway, NJ, p. 338 (1998).

Kjekshus, A., *Acta Chem. Scand.* **15**, 678 (1961).

Kroll, W., *Transact. Electrochem. Soc.* **78**, 35 (1940).

Kruth, J. P., M. C. Leu, and T. Nakagawa, *CIRP Annals-Manufacturing Technol.* **47**, 525 (1998).

Kumar, S. R. S., A. Alyamani, J. W. Graff, T. M. Tritt, and H. N. Alshareef, *J. Mater. Sci.* **26**, 1836 (2011).

Lei, Y., W. S. Gao, R. Zheng, Y. Li, R. D. Wan, W. Chen, L. Q. Ma, H. W. Zhou, and P. K. Chu, *J. Alloys Compd.* **806**, 537 (2019).

Le Tonquesse, S., E. Alleno, V. Demange, V. Dorcet, L. Joanny, C. Prestipino, O. Rouleau, and M. Pastural, *J. Alloys Compd.* **796**, 176 (2019).

Liang, T., X. Su, Y. Yan, G. Zheng, Q.-J. Zhang, H. Chi, X. F. Tang, and C. Uher, *J. Mater. Chem A* **2**, 17914 (2014).

Lim, J. R., G. J. Snyder, C.-K. Huang, J. A. Herman, M. A. Ryan, and J.-P. Fleurial, *Proc. 21th Int. Conf. on Thermoelectrics*, IEEE Catalog Number 02TH8657, Piscataway, NJ, p. 535 (2002).

Liu, H., J.-Y. Wang, X. Hu, L.-X. Li, F. Gu, S.-R. Zhao, M.-Y. Gu, R. I. Boughton, and M.-H. Jiang, *J. Alloys Compd.* **334**, 313 (2002).

Lonsky, M., S. Heinz, M. V. Daniel, M. Albrecht, and J. Müller, *J. Appl. Phys.* **120**, 142101 (2016).

Mandrus, D., A. Magliori, T. W. Darling, M. F. Hundley, E. J. Peterson, and J. D. Thompson, *Phys. Rev. B* **52**, 4926 (1995).

Mao, Y., Y. Yan, K. Wu, H. Xie, Z. Xiu, J. Yang, Q. Zhang, C. Uher, and X. F. Tang, *RSC Adv.* **7**, 21439 (2017).

Martin, J. H., B. D. Yahata, J. M. Hundley, J. A. Mayer, T. A. Schaedler, and T. M. Pollock, *Nature* **549**, 365 (2017).

Martin-Gonzales, M. S., A. L. Prieto, R. Gronsky, T. Sands, and A. M. Stacy, *J. Electrochem. Soc.* **149**, C546 (2002).

Merzhanov, A. G., *Adv. Mater.* **4**, 2945 (1992).

Merzhanov, A. G., in *Combustion and Plasma Synthesis of High-Temperature Materials*, ed. Z. A. Munir and J. B. Holt, VCH, Weinheim (1990).

Mi, J. L., T. J. Zhu, X. B. Zhao, and J. Ma, *J. Appl. Phys.* **101**, 054314 (2007).

Möchel, A., I. Sergueev, N. Nguyen, G. J. Long, F. Grandjean, D. C. Johnson, and R. P. Hermann, *Phys. Rev. B* **84**, 064302 (2011).

Moore, J. J. and H. Feng, *Prog. Mater. Sci.* **39**, 275 (1995).

Morimura, T., H. Kitagawa, M. Hasaka, and S. Kondo, *Proc. 16the Int. Conf. Thermoelectrics*, IEEE Catalog Number 97TH8291, Piscataway, NJ, p. 356 (1997).

Munir, Z. A. and U. Anselmi-Tamburini, *Mater. Sci. Rep.* **3**, 277 (1989).

Nakagawa, H., H. Tanaka, A. Kasama, K. Miyamura, H. Masumoto, and K. Matsubara, *Proc. 15th Int. Conf. on Thermoelecrics*, IEEE Catalog Number 96TH8169, Piscataway, NJ, p. 117 (1996).

Odile, J. P., S. Soled, C. A. Castro, and A. Wold, *Inorg. Chem.* **17**, 283 (1978).

Quach, D. V., R. Vidu, J. R. Groza, and P. Stroeve, *Ind. Eng. Chem. Res.* **49**, 11385 (2010).

Rao, K. J., B. Vaidhyanathan, M. Ganguli, and P. A. Ramakrishnan, *Chem. Mater.* **11**, 882 (1999).

Rogl, G., L. Zhang, P. Rogl, A. Grytsiv, M. Falmbigl, D. Rajs, M. Kriegisch, H. Müller, E. Bauer, J. Koppensteiner, W. Schranz, M. Zehetbauer, Z. Henkie, and M. B. Maple, *J. Appl. Phys.* **107**, 043507 (2010).

Rundqvist, S. and E. Larsson, *Acta Chem. Scand.* **13**, 551 (1959).

Rundqvist, S. and N.-O. Ersson, *Ark. Kemi* **30**, 103 (1968).

Sadana, Y. N. and R. Kumar, *Surf. Technol.* **11**, 37 (1980).

Savchuk, V., A. Boulouz, S. Chakraborty, J. Schumann, and H. Vinzelberg, *J. Appl. Phys.* **92**, 5319 (2002).

Savchuk, V., J. Schumann, B. Schüpp, G. Behr, N. Mattern, and D. Soupel, *J. Alloys Compd.* **351**, 248 (2003).

Schüpp, B., I. Bächer, M. Hecker, N. Mattern, V. Savchuk, and J. Schumann, *Thin Solid Films* **434**, 75 (2003).

Sekine, C., H. Saito, T. Uchiumi, A. Sakai, and I. Shirotani, *Solid State Commun.* **106**, 441 (1998).

Sekine, C., T. Uchiumi, I. Shirotani, K. Matsuhira, T. Sakakibara, T. Goto and T. Yagi, *Phys. Rev. B* **62**, 11581 (2000).

Sekine, C. and Y. Mori, *Jpn. J. Appl. Phys.* **56**, 05FA09 (2017).

Sellinschegg, H., D. C. Johnson, G. S. Nolas, G. A. Slack, S. B. Schujman, F. Mohammed, T. M. Tritt, and E. Nelson, *Proc. 17th Int. Conf. Thermoelectrics*, IEEE Catalog Number 98TH8365, Piscataway, NJ, p. 338 (1998).

Shirotani, I., Y. Shimaya, K. Kihou, C. Sekine, and T. Yagi, *J. Solid State Chem.* **174**, 32 (2003).

Smalley, A. L. E., S. Kim, and D. C. Johnson, *Chem. Mater.* **15**, 3847 (2003).

Su, X., F. Fu, Y. Yan, G. Zheng, T. Liang, Q. Zhang, X. Cheng, D. Yang, H. Chi, X. F. Tang, Q.-J. Zhang, and C. Uher, *Nat. Commun.* **5**, 4908 (2014).

Suzuki, A., *Jpn. J. Appl. Phys.* **42**, 2843 (2003).

Tafti, M. Y., M. Saleem, A. Jacquot, M. Jägle, M. Muhammed, and M. S. Toprak, *Mat. Res. Soc. Symp. Proc.* Vol. 1543, 105 (2013).

Takizawa, H., K. Miura, M. Ito, T. Suzuki, and T. Endo, *J. Alloys Compd.* **282**, 79 (1999).

Takizawa, H., K. Okazaki, K. Uhedu, T. Endo, and G. S. Nolas, *Mater. Res. Soc. Symp. Proc.* **691**, 37 (2002).

Tan, G., W. Liu, S. Wang, Y. Yan, H. Li, X. F. Tang, and C. Uher, *J. Mater. Chem A* **1**, 12657 (2013).

Tanaka, K., Y. Kawahito, Y. Yonezawa, D. Kikuchi, H. Aoki, K. Kuwahara, M. Ichihara, H. Sugawara, Y. Aoki, and H. Sato, *J. Phys. Soc. Jpn.*, **76**, 103704 (2007)

Torikachvili, M. S., J. W. Chen, Y. Dalichaouch, R. P. Czuertin, M. W. McElfresh, C. Rossel, M. B. Maple, and G. P. Meisner, *Phys. Rev. B* **36**, 8660 (1987).

Uchida, H., V. Crnko, H. Tanaka, A. Kasama, and K. Matsubara, *Proc. 17th Int. Conf. Thermoelectrics*, IEEE Catalog Number 98TH8365, Piscataway, NJ, p. 330 (1998).

Vanysek, P., in *Handbook of Chemistry and Physics*, 93rd edition, ed. W. M. Hayes, Chemical Rubber Company, p. 5 (2012).

Venkatasubramanian, R., E. Siivola, T. Colpitt, and B. O'Quinn, *Nature* **413**, 597 (2001).

Wang, M., Y. Zhang, and M. Muhammed, *Nanostr. Mater.* **12**, 237 (1999).

Wang, S., G. Tan, W. Xie, G. Zhang, H. Li, J. Yang, and X. F. Tang, *J. Mater. Chem.* **22**, 20943 (2012).

Williams, J. R., M. Johnson, and D. C. Johnson, *J. Amer. Chem. Soc.* **123**, 1645 (2001).

Xie, J., X. B. Zhao, J.-L. Mi. G.-S. Cao, and J.-P. Tu, *J. Zhejiang Univ. Science* **5**, 1504 (2004).

Yadav, S., B. S. Yadav, S. Chaudhary, and D. K. Pandya, *RSC Adv.* **7**, 20336 (2017).

Yamamoto, H., M. Morishita, Y. Mizuta, and A. Masubuchi, *Surf. Coat. Technol.* **206**, 3415 (2012).

Yan, Y., H. Q. Ke, J. Yang, C. Uher, and X. F. Tang, *ACS Appl. Mater. Inter.* 10, 13669 (2018).

Yang, J. Y., Y. H. Chen, J. Y. Peng, X. L. Song, W. Zhu, J. F. Su, and R. Chen, *J. Alloys Compd.* **375**, 229 (2004).

Yang, J. Y., Y. H. Chen, W. Zhu, J. Y. Peng, S. Bao, X. Fan, and X. Duan, *J. Solid State Chem.* **179**, 212 (2006).

Zeipl, R., J. Walachová, J. Lorinčik, S. Leshkov, M. Josieková, M. Jelinek, T. Kocourek, K. Jurek, J. Navrátil, L. Beneš, and T. Plecháček, *J. Vac. Sci. Technol. A* **28**, 523 (2010).

Zhang, J., B. Xu, L.-M. Wang, D. Yu, Z. Liu, J. He, and Y. Tian, *Appl. Phys. Lett.* **98**, 072109 (2011).

Zhao, L., B. Zhang, J. Li, H. Zhang, and W. Liu, *Solid State Sci.* **10**, 651 (2008).

Zheng, Z.-H., P. Fan, Y. Zhang, J.-T. Luo, Y. Huang, and G.-X. Liang, *J. Alloys Compd.* **639**, 74 (2015).

Zheng, Z.-H., M. Wei, F. Li, J.-T. Luo, H.-L. Ma, G.-X. Liang, X.-H. Zhang, and P. Fan, *J. Mater. Sci: Mater. Electron.* **28**, 17221 (2017a).

Zheng, Z.-H., M. Wei, F. Li, J.-T. Luo, G.-X. Liang, H.-L. Ma, X.-H. Zhang, and P. Fan, *Coatings* **7**, 205 (2017b).

Zheng, Z.-H., F. Li, Y.-Z. Li, P. Fan, J.-T. Luo, G.-X. Liang, B. Fan, and A.-H. Zhong, *Thin Solid Films* **632**, 88 (2017c).

Zheng, Z.-H., F. Li, J.-T. Luo, G.-X. Liang, H.-L. Ma, X.-H. Zhang, and P. Fan, *J. Alloys Compd.* **732**, 958 (2017d).

Zheng, Z.-H., M. Wei, J.-T. Luo, F. Li, G.-X. Liang, Y. Liang, J. Hao, H.-L. Ma, X.-H. Zhang, and P. Fan, *Inorg. Chem. Front.* **5**, 1409 (2018).

Zobrina, B. N. and L. D. Dudkin, *Sov. Phys.-Solid State* **1**, 1668 (1960).

3 Electronic Energy Band Structure

Although considerable insight into the electronic structure of a material can be gained from its crystal structure and by applying appropriate models of the chemical bonding topologies, for the detailed picture and understanding of electronic bands, including the position of the band edges and the band dispersion, reliable computations are indispensable. During the past two decades, the development of efficient algorithms coupled with the progress in computing hardware has reached a point where computational packages have become widely available and accessible and even experimentally trained graduate students and postdocs with some training can make great use of such packages, as was the case of several of my own students. In this context, one might question the need to discuss earlier theoretical approaches aiming to describe the band structure of skutterudites. The point is that, even though lacking the state-of-the-art sophistication, the foundational developments more or less captured the key features of the electronic structure, and the later efforts merely fine-tuned the picture and clarified certain details. Moreover, it would be grossly unfair not to list the pioneering works as they provided important guidance to experimentalists to focus their search in areas that appeared most fruitful from the perspective of achieving the optimal material properties. In that sense, band structure calculations have truly shaped the development of skutterudites as one of the most prospective novel thermoelectrics.

3.1 BAND STRUCTURE OF BINARY SKUTTERUDITES

The key structural elements of MX_3 binary skutterudites, which should be reflected in any reliable band structure calculations, are the octahedral coordination of the transition metal M and the near-square planar configuration of the pnicogen X_4 rings. In describing the band structure of skutterudites, the theoretical papers use particular terminology, such as the t_{2g} and e_g block bands. For readers who might not be familiar with such terms or who forgot how such bands arise, a brief review is presented of the key points of the crystal field theory (CFT) that describe the breaking of orbital degeneracy in transition metal ions in the presence of ligands (in our case pnicogen atoms X). CFT was developed by Bethe in 1929 and extended to include covalent bonding by van Vleck three years later (1932). In its rather naïve but easily understandable form, the metal ions are taken as point charges q_1 while the ligands have charges q_2. The bond energy between the two is then simply $E \propto q_1 q_2 / r$, where r is the separation between the metal ion and the ligand.

A single transition metal ion has five d-orbitals, all with the same energy. When ligands with a spherically distributed cloud of electrons approach the metal ion, the energy of all d-orbitals is raised by the same amount on account of the increased electrostatic energy. However, when the electrons of the ligand are redistributed in such a way that electrons are placed at vertices of an octahedron, such as happens in the skutterudite's MX_6 complex, some d-orbitals experience stronger repulsion from the electrons of the ligand, depending on the orientation of the d-orbital's wavefunction lobes. The five d-orbitals are designated d_{xy}, d_{yz}, d_{xz}, d_{z^2}, $d_{x^2-y^2}$, and are schematically pictured in Figure 3.1. Four of the orbitals are of the "clover-leaf" shape and one (the d_{z^2} orbital) can be viewed as consisting of a dumbbell with a doughnut in the middle. Drawing attention to the orientation of the lobes of each d-orbital with respect to the axes x, y, and z, one notes that the lobes of orbitals d_{xy}, d_{yz}, and d_{xz} lie between the respective axes, while the orbitals d_{z^2} and $d_{x^2-y^2}$ have their lobes along the z-axis and along the x- and y-axes, respectively. Assuming the octahedrally-shaped ligand has its axes oriented along the x-, y-, and z-directions, the orbitals d_{z^2} and $d_{x^2-y^2}$ experience a strong

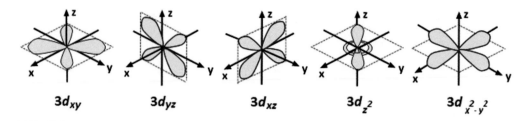

FIGURE 3.1 The shape of 3d orbitals. Note, the lobes of orbitals $d_{xy}, d_{yz}, and\ d_{xz}$ fall between the x-, y- and z-axes, while orbitals d_{z^2} and $d_{x^2-y^2}$ have the lobes in the direction of the axes.

repulsion and their energy is raised (the orbitals become more destabilized) because the lobes of these two orbitals are in the direction where electrons of the ligand are located and approach closer to them. On the other hand, because the orbitals d_{xy}, d_{yz}, and d_{xz} have their lobes between the axes, these orbitals do not interact strongly with the octahedrally distributed electrons of the ligand, and their energy is actually lowered (orbitals become more stable) compared to their original energy. The net effect of the interaction is a split in the energy of d-orbitals, as depicted schematically in Figure 3.2. The energy of the orbitals d_{z^2} and $d_{x^2-y^2}$ is raised and the energy of orbitals d_{xy}, d_{yz}, and d_{xz} is lowered so that they are separated by energy Δ_0, of the same order of magnitude as the energy of a chemical bond. The three lower energy orbitals are referred to as the t_{2g} block while the two higher energy orbitals are the e_g block. Since the barycenter energy of the d-orbitals before and after the octahedral field of ligands was applied must be the same, the e_g block of orbitals is raised by 3/5 Δ_0, while the energy of the t_{2g} block of orbitals is decreased by −2/5 Δ_0.

According to the Pauli principle, the five d-orbitals can accommodate up to 10 electrons. The filling of the d-orbitals is governed by Hund's rules. For transition metals with one to three electrons, d^1–d^3, all electrons occupy the t_{2g} block. Transition metals with four to seven electrons, d^4–d^7, can accommodate d-electrons in two different ways: either in the so-called high spin configuration or in the low spin configuration, depending on the relative strength of the spin-pairing energy P with respect to the splitting energy Δ_0. As an example, I illustrate this for the case of Co, the transition element of the most interest when discussing skutterudites.

Co has nine valence electrons (d^7s^2), three of which (the two s-electrons and one d-electron) engage in bonding with the six neighboring pnicogen atoms (Sb in the case of $CoSb_3$). This leaves cobalt in the 3+ state with six nonbonding electrons that distribute over the d-orbitals. In principle, two scenarios can happen, depicted in Figure 3.3. In Figure 3.3(a), all five d-orbitals are occupied by at least one electron and the sixth electron goes on the lowest energy level, i.e., on one of the t_{2g} states.

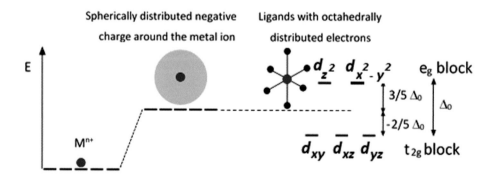

FIGURE 3.2 Spherically distributed charge of a ligand merely raises energy of all d-orbitals but does not split them. Octahedral field of a ligand splits the d-orbitals into a t_{2g} block and the e_g block separated in energy by Δ_0.

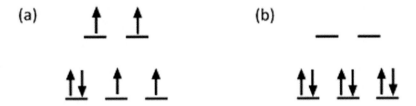

(a)   ↑ ↑   (b)   __ __

  ↑↓ ↑ ↑   ↑↓ ↑↓ ↑↓

FIGURE 3.3   Electron spins of a Co^{3+} ion distributed over five d-orbitals in (a) the high spin state, and (b) the low spin state of $CoSb_3$.

In Figure 3.3(b), all electrons are spin-paired on t_{2g} states. The latter scenario is the actual situation of the Co^{3+} ion in binary skutterudites. In this case, the penalty paid by pairing the spins is smaller than the energy needed to promote electrons to the e_g block of states. How do we know that this, indeed, is the case in skutterudites? In configuration (a), there are four unpaired electron spins and the structure would inevitably have a magnetic moment. In reality, $CoSb_3$ and all binary skutterudites are diamagnetic substances and thus cannot have any unpaired spins. Consequently, transition metals M (M = Co, Rh, Ir) in binary skutterudites have all spins paired and are in the low spin d^6 state. Similar to transition metals with one to three electrons, transition metals with eight to ten electrons, d^8–d^{10}, have again uniquely defined spin configurations.

Apart from the octahedrally coordinated transition metal atom M, the other key structural feature of skutterudites is the planar rectangular pnicogen rings X_4 forming linear arrays along the x-, y-, and z-crystallographic directions with the neighboring rings orthogonal to each other, as depicted in Figure 3.4. Each pnicogen ring carries a charge of −4. As already noted in Section 1.1, each pnicogen atom (ns^2np^3) has five valence electrons. Two of them engage in covalent bonds of σ-character with its two pnicogen neighbors on the ring. The remaining three electrons bond with the two nearest transition metal atoms M, giving rise to hybrid d^2sp^3 bonds.

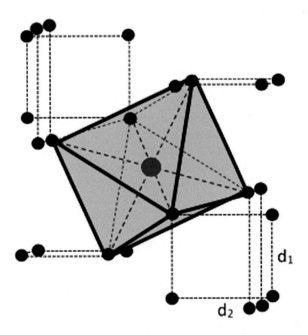

FIGURE 3.4   Transition metal M (blue circle) octahedrally coordinated by pnicogen atoms X (black circles). Each vertex pnicogen atom is part of the mutually orthogonal array of X_4 pnicogen rings. The short, d_1, and the long, d_2, bond distances between pnicogen atoms on the ring are indicated.

Most of the reliable band structure calculations are being performed using first principles (*ab-initio*) methods based on the Density Functional Theory (DFT), which is perhaps the most versatile quantum mechanical-based computational method to study electronic properties of many-body systems, such as condensed matter phases. The ground work for DFT was laid by Hohenberg and Kohn (1964), who formulated their ideas in terms of two theorems. The essence of the first H.-K. theorem is a statement that the ground state properties of a many-body system are uniquely determined by the spatial distribution of an electron density. The second H.-K. theorem defines the system's energy functional and shows that the correct ground state electron density minimizes this energy functional. An important step toward practical realization of DFT was taken by Kohn and Sham (1965), who derived self-consistent single-particle equations that basically showed that an intractable many-body problem of interacting electrons in a static external potential can be reduced to a manageable problem of non-interacting electrons moving in an effective potential consisting of the external potential and the Coulomb interactions between the electrons. Walter Kohn shared the Nobel Prize for Chemistry in 1998 for his contributions to the development of DFT.

The major challenge for DFT is the exchange and correlations energies, two purely quantum-mechanical phenomena that have no equivalence in the realm of classical physics. The exchange energy is a direct consequence of the Pauli exclusion principle that dictates that no two electrons with the same spin can occupy the same state (orbital). In other words, electrons with the same spin are kept apart. The exchange energy thus lowers the energy and reduces the usual Coulomb repulsion. Correlation energy is a measure of how much the movement of one electron is influenced by the presence of all other electrons, i.e., how its spatial position is correlated to the spatial positions of neighboring electrons. Correlations are a consequence of the collective motion of electrons to screen and decrease the Coulomb interaction. Unlike the exchange energy, the correlation energy is more important for electrons with opposite spins because they are more likely to be found in nearby locations. While well defined, the exchange and correlation energies are not known *a priori* and must be approximated. How well such approximations match the exact value determines the quality of a DFT calculation. The simplest and often used approximation is referred to as the Local Density Approximation (LDA). Here, one makes use of the *local* exchange and correlation energies and treats the potential as that of a uniform electron gas with the local density $n(r)$. LDA is a major simplification of the problem, and the penalty one pays for this simplification is an underestimation of the exchange energy and overestimation of the correlation energy. To correct for it, one can carry out an expansion in terms of the gradient of the electron density, which takes into account the actual inhomogeneous electron density. This is referred to as the General Gradient Approximation (GGA), Perdew et al. (1992), and is considered a refinement and improvement over the LDA approximation. There are different variants of both methods, depending on the choice of specific basis set expansions of the wave function in computations (plane waves, muffin-tin orbitals, local orbitals, etc.). A particularly serviceable form of DFT uses a plane-wave-based algorithm implemented in the Vienna *Ab Initio* Simulation Package (VASP), Blaha et al. (1990). Several excellent monographs and review articles deal with the topic of DFT, the use of various exchange-correlation functionals, and basis set expansions of wave functions, and the interested reader is referred to them for details (Dreizler and Gross 1990, Singh 1994, Stoll and Steckel 2009, Giustin 2014, and Engel and Dreizler 2018). With this very brief introduction to the DFT-based computations of the electronic energy band structure, we now turn to band structure calculations specifically relevant to skutterudites.

Until the early 1990s, self-consistent band structure calculations for skutterudites were considered too difficult because of their complex crystal structure and a relatively open unit cell. The first attempt to do so was made by Singh and Pickett (1994). The calculations were carried out within the local density approximation (LDA) using an extended general potential linearized augmented-plane-wave method (LAPW). The target compounds were $IrSb_3$, $CoSb_3$, and $CoAs_3$, and the calculations were performed with experimental crystal structure parameters. The spin-orbit interaction was included only for $IrSb_3$, on account of the heavy mass of Ir, and neglected for the other two skutterudites. The exchange-correlation energy was modeled with the Hedin-Lundqvist functional, one of the early and often used functionals, Hedin and Lundqvist (1971). The resulting

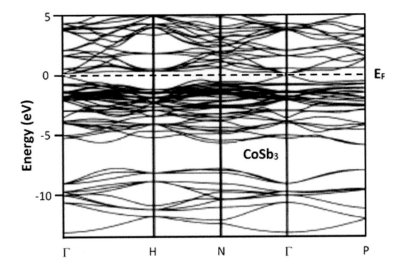

FIGURE 3.5 Electronic band structure of CoSb₃ reproduced from D. J. Singh and W. E. Pickett, *Physical Review B* 50, 11235 (1994). With permission from the American Physical Society.

band structure is shown in Figure 3.5. There are two important aspects of the calculation results: a clear separation between the valence and conduction bands in all three skutterudites, often referred to as a pseudogap, and a single band crossing the gap and touching (or nearly so in CoSb₃) the triply degenerate conduction band minimum at Γ. Both the valence and the conduction bands are derived from hybridized combinations of transition metal d-states and pnicogen p-states. The p-derived component of the Density of States (DOS) is very flat and symmetrical about the pseudogap, while the d-component of the DOS is peaked below the pseudogap, particularly for Co skutterudites, due to the more spatially localized Co $3d$ orbitals relative to Ir $5d$ orbitals. Direct values of the pseudogap at Γ are: 1.21 eV for IrSb₃, 0.80 eV for CoSb₃, and 0.95 eV for CoAs₃. The above values are presumably those one would obtain in measurements of the optical gap. Indeed, the optical gap measured by Slack and Tsoukala (1994) for IrSb₃ of 1.4 eV is not too far from the direct gap of the IrSb₃ compound.

The single band breaching the pseudogap has quite remarkable properties. Again, the band is of hybridized transition metal d and pnicogen p character, with the latter gaining more importance away from Γ. In IrSb₃ and CoAs₃, the band touches the conduction band edge at Γ, predicting the two skutterudites to be zero-gap semiconductors. In CoSb₃, the calculations yield a small direct gap of 50 meV (not really distinguishable in Figure 3.5). While the values of the band gap calculated with LDA are notoriously underestimated, the truly unique feature of the single band crossing the gap is its dispersion. While in CoAs₃ it has the usual parabolic shape, in IrSb₃ and CoSb₃ it is parabolic for only about 10^{-5} of the zone very close to point Γ. For wave vectors larger than 1–2% of the distance to the zone boundary (corresponding to a hole concentration of about 3×10^{16} cm⁻³), the band acquires a distinctly linear shape. The slopes of the linearly dispersing region are $\alpha = -3.45$ eVÅ for IrSb₃ and -3.10 eVÅ for CoSb₃. Because it is just a single band in a unit cell containing many atoms, the imprint of the band on the DOS in the pseudogap is very small. However, the band has a significant impact on the transport properties, with the linear dispersion dominating because the carrier densities in skutterudites are orders of magnitude larger than 3×10^{16} cm⁻³. The functional form of several transport parameters is consequently altered. For instance, near the band edge, the density of states is a quadratic function of energy ε rather than the usual three-dimensional form of $\varepsilon^{1/2}$; the carrier concentration as a function of the Fermi energy becomes cubic instead of the usual 3/2 power; the normally doping-independent carrier mobility under the constant relaxation time approximation turns out to be doping dependent and proportional to $n^{-1/3}$, and the inverse mass tensor $\nabla\nabla\varepsilon_\kappa$ becomes entirely off-diagonal, implying that the transport mass perpendicular to the Fermi surface is infinite. Fortunately, the carrier density as a function of the wavevector k is unchanged and one can continue using Hall measurements to determine the carrier concentration as in the case of usual parabolic bands. From the perspective of thermoelectricity, perhaps the most impacting difference between the linear and quadratic band shapes concerns the form of the degenerate Seebeck coefficient within the constant relaxation time approximation. It now becomes

$$S = -\left(\frac{2\pi k_B^2 T}{3e\,\alpha}\right)\left(\frac{\pi}{3n}\right)^{1/3},\tag{3.1}$$

i.e., it has a weaker carrier concentration dependence than for the usual parabolic band, $\left(\frac{\pi}{3n}\right)^{2/3}$.

The weaker doping dependence of the Seebeck coefficient and thus the power factor is unfortunate as it makes optimization of the carrier concentration more challenging compared to an ordinary semiconductor with parabolic bands.

Following the work of Singh and Pickett, band structure calculations of binary skutterudites have been performed numerous times using similar LDA approximations as well as various modifications of GGA. While the results differ in details, the main points, namely a clearly defined pseudogap separating the valence and conduction band manifolds and a single linearly dispersing band in IrSb$_3$ and CoSb$_3$ have been reproduced. To better appreciate contributions of Sb p-states and Co 3d-states to the band structure of CoSb$_3$, I include here more recent calculations of the band structure by Pardo et al. (2012) carried out by two distinct all-electron full-potential codes, FPLO-9 (Koepernik and Eschrig 1999) and WIEN2k, using the augmented plane wave + local orbital method, both returning consistent results. Figure 3.6(a) highlights with *fatbands* the Sb p-character of the bands and Figure 3.6(b) emphasizes the Co 3d contributions to the bands.

Sofo and Mahan (1999) pointed out that the band structure of CoSb$_3$ is exactly what one would expect of a small band-gap semiconductor that can be described by a two-band Kane model with the dispersion of the form

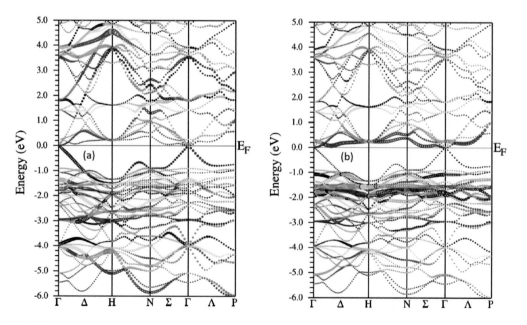

FIGURE 3.6   Band structure of CoSb$_3$ highlighting (by fatbands) the contributions of (a) Sb p-states and (b) Co 3d-states at various energies within the Brillouin zone. Reproduced from V. Pardo et al., *Physical Review B* 85, 214531 (2012). With permission from the American Physical Society.

$$\frac{\hbar^2 k^2}{2m^*} = \varepsilon_k \left(1 + \frac{\varepsilon_k}{E_g} \right), \tag{3.2}$$

where m^* is the effective mass at the band edge, and E_g is the band gap, Kane (1975). Regardless of using LDA or GGA, their DFT calculations based on the LAPW yielded the direct band gap at Γ of $E_g = 0.22$ eV for $CoSb_3$, some four times larger value than the band gap determined by Singh and Pickett. However, the value was obtained with the positional parameters y and z of the pnicogen atom that minimize the energy rather than the experimental values used in calculations by Singh and Pickett. When Sofo and Mahan carried out the calculations with the experimental values of y and z, they obtained the same small band gap of 50 meV as Singh and Pickett. The calculations revealed an extreme sensitivity of the band gap on the values of the positional parameters used, i.e., on the exact shape and distances d_1 and d_2 in the pnicogen X_4 ring. Sofo and Mahan have noted that such extreme sensitivity of the band gap is in accord with the Jung et al. (1990) chemical picture of the structure, where the top of the valence band at Γ corresponds to the most anti-bonding combination of the π_4 orbitals of the ring. A very small change in the position of the Sb atom drastically changes the energy of this molecular orbital, i.e., alters the band gap.

Llunell et al. (1996) explored the band structure of two phosphide-based skutterudites, CoP_3 and the controversial NiP_3. They used LDA with the linear muffin-tin orbital (LMTO) expansion method, a technique that is particularly suited to compact crystal structures. To make it relevant to the rather open environment of skutterudites, the authors introduced interstitial spheres to force small radii on the atom-centered spheres so that they would not overlap substantially. They used experimental lattice parameters a (7.073 Å for CoP_3 and 7.819 Å for NiP_3) and the same sphere radii for both compounds. The band structure of CoP_3 is similar to antimony-based binary skutterudites, characterized by a substantial pseudogap (1.26 eV) separating the valence and conduction band manifolds and a single band crossing the pseudogap that nearly touches the conduction band at the center of the Brillouin zone. The indirect gap with a conduction band minimum at point H is 0.07 eV, and the direct gap at Γ is 0.28 eV. Again, similar to antimonide skutterudites, the single pseudogap-crossing band becomes linear for wave vectors of about 5% the distance to the zone boundary. The calculations confirm the argument of Jung et al. that the band is formed by π-type antibonding orbitals of the X_4 pnicogen rings, and its dispersion arises as a consequence of the mixing of these orbitals with transition metal p-orbitals.

The inherent problem with using the linear muffin-tin orbital sphere approximation (Llunell et al. 1996, Zhukov 1996), to describe open crystalline environments, such as those represented by skutterudites, has been documented by subsequent calculations by Fornari and Singh (1999). In their work, Fornari and Singh used a more appropriate full-potential LAPW method in the DFT calculations with the Hedin-Lundqvist parameterization of the exchange-correlation LDA functional. The results are depicted in Figure 3.7 and show clearly that CoP_3 has a metallic rather than semiconducting character, as the pseudogap near the Fermi energy is crossed by a single band of mostly phosphorus p origin. This is the same valence band seen in $CoSb_3$, $CoAs_3$, and $IrSb_3$, but here it also crosses the conduction bands above the Fermi level. However, compared to $CoSb_3$ and $IrSb_3$, it does not seem to have the distinctly linear dispersion. The metallic nature of CoP_3 makes it much less appealing for thermoelectricity unless one finds a way to modify the band structure so that a gap opens. In Section 3.3, we shall see that filling the CoP_3 structure will just do the trick.

Regarding NiP_3, the calculations by Llunell et al. (1996) confirmed the metallic nature of the skutterudite, expected based on the fact that Ni provides one more electron per formula unit, and the extra electron occupies the bottom of the conduction band. The more strongly dispersing bands smear out and decrease the pseudogap (some semblance of it with a magnitude of about 0.6 eV can be noted in Figure 3.8). An interesting finding is that the projected density of states for transition metal atoms Co and Ni has a dominant peak just below the pseudogap. This implies that the d-states of the transition metals are fully occupied and provides a solid argument for a substantially covalent character of bonding in binary phosphide skutterudites. The projected density-of-states of phosphorus gradually dominates above the pseudogap and has a distinct peak about 4 eV above the

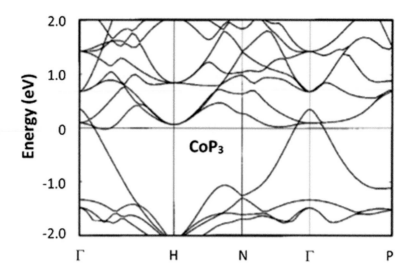

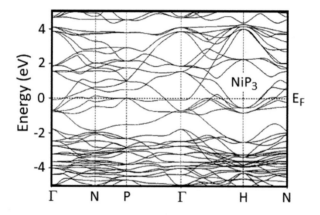

FIGURE 3.7   Electronic structure of CoP_3 in the vicinity of the Fermi energy. Note a distinctly metallic nature of the structure as the highest valence band crosses conduction bands above the Fermi level. Reproduced from M. Fornari and D. J. Singh, *Physical Review B* 59, 9722 (1999). With permission from the American Physical Society.

FIGURE 3.8   Electronic band structure of NiP_3, showing a metallic state but also some semblance of the pseudogap separating the valence and conduction manifolds. Reproduced from M. Llunell et al., *Physical Review B* 53, 10605 (1996). With permission from the American Physical Society.

Fermi level. With the calculated ratio of phosphorus to nickel contributions to the DOS at the Fermi level of 51.7 : 13.7 states/eV/unit cell, one arrives at a surprising conclusion that the bands responsible for electrical conduction in NiP_3 are centered on the phosphorus rather than nickel sublattice. Specifically, the states responsible are mostly nonbonding and π-type bonding orbitals of the pnicogen rings. These orbitals are coupled to neighboring rings in all three crystallographic directions through interactions with the s and p orbitals of the transition metal atoms.

   Quite apart from the extreme sensitivity of the band gap to the values of the positional parameters of the pnicogen atoms, DFT calculations and, in particular the LDA method, tend to underestimate the magnitude of the band gap. The disparity is reflected in a plethora of conflicting results reported in the literature on the size of the band gap and even whether it is positive (semiconducting system) or negative (semimetallic system). To illustrate the point, I have collected in Table 3.1 the theoretically predicted band gaps of binary skutterudites, which, for each particular binary compound, are presented in chronological order.

TABLE 3.1

Theoretical Band Gaps of Binary Skutterudites Computed by DFT with Various Functionals and Wave Function Expansions

Skutterudite	Band Gap (eV)	Functional	Comments	Reference
$CoSb_3$	0.050	LDA		Singh & Pickett (1994)
	0.080	LDA, LMTO		Akai et al. (1997)
	0.220	LDA	y, z minim. energy	Sofo & Mahan (1999)
	0.050	LDA	experimental y, z	Sofo & Mahan (1999)
	0.110	GGA, FPLAPW	with spin-orbit	Koga et al. (2001)
	0.195	LDA, CASTEP	unrelaxed y and z	Lefebvre et al. (2001)
	0.330	LDA, CASTEP	relaxed y and z	Lefebvre et al. (2001)
	0.003	LDA, FPLAPW		Takegahara & Harima (2003a)
	0.070	KKR with LDA	refined y, z, a	Wojciechowski et al. (2003)
	0.170	KKR with LDA	minimized y, z, a	Wojciechowski et al. (2003)
	0.105	B3PW with CRYSTAL98		Bertini et al. (2003)
	0.140	LDA, LMTO		Kurmaev et al. (2004)
	0.118	GGA		Koga et al. (2005)
	0.140	GGA, FPLAPW		Chaput et al. (2005)
	0.240	LDA, ABINIT p		Ghosez & Veithen (2007)
	0.660	B3PW with CRYSTAL98	optimized structure	Bertini & Cenedese (2007)
	0.150	PBE, FPLAPW	without spin-orbit	Wei et al. (2009)
	0.120	PBE, FPLAPW	with spin-orbit	Wei et al. (2009)
	0.400	GGA, CASTEP		Lu et al. (2010)
	0.260	LDA, VASP		Park & Kim (2010)
	0.225	LDA, Quant. Espresso	pressure P=0	Wee et al. (2010)
	0.318	LDA, Quant. Espresso	P = 33.6 kbar	Wee et al. (2010)
	0.134	LDA, Quant. Espresso	P = −28.3 kbar	Wee et al. (2010)
	0.620	GGA, VASP		Li et al. (2012)
	0.036	GGA, FPLAPW		Aliabad et al. (2012)
	0.220	LDA, PZ	relaxed y and z	Volja et al. (2012)
	0.129	various DFT's		Hammerschmidt et al. (2013)
	0.226	GGA, PBEsol	without spin-orbit	Sharma & Pandey (2014)
	0.188	GGA, PBEsol	with spin-orbit	Sharma & Pandey (2014)
	0.230	GGA, Quantum Espresso		Tang et al. (2015)
	0.335	GWA, ABINIT		Khan et al. (2015)
	0.580	TB-mBJ		Khan et al. (2015)
	0.610	nTB-mBJ		Khan et al. (2015)Hu et al. (2017)
	0.23	optB86b-vdW		Kolezynski & Szczypka (2017)
	0.27	GGA, FPLAPW		Tu et al. (2019)
	0.17	GGA, VASP		
	0.352	Meta-GGA	with spin-orbit	Yang et al. (2019)
	0.155	GGA, VASP		Isaacs & Wolverton (2019)
$RhSb_3$	Zero gap	GGA, FPLAPW		Akai et al. (1999)
	Zero gap	LDA, LAPW		Fornari & Singh (1999)
	Metal	LDA, FPLAPW		Takegahara & Harima (2003a)
	Zero gap	WIEN2k, PW		Pardo et al. (2012)
	0.165	GWA, ABINIT		Khan et al. (2015)
	0.170	TB-mBJ		Khan et al. (2015)

(*Continued*)

TABLE 3.1 (CONTINUED)

Theoretical Band Gaps of Binary Skutterudites Computed by DFT with Various Functionals and Wave Function Expansions

Skutterudite	Band Gap (eV)	Functional	Comments	Reference
	0.730	nTB-mBJ		Khan et al. (2015)
	Zero gap	WIEN2k		Kolezynski & Szczypka (2017)
IrSb$_3$	Zero gap	LDA		Singh & Pickett (1994)
	0.103	GGA, FPLAPW	without spin-orbit	Akai et al. (1999)
	0.060	GGA, FPLAPW	with spin-orbit	Akai et al. (1999)
	0.075	LDA, FPLAPW		Takegahara & Harima (2003a)
	0.080	WIEN2k, PW	with spin-orbit	Pardo et al. (2012)
	0.086	GWA, ABINIT		Khan et al. (2015)
	0.870	TB-mBJ		Khan et al. (2015)
	0.870	nTB-mBJ		Khan et al. (2015)
	0.130	PBE, VASP	without spin-orbit	Yin et al. (2017)
	0.104	Meta-GGA	without spin-orbit	Yang et al. (2019)
CoAs$_3$	Zero gap	LDA		Singh & Pickett (1994)
	Zero gap	LDA, FPLAPW		Takegahara & Harima (2003a)
	0.045	GWA, ABINIT		Khan et al. (2015)
	0.650	TB-mBJ		Khan et al. (2015)
	0.650	nTB-mBJ		Khan et al. (2015)
RhAs$_3$	Zero gap	LDA, FPLAPW		Takegahara & Harima (2003a)
	0.214	GWA, ABINIT		Khan et al. (2015)
	0.380	TB-mBJ		Khan et al. (2015)
	0.770	nTB-mBJ		Khan et al. (2015)
IrAs$_3$	0.156	LDA, FPLAPW		Takegahara & Harima (2003a)
	0.635	GWA, ABINIT		Khan et al. (2015)
	0.900	TB-mBJ		Khan et al. (2015)
	1.380	nTB-mBJ		Khan et al. (2015)
CoP$_3$	0.280	LDA, LMTO		Llunell et al. (1996)
	0.070	LMTO-ASA		Zhukov (1996)
	Metal	LDA, LAPW		Fornari and Singh (1999)
	Metal	FPLAPW		Harima (2000)
	Metal	LDA, FPLAPW		Takegahara & Harima (2003a)
	Metal	GGA, VASP		Løvvik & Prytz (2004)
	0.592	GWA, ABINIT		Khan et al. (2015)
	Zero gap	TB-mBJ		Khan et al. (2015)
	0.420	nTB-mBJ		Khan et al. (2015)
RhP$_3$	Metal	LDA, FPLAPW		Takegahara & Harima (2003a)
	0.123	GWA, ABINIT		Khan et al. (2015)
	Metal	TB-mBJ		Khan et al. (2015)
	Metal	nTB-mBJ		Khan et al. (2015)
IrP$_3$	0.136	LDA, FPLAPW		Takegahara & Harima (2003a)
	0.119	GWA, ABINIT		Khan et al. (2015)

(Continued)

TABLE 3.1 (CONTINUED)

Theoretical Band Gaps of Binary Skutterudites Computed by DFT with Various Functionals and Wave Function Expansions

Skutterudite	Band Gap (eV)	Functional	Comments	Reference
	0.610	TB-mBJ		Khan et al. (2015)
	0.610	nTB-mBJ		Khan et al. (2015)
NiP_3	Metal	LDA, LMTO		Llunell et al. (1996)
	Metal	LDA, FPLAPW		Takegahara & Harima (2003a)

FPLAPW stands for full potential linearized augmented plane waves; KKR means the Korringa-Kohn-Rostoker method; PBE represents the Perdew, Burke and Ernzerhof functional described in *Physical Review Letters* **77**, 3865 (1996); PBEsol is its modified version presented in *Physical Review Letters* **100**, 136406 (2008); the Perdew-Zunger (PZ) functional is discussed in *Physical Review B* **23**, 5048 (1981); PW stands for the Perdew-Wang functional (sometimes designated as PW91) described in *Physical Review B* **45**, 13244 (1992); and B3PW is a hybrid functional published in *Journal of Chemical Physics* **98**, 5648 (1993). VASP means the Vienna *Ab Initio* Simulation Package code described in *Physical Review B* **54**, 11169 (1996); CASTEP is an acronym for the Cambridge Serial Total Energy Package; CRYSTAL98 is a program presented in *Z. Kristallogr.* **220**, 571 (2005); and ABINIT is a package described in *Comput. Mater. Sci.* **25**, 478 (2002). Quantum Espresso is a software suite for *ab initio* calculations of electronic structures and is described in *J. Phys.: Condens. Matter* **21**, 395502 (2009). The more advanced version of the software is published in *J. Phys.: Condens. Matter* **29**, 465901 (2017). WIEN2k is a Fortran-based program for quantum-mechanical calculations of solids originally developed by P. Blaha and K. Schwarz at the Vienna University of Technology and is described in *Comput. Mater. Sci.* **28**, 259 (2003). HF stands for the Hartree-Fock method; GWA represents the Green's Function Approximation to compute the exchange-correlation energy described in *Comput. Mater. Sci.* **25**, 478 (2002); TB-mBJ and nTB-mBJ stand for the regular and non-regular Tran and Blaha-modified Becke-Johnson method introduced in *Physical Review Letters* **102**, 226401 (2009) and recently modified in *Physical Review B* **85**, 155109 (2012); TSPACE is a Fortran Program for Space Group developed by A. Yanase, Shobako, Tokyo, 1995; LMTO stands for Linear Muffin-Tin Orbital method described in *Physical Review Letters* **53**, 2571 (1984), and optB86b-vdW indicates a potential described in *J. Phys.: Condens. Matter* **22**, 022201 (2010). Data assembled from the literature.

Indeed, the scatter among the theoretical band gaps is enormous. One might argue that the more recent calculations would be more reliable, but that is not supported by entries in Table 3.1. Moreover, it is really difficult to say, which calculations are more in line with the experimentally determined band gaps as the latter are also all over the place, as shown in Table 3.2. The problem here is that the optical measurements of the band gap (mostly absorption edge) preferentially sense the pseudogap rather than the actual band gap, because the footprint left behind by a single linear band that crosses the pseudogap is, in terms of its DOS, too small to resolve. The alternative way of determining the band gap from high temperature electrical conductivity measurements via

$$\sigma = \sigma_0 e^{-\frac{E_g}{2kT}} \tag{3.3}$$

is also not as sharp a probe as one would wish.

Given the rather unsatisfactory state of affairs, two theoretical studies attempted to shed light on the problem in a more systematic manner. In the first one, Takegahara and Harima (2003a) used the identical FPLAPW calculations to make a relative comparison of band structures of all 10 binary skutterudites, MX_3 (M = Co, Rh, Ir; X = P, As, Sb) including NiP_3, and documented the presence of a substantial pseudogap between the valence and conduction bands of 0.7 eV to 1.4 eV in all binary compounds. Moreover, both the valence and conduction bands have contributions from the transition metal *d*-states and pnicogen *p*-states, the latter being uniformly distributed while the former is

TABLE 3.2

Experimental Band Gaps for Binary Skutterudites

Skutterudite	Band Gap (eV)	Technique	Comments	Reference
$CoSb_3$	0.50	High-T res.		Dudkin & Abrikosov (1956)
	0.55	High-T res.		Caillat et al. (1995a)
	0.05	Resistivity		Mandrus et al. (1995)
	0.35	High-T res.		Matsubara et al. (1994)
	0.6-0.7	High-T res.		Sharp et al. (1995)
	~0.10	Resistivity		Morelli et al. (1995)
	0.56	Resistivity		Caillat et al. (1996)
	0.035	SdH	sc	Rakoto et al. (1999)
	0.035	SdH	sc, 2 band fit	Arushanov et al. (2000)
	0.031	SdH	sc, 3 band fit	Arushanov et al. (2000)
	0.05	e-tunneling		Nagao et al. (2000)
	0.03-0.04	UPS		Ishii et al. (2002)
$RhSb_3$	Metal	Optical		Kliche & Bauhofer (1987)
	Metal	Resistivity		Kliche & Bauhofer (1987)
	0.80	Resistivity		Caillat et al. (1996)
$IrSb_3$	1.4	Reflectivity		Slack & Tsoukala (1994)
	1.18	Resistivity		Caillat et al. (1995b)
$CoAs_3$	0.18	Resistivity	Hot-pressed	Kliche & Bauhofer (1988)
	0.18	Optical	Hot-pressed	Kliche & Bauhofer (1988)
	0.69	Resistivity		Caillat et al. (1995c)
$RhAs_3$	>0.85	Resistivity		Caillat et al. (1995c)
CoP_3	0.43			Biltz & Heimbrecht (1939)
NiP_3	Metal			Rundqvist & Larsson (1959)

SdH stands for Shubnikov-de Haas measurements and UPS for Ultraviolet Photoemission Spectroscopy.
Data assembled from the literature.

hybridized with p-states of the pnicogen and peaks just below the Fermi level. This is particularly pronounced for Co-based skutterudites due to more spatially localized Co $3d$ orbitals compared to Rh $4d$ and Ir $5d$ orbitals. The real band gap is obtained for $CoSb_3$ (0.003 eV) and all Ir-based binary skutterudites: $IrSb_3$ (0.075 eV), $IrAs_3$ (0.156 eV), and IrP_3 (0.136 eV). In $CoAs_3$ and $RhAs_3$, the top of the valence band touches the bottom of the conduction band at Γ, and the remaining binary skutterudites, except NiP_3 that is a metal, develop a small overlap, i.e., are semimetals. Skutterudites with heavier pnicogen atoms tend to have narrower valence bandwidth: 7.6 – 8.3 eV for MP_3, 6.8 – 7.2 eV for MAs_3, and 5.7 – 6.3 eV for MSb_3. On the other hand, the valence bandwidth increases as one goes from M–$3d$ to M–$4d$ and to M–$5d$ orbitals.

The other study by Hammerschmidt et al. (2013) took one particular skutterudite, $CoSb_3$, and applied different, mostly DFT-based, electronic band structure calculations to look at differences arising from the use of various density functionals, basis sets, and electronic codes. The plane wave-based DFT calculations employed the VASP 4.6 program package with PBE functionals. Gaussian-type atomic basis set computations were carried out using DFT and Hartree-Fock (HF) theory and were implemented in the CRYSTAL09 program package, Dovesi et al. (2009). DFT calculations utilized five different density functionals; one based on LDA, two generalized GGA functionals (PBE and Perdew-Wang), and two hybrid functionals B3PW, Becke (1993), and B3LYP, Lee et al. (1988). The outcome of this extensive computational work is presented in Table 3.3, which lists the computed lattice parameter a_0, the positional parameters y and z of the Sb atom optimized with various functionals and the HF theory, the bulk modulus B_0 and its pressure derivative B_0', the band gap E_g, and the cohesive energy E_{coh}.

The lattice parameter of $CoSb_3$ computed using the various DFT functionals is quite close to its experimental value with the spread of less than 1.6%. The agreement in the positional parameters is even closer. There is very little difference between computations carried out with the plane wave basis set and the atomic basis set. On the other hand, what clearly stands out are HF calculations, which overestimate the lattice parameter by some 4% compared to the experimental value.

The bulk moduli were computed by least-squares fitting of B_0 and B_0' to the Murnaghan equation of state, Murnaghan (1944), Erk et al. (2011),

$$E(V) - E(V_0) = B_0 V_0 \frac{1}{B_0'} \left\{ \frac{V}{V_0} - 1 + \frac{1}{B_0' - 1} \left[\left(\frac{V_0}{V} \right)^{B_0' - 1} - 1 \right] \right\},$$ (3.4)

for at least seven data points around the equilibrium volume V_0. Both the equilibrium volume and the equilibrium energy $E(V_0)$ were taken directly from the structure optimization. Assessing the success of computations based on how the values of the bulk modulus agree with the experimental bulk modulus is somewhat tricky as the two existing experimental values deviate by 10%, roughly the spread in the computed values of the bulk moduli. However, assuming that the experimental values, indeed, fall in the range of 80–90 0 GPa, the DFT-based computations, except for the

TABLE 3.3

Lattice parameter, positional parameters y and z of Sb optimized with different functionals and the HF theory. Also included is the computed bulk modulus, its pressure derivative B_0' and the band gap at Γ

Method	Type	a_0 (Å)	y	z	B_0(GPa)	B_0'	E_g (eV)
HF	C	9.3975	0.3442	0.1488	124.2	17.69	4.517
LDA	C	8.9244	0.3325	0.1597	114.2	9.37	0.221
PBE	C	9.1122	0.3334	0.1590	91.8	3.94	0.109
PBE	V	9.1152	0.3334	0.1596	80.6	3.18	0.129
PW91	C	9.1140	0.3333	0.1589	92.0	10.01	0.089
B3LYP	C	9.1819	0.3355	0.1566	89.6	3.93	0.572
B3PW	C	9.0401	0.3351	0.1574	97.4	15.79	0.756
Exp [4.a]		9.0385	0.3354	0.15788	81 [4.b]	6 [4.b]	0.03 [4.c]
Exp [4.d]		9.03573	0.3348	0.1570	90 [4.e]		0.05 [4.f]

Source: The table is modified from L. Hammerschmidt et al., *Physica Status Solidi A* **210**, 131 (2013). With permission from John Wiley and Sons.

C and V stand for results obtained from CRYSTAL09 and VASP4.6, respectively.

[4.a] are data from T. Schmidt et al., *Acta Cryst. C* **43**, 1678 (1978).

*[4.b] are data from I. Shirotani et al., J. Phys.: Condens. Matter **16**, 7853 (2004).*

[4.c] are data from E. Arushanov et al., *Phys. Rev. B* **61**, 4672 (2001).

*[4.d] are data from D. Mandrus et al., Phys. Rev. B **52**, 4926 (1995).*

[4.e] are data from C. Recknagel et al., *Sci. Technol. Adv. Mater.* **8**, 357 (2007).

[4.f] are data from J. Nagao et al., *Appl. Phys. Lett.* **76**, 3436 (2000).

LDA functional, which significantly overestimates the bulk modulus, are in reasonable agreement. Again, the HF theory fails here, yielding some 30% overestimation with respect to the larger of the two experimental values. The pressure derivative of the bulk modulus is, obviously, very sensitive to the computational approach with the HF theory results being particularly exceptional. On the other hand, computations of B_0' with the PBE functional using VASP and CRYSTAL are fairly close to each other, again demonstrating that the choice of the functional influences ground-state properties far more than the choice of the basis set.

Turning our attention to the electronic properties and the band gap in particular, we note two distinct trends in the computations of Hammerschmidt et al. (2013). The results agree well with the previous calculations by Singh and Pickett (1994) and Sofo and Mahan (1999) using LDA, by Lefebvre-Davos et al. (2001) and Kurmaev et al. (2004) using PBE, and by Pardo et al. (2012) using PW density functionals. Yet, the use of hybrid functionals results in band gaps that are 5–7 times larger, and the HF theory yields a clearly unrealistic band gap of some 4.5 eV. The reason why the two hybrid functionals B3PW and B3LYP give rise to such distinctly larger band gaps compared to DFT functionals is the content of the HF exact exchange mixed into the functional expression, which shifts the weight of Co d-states and Sb p-states in the DOS near the Fermi level.

Hammerschmidt et al. also computed the band gap as a function of the lattice parameter for all functionals and the HF method. The results are plotted in Figure 3.9 and show a distinctly decreasing band gap as the lattice parameter increases. However, as the band gap gradually closes, the shape of the bands and the DOS remain substantially the same. This is somewhat surprising since, as the overlap between the π_4 antibonding orbitals of the Sb_4 ring decreases, Sofo and Mahan (1999), one would expect the highest valence band to become less dispersive than the results indicate. The computations also confirmed the extreme sensitivity of the band gap on the exact value of the positional parameters y and z, i.e., on the degree of stretching and squeezing of the rectangular Sb_4 ring, seen previously by Sofo and Mahan (1999).

Subsequent electronic structure calculations of all binary skutterudites by Khan et al. (2015) used the latest versions of the density functionals based on the regular Becke and Johnson (2006) functionals modified by Tran and Blaha (2009) and designated as TB-mBJ, and also its non-regular TB-mBJ version, Koller et al. (2012). The data are presented as part of the entries in Table 3.1. By selectively picking the experimental values of the band gap (from the many presented in Table 3.2)

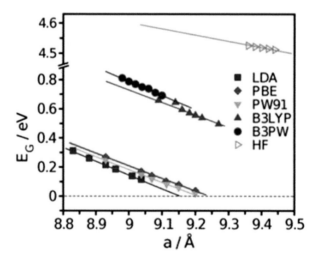

FIGURE 3.9 Band gap as a function of the lattice parameter computed with various DFT functionals. Reproduced from L. Hammerschmidt et al., *Physica Status Solidi A* 210, 131 (2013). With permission from John Wiley and Sons.

that most closely matched their calculated values, Khan et al. (2015) declared that the non-regular TB-mBJ functional reproduces the electronic structure of the binary skutterudites most faithfully. The correct experimental value of the band gap and justification of its selection were left mute. The most recent band structure calculations by Yang et al. (2019) employed the SCAN method of mega-GGA, Sun et al. (2015), within the VASP package and resulted in the band gap of 0.352 eV for $CoSb_3$ and 0.104 eV for the band gap of $IrSb_3$, both included in Table 3.1.

To appreciate the structure of the bands near the band edges, Figure 3.10 shows the band structure of $CoSb_3$ calculated recently by Hu et al. (2017) in the proximity of the Fermi energy. When the spin-orbit interaction is neglected, the conduction band edge directly above the linearly dispersing valence band is triply degenerate at the Γ-point but separates immediately into three distinct bands along the Γ–H direction, and into a doubly degenerate set of parabolic bands and a single linearly dispersing band, mirroring the valence band, along the Γ–P direction. However, when the spin-orbit interaction is turned on, the degeneracy at Γ is lifted, except for the two lower bands, as depicted in the inset on the right-hand side of the figure. The two lower conduction bands comprise a heavy conduction band and a light conduction band, and the third, split-off band, lies at Γ only 0.02 eV above and remains quasi-linear even with the spin-orbit interaction. The spin-orbit interaction slightly reduces the band gap to 0.18 eV.

Another interesting feature of the band structure of $CoSb_3$ pointed out previously by Tang et al. (2015) and replicated in the band structure calculations shown in Figure 3.10 is the existence of the second conduction band minimum (labeled as 'valley') along the Γ–N direction. Tang et al. and Hu et al. place the second conduction band minimum only about 0.11 eV and 0.13 eV, respectively, above the conduction bands at Γ. In fact, making careful measurements of the temperature-dependent optical absorption from 300 K to 673 K, Tang et al. have shown that the direct $E_{g,\Gamma-\Gamma}$ and the indirect $E_{g,\Gamma-valley}$ band gaps are temperature dependent in such a way that they converge at T_{cvg}

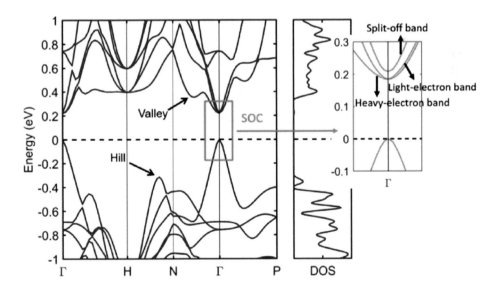

FIGURE 3.10 Band structure and total density of states of $CoSb_3$ in the vicinity of the Fermi energy set at 0 eV at the top of the valence bands. With no spin-orbit interaction at play, the conduction band edge at Γ is triply degenerate. Note a "valley" band located at 0.13 eV above the conduction band minimum and a "hill" band located at 0.32 eV below the valence band maximum. The inset on the right depicts the valence and conduction band edges around the zone center Γ calculated including spin-orbit coupling (SOC). The conduction band edge in this case consists of doubly degenerate heavy and light conduction bands and a split-off band located merely 0.02 eV above. Reproduced from C. Z. Hu et al., *Physical Review B* 95, 165204 (2017). With permission from the American Physical Society.

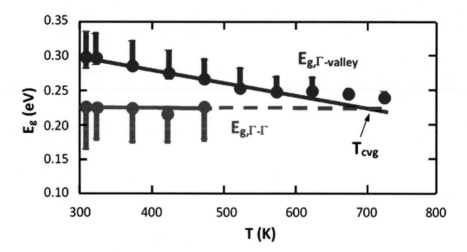

FIGURE 3.11 Temperature dependence of the direct ($E_{g,-}$) and indirect ($E_{g,-valley}$) band gaps for $CoSb_3$. The band convergence is estimated to occur at $T_{cvg} \approx 700$ K. Redrawn from the data of Y. Tang et al., *Nature Materials* 14, 1223 (2015). With permission from Springer Nature.

≈ 800 K, Figure 3.11. Since the second ("valley") conduction minimum is at a point along the Γ-N direction, the symmetry implies it has valley degeneracy $N_v = 12$. Including the valley degeneracy of 3 from the conduction band at Γ, the total valley degeneracy (number of electron carrier pockets) of 15 is exceptionally large upon band convergence. Such a band structure aspect dramatically improves the thermoelectric performance by increasing the electrical conductivity with no penalty to the Seebeck coefficient. It would be interesting to explore the temperature dependence (or doping dependence) of the second ('hill') valence band maximum with respect to that of the valence band at Γ. Perhaps there might also be a possibility to induce the band convergence of these two valence bands, even though they are separated by a much larger 0.34 eV.

As described in Section 1.1 and illustrated in Figure 1.4, the skutterudite structure with its characteristic near-square rings of pnicogen atoms can be viewed as evolving from the more symmetric ReO_3 structure, which, in turn, is a substructure of the perovskite ATX_3 structure where the atom A is missing. Smith et al. (2011) and subsequently Pardo et al. (2012) explored the formation of the band gap as the structure transforms from a highly metallic perovskite to a semiconducting skutterudite. The transformation can be expressed in terms of the positional parameters of the pnicogen atom y and z as

$$y'(s) = \frac{1}{4} + s\left(y - \frac{1}{4}\right) \text{ and } z'(s) = \frac{1}{4} + s\left(z - \frac{1}{4}\right), \tag{3.5}$$

where s is a continuous variable. The initial perovskite structure corresponds to $s = 0$, while the final skutterudite structure forms when $s = 1$. Their calculations using all-electron full-potential codes (WIEN2k and FPLO-9) revealed that the separation of the conduction and valence band manifolds starts to be notable for $s = 0.75$. The formation of the band gap and pseudogap was clearly discerned at $s = 0.9$, and the linearly dispersing valence band at Γ starts to dominate the valence bands at $s = 0.95$. At $s = 1$, the valence band is clearly a partner to the triply degenerate conduction bands at Γ, giving rise to a small direct band gap, i.e., to the formation of occupied bonding bands and unoccupied antibonding bands. Thus, the band gap arises only toward the end point of the transformation from the perovskite to the skutterudite structure.

As originally pointed out by Singh and Pickett (1994), the valence band remains linear very close to the Γ point, but in the immediate vicinity of the zone center ($k = 0$), it acquires a quadratic form. The calculations of Smith et al. and Pardo et al. indicated that with a very small uniaxial strain applied to the pnicogen ring, and its strength tuned by the same parameter s used in Eq. 3.5 above (in practice, this could be a small lattice mismatch between a $CoSb_3$ film and the substrate the film is deposited on), the linear dispersion can extend right up to the zone center Γ, at least along the <111> direction. When this happens and the band gap closes, the value of the parameter s marks a critical point, referred to as the Dirac point, and the linear band becomes known as the Dirac band. Computations by Smith et al. (2011) indicate that the onset of the Dirac point happens precisely at the critical value of $s_c = 1.023$. In the subsequent paper by Pardo et al. (2012), the critical value of s_c is slightly lower at 1.020 as a consequence of including spin-orbit coupling. The situation is nicely depicted in Figure 3.12, where the progression of the relevant bands near Γ is plotted for three values of the parameter s that represents the degree of distortion of the pnicogen ring.

Due to the band inversion and the parity of the bands at Γ, the transformation as the parameter s is varied from 1.020 to 1.025 is from a trivial insulator to one having the topological character. However, because the degeneracy at the point Fermi surface at Γ remains, the transformed state is strictly a topological zero-gap semiconductor rather than the topological insulator.

Regarding other binary antimonide skutterudites, the calculations of Pardo et al. indicate that, for a range of lattice parameters around its experimental value, $RhSb_3$ has an inverted band structure with a point Fermi surface, i.e., $RhSb_3$ is a zero-gap topological semimetal. Assuming the absence of spin-orbit coupling (SOC), the single valence band at Γ is actually 23 meV above

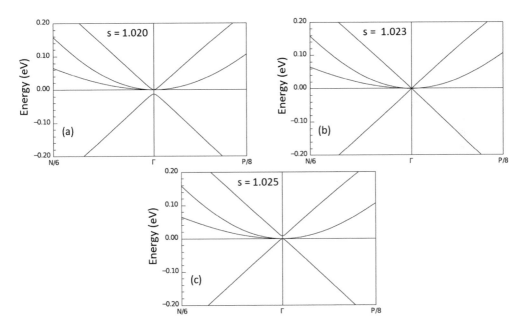

FIGURE 3.12   Band structure of $CoSb_3$ near the critical point as a function of the parameter s expressing minor changes in the positional parameters of Sb. (a) s = 1.020, a situation just before the critical point is reached with a small direct band gap between the triply degenerate conduction band and the near-linear valence band; (b) s_c = 1.023, at the critical point the valence band becomes exactly of the Dirac type with the linear dispersion even at the zone center. The Dirac bands are degenerate with two conventional quadratic and rather massive conduction bands; (c) s = 1.025, a situation depicting a zero-gap semiconductor with the near-linear singlet conduction band now lying above the triply degenerate bands at the Fermi level. N/6 and P/8 indicate fractional distances along the Γ-N and Γ-P directions, respectively. Redrawn and modified from J. C. Smith et al., *Physical Review Letters* 106, 056401 (2011), with permission from the American Physical Society.

the triply degenerate conduction band that pins the Fermi level. With the SOC switched on, the Fermi level is pinned to a doubly degenerate conduction band, as the third band is split off by the interaction. When a tetrahedral distortion is applied, the degeneracy is lifted and a topological insulator results for c/a > 1, with no assistance from any sublattice strain. As shown by Wee et al. (2010), applying pressure to binary skutterudites tends to increase the band gap, and the same would be expected to happen in $RhSb_3$. However, Pardo et al. demonstrated that the gap in this skutterudite does not even open at high pressure because the energy of other Rh $4d$ bands decreases with pressure, and the bands cross the Fermi level along the Γ-N direction, making the structure a metal. In $IrSb_3$, on account of the heavier mass of Ir, one might expect the SOC to be even stronger. However, because the linearly dispersing valence band in this case has essentially no metallic character, it is immune to spin-orbit interaction, and the calculations return a band gap of about 80 meV, and thus no topological character. As in the case of $CoSb_3$, one would need the assistance of internal and tetrahedral strain to drive $IrSb_3$ into a state with an inverted band structure.

Topological states of matter have generated a tremendous interest during the past 15 years, Kane and Mele (2005), Moore and Balents (2007), Roy (2009). Among the best examples are, in fact, high-performance thermoelectric materials. Most notable among them are the tetradymite-type structures of Bi_2Te_3, Sb_2Te_3 and Bi_2Se_3 (Zhang et al. 2009), alloys of Bi with Sb (Fu and Kane 2007), certain half-Heusler alloys (Chadov et al. 2010), and now even suitably distorted skutterudite structures. Why should thermoelectric materials be such impressive examples of a topological insulator? The reason is the common presence of heavy elements, which in thermoelectric materials underpins the desirable low lattice thermal conductivity, while in topological insulators gives rise to a strong spin-orbit interaction that leads to the required electronic band inversions. Whether one can make an outstanding thermoelectric material based on a detailed understanding of topological insulators has been debated, see, e.g., Müchler et al. (2013) and Yang (2017). I personally think that, for power generation applications, it is unlikely that thermoelectric energy conversion would benefit from topological features because the appropriate thermoelectric materials for such applications are rather heavily doped semiconductors with the Fermi level situated in the conduction or the valence band, whereas topological insulators require the Fermi level to be within the band gap. However, for cooling applications, where the requirement for the carrier degeneracy is more relaxed, perhaps a contribution from the surface state transport that is protected by time-reversal symmetry (meaning that non-magnetic impurities cannot cause undesirable elastic back-scattering of the charge carriers) might boost the thermoelectric performance.

3.1.1 EFFECT OF PRESSURE ON THE BAND STRUCTURE OF BINARY SKUTTERUDITES

In Section 1.1.4, it was shown that binary skutterudites are quite stable to pressures of about 30 GPa, at which point the so-called pressure-induced self-insertion reaction takes place. Kolezynski and Szczypka (2017) calculated the effect of pressure on the band structure of $CoSb_3$ and $RhSb_3$ and noted a nearly linearly increasing band gap with pressure. The effect is caused by the upshift of the Γ-point conduction bands. However, at the same time, the band edge of the secondary conduction band located along the Γ–N direction (see Figure 3.10), moves down relative to the Γ-point conduction bands, and at about 10 GPa, the band edges cross. One thus observes a pressure-driven band convergence and, at higher pressures, the originally secondary conduction band becomes the global minimum, converting $CoSb_3$ from a direct band gap to an indirect band gap semiconductor. Band gap movements with pressure in $CoSb_3$ and $IrSb_3$ were evaluated recently by Yang et al. (2019) by using the SCAN method of mega-GGA, Sun et al. (2015, 2016), within VASP. The authors noted gradually increasing band gaps up to about 30 GPa where they reached values of 0.773 eV for $CoSb_3$ and 0.645 eV for $IrSb_3$. At higher pressures, the band gaps slowly decreased. Interestingly, the pressure of 30 GPa is close to the critical pressure for the pressure-induced self-insertion reaction estimated for $CoSb_3$ by Matsui et al. (2010). Moreover, Yang et al. confirmed that the direct band gap characterizing the band structure of $CoSb_3$ and $IrSb_3$ at ambient pressure changes into an indirect band gap, this time at 6 GPa for $CoSb_3$ and at 10 GPa for $IrSb_3$. Again, the band movement is primarily due to the upshift of the conduction bands at the Γ point, while the valence band at the Γ point seems to be immune to the pressure. The position of the band edges in $CoSb_3$ at three different pressures is indicated in Figure 3.13.

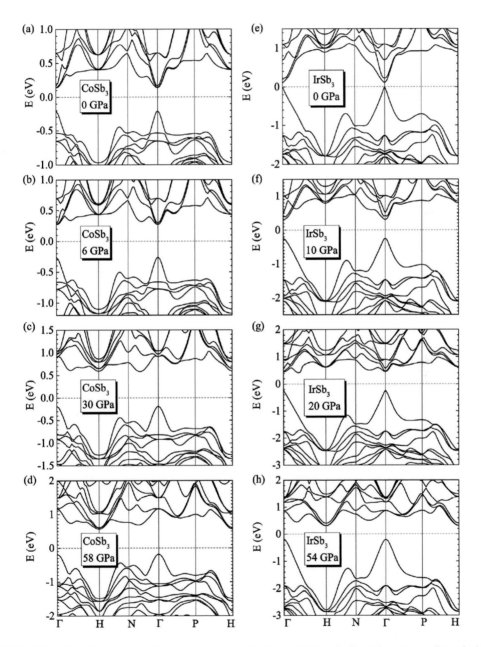

FIGURE 3.13 Calculated electronic band structure of $CoSb_3$ and $IrSb_3$ at indicated pressures along the high symmetry points within the first Brillouin zone. Note a progressively larger band gap with pressure and the crossover from the direct to the indirect band gap at 6 GPa for $CoSb_3$ and at 10 GPa for $IrSb_3$. Reproduced from X. X. Yang et al., *Physical Chemistry Chemical Physics* 21, 851 (2019). With permission from the Royal Society of Chemistry.

3.2   BAND STRUCTURE OF TERNARY SKUTTERUDITES

Given the extreme sensitivity of the band structure of binary skutterudites on the positional parameters y and z of the pnicogen atom in the X_4 ring structure, one expects significant modifications in the electronic bands of ternary skutterudites in which the pnicogen atom is substituted by a pair of elements from Columns 14 (Ge, Sn) and 16 (S, Se, Te). It is not only that the previously homogeneous environment of the X_4 ring turns heterogeneous with the two different ions in *trans* (opposite) position to each other, but also that the rings are no longer rectangular, with the dihedral angle reduced from the original 90.0°, depending on the particular structure. The bonds acquire a more ionic character as charge transfers between ions in the unit cell.

The first DFT *ab-initio* computations of ternary skutterudites were carried out by Bertini and Cenedese (2007) for $CoSn_{1.5}Te_{1.5}$ using CRYSTAL98 code with the B3PW (GGA) functional. The electronic energy band calculations identified this ternary skutterudite as a semiconductor with the band gap at Γ about 22% smaller (0.51 eV) than the band gap of $CoSb_3$ (0.66 eV) computed using the same code. An interesting outcome of the study was a considerably enhanced charge transfer (and therefore ionicity of the bonds) between atoms in the unit cell of $CoSn_{1.5}Te_{1.5}$ compared to that of $CoSb_3$, see Figure 3.14. While in both structures Co carries a net negative charge, in $CoSn_{1.5}Te_{1.5}$ its charge is significantly larger at −1.232 compared to the Co charge in $CoSb_3$ where it is merely −0.298. By the same token, the positive charge on the Sn_2Te_2 ring is +1.636 while on the Sb_4 ring it is only +0.372. As I have already pointed out, the interest in ternary skutterudites as thermoelectric materials has stemmed primarily from the greater mass disorder leading to a lower lattice thermal conductivity compared to binary skutterudites. The much enhanced charge transfer in ternary skutterudites arising from the work of Bertini and Cenedese might modify the frequency of phonon modes of the ring and provide an additional pathway for a reduction of the lattice thermal conductivity.

Detailed DFT-based *ab-initio* computations using the Perdew-Zunger LDA exchange-correlation functional by Volja et al. (2012) focused on six ternary skutterudites: $CoGe_{1.5}S_{1.5}$, $CoGe_{1.5}Se_{1.5}$, $CoGe_{1.5}Te_{1.5}$, $CoSn_{1.5}S_{1.5}$, $CoSn_{1.5}Se_{1.5}$, and $CoSn_{1.5}Te_{1.5}$. The calculated band structures are displayed in Figure 3.15, with the result for $CoSb_3$ plotted in solid red traces for comparison. Inspecting the band structures in Figure 3.15, one notes the much larger (perhaps by as much as a factor of 2–3) band gaps of ternary skutterudites compared to $CoSb_3$. This includes $CoSn_{1.5}Te_{1.5}$ that Bertini and Cenedese claimed to have the band gap some 22% smaller than $CoSb_3$. In computations of Volja et al., the band gap of ternary skutterudites varies between 0.41 eV in $CoSn_{1.5}Se_{1.5}$ and 0.61 eV in $CoGe_{1.5}S_{1.5}$, while that of $CoSb_3$ is 0.22 eV. The larger band gaps in ternary skutterudites arise chiefly from the $t_{2g}-e_g$ derived manifold splitting and the flatter band dispersion due to the more ionic bonding than in binary skutterudites. Just as in binary antimonide skutterudites, there is a single band linearly dispersing from the t_{2g} valence manifold toward the Γ point in all six ternary skutterudites studied. However, compared to $CoSb_3$, the linear dispersion is weaker. Interestingly, the second highest valence bands in all ternaries possess a multivalley character and lie much

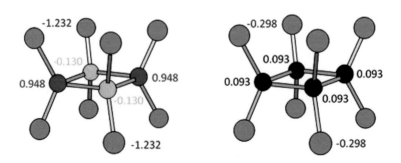

FIGURE 3.14   Computed net atomic charges for $CoSn_{1.5}Te_{1.5}$ (left panel) and $CoSb_3$ (right panel). Co atoms are in blue, Sb atoms in black, Sn atoms in red, and Te atoms in orange. Redrawn from L. Bertini and S. Cenedese, *Physica Status Solidi (RRL)* 1, 244 (2007). With permission from John Wiley and Sons.

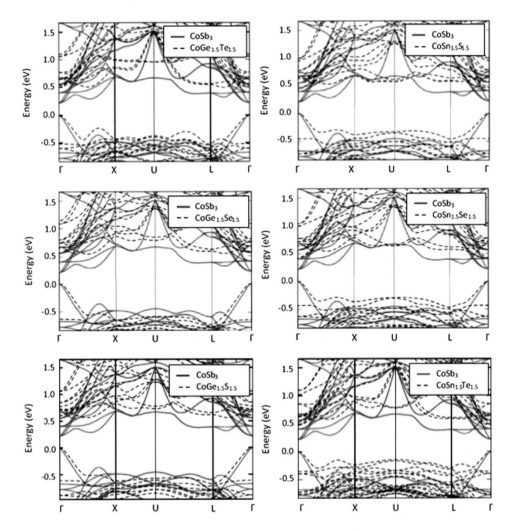

FIGURE 3.15 First principles band structures (dashed black curves) of the ternary skutterudites compared to the band structure of CoSb$_3$ plotted in solid red traces. Reproduced from D. Volja et al., *Physical Review B* 85, 245211 (2012). With permission from the American Physical Society.

closer to the top, nearly-linearly dispersing valence band, particularly in CoSn$_{1.5}$Te$_{1.5}$. Reducing the valence band separation further, so that the heavy carriers of the second highest band would contribute to the transport, should benefit the Seebeck effect. Values of the effective masses of holes evaluated by Volja et al. turned out to be 0.196 m$_e$ (CoGe$_{1.5}$S$_{1.5}$), 0.169 m$_e$ (CoGe$_{1.5}$Te$_{1.5}$), and 0.134 m$_e$ (CoSn$_{1.5}$Te$_{1.5}$), in all cases significantly larger than the value of ~0.07 m$_e$ reported for holes in CoSb$_3$ by Caillat et al. (1996) and by Arushanov et al. (1997). Isoelectronic substitutions on the pnicogen ring also lead to modifications of the lowest conduction bands that attain non-equivalent minima in Γ–L and Γ–X directions that may provide carriers with large effective mass upon appropriate n-type doping.

In a paper by Zevalkink et al. (2015), the authors focused on the electronic structure of MSn$_{1.5}$Te$_{1.5}$ with M = Co, Rh, Ir, i.e., including the two heavier and more expensive transition metals Rh and Ir. Using Quantum Espresso software with the PBE form of the GGA functional, they computed the electronic structure. The experimental lattice and positional parameters were used as initial inputs, from which the theoretical minimum energy lattice parameters were determined. The positional parameters *y* and *z* were relaxed using the Broyden-Fletcher- Goldfarb-Shanno method (see, e.g., Fletcher (1987)) until the total force on each atom fell below 0.01 eV/Å.

Even though the structures included a heavy Ir-based ternary, they found that spin-orbit coupling was not critical and thus was omitted. Similar to the calculations of Volja et al., the authors found rather large band gaps for all three ternary skutterudites: 0.43 eV for $CoSn_{1.5}Te_{1.5}$, 0.19 eV for $RhSn_{1.5}Te_{1.5}$, and 0.41 eV for $IrSn_{1.5}Te_{1.5}$. In particular, the value for $CoSn_{1.5}Te_{1.5}$ matches well with that obtained by Volja et al. Furthermore, the valence band is dominated by a single, nearly linearly dispersing band at Γ, and the energy separation to the next highest valence band is dramatically reduced compared to the case of binary skutterudites. Specifically, in $CoSn_{1.5}Te_{1.5}$, the energy separation is merely 0.2 eV, providing a tantalizing possibility for multi-band conduction in p-type forms of these ternary skutterudites. The conduction band edge at Γ consists of two degenerate bands and a third band lying slightly higher by about 0.04 eV for $CoSn_{1.5}Te_{1.5}$ and up to 0.12 eV for $IrSn_{1.5}Te_{1.5}$. Again, these are quite small energy separations that might be explored for multiband conduction, this time in n-type doped ternary skutterudites.

In Figure 3.16 are reproduced the partial density of states to appreciate the contribution of the transition metal M = Co, Rh, Ir in the three ternary skutterudites $MSn_{1.5}Te_{1.5}$. The lowest energy region of the DOS up to about −6 eV is dominated by contributions from bonding s-orbitals of Sn

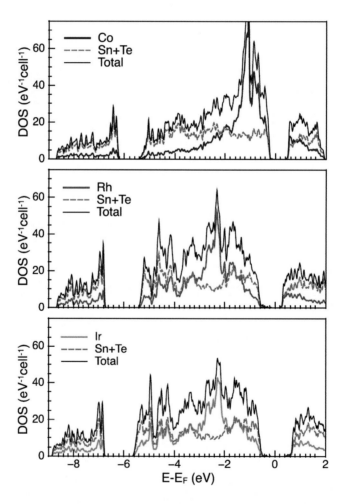

FIGURE 3.16 Partial density of states comparing the influence of the M = Co, Rh, Ir atom in $MSn_{1.5}Te_{1.5}$ ternary skutterudites. The energy $E - E_F = 0$ is set equal to the valence band maximum. Reproduced from A. Zevalkink et al., *Journal of Applied Physics* 118, 035107 (2015). With permission from the American Institute of Physics.

and Te. In the energy interval from −6 eV to 0 eV, the DOS is chiefly due to t_{2g} bonding states arising from the M(Sn,Te)$_6$ distorted octahedron, in which p-orbitals of Sn and Te hybridize with d-orbitals of the transition metal M. The conduction band minimum above 0 eV is formed by antibonding states of these hybridized bands. As noted before, the $3d$ orbitals of Co are typically more spatially localized than are $4d$ and $5d$ orbitals of Rh and Ir, respectively. This is reflected in a relatively less dispersive Co contribution between −2 eV and 0 eV compared to Rh and Ir. Moreover, different Pauling electronegativities of the M element (1.88 for Co, 2.28 for Rh, and 2.20 for Ir) also play a role regarding a relative position of the M electronic states with respect to those of Sn and Te. On account of the smaller electronegativity, the peak density of the electronic states of Co lies about 1.2 eV higher in energy relative to those of Rh and Ir, and the corresponding DOS near the band gap in CoSn$_{1.5}$Te$_{1.5}$ ternary skutterudites is much larger than that of the two heavier structures.

Physical properties of a large collection of alkaline earth-filled ternary skutterudites were calculated by Bang et al. (2016) using DFT and DFTP (Density Functional Perturbation Theory) with the Quantum-ESPRESSO package. The data sets include the lattice parameters, unit cell volumes, band gaps, the estimated power factors, as well as frequencies of the lowest optic phonons.

3.3   BAND STRUCTURE OF FILLED SKUTTERUDITES

As described in Section 1.3, filled skutterudites form from binary skutterudites upon filling the structural voids at the body-centered position of the cubic skutterudite lattice, the 2a sites using Wyckoff notation. It is perhaps surprising that the early interest in filled skutterudites was focused on phosphide-based skutterudites rather than on the antimonide-based forms. The reason, however, should be clear; throughout the 1980s and very early 1990s, the greatest fascination was with the magnetic state of Fe in LaFe$_4$P$_{12}$ and, above all, the rather high superconducting transition temperature of the compound and other filled phosphide skutterudites (see Table 1.3). At that time, no one had any notion that filled skutterudites would a few years later become *bona fide* thermoelectric materials with antimonides leading the way. Thus, the original calculations of the band structure of filled skutterudites were made for phosphide skutterudites by Jung et al. (1990) using the tight-binding method within the extended Hückel framework, Hoffmann (1963) and Whangbo and Hoffmann (1978). The authors computed the dispersion relation and the density of states (DOS) for [Fe$_4$P$_{12}$]$^{3-}$, the primitive unit cell that is the fundamental structural block of LaFe$_4$P$_{12}$ but entirely omitted the influence of La^{3+} ions. The essential outcome of these calculations was that the t_{2g} and e_g block bands are separated by a rather large gap (~ 2.7 eV) with the Fermi level cutting the highest occupied valence band so that it is half-filled, in other words the structure was a metal. As all skutterudites filled with the lanthanoid 3+ species have the lattice building blocks [T$_4$P$_{12}$]$^{3-}$ that are isoelectronic and isostructural with the [Fe$_4$P$_{12}$]$^{3-}$ framework, they all should be metals. On the other hand, the [Fe$_4$P$_{12}$]$^{4-}$ lattice and its isoelectronic [M$_4$X$_{12}$] lattices have the t_{2g} block bands completely filled, and the structure should be semiconducting. This is the case of a 4+ filler ion, such as U^{4+} and Th^{4+}, or a strictly tetravalent form of Ce^{4+}.

As Figure 3.17 indicates, the Fermi level for [Fe$_4$P$_{12}$]$^{3-}$ falls in the energy region where the DOS is small (~ 5 electrons/eV/formula unit) and where the Fe $3d$-orbital character is negligible. The low value of the DOS near the Fermi level is in accord with the experimentally estimated value of about four electrons/eV/formula unit by Shenoy et al. (1982). Consequently, the highest occupied band has mainly phosphorus character. Interestingly, since electrons near the Fermi level are primarily responsible for superconductivity, the superconducting properties of LaFe$_4$P$_{12}$ observed by Meisner (1981) had to be governed by the electrons of its phosphorus sublattice. It is, however, important to keep in mind that the computed band structure refers strictly to the [Fe$_4$P$_{12}$]$^{3-}$ polyanion and entirely neglects the influence of the filler ions, a rather unreasonable proposition that limits the conclusions of the paper.

The first DFT calculations of filled skutterudites (CeFe$_4$P$_{12}$ and CeFe$_4$Sb$_{12}$) were made by Nordström and Singh (1996) using the FP-LAPW method with the Hedin-Lundqvist version of

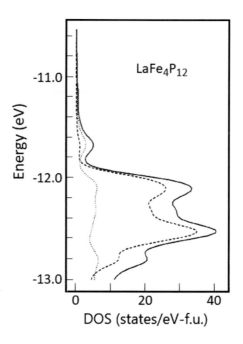

FIGURE 3.17 Total and projected density of states for the t_{2g} block bands of LaFe$_4$P$_{12}$. The total DOS is shown by the solid line, the contribution of Fe 3d orbitals by the dashed line, and the contribution of phosphorus 3s/3p orbitals by the (faint) dotted line. Redrawn from D. Jung et al., *Inorganic Chemistry* 29, 2252 (1990). With permission from the American Chemical Society.

the LDA functional. The calculations included spin-orbit interaction, fully respected the presence of Ce fillers, and were performed with the experimental values of the lattice parameter. While the positional parameters of phosphorus in CeFe$_4$P$_{12}$ were already known from X-ray diffraction studies by Grandjean et al. (1984) ($y = 0.3522$ nm and $z = 0.1501$ nm), no such data were available for CeFe$_4$Sb$_{12}$. Consequently, the authors determined the parameters by minimizing the total energy while keeping the volume fixed at the experimental value. The resulting $y = 0.333$ nm and $z = 0.163$ nm were then used in the calculations of the band structure. By the way, both sets of the positional parameters implied that the filled skutterudites possess near square pnicogen rings, significantly more so than in the case of binary skutterudites, as the Oftedal relation, $y + z = \frac{1}{2}$ was followed quite closely (0.5032 and 0.496 for the two compounds, respectively). The key question explored in the computations was the valence state of the Ce filler and whether its presence in the structure would result in a metallic or semiconducting state. A truly tetravalent Ce^{4+} would be expected to yield a semiconducting structure as Ce^{4+}[Fe$_4$X$_{12}$]$^{4-}$ would achieve the valence electron count of 72. If, however, the valence state of Ce was 3+, the filled skutterudite should be a metal. The band structures, presented here for energies close to the Fermi level, are reproduced in Figure 3.18(a) and 3.18(b). As is obvious, both filled skutterudites have a band gap: 0.34 eV for the phosphorus-based skutterudite and a smaller one of 0.10 eV for the antimonide-based skutterudite. The band gap forms right below the flat spin-split Ce 4f bands that, according to the detailed site-projected DOS, are slightly hybridized with Fe 3d and pnicogen p states. The valence band top is formed from hybridized Fe 3d and pnicogen p states. Apart from different sizes of the band gap, CeFe$_4$P$_{12}$ shows two closely lying parabolic bands at the valence band edge at Γ, while in CeFe$_4$Sb$_{12}$ there is only one. The authors drew attention to various valence and conduction bands of the two structures in terms of the different contributions of the filler, transition metal and pnicogen states. For instance, based on the symmetry and orbital character, the second highest valence band in the phosphide and the highest valence band in the antimonide both have the dominant Fe 3d character (~ 80%) that at Γ is slightly

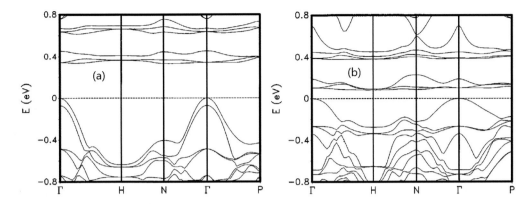

FIGURE 3.18 Band structure of (a) $CeFe_4P_{12}$ and (b) $CeFe_4Sb_{12}$ near the Fermi level. Reproduced from L. Nordström and D. J. Singh, *Physical Review B* 53, 1103 (1996). With permission from the American Physical Society.

hybridized with pnicogen p states but no contribution from Ce $4f$ states. In contrast, the topmost phosphide valence band is closely related to the fourth highest valence band in the antimonide, both having the predominantly Ce $4f$ character at Γ, which, however, weakens and becomes mostly Fe $3d$ as the zone boundary at N is approached. The respective effective masses of the valence bands at Γ are 0.8 m_e for the phosphide and 2.2 m_e for the antimonide. This is to be contrasted with the effective masses of electrons that are 10–20 m_e for $CeFe_4P_{12}$ and 6–8 m_e for $CeFe_4Sb_{12}$, arising from the flat conduction bands. The results underscored the strong influence of filler species on the band structure near the Fermi level.

The question concerning the tetravalent or trivalent nature of Ce, i.e., whether the $4f$ level is unoccupied, f^0, or has the occupancy of one, f^1, was resolved by comparing overlapping ionic charge densities that agreed most closely with the self-consistently derived result. Although the ionic charges are difficult to establish accurately in compounds that do not have strictly ionic bonding, the calculations left no doubt that the best match was obtained when Ce was trivalent, i.e., the $4f$ level was occupied with one electron. To explain the existence of the band gap under such circumstances, the authors postulated the hybridization of Ce $4f$ states with Fe $3d$ and pnicogen p states.

It is to be noted that the computation results yielded a somewhat unusual situation: small band gap semiconductors with very heavy band masses! The authors pointed out that since the LDA tends to underestimate the intra-atomic correlation of Ce $4f$ electrons and thus overestimates the hybridization, one arrives at a rather surprising finding that the LDA computations actually overestimate rather than underestimate the band gaps. Consequently, it is not a big surprise that the theoretically predicted semiconducting state for $CeFe_4Sb_{12}$ has been contradicted by the experimentally observed metallic nature of conduction. On the other hand, the larger band gaps of $CeFe_4P_{12}$ and $CeFe_4As_{12}$ are supported experimentally, and the two structures are semiconductors (Grandjean et al. 1984).

Subsequently, Singh and Mazin (1999) revisited the case of La fillers and examined their influence on the band structure of $La(Fe,Co)_4Sb_{12}$ for Co fractions of 0, 25%, and 50%. Calculations were again performed by the FPLAPW method and treated the upper core states of La consistently with the valence states. The calculated bands near the Fermi level differed little as the content of Co was varied, Figure 3.19 reproduces the virtual band structure for $LaFe_3Co_1Sb_{12}$. The band structure shows an indirect gap between the parabolic valence band edge at Γ (effective mass about 0.2 m_e) and the parabolic conduction band edge at N of about 0.56 eV and a direct gap of about 0.67 eV. Both the conduction and valence band edges have mostly Sb p character. There are also rather flat, i.e., heavy mass, bands of the Fe/Co d origin above and below the gap. Among them is the second highest valence band at Γ derived from the Fe/Co t_{2g} states, which lies only about 0.10 eV below the top valence band and contains heavy holes with the effective mass ~3 m_e. The proximity of two valence

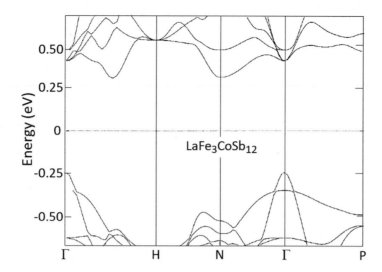

FIGURE 3.19   Virtual band structure of $LaFe_3Co_1Sb_{12}$ near the Fermi level. Redrawn from D. J. Singh and I. I. Mazin, *Physical Review B* 56, R1650 (1997). With permission from the American Physical Society.

bands with vastly different effective masses might benefit the Seebeck effect at high temperatures. It is clear though, just as was the case of Ce fillers, that the presence of La strongly alters the band structure away from that of the binary skutterudite.

In Section 3.1, discussing the band structure of CoP_3 computed by Fornari and Singh (1999), I noted that the crossing of valence and conduction bands in CoP_3 makes it challenging to develop efficient thermoelectric materials based on this skutterudite unless one modifies the structure to open a gap. In the same paper, Fornari and Singh have shown that filling, such as with La forming $LaFe_4P_{12}$, might be such an approach. Indeed, as shown by the density of states and the band structure near the Fermi level in Figure 3.20, a strong repulsion of La f-resonance states lying about 3 eV above the Fermi level drives the previously crossing valence band down and a small gap of E_g = 0.098 eV opens. Unfortunately, the band edge has not fallen low enough to be close to the heavy valence bands lying beneath to benefit the Seebeck coefficient with heavy carriers.

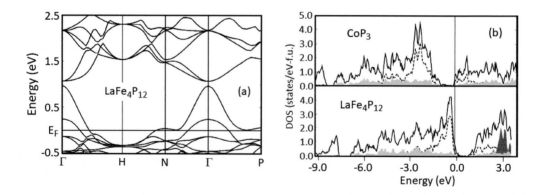

FIGURE 3.20   (a) Band structure of $LaFe_4P_{12}$ near the Fermi level. (b) Density of states of CoP_3 (upper panel) and of $LaFe_4P_{12}$ (lower panel). The solid line indicates the total density of states, light gray regions represent phosphorus p states, the dashed line traces the contribution of Co d states, and the resonant La f component is in dark gray shade. Adapted from M. Fornari and D. J. Singh, *Physical Review B* 59, 9722 (1999). With permission from the American Physical Society.

Harima and Takegahara (2003a) carried out DFT computations of the $LaFe_4X_{12}$ filled skutterudites for X = P, As, and Sb using the FPLAPW method to evaluate the band progression from the phosphide to the arsenide and to the antimonide structure. The valence band manifolds of all three filled skutterudites are similar, but the density of states at the Fermi level dramatically increases from the phosphide (13.97 states/eV) to the arsenide (18.16 states/eV) to the antimonide (25.2 states/eV). This is primarily due to the elevated t_{2g} bands associated with Fe d-orbitals that do not seem to hybridize with the pnicogen p-states and contribute progressively larger peaks in the DOS as the interatomic distance La-Fe increases more than the La-X distance.

Clearly, all three filled skutterudites in the calculations are predicted to be p-type metals possessing a closed Fermi surface near Γ and a nesting-type Fermi surface in $LaFe_4P_{12}$ that progressively shrinks in $LaFe_4As_{12}$ and $LaFe_4Sb_{12}$.

Since Yb-filled skutterudites are of considerable interest for thermoelectricity, I wish to point out interesting recent DFT computations by Isaacs and Wolverton (2019) comparing electronic bands in $CoSb_3$ with the bands of hypothetical fully filled $YbCo_4Sb_{12}$, both computed with fully-relaxed structures. Referring to Figure 3.21, the filled skutterudite reveals additional conduction bands, marked by blue vertical arrows. In fact, three of these bands have edges at or below the band edge of the degenerate Γ-point conduction bands, indicating that at some content y of Yb the conduction bands will converge. While such filled skutterudites are clearly metals with the Fermi level deep inside the conduction bands (marked by a horizontal arrow), the separation between the conduction and valence band manifolds has significantly increased compared to $CoSb_3$. Moreover, the effective masses of both holes and electrons increase upon Yb filling as the bands become somewhat less dispersive. The band convergence in $Yb_yCo_4Sb_{12}$ skutterudites, as a result of the increased density of electrons populating antibonding states, has been noted previously by Hanus et al. (2017) based on their DFT calculations augmented by structural changes observed in $Yb_yCo_4Sb_{12}$ via temperature-dependent synchrotron powder x-ray diffraction measurements.

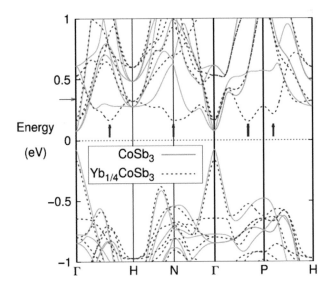

FIGURE 3.21   Electronic band structure of $Yb_yCo_4Sb_{12}$ for the fully-relaxed endmembers, y = 0 and y = 1. Energy is plotted with respect to the mid-gap at Γ. The Fermi energy for $YbCo_4Sb_{12}$ is indicated by the horizontal arrow. Note four conduction bands (marked by vertical arrows) in $YbCo_4Sb_{12}$, at least three of which have band edges at or below the edge of the Γ-point conduction bands, revealing that the band convergence happens at some content y of Yb. Reproduced from E. B. Isaacs and C. Wolverton, *Chemistry of Materials* 31, 6154 (2019). With permission from the American Chemical Society.

Band structure calculations have been carried out for many filled skutterudites. The interest was driven not only by their potential use in thermoelectrics but also for their fascinating magnetic and superconducting properties. I list here many of the references and, for the benefit of readers interested in specific filled skutterudites, I indicate the targeted material being described: Singh (2002), Takegahara and Harima (2002) [LaFe$_4$Sb$_{12}$]; Harima (1998) [LaFe$_4$P$_{12}$]; Harima and Takegahara (2003a) [LaFe$_4$X$_{12}$, X = P, As, Sb]; Harima and Takegahara (2003b) [PrRu$_4$P$_{12}$, PrFe$_4$P$_{12}$, LaFe$_4$P$_{12}$]; Saha et al. (2002), Harima (2008) [LaRu$_4$P$_{12}$]; Harima and Takegahara (2002a), Harima et al. (2002) [perfect nesting in PrRu$_4$P$_{12}$]; Harima and Takegahara (2002b) [LaOs$_4$Sb$_{12}$]; Akai et al. (2002), Takegahara and Harima (2002) [YbCo$_4$Sb$_{12}$, YbFe$_4$Sb$_{12}$]; H. Sugawara et al. (2002), Harima and Takegahara (2005) [PrOs$_4$Sb$_{12}$]; Takegahara and Harima (2003b) [ThFe$_4$P$_{12}$]; Takegahara and Harima (2008) [SmOs$_4$Sb$_{12}$]; Yan et al. (2012) [CeOs$_4$As$_{12}$, CeOs$_4$Sb$_{12}$]; Nieroda et al. (2014) [Ag$_x$Co$_4$Sb$_{12}$]; Ram et al. (2014) [LaRu$_4$X$_{12}$, X = P, As, Sb]; Xing et al. (2015) [LaFe$_4$X$_{12}$, NaFe$_4$X$_{12}$, X = P, As, Sb]; Luo et al. (2015) [LaT$_4$Sb$_9$Sn$_3$]; Shankar et al. (2017) [EuRu$_4$As$_{12}$]; Hu et al. (2017) [La$_x$Co$_4$Sb$_{12}$]; Qi et al. (2017) [Ba$_x$Ir$_4$As$_{12}$]; Tütüncü et al. (2017) [LaRu$_4$P$_{12}$, LaRu$_4$As$_{12}$].

3.4 BAND STRUCTURE OF SKUTTERUDITES WITH THE [Pt$_4$Ge$_{12}$] FRAMEWORK

The papers reporting the discovery of skutterudites having the [Pt$_4$Ge$_{12}$] framework included DFT calculations of the total density of states as well as contributions of each constituent element to the DOS. Often, the calculations were carried out with both nonrelativistic and fully relativistic treatments, the latter taking care of the spin-orbit interaction. Most of the studies were carried out using the Vienna *ab initio* simulation package (VASP) described by Kresse and Furthmuller (1996a,b). The density of states of the first-reported BaPt$_4$Ge$_{12}$ and SrPt$_4$Ge$_{12}$ skutterudites by Bauer et al. (2007), computed with the exchange-correlation functional within LDA, are shown in Figure 3.22. Panel (e) depicts the DOS of SrPt$_4$Ge$_{12}$ over a wider range of energies, calculated by Rosner et al. (2009) using DFT within LDA employing the full-potential local-orbital (FPLO) code. The exchange and correlation functional was that of Perdew and Wang (1992).

As follows from Figure 3.22, *p*-like states of Ge overwhelmingly dominate the DOS near the Fermi energy, with a small contribution of Pt 5*d*-states. Both Ba and Sr fillers have an insignificant impact on the DOS near E_F, implying that physical properties, such as superconductivity, are tied

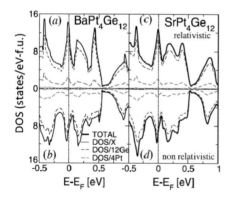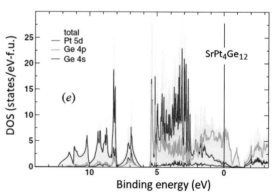

FIGURE 3.22 Density of states of BaPt$_4$Ge$_{12}$ computed using (a) a fully relativistic treatment and (b) computed using a standard nonrelativistic treatment. Panels (c) and (d) show the same for SrPt$_4$Ge$_{12}$. In each case, the contribution of the constituent elements (1 Ba and Sr atom, 4 Pt atoms, and 12 germanium atoms) to the DOS is shown in color and dashed curves. The Fermi energy is set at $E - E_F = 0$. Panel (e) depicts the DOS of SrPt$_4$Ge$_{12}$ over a wider range of energies. The Fermi energy is shown by a vertical line. Panels (a) − (d) are reproduced from E. Bauer et al., *Physical Review Letters* 99, 217001 (2007), while panel (e) is modified from H. Rosner et al., *Physical Review B* 80, 075114 (2009). With permission from the American Physical Society.

to the properties of the [Pt$_4$Ge$_{12}$] framework rather than to the choice of the filler. The DOS of BaPt$_4$Ge$_{12}$ at the Fermi level in panel (a) is 13.2 states eV^{-1} (f.u.)$^{-1}$, while for the SrPt$_4$Ge$_{12}$ compound in panel (c), it is 12.1 states eV^{-1} (f.u.)$^{-1}$. It is interesting to note that the spin-orbit interaction tunes the peak in the DOS to a near coincidence with the Fermi energy. It also implies that a modest doping might be able to shift the peak to a perfect overlap with E_F, which would increase the DOS at E_F and give rise to a large Seebeck coefficient.

The density-of-states of LaPt$_4$Ge$_{12}$ and PrPt$_4$Ge$_{12}$, a part of the paper announcing the discovery of the above two skutterudites by Gumeniuk et al. (2008), is shown in Figure 3.23.

The low energy bands below −6 eV are formed predominantly by Ge 4s states. Pt 5d states are found chiefly between −5.5 eV and −2.5 eV and hybridize with Ge 4p orbitals. At the Fermi level, the dominant contribution comes from Ge 4p bands with rare earth fillers La and Pr again making an insignificant contribution, just as was the case of alkaline earth Ba and Sr fillers. The computed DOS at the Fermi energy for LaPt$_4$Ge$_{12}$ is 13.4 states eV^{-1}(f.u.)$^{-1}$, and for PrPt$_4$Ge$_{12}$ it is a bit smaller at 9.3 states eV^{-1}(f.u.)$^{-1}$, the values comparable to those of BaPt$_4$Ge$_{12}$ and SrPt$_4$Ge$_{12}$. A similar DOS was computed for EuPt$_4$Ge$_{12}$ by Grytsiv et al. (2008) with the value of 9.2 states eV^{-1}(f.u.)$^{-1}$ at the Fermi energy. Bader's analysis,[*] Bader (1990), indicated that about 1.3 electrons were transferred from Eu to Pt, similar to the amount of charge transfer estimated in BaPt$_4$Ge$_{12}$ and SrPt$_4$Ge$_{12}$. The charge transfer is what stabilizes the [Pt$_4$Ge$_{12}$]-based skutterudites. Isosurfaces of the electron density (localized electronic function with the electronic density of 0.28 electrons per Å^3 for the states of the DOS peak at the Fermi energy) are shown in Figure 3.24. Particularly notable are tube-like regions in the left panel indicating strongly directional covalent Ge–Ge bonds, while spherical regions around Pt formed by Pt 5d states suggest metal-like charge distribution. The metal-like

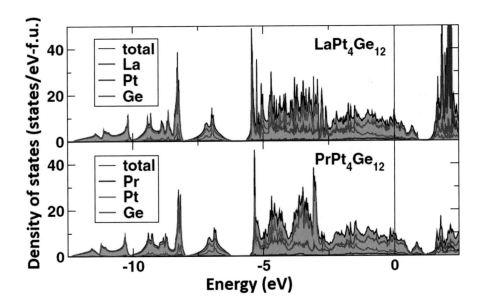

FIGURE 3.23   Total and atom resolved electronic density-of-states for LaPt$_4$Ge$_{12}$ and PrPt$_4$Ge$_{12}$. Modified from R. Gumeniuk et al., *Physical Review Letters* 100, 017002 (2008). With permission from the American Physical Society.

---

[*] Since individual atomic charges in a molecule are not observable in the quantum mechanical sense, Bader's analysis is a technique that delineates atoms and their charge in a molecule. The surfaces separating atoms in a molecule are constructed by connecting minima in the charge density. In other words, at any point on the Bader dividing boundary, the gradient of the electron density perpendicular to the boundary is zero. The charge associated with the atom is then obtained by integrating the electron density within the Bader region.

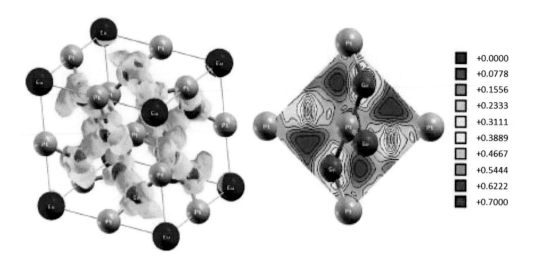

■	+0.0000
■	+0.0778
■	+0.1556
□	+0.2333
□	+0.3111
□	+0.3889
■	+0.4667
■	+0.5444
■	+0.6222
■	+0.7000

FIGURE 3.24 Graphical representation of isosurfaces of the localized electronic function in $EuPt_4Ge_{12}$ with the electron density of 0.28 electrons per $Å^3$. Eu atoms in blue, Ge atoms in purple, and Pt atoms in gray. Strongly directional covalent Ge-Ge bonds show as forming tubes, while the Pt $5d$ states are spherically symmetric around Pt. The right panel is a contour plot depicting the relatively weak covalent bonds between Pt and Ge, compared to the Ge-Ge bonds. Adapted from A. Grytsiv et al., *Journal of the Physical Society of Japan* 77, Suppl. A, 121 (2008). With permission from the Physical Society of Japan.

regions in the DOS near the Fermi energy were interpreted by the authors as proof that the Zintl concept is not applicable in the case of $[Pt_4Ge_{12}]$-based skutterudites, in contrast to its relevance for pnicogen-based skutterudites.

Subsequent DFT calculations of the band structure of $LaPt_4Ge_{12}$ and $PrPt_4Ge_{12}$ by Sharath Chandra et al. (2016) using the FPLAPW method with the PBESol exchange-correlation functional, Perdew et al. (2008), and Tütüncü et al. (2017), who used the Quantum Espresso package, Giannozzi et al. (2009), with the exchange-correlation energy estimated by GGA of Perdew, Burke, and Ernzerhof (1996) in their band structure calculations of $LaPt_4Ge_{12}$, returned essentially the same structure of the DOS. In their DFT-based computations carried out by VASP with the PBE exchange-correlation potential, Humer et al. (2013) expanded their study from $LaPt_4Ge_{12}$ to include $LaPt_4Ge_7Sb_5$, which drove a good metal with a fairly large DOS toward an insulating state. Gumeniuk et al. (2010) carried out band structure calculations for $SmPt_4Ge_{12}$ using the same approach as in their study of $LaPt_4Ge_{12}$ and found a very similar DOS, apart from the contribution of strongly correlated $4f$ electrons and a tiny shift toward higher energy.

The last band structure I want to mention is that of $ThPt_4Ge_{12}$, described in several publications (Bauer et al. 2008, Tran et al. 2009a,b, and Galvan 2009). The results of FPLO calculations by Tran et al. (2009a) within LDA using the Perdew and Wang functional for exchange and correlation, and with fully relativistic treatment of spin-orbit interaction, are displayed in Figure 3.25. The lowest energy bands (down to −12 eV) are dominated by Ge $4s$-like states. The majority of Pt $5d$ states occupy the range from −6 eV to −2.5 eV, and their presence diminishes for energies above −3 eV. At about −2.5 eV, there is a distinct change in the shape of the bands from d-like (flat) to p-like (more undulating). Below the Fermi level, the states are dominantly occupied by Ge $4p$ electrons. As the right panel in Figure 3.25 indicates, the Fermi level is crossed by three bands, suggesting that $ThPt_4Ge_{12}$ is a multiple-conduction skutterudite, and the beloved single parabolic band model is unlikely to describe the transport data faithfully.

The DOS of $ThPt_4Ge_{12}$, computed by Tran et al. (2009b) using FPLAPW and employing the EXCITING code, Dewhurst et al., is shown in Figure 3.26. While the Ge $4p$ states, again, make the dominant contribution to the DOS at the Fermi level, the $5f$ states of the Th filler are, in this

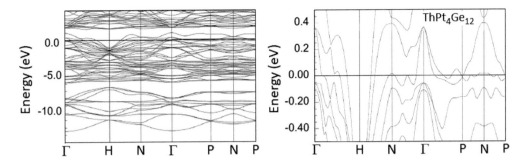

FIGURE 3.25   Electronic bands in $ThPt_4Ge_{12}$. The right panel shows an expanded region near the Fermi energy set at zero. Reproduced from V. H. Tran et al., *Physical Review B* 79, 054520 (2009). With permission from the American Physical Society.

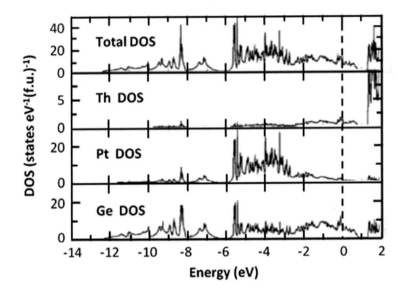

FIGURE 3.26   Calculated total and partial densities of states in $ThPt_4Ge_{12}$. Adapted from V. H. Tran et al., *Physical Review B* 79, 144510 (2009). With permission from the American Physical Society.

case, not negligible, accounting for about 8.7% of all 9.75 states $eV^{-1}(f.u.)^{-1}$. The DOS of $ThPt_4Ge_{12}$ computed by Bauer et al. (2008) using the VASP package, and including the spin-orbit interaction, yielded a similar result with 9.63 states $eV^{-1}(f.u.)^{-1}$ at the Fermi energy. Moreover, the authors have shown that the spin-orbit interaction in $ThPt_4Ge_{12}$ does not alter the DOS in any major way.

The band structure of $ThPt_4Ge_{12}$ computed by Galvan (2009) was carried out using the tight-binding method within the extended Hückel framework, and the results deviate significantly from the DFT-based computations, yielding a sharp Th peak very near the Fermi energy with Th 5f states accounting for 20% of all states and little contributions from Ge p states and Pt d states.

3.5   BAND STRUCTURE OF SKUTTERUDITES FILLED WITH ELECTRONEGATIVE FILLERS

In their pioneering work with the iodine-filled $[Rh_4Sb_{12}]$ framework, Fukuoka and Yamanaka (2010) used LAPW calculations (WIEN2k code) to obtain the band structure of IRh_4Sb_{12}, DOS, and the

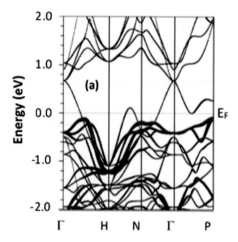
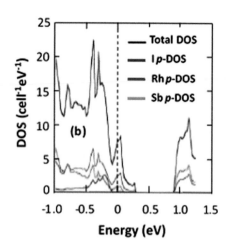

FIGURE 3.27 (a) Band structure of IRh_4Sb_{12} for paths along symmetry points of the Brillouin zone. Bands arising from iodine are shown in fat curves. (b) Total density-of-states of IRh4Sb$_{12}$ and contributions of individual elements to the DOS near the Fermi energy set at zero. Iodine p-states lie below the Fermi level. Adapted and redrawn from H. Fukuoka and S. Yamanaka, *Chemistry of Materials* 22, 47 (2010). With permission from the American Chemical Society.

element-projected DOS. As shown in Figure 3.27(a), the band structure features the familiar nearly linearly dispersing Γ-point valence band, so typical of binary $RhSb_3$ and $CoSb_3$. However, unlike in the above binary skutterudites, where the Fermi level was at the top of the single Γ-point valence band, in IRh_4Sb_{12} the Fermi level crosses this band along the path from Γ to H, from H to N, and from N to Γ, making IRh_4Sb_{12} a metal dominated by holes. The thick black traces in Figure 3.27a indicate bands contributed by iodine. Evidently, iodine makes no contribution to low-lying conduction bands and its p-orbitals fall below the Fermi level. The DOS for IRh_4Sb_{12} and the respective contributions of I, Rh, and Sb are shown in Figure 3.27(b). The conducting electron bands consist chiefly of p-orbitals of Sb, and p- and d-orbitals of Rh.

A few years later, supporting their experimental studies on iodine-filled $CoSb_3$, Li et al. (2014) computed the band structure and DOS of ICo_4Sb_{12}, and confirmed that, just as in $RhSb_3$, iodine has little effect on $CoSb_3$, except shifting the Fermi level down so that it intersects more bands than just the top-lying valence band and enhances the DOS near the Fermi energy compared to $CoSb_3$.

The density of states of Br and S, the two recently discovered electronegative fillers in the $[Co_4Sb_{12}]$ framework and its partly Te-substituted framework $[Co_4Sb_{11.5}Te_{0.5}]$, were computed by DFT using the VASP package utilizing the GGA functional by Duan et al. (2016). The total density of states and the contribution due to Br, respectively, S are depicted in Figures 3.28(a) and 3.28(b). Again, both electronegative fillers make a modest contribution to the DOS only below the Fermi energy. However, while Br-filled $CoSb_3$ is a p-type metal, the S-filled $[Co_4Sb_{11.5}Te_{0.5}]$ framework is an n-type metal as a result of Te doping. Instead of Te doping on the pnicogen site, skutterudites with electronegative fillers, such as S and Se, can also be charge compensated by replacing some fraction of Co with Ni, Pd, and Pt, Li et al. (2019) and Tu et al. (2019). The DFT VASP calculations by the latter authors also indicated a significant effect of Ni, Pd, and Pt on the conduction band edge, while the valence band was immune to the presence of the dopants. The triply degenerate conduction band at Γ was split upon the presence of Ni, Pd, and Pt to a different degree, decreasing from Ni- to Pd- to Pt-doped structure, as shown in Figure 3.29, with a consequent reduction in the band gap. In fact, only the Pt-doped structure maintained a small band gap (0.05 eV), while the band gaps were essentially closed when Ni and Pd substituted for Co. The authors proposed that the splitting of the conduction band edge reflects different degree of anti-bonding strength between the doping atoms and Sb.

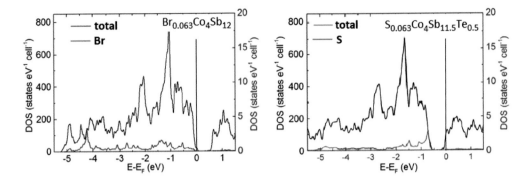

FIGURE 3.28 (a) The total density-of-states of $Br_{0.063}Co_4Sb_{12}$ and a contribution due to Br. (b) The total density-of-states of $S_{0.063}Co_4Sb_{11.5}Te_{0.5}$ and a contribution due to S. In both cases, the Fermi energy is indicated by a dashed line. Adapted from supplementary information of a paper by B. Duan et al., *Energy & Environmental Science* 9, 2090 (2016). With permission from the Royal Society of Chemistry.

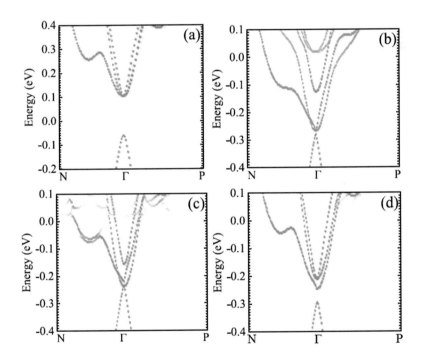

FIGURE 3.29 Band structure focusing on the conduction and valence bands at Γ point. (a) Stoichiometric Co_4Sb_{12}, (b) $Co_{3.5}Ni_{0.5}Sb_{12}$, (c) $Co_{3.5}Pd_{0.5}Sb_{12}$, and (d) $Co_{3.5}Pt_{0.5}Sb_{12}$. Adapted from Z. K. Tu et al., *AIP Advances* 9, 045325 (2019). I wish to thank Prof. L. L. Xi for allowing me to use the above figure that was published as part of a paper under the Creative Common Attribute license.

3.6 BENEFITS OF ACCURATE COMPUTATIONS OF ELECTRONIC BANDS

With the improved sophistication of DFT computing packages, the electronic band structure calculations have evolved from *a posteriori* justification of experimental results to highly accurate predictive assessment of physical properties. In the case of skutterudites, the band structure calculations have been indispensable in explaining exotic heavy fermion superconductivity in $PrOs_4Sb_{12}$, and verifying various magnetically ordered phases featured in rare-earth filled skutterudites. For thermoelectricity, the detailed description of the electronic bands in the vicinity of the Fermi energy revealed road maps for achieving the desirable convergence of electronic bands (Tang et al. 2015), which increases the number of carrier pockets and thus enhances the Seebeck coefficient. Moreover, the shape and structure of the bands inform about the effective mass of charge carriers, and point out whether n-type or p-type doping is more effective in achieving the high power factor. Calculations also suggested that certain filler species, such as Ag, may give rise to sharp peaks (resonant states) in the density-of-states near the Fermi energy that might further benefit the power factor (Nieroda et al. 2014). *Ab initio* calculations were instrumental in identifying defect structures that form in binary skutterudites as a consequence of a slight non-stoichiometry in the starting composition (Park and Kim 2010). As the defect states are charged, they affect the transport properties. Furthermore, simulations also clarified the dual role of Column 13 elements (Ga and In) that can enter voids (in small quantities) as well as substitute on the site of Sb, playing a very different role in their two respective locations (Tang et al. 2014). DFT-based computations have also revealed how electronic bands respond to external stimuli, such as applied pressure (Ram et al. 2014). Finally, calculations by Smith et al. (2011) have shown that a trivial insulator $CoSb_3$ can transition to a topological point Fermi surface, and subsequent work by Pardo et al. (2012) indicated that with slight band structure modifications, skutterudites can reveal topological features. Although no new family of efficient thermoelectric materials has as yet been discovered based on the predictive power of electronic band structure calculations, many thermoelectric structures, including slutterudites, have benefitted from the detailed understanding of electronic bands near the Fermi level.

REFERENCES

Akai, K., H. Kurisu, T. Shimura, and M. Matsuura, *Proc. 16th Int. Conf. on Thermoelectrics*, IEEE Catalog Number 97TH8291, Piscataway, NJ, p. 334 (1997).

Akai, K., K. Oshiro, and M. Matsuura, *Proc. 18th Int. Conf. on Thermoelectrics*, IEEE Catalog Number 99TH8407, Piscataway, NJ, p. 444 (1999).

Akai, K., K. Koga, K. Oshiro, and M. Matsuura, *Proc. 20th Int. Conf. on Thermoelectrics*, IEEE Catalog Number 01TH8589, Piscataway, NJ, p. 93 (2002).

Aliabad, H. A. R., M. Ghazanfari, I. Ahmad, and M. A. Saeed, Comp. Mater. Sci. **65**, 509 (2012).

Arushanov, E., K. Fess, W. Kaefer, C. Kloc, and E. Bucher, *Phys. Rev. B* **56**, 1911 (1997).

Arushanov, E., M. Respaud, H. Rakoto, J. M. Broto, and T. Caillat, *Phys. Rev. B* **61**, 4672 (2000).

Bader, R., in *Atoms in Molecules: A Quantum Theory*, Oxford University Press, NY (1990).

Bang, S., D.-H. Wee, A. Li, M. Fornari, and B. Kozinski, *J. Appl. Phys.* **119**, 205102 (2016).

Bauer, E., A. Grytsiv, X.-Q. Chen, N. Melnychenko-Koblyuk, G. Hilscher, H. Kaldarar, H. Michor, E. Royanian, G. Giester, M. Rotter, R. Podloucky, and P. Rogl, *Phys. Rev. Lett.* **99**, 217001 (2007).

Bauer, E., X.-Q. Chen, P. Rogl, G. Hilscher, H. Michor, E. Royanian, R. Podloucky, G. Giester, O. Sologub, and A. P. Gonçalves, *Phys. Rev. B* **78**, 064516 (2008).

Becke, A. D., *J. Chem. Phys.* **98**, 5648 (1993).

Becke, A. D. and E. R. Johnson, *J. Chem. Phys.* **124**, 221101 (2006).

Bertini, L., C. Stiewe, M. Toprak, S. Williams, D. Platzek, A. Mrotzek, Y. Zhang, C. Gatti, E. Müller, M. Muhammed, and M. Rowe, *J. Appl. Phys.* **93**, 438 (2003).

Bertini, L. and S. Cenedese, Phys. Stat. Solidi (*RRL*) **1**, 244 (2007).

Bethe, H., *Ann. Physik* **5**, 133 (1929).

Biltz, W. and M. Heimbrecht, *Z. Anorg. Allgem. Chem.* **241**, 349 (1939).

Blaha, P., K. Schwarz, P. Sorantin, and S. B. Trickey, Comp. Phys. Commun. **59**, 399 (1990).

Caillat, T., A. Borshchevsky, and J.-P. Fleurial, *Proc. 13th Int. Conf. on Thermoelectrics*, Kansas City, MO, 1994, ed. B. Mathiprakasam, American Institute of Physics Press No. 316, p. 31 (1995a).

Caillat, T., A. Borshchevsky, and J.-P. Fleurial, *Proc. 13th Int. Conf. on Thermoelectrics*, Kansas City, MO, 1994, ed. B. Mathiprakasam, American Institute of Physics Press No. 316, p. 58 (1995b).

Caillat, T., J.-P. Fleurial, and A. Borshchevsky, *Proc. 30th* Intersoc. Ener. Conv. Engin. *Conf.*, Orlando, FL, August 3–7 (1995c).

Caillat, T., A. Borshchevsky, and J.-P. Fleurial, *J. Appl. Phys.* **80**, 4442 (1996).

Caillat, T., J.-P. Fleurial, and A. Borshchevsky, *J. Crystal Growth* **166**, 722 (1996).

Chadov, S., X. Qi, J. Kübler, G. H. Fecher, C. Felser, and S.-C. Zhang, *Nat. Mater.* **9**, 541 (2010).

Chaput, L., P. Pecheur, J. Tobola, and H. Scherrer, *Phys. Rev. B* **72**, 085126 (2005).

Dewhurst, J. K., S. Sharma, and C. Ambrosch-Draxl, EXCITING Code Version 0.9.224, http://exciting .sourceforge.net.

Dovesi, R., V. R. Saunders, R. Roetti, R. Orlando, C. M. Zicovich-Wilson, F. Pascale, B. Civalleri, K. Doll, N. M. Harrison, I. J. Bush, P. D'Arco, and M. Llunell, CRYSTAL09, User's manual (University of Torino), 2009.

Dreizler, R. M. and E. K. U. Gross, *Density Functional Theory*, Springer-Verlag, Berlin (1990).

Duan, B., Jiong Yang, J. R. Salvador, Y. He, B. Zhao, S. Wang, P. Wei, F. S. Ohuchi, W. Q. Zhang, R. P. Hermann, O. Gourdon, S. X. Mao, Y. W. Cheng, C. M. Wang, J. Liu, P. C. Zhai, X. F. Tang, Q. J. Zhang, and Jihui Yang, *Energy Environ. Sci.* **9**, 2019 (2016).

Dudkin, L. D. and N. K. Abrikosov, *Sov. Phys. Solid State* **1**, 29 (1956).

Engel, E. and R. M. Dreizler, *Density Functional Theory: An Advanced Course*, Springer (2018).

Erk, C., L. Hammerschmidt, D. Andrae, B. Paulus, and S. Schlecht, *Phys. Chem. Chem. Phys.* **13**, 6029 (2011).

Fletcher, R., *in Practical methods of optimization*, 2nd edition, John Wiley & Sons, NY (1987).

Fornari, M. and D. J. Singh, *Phys. Rev. B* **59**, 9722 (1999).

Fu, L. and C. L. Kane, *Phys. Rev. B* **76**, 045302 (2007).

Fukuoka, H. and S. Yamanaka, *Chem. Mater.* **22**, 47 (2010).

Galvan, G. H., *J. Supercond. Nov.* Magn. **22**, 367 (2009).

Ghosez, P. and M. Veithen, *J. Phys.: Condens. Matter* **19**, 096002 (2007).

Giannozzi, P., S. Baroni, N. Bonini, M. Calandra, R. Car, C. Cavazzoni, D. Ceresoh, G. L. Chiarotti, M. Cococcioni, I. Dabo, A. D. Corso, S. de Gironcoli, S. Fabris, G. Fratesi, R. Gebauer, U. Gerstmann, C. Gougoussis, A. Kokalj, M. Lazzeri, L. Martin-Samos, N. Marzari, F. Mauri, R. Mazzarello, S. Paolini, A. Pasquarello, L. Paulatto, C. Sbraccia, S. Scandolo, G. Sclauzero, A. P. Seitsonen, A. Smogunov, P. Uman, and R. M. Wentzcovich, *J. Phys.: Condens. Matter* **21**, 395502 (2009).

Giustin, F. *Materials Modelling Using Density Functional Theory*, Oxford University Press (2014).

Grandjean, F., A. Gérard, D. J. Braun, and W. Jeitschko, *J. Phys. Chem. Solids* **45**, 877 (1984).

Grytsiv A., X.-Q. Chen, N. Melnychenko-Koblyuk, P. Rogl, E. Bauer, G. Hilscher, H. Kaldarar, H. Michor, E. Royanian, R. Podloucky, M. Rotter, and G. Giester, *J. Phys. Soc. Jpn.* **77**, 121 (2008).

Gumeniuk, R., W. Schnelle, H. Rosner, M. Nicklas, A. Leithe-Jasper, and Y. Grin, *Phys. Rev. Lett.* **100**, 017002 (2008).

Gumeniuk, R., M. Schöneich, A. Leithe-Jasper, W. Schnelle, M. Nicklas, H. Rosner, A. Ormeci, U. Burkhardt, M. Schmidt, U. Schwarz, M. Ruck, and Y. Grin, *New J. Phys.* **12**, 103035 (2010).

Hammerschmidt, L., S. Schlecht, and B. Paulus, *Phys. Stat. Solidi A* **210**, 131 (2013).

Hanus, R., X. Y. Guo, Y. L. Tang, G. D. Li, G. J. Snyder, and W. G. Zeir, *Chem. Mater.* **29**, 1156 (2017).

Harima, H., *J. Mag. Magn. Mater.* **177–181**, 321 (1998).

Harima, H., *Progr. Theor. Phys. Suppl.* No. 138, 117 (2000).

Harima, H., *J. Phys. Soc. Jpn.*, **77**, 114 (2008).

Harima, H., K. Takegahara, K. Ueda, and S. H. Curnoe, *Acta Phys. Pol. B* **34**, 1189 (2002).

Harima, H. and K. Takegahara, *Physica B* **312–313**, 843 (2002a).

Harima, H. and K. Takegahara, *Physica C* **388–389**, 555 (2002b).

Harima, H. and K. Takegahara, *Physica B* **328**, 26 (2003a).

Harima, H. and K. Takegahara, *J. Phys.: Condens. Matter* **15**, S2081 (2003b).

Harima, H. and K. Takegahara, *Physica B* **359–361**, 920 (2005).

Hedin, L. and B. I. Lundqvist, *J. Phys. C* **4**, 2064 (1971).

Hoffmann, R., *J. Chem. Phys.* **39**, 1397 (1963).

Hohenberg, P. and W. Kohn, Phys. Rev. B **136**, 864 (1964).

Hu, C. Z., X. Y. Zeng, Y. F. Liu, M. H. Zhou, H. J. Zhao, T. M. Tritt, J. He, J. Jakowski, P. R. C. Kent, J. S. Huang, and B. G. Sumpter, *Phys. Rev. B* **95**, 165204 (2017).

Humer, S., E. Royanian, H. Michor, E. Bauer, A. Grystiv, M. X. Chen, R. Podloucky, and P. Rogl, in *New Materials for Thermoelectric Applications: Theory and Experiment*, NATO Science for Peace and Security Series B: Physics and Biophysics, ed. V. Zlatic and A. Hewson, Springer Science + Business Media, Dordrecht, Ch. 9, p. 115 (2013).

Isaacs, E. B. and C. Wolverton, *Chem. Mater.* **31**, 6154 (2019).

Ishii, H., K. Okazaki, A. Fujimori, Y. Nagamoto, T. Koyanagi, and J. O. Sofo, *J. Phys. Soc. Japan* **71**, 2271 (2002).

Jung, D., M.-H. Whangbo, and S. Alvarez, *Inorg. Chem.* **29**, 2252 (1990).

Kane, C. and E. Z. Mele, *Phys. Rev. Lett.* **95**, 146802 (2005).

Kane, E. O., in *Semiconductors and Semimetals*, ed. R. K. Willardson and A. C. Beer, Academic Press, NY, Vol. 1. Chap. 3 (1975).

Khan, B., H. A. R. Aliabad, Saifulla, S. Jalali-Asadabadi, I. Khan, and I. Ahmad, *J. Alloys Compd.* **647**, 364 (2015).

Kliche, G. and W. Bauhofer, *Mater. Res. Bull.* **22**, 551 (1987).

Kliche, G. and W. Bauhofer, *J. Phys. Chem. Solids* **49**, 267 (1988).

Koepernik, K. and H. Eschrig, *Phys. Rev. B* **59**, 1743 (1999).

Koga, K., K. Akai, K. Oshiro, and M. Matsuura, *Proc. 20th Int. Conf. on Thermoelectrics*, IEEE Catalog Number 01TH8589, Piscataway, NJ, p. 105 (2001).

Koga, K., K. Akai, K. Oshiro, and M. Matsuura, *Phys. Rev. B* **71**, 155119 (2005).

Kohn, W. and L. D. Sham, *Phys. Rev. B* **140**, A1133 (1965).

Kolezynski, A. and W. Szczypka, *J. Alloys Compd.* **691**, 299 (2017).

Koller, D., F. Tran, and P. Blaha, *Phys. Rev. B* 85, 155109 (2012).

Kurmaev, E. Z., A. Moewes, I. R. Shtein, L. D. Finkelstein, A. L. Ivanovski, and H. Anno, *J. Phys.: Condens. Matter* **16**, 979 (2004).

Lee, C., W. Yang, and R. G. Parr, *Phys. Rev. B* **37**, 785 (1988).

Lefebvre-Devos, I., M. Lassalle, X. Wallart, J. Olivier-Fourcade, L. Monconduit, and J. Jumas, *Phys. Rev. B* **63**, 125110 (2001).

Li, J. L., B. Duan, H. J. Yang, H. T. Wang, G. D. Li, J. Yang, G. Chen, and P. C. Zhai, *J. Mater. Chem. C* **7**, 8079 (2019).

Li, S., X. Jia, and H. Ma, *Chem. Phys. Lett.* **549**, 22 (2012).

Li, X. D., B. Xu, L. Zhang, F. F. Duan, X. L. Yan, J. Q. Yang, and Y. J. Tian, *J. Alloys Compd.* **615**, 177 (2014).

Llunell, M., P. Alemany, S. Alvarez, and V. P. Zhukov, *Phys. Rev. B* **53**, 10605 (1996).

Løvvik, O.M. and Ø. Prytz, *Phys. Rev. B* **70**, 195119 (2004).

Lu, P.-X., Q.-H. Ma, Y. Li, and X. Hu, *J. Mag. Mag. Mater.* **322**, 3080 (2010).

Luo, H. X., J. W. Krizan, L. Muechler, N. Haldolaarachchige, T. Klimczuk, W. W. Xie, M. K. Fuccillo, C. Felser, and R. J. Cava, *Nat. Commun.* **6**, 6489 (2015).

Mandrus, D., A. Migliori, T. W. Darling, M. F. Hundley, E. J. Peterson, and J. D. Thompson, *Phys. Rev. B* **52**, 4926 (1995).

Matsui, K., J. Hayashi, K. Akahira, K. Ito, K. Takeda, and C. Sekine, *J. Phys.: Conf. Series* **215**, 012005 (2010).

Meisner, G. P., *Physica B* **108**, 763 (1981).

Moore, J. E. and L. Balents, *Phys. Rev. B* **75**, 121306(R) (2007).

Morelli, D. T., T. Caillat, J.-P. Fleurial, A. Borshchevsky, J. Vandersande, B. Chen, and C. Uher, *Phys. Rev. B* **51**, 9622 (1995).

Müchler, L., F. Casper, B. Yan, S. Chodov, and C. Felser, *Phys. Stat. Solidi RRL* **7**, 91 (2013).

Murnaghan, F. D., Proc. Natl. Acad. Sci. USA **30**, 244 (1944).

Nagao, J., M. Ferhat, H. Anno, K. Matsubaru, E. Hatta, and K. Mukasa, *Appl. Phys. Lett.* **76**, 3436 (2000).

Nieroda, P., K. Kutorasinski, J. Tobola, and K. T. Wojciechowski, *J. Electron. Mater.* **43**, 1681 (2014).

Nordström, L. and D. J. Singh, *Phys. Rev. B* **53**, 1103 (1996).

Pardo, V., J. C. Smith, and W. E. Pickett, *Phys. Rev. B* **85**, 214531 (2012).

Park, C.-H. and Y.-S. Kim, *Phys. Rev. B* **81**, 085206 (2010).

Perdew, J. P., J. A. Chevary, S. H. Vasko, K. A. Jackson, M. R. Pederson, D. J. Singh, and C. Fiolhais, *Phys. Rev. B* **46**, 6671 (1992).

Perdew, J. P., K. Burke, and M. Ernzerhof, *Phys. Rev. Lett.* **77**, 3865 (1996).

Perdew, J. P., A. Ruzsinszky, G. I. Csonka, O. A. Vydrov, G. E. Scuseria, L. A. Constantin, X. Zhou, and K. Burke, *Phys. Rev. Lett.* **100**, 136406 (2008).

Qi, Y. P., H. C. Lei, J. G. Guo, W. J. Shi, B. H. Yan, C. Felser, and H. Hosono, *J. Am. Chem. Soc.* **139**, 8106 (2017).

Rakoto, H., M. Respaud, J. M. Broto, E. Arushanov, and T. Caillat, *Physica B* **269**, 13 (1999).

Ram, S., V. Kanchana, and M. C. Valsakumar, *J. Appl. Phys.* **115**, 093903 (2014).

Rosner, H., J. Gegner, D. Regesch, W. Schnelle, R. Gumeniuk, A. Leithe-Jasper, H. Fujiwara, T. Haupricht, T. C. Koethe, H.-H. Hsieh, H.-J. Lin, C. T. Chen, A. Ormeci, Y. Grin, and L. H. Tjeng, *Phys. Rev. B* **80**, 075114 (2009).

Roy, R., *Phys. Rev. B* **79**, 195322 (2009).

Rundqvist, S. and E. Larsson, *Acta Chem. Scand.* **13**, 551 (1959).

Saha, S. R., H. Sugawara, R. Sakai, Y. Aoki, H. Sato, Y. Inada, H. Shishido, R. Settai, Y. Onuki, and H. Harima, *Physica B* **328**, 68 (2002).

Shankar, A., D. P. Rai, Sandeep, M. P. Ghimire, and R. K. Thapa, *Indian J. Phys.* **91**, 17 (2017).

Sharath Chandra, L. S., M. K. Chattopadhyay, S. B. Roy, and S. K. Pandey, *Phil. Mag.* **96**, 2161 (2016).

Sharma, S. and S. K. Pandey, *Comp. Mater. Sci.* **85**, 340 (2014).

Sharp, J. W., E. C. Jones, R. K. Williams, P. M. Martin, and B. C. Sales, *J. Appl. Phys.* **78**, 1013 (1995).

Shenoy, G. K., D. R. Nokes, and G. P. Meisner, *J. Appl. Phys.* **53**, 2628 (1982).

Singh, D. J., *Planewaves, Pseudopotentials and the LAPW Method*, Kluwer Academic, Boston, 1994.

Singh, D. J. and W. E. Pickett, *Phys. Rev. B* **50**, 11235 (1994).

Singh, D. J. and I. I. Mazin, *Phys. Rev. B* **56**, R1650 (1999).

Singh, D. J., *Mat. Res. Soc. Symp. Proc.*, Vol. 691, 15 (2002).

Slack, G. A. and V. G. Tsoukala, *J. Appl. Phys.* **76**, 1665 (1994).

Smith, J. C., S. Banerjee, V. Pardo, and W. E. Pickett, *Phys. Rev. Lett.* **106**, 056401 (2011).

Sofo, J. O. and G. D. Mahan, Mater. Res. Symp. Proc., Vol. **545**, p. 315 (1999).

Stoll, D. and J. A. Steckel, *Density Functional Theory: A Practical Introduction*, John Wiley & Sons, Inc. (2009).

Sugawara, H., S. Osaki, S. R. Saha, Y. Aoki, H. Sato, Y. Inada, H. Shishido, R. Settai, Y. Onuki, H. Harima, and K. Oikawa, *Phys. Rev. B* **66**, 220504 (2002).

Sun, J., A. Ruzsinszky, and J. P. Perdew, *Phys. Rev. Lett.* **115**, 036402 (2015).

Sun, J., R. C. Remsing, Y. Zhang, Z. Sun, A. Ruzsunszky, H. Peng, Z. yang, A. Paul, U. Waghnare, X. Wu, M. L. Klein, and J. P. Perdew, *Nat. Chem.* **8**, 831 (2016).

Takegahara, K. and H. Harima, *J. Phys. Soc. Jpn.* **71**, 240 (2002).

Takegahara, K. and H. Harima, *Physica B* **328**, 74 (2003a).

Takegahara, K. and H. Harima, *Physica B* **329–333**, 464 (2003b).

Takegahara, K. and H. Harima, *J. Phys. Soc. Jpn.* **77**, 193 (2008).

Tang, Y. L., Y. T. Qiu, L. Xi, X. Shi, W. Q. Zhang, L. D. Chen, S.-M. Tseng, S.-W. Che, and G. J. Snyder, *Energy Environ. Sci.* **7**, 812 (2014).

Tang, Y. L., Z. M. Gibbs, L. A. Agapito, G. Li, H.-S. Kim, M. B. Nardelli, S. Curtarolo, and G. J. Snyder, *Nat. Mater.* **14**, 1223 (2015).

Tran, F. and P. Blaha, *Phys. Rev. Lett.* **102**, 226401 (2009).

Tran, V. H., D. Kaczorowski, W. Miller, and A. Jezirski, *Phys. Rev. B* **79**, 054520 (2009a).

Tran, V. H., B. Nowak, A. Jezirski, and D. Kaczorowski, *Phys. Rev. B* **79**, 144510 (2009b).

Tu, Z. K., X. Sun, X. Li, R. X. Li, L. L. Xi, and J. Yang, *AIP Adv.* **9**, 045325 (2019).

Tütüncü, H. M., E. Karaca, and G. P. Srivastava, *Phys. Rev. B* **95**, 214514 (2017).

van Vleck, J. H., *Phys. Rev.* **41**, 208 (1932).

Volja, D., B. Kozinsky, A. Li, D. Wee, N. Marzari, and M. Fornari, *Phys. Rev. B* **85**, 245211 (2012).

Wee, D., B. Kozinsky, N. Marzari, and M. Fornari, *Phys. Rev. B* **81**, 045204 (2010).

Wei, W., Z. Y. Wang, L. L. Wang, H. J. Liu, R. Xiong, J. Shi, and X. F. Tang, *J. Phys. D: Appl. Phys.* **42**, 115403 (2009).

Whangbo, M.-H. and R. Hoffmann, *J. Am. Chem. Soc.* **100**, 6093 (1978).

Wojciechowski, K. T., J. Tobola, and J. Leszczynski, *J. Alloys Compd.* **361**, 19 (2003).

Xing, G. Z., X. F. Fan, W. T. Zheng, Y. M. Ma, H. L. Shi, and D. J. Singh, *Sci. Rep.* **5**, 10782 (2015).

Yan, B., L. Müchler, X.-L. Qi, S.-C. Zhang, and C. Felser, *Phys. Rev. B* **85**, 165125 (2012).

Yang, J., in *Materials Aspects of Thermoelectricity*, ed. C. Uher, Ch. 21, CRC Press, Taylor & Francis, Boca Raton, FL (2017).

Yang, X. X., Z. H. Dai, Y. C. Zhao, W. C. Niu, J. Y. Liu, and S. Meng, *Phys. Chem. Chem. Phys.* **21**, 851 (2019).

Yin, Y., Y. Huang, Y. Wu, G. Chen, W.-J. Yin, S.-H. Wei, and X. Gong, *Chem. Mater.* **29**, 9429 (2017).

Zevalkink, A., K. Star, U. Aydemir, G. J. Snyder, J.-P. Fleurial, S. Bux, T. Vo, and P. von Allmen, *J. Appl. Phys.* **118**, 035107 (2015).

Zhang, H., C.-X. Liu, X.-L. Qi, X. Dai, Z. Fang, and S.-C. Zhang, *Nat. Phys.* **5**, 438 (2009).

Zhukov, V. P., *Phys. Stat. Solidi* **38**, 90 (1996).

4 Electronic Transport Properties of Skutterudites

4.1 INTRODUCTION

Transport phenomena in solids primarily include such processes as the flow of charge and flow of heat. Typically, the interest is in steady-state flow, i.e., the flow established as a combined effect of external driving forces (electric fields and thermal gradients) and internal scattering processes that tend to restore the system to equilibrium. A steady state therefore should be distinguished from the equilibrium state, in which external driving forces are absent. Transport theory is thus a branch of nonequilibrium statistical mechanics. However, the deviations from the truly equilibrium state are usually rather small since we assume that the driving forces are not too large. The questions then are how the driving forces and the scattering processes are interrelated, how the population of the species (charge carriers and phonons) evolves as a function of time, and how the steady state manifests in terms of measurable transport parameters.

The first attempt to systematically describe transport properties of solids (metals in particular) is due to Drude (1900), who envisioned electrons (particles discovered by J. J. Thomson 1897, just three years earlier), as the free point charge entities. Endowed with the classical Maxwell-Boltzmann velocity distribution, the electrons constitute the electric current when subjected to an external electric field. In spite of the obvious shortcomings associated with the classical description of the problem, the intuitive form of expressions derived by Drude has survived the test of time and was confirmed some 30 years later by Sommerfeld (1928) in his treatment of the conduction process. Sommerfeld recognized that electrons are fermions, i.e., particles with the spin one-half, the population of which is described by the Fermi-Dirac distribution function. The distribution stems from the fact that fermions must obey the Pauli exclusion principle, which states that no more than two electrons (with opposite spins) can occupy the same quantum level. The Sommerfeld theory was a major step forward and avoided some gross shortcomings of the classical treatment (the most glaring being the value of the electronic specific heat). But, just as the theory of Drude, it considered electrons as free particles limited only by occasional interactions with static ions that maintained the overall charge neutrality of the metal. What was not yet recognized was the influence of the periodic arrangement of ions in a crystalline lattice. Since ions sit at regular periodic lattice positions, they generate a periodic potential. The existence of such a periodic potential in which electrons move dramatically alters the energy states of the system. The crucial role of lattice periodicity was recognized by Bloch (1928), who showed that it imposes a modulated plane wave form (Bloch wave) as a solution of the one-electron Schrodinger equation, with the modulating function having the same periodicity as the crystalline lattice. Moreover, picturing the lattice potential as a periodic array of potential wells, Kronig and Penney (1930) demonstrated that the energy spectrum of an electron moving in such potential breaks into allowed and forbidden energy bands. These developments provided the first consistent explanation why among the solids we have metals on one hand and insulators on the other. Since no more than two electrons can occupy the same state (one with spin up and one with spin down), the electrons fill a range of energy states. The highest occupied state is referred to as the Fermi level. If the Fermi level falls within the allowed energy band, the structure is a metal. If, however, the electrons fill the entirety of the allowed states in the energy band, and there are no electrons in the next higher energy band, such a configuration leads to an insulator, since there are no states available into which an electric field could move an electron. Thus, fully filled and completely empty bands result in zero electrical conductivity. Formally, the effect of periodicity and the ensuing band formation is taken into account by replacing the mass of an electron, m_e, in the Sommerfeld treatment with the so-called effective electron mass, m^*.

What about semiconductors? The short answer is that semiconductors are structures with a rather modest size of the energy gap. We distinguish two kinds of semiconductors:

An *intrinsic semiconductor* is typically a very pure structure with a rather small band gap, where electrons fully occupy the last filled band (the valence band), and the next higher unoccupied band (the conduction band) is entirely empty, i.e., there is no conduction at T = 0. As the temperature increases, the energy of electrons gained from the lattice may become comparable or even exceed the band gap, and electrons start to populate the conduction band, leaving behind in the valence band an equal number of missing electrons, i.e., effectively positive entities called holes. In all respects (except that they have an opposite charge and likely different effective mass), the holes respond to the applied electric field as electrons, and both contribute to the transport process.

An *extrinsic semiconductor* usually has a larger band gap and contains a number of impurities that can either give off electrons to the conduction band (donors) or accept electrons from the valence band (acceptors). The impurity levels are located in the band gap and can be either fairly close to the respective conduction or valence band edges (shallow impurities) or near the middle of the band gap (deep impurities). The conduction process is determined by the density of the donors (or acceptors), their separation from the conduction (valence) band edge, and the temperature.

A schematic description of a metal, semimetal, intrinsic semiconductor, and extrinsic n-type semiconductor is depicted in Figure 4.1. A semimetal can be viewed as a semiconductor with a negative band gap, i.e., overlapping conduction and valence bans. The semiconductor shown in Figure 4.1c is assumed to be the direct band gap semiconductor, where an electron transitions in a vertical fashion at constant momentum. In an indirect semiconductor, the conduction band edge does not lie directly above the valence band edge but is offset, and the transition requires assistance of a phonon to satisfy the crystal momentum. The energy of a charge carrier is usually measured from the respective band edge taken positive going into the band.

In the formal assessment of the effectiveness of a thermoelectric material via the dimensionless thermoelectric figure of merit, $ZT = S^2\sigma T/\kappa$, there are three key transport parameters that must be specified: the electrical conductivity, σ, the thermal conductivity, κ, consisting of the electronic thermal conductivity, κ_e, and the lattice contribution, κ_L, and the Seebeck coefficient, S. Since two distinct species (charge carriers and lattice phonons) compose the thermal conductivity, it is important to know their respective contributions. The electronic portion is usually assessed by invoking

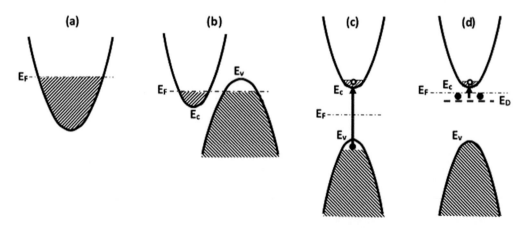

FIGURE 4.1 Schematic representation of a (a) metal, (b) semimetal, (c) intrinsic semiconductor, and (d) extrinsic n-type semiconductor. The Fermi energy and the respective band edges are indicated.

the Wiedemann-Franz relation that ties the electronic thermal conductivity to the electrical conductivity *via* the Lorenz number L, $\kappa_e = \sigma L T$. The lattice thermal conductivity then follows from $\kappa_L = \kappa - \kappa_e$. In addition, it is essential to know the carrier concentration in a material, which is conveniently provided by measurements of the Hall effect. Although a magnetic field might, in certain cases, enhance the thermoelectric performance, essentially all measurements of thermoelectric properties and all TE device operations are carried out in zero magnetic field. I will therefore not consider here the influence of the magnetic field on the transport parameters except, when considering the Hall effect.

Transport processes can be tackled by different approaches. Perhaps the most straightforward and versatile treatment is based on the Boltzmann Transport Equation (BTE), which monitors how the carrier distribution (Fermi-Dirac function) evolves as a result of the action of an external electric field and thermal gradient, and the scattering processes that try to re-establish thermal equilibrium. Beyond its relative simplicity, the BTE yields expressions for transport parameters that are readily compared with the experimental results. The actual derivation of the Boltzmann transport equation, together with all relevant expressions for the transport parameters under arbitrary carrier degeneracy, are given in online appendix. The two most important transport parameters that describe the nature of charge carrier transport in any material are the electrical conductivity and the Seebeck coefficient. Therefore, most of the discussion of this chapter will focus on these two parameters. The same two parameters happen to be particularly relevant to thermoelectric phenomena, where a product of the electrical conductivity and the square of the Seebeck coefficient, σS^2, called the power factor, provides the pivotal input regarding the contribution of electrons (and/or holes) to thermoelectric energy conversion.

4.2 CONDUCTION IN A SINGLE PARABOLIC BAND

In a single-band conductor, one assumes that there is a single pocket of charge carriers (either electrons or holes) responsible for transport. The energy-wavevector relation of a parabolic band is of the form

$$E(k) = \frac{\hbar^2 k^2}{2m^*}, \tag{4.1}$$

where m^* is the effective mass of the electron. As already stated, the distribution of electrons is specified by the Fermi-Dirac distribution function,

$$f^0(E,T) = \frac{1}{e^{(E-\xi)/k_B T} + 1}, \tag{4.2}$$

where ξ is the Fermi energy and k_B is the Boltzmann constant. For energies much smaller than the Fermi energy, $E - \xi \ll 0$, the Fermi function attains the value of unity. An opposite inequality brings the Fermi function to zero. The transition from zero to unity takes place over a narrow range of energies on the order of $2k_B T$ around the Fermi level, and becomes sharper as the temperature decreases. In fact, at $T = 0$, it happens at $E = \xi$.

To obtain the total number of electrons participating in the transport process, apart from the probability of an electron occupying the state in Eq. 4.2, we also need to know how densely the electron states are distributed with energy, i.e., how many allowed electron states there are in the energy range between E and $E + dE$. The so-called electron density of states $D(E)$ in a three-dimensional space is given by

$$D(E)dE = \frac{4\pi(2m^*)^{3/2}}{h^3} E^{1/2} dE. \tag{4.3}$$

Equation 4.3 follows immediately from the fact that electrons within the range dE of energy E are electrons within a spherical shell of thickness dk in k-space. The number of such electrons is equal to the volume of the shell, $4\pi k^2 dk$, multiplied by the density of states in k-space, $V / (2\pi)^3$, i.e., $V 4\pi k^2 dk / (2\pi)^3$. Dividing by the volume V, and substituting for k^2 and dk from Eq. 4.1, yields the desired electron density of states.

The total number of electrons n in the system is then

$$n = \int_0^\infty f^0(E,T) D(E) dE = 2 \left(\frac{2\pi m^* k_B T}{h^2} \right)^{3/2} \frac{2}{\sqrt{\pi}} F_{1/2}(\eta). \tag{4.4}$$

Here, $F_{1/2}(\eta)$ is one of the Fermi integrals defined as

$$F_n(\eta) = \int_0^\infty \frac{\varepsilon^n}{e^{\varepsilon - \eta} + 1} d\varepsilon, \tag{4.5}$$

where η is the reduced Fermi energy, $\eta = \xi / k_B T$.

I note that because a hole in the valence band signifies a missing electron, the Fermi-Dirac distribution function for holes is simply equal to

$$1 - f^0(E,T) = \frac{1}{e^{(\xi - E)/k_B T} + 1}, \tag{4.6}$$

and the total number of holes is given by

$$p = \int_0^\infty \left(1 - f^0(E,T)\right) D(E) dE. \tag{4.7}$$

In general, we distinguish between degenerate and non-degenerate carrier systems. The degeneracy here implies that a semiconductor is heavily doped so that its transport behavior mimics that of a metal rather than a semiconductor, e.g., its electrical conductivity decreases rather than increases with the increasing temperature. The carrier degeneracy is closely tied with the value of the exponent $(E - \xi)/k_B T$ in Eq. 4.2 in the case of electrons, or the value of $(\xi - E)/k_B T$ in Eq. 4.6 in the case of holes. When $(E - \xi)/k_B T \gg 1$, or when $(\xi - E)/k_B T \gg 1$ in the case of holes, the Fermi-Dirac distribution function turns into the classical Maxwell-Boltzmann function, and the system is considered nondegenerate. The opposite inequalities indicate degenerate systems of electrons, respectively holes.

The expression in Eq. 4.4 for an arbitrarily degenerate electron system contains the Fermi integral that can be expanded and taking the first term in the case of strong degeneracy, while for the non-degenerate carriers the Fermi integral is approximated by a Γ-function. The resulting respective electron carrier concentrations are

$$n_{sd} = \frac{8}{3\sqrt{\pi}} \left(\frac{2\pi m^* k_B T}{h^2} \right)^{3/2} \eta^{3/2} \text{ for the case of strong degeneracy}, \tag{4.8}$$

$$n_{nd} = 2 \left(\frac{2\pi m^* k_B T}{h^2} \right)^{3/2} e^{\eta} \quad \text{for the non-degenerate case.} \tag{4.9}$$

The appropriate expressions for the electrical conductivity, electronic thermal conductivity, and the Seebeck coefficient are given in online appendix.

4.3 TWO-BAND CONDUCTION, BIPOLAR THERMAL CONDUCTIVITY

In Section 4.2, we assumed that the conduction proceeds via a single parabolic band of charge carriers. However, semimetals, such as Bi, by definition possess equal numbers of electrons and holes. Furthermore, semiconductors at elevated temperatures feature majority and minority charge carriers due to the increased thermal energy exciting carriers across the band gap. The conduction process therefore requires consideration of more than a single carrier and leads to the concept of two-band conduction, the most frequently encountered multicarrier situation. I note that the two-band conduction model also encompasses a situation where the two carriers have the same sign but different effective masses, but I consider here specifically a system of electrons and holes.

To develop the relevant relations, one starts with carriers labeled 1 (electrons) and 2 (holes) that each give rise to the current density in the presence of an electric field and temperature gradient. For simplicity, the electric field E_x and the temperature gradient dT/dx are assumed along the x-axis only, yielding

$$J_{1,x} = \sigma_1 \left(E_x - S_1 \frac{dT}{dx} \right) \quad \text{and} \quad J_{2,x} = \sigma_2 \left(E_x - S_2 \frac{dT}{dx} \right). \tag{4.10}$$

Assuming zero temperature gradient, the total current density is $J_x = J_{1,x} + J_{2,x} = (\sigma_1 + \sigma_2) E_x$, and thus the electrical conductivity is

$$\sigma = \sigma_1 + \sigma_2. \tag{4.11}$$

Setting the overall current density to zero, which implies that the current densities of the two types of carriers are equal and opposite, leads to $(\sigma_1 + \sigma_2) E_x = (\sigma_1 S_1 + \sigma_2 S_2) \frac{dT}{dx}$, from which the Seebeck coefficient follows as

$$S = \frac{E_x}{dT / dx} = \frac{\sigma_1 S_1 + \sigma_2 S_2}{\sigma_1 + \sigma_2}. \tag{4.12}$$

The corresponding heat current densities due to the two charge carriers are

$$Q_{1,x} = S_1 T J_{1,x} - \kappa_{1,q} \frac{dT}{dx}, \quad \text{and} \quad Q_{2,x} = S_2 T J_{2,x} - \kappa_{2,q} \frac{dT}{dx}, \tag{4.13}$$

where the respective Peltier coefficients have been replaced by the terms $S_1 T$ and $S_2 T$ with the aid of the Kelvin relation. As before, the thermal conductivity is determined in the absence of the electric current density, i.e., $J_{1,x} = -J_{2,x}$. Using this in Eq. 4.10, we obtain, after simple algebra,

$$J_{1,x} = -J_{2,x} = \frac{\sigma_1 \sigma_2}{\sigma_1 + \sigma_2} (S_2 - S_1) \frac{dT}{dx}. \tag{4.14}$$

Substituting for the respective current densities in Eq. 4.13 yields

$$Q_x = Q_{1,x} + Q_{2,x} = - \left[\kappa_{1,q} + \kappa_{2,q} + \frac{\sigma_1 \sigma_2}{\sigma_1 + \sigma_2} (S_2 - S_1)^2 T \right] \frac{dT}{dx}, \tag{4.15}$$

from which the thermal conductivity associated with the charge carriers follows as

$$\kappa_q = -\frac{Q_x}{dT/dx} = \kappa_{1,q} + \kappa_{2,q} + \frac{\sigma_1\sigma_2}{\sigma_1+\sigma_2}\left(S_2 - S_1\right)^2 T. \qquad (4.16)$$

Equation 4.16 reveals that the overall electronic thermal conductivity of a system consisting of two types of carriers, is not just an addition of the heat conductivity contributions of the two carrier species, but it is augmented by the term $\frac{\sigma_1\sigma_2}{\sigma_1+\sigma_2}\left(S_2 - S_1\right)^2 T$, which can be quite large when the carriers have an opposite sign and increases with the temperature. The physical origin of this term stems from the fact that in a multicarrier system the Peltier heat can flow even in the absence of the overall current density. The thermal conductivity contribution is referred to as the *bipolar thermal conductivity* (Price 1955, 1956). The reader should note that if a material contains both electrons and holes, it is a double jeopardy from the perspective of thermoelectricity: the competing contributions of electrons and holes will make not only the Seebeck coefficient smaller, see Eq. 4.12, but a large bipolar thermal conductivity may also far exceed the contribution of each individual species, dramatically degrading the thermoelectric performance.

4.4   THE ROLE OF EFFECTIVE MASS

In the expressions describing various transport parameters, the effective mass of charge carriers plays a prominent role. However, this effective mass has a somewhat different meaning depending on where it is used. Specifically, it is important to distinguish between the transport (inertial) mass, m_t^*, that enters in the electrical conductivity, and the effective mass that enters from considerations of the distribution of energy states, i.e., the density-of-states effective mass. Moreover, in multi-valley semiconductors, one must distinguish between the effective mass of a single valley (carrier pocket), m_b^*, and the effective mass representing all energy surfaces, the so-called density-of-states effective mass, m_{DOS}^*.

The motion of a free electron is governed by the dispersion relation $E = \frac{\hbar^2 k^2}{2m_e}$, where E is the energy, k is the wave vector, and m_e is the free electron mass. A question is, what happens when an electron moves in a periodic potential of the crystal lattice? In this case, the electron is subjected to various forces $= m_t a = \frac{\partial p}{\partial t} = m_t \frac{\partial v_g}{\partial t} = \hbar \frac{\partial k}{\partial t}$, where $v_g = \frac{1}{\hbar}\frac{\partial E}{\partial k}$ is its group velocity. Differentiating the group velocity with time, $\frac{\partial v_g}{\partial t} = \frac{1}{\hbar}\frac{\partial^2 E}{\partial t \partial k} = \frac{1}{\hbar}\frac{\partial^2 E}{\partial k^2}\frac{\partial k}{\partial t}$, isolating $\frac{\partial k}{\partial t}$ from the above equation, and substituting it back into the Newton law equation, one notes that the electron (or hole) moves freely, but with an effective mass equal to

$$m_t^* = \hbar^2 \left(\frac{\partial^2 E}{\partial k^2}\right)^{-1}. \qquad (4.17)$$

The above equation probes the curvature of the energy band, and the effective mass is referred to as the transport (inertial) effective mass. As written above, it is assumed that the energy surface is isotropic (spherically symmetric) and is located at the center of the Brillouin zone (Γ-point) so that only a single value of the effective mass arises. The effective mass m_t^* is then independent of energy and the carrier concentration. Spherically symmetric energy surfaces are rarely encountered in real semiconductors, and a more realistic is an anisotropic energy surface of ellipsoidal shape. Then, along the principal axes of such an ellipsoid, the masses are m_1^*, m_2^*, and m_3^*, and the transport mass becomes

$$\frac{1}{m_t^*} = \frac{1}{3}\left(\frac{1}{m_1^*} + \frac{1}{m_2^*} + \frac{1}{m_3^*}\right). \tag{4.18}$$

Clearly, an electron (or hole) with a heavy mass has a difficulty to respond rapidly to changes in the applied electric field, and the electrical conductivity is thus inversely proportional to the transport effective mass.

The other mass found in the expressions of some transport parameters, e.g., the Seebeck coefficient, is the density-of-states effective mass that enters via the density of the charge carriers. For a strictly isotropic energy surface of a parabolic band, the band mass, m_b^*, is identical to the transport effective mass m_t^*. However, the two masses are not the same for an anisotropic energy surface. For a single valley, the density-of-states effective mass m_b^* is a geometrical average of the masses along the principal axes of an ellipsoid,

$$m_b^* = \left(m_1^* m_2^* m_3^*\right)^{1/3}. \tag{4.19}$$

If the energy surface is not located in the Brillouin zone center, the symmetry of the system leads to N_v symmetry-related degenerate valleys, each with the same mass m_b^*. In this case, the total density-of-states effective mass, m_{DOS}^*, is given by

$$m_{DOS}^* = N_v^{2/3} m_b^*. \tag{4.20}$$

The masses m_t^* and m_{DOS}^* can be quite different, and this benefits thermoelectricity. The density-of-states effective mass enhanced by the number of valleys N_v obviously increases the Seebeck coefficient without any penalty to the electrical conductivity that depends on m_t^* and not on m_{DOS}^*. Even a single highly anisotropic energy surface is advantageous from the perspective of thermoelectricity, as is easily seen by considering the often modeled case of an ellipsoid of revolution. Taking its longitudinal effective mass as $m_{\parallel}^*$ and its transverse mass as $m_{\perp}^*$, the transport and density-of-states effective masses become $m_t^* = 3\left(\frac{1}{m_{\parallel}^*} + \frac{2}{m_{\perp}^*}\right)^{-1}$ and $m_b^* = \left(m_{\parallel}^* m_{\perp}^{*2}\right)^{1/3}$. Forming the ratio m_b^*/m_t^*, we obtain

$$\frac{m_b^*}{m_t^*} = \frac{m_{\perp}^* + 2m_{\parallel}^*}{3m_{\parallel}^{*2/3} m_{\perp}^{*1/3}} = \frac{a^{2/3}}{3} + \frac{2a^{-1/3}}{3}, \tag{4.21}$$

where in the last term I used $a = \frac{m_{\perp}^*}{m_{\parallel}^*}$. A plot of $\frac{m_b^*}{m_t^*}$ as a function of a is shown in Figure 4.2.

It is obvious that a highly prolate or oblate ellipsoid leads to a large ratio m_b^*/m_t^*, and therefore to a large Seebeck coefficient without affecting the electrical conductivity. Of course, with multivalley energy surfaces, the benefit would be further enhanced by the factor $N_v^{2/3}$.

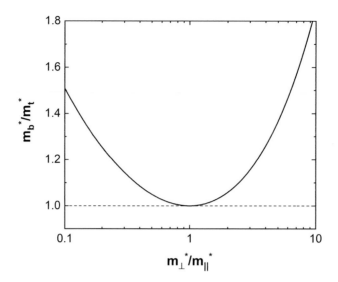

FIGURE 4.2   Ratio of the density-of-states mass, m_b^*, to the transport (inertia) mass, m_i^*, as a function of the ratio of the transverse and longitudinal masses, $m_\perp^*/m_\parallel^*$, of an ellipsoid of revolution. The figure highlights the advantage of having an anisotropic energy surface, either highly prolate or highly oblate.

4.5   NONPARABOLIC BANDS

So far, we have assumed that the energy of a charge carrier is a quadratic function of the wavevector, i.e., that the band has the shape of a parabola. While this assumption holds very well close to the band edge, in small band-gap semiconductors away from the band edge, the band can attain a distinctly nonparabolic shape. If so, the effective mass is no longer constant and depends on the energy. This complication is often dealt with by invoking a model due to Kane (1957) or its modified forms developed by Ravich et al. (1970). In the latter case, the energy dispersion of charge carriers is written as

$$E\left(1+\frac{E}{E_g}\right) = \frac{\hbar^2 k_T^2}{2m_{T0}^*} + \frac{\hbar^2 k_L^2}{2m_{L0}^*},$$
(4.22)

where k_T and k_L are the transverse and longitudinal components of the wavevector, respectively, m_{T0}^* and m_{L0}^* are the transverse and longitudinal components of the effective mass tensor near the band extremum, and E_g is the band gap. The actual energy dependence of the effective mass tensor becomes

$$m_i^* = m_{i0}^*\left(1+2\frac{E}{E_g}\right).$$
(4.23)

Appropriate modifications of the density of states, carrier concentrations, and transport parameters applicable for nonparabolic bands have been worked out by Kolodziejczak (1961), Ravich et al. (1970), and Zawadzki (1974).

4.6   RELAXATION TIME/SCATTERING MECHANISMS OF CHARGE CARRIERS

As charge carriers are subjected to an external electric field or thermal gradient, electrons (holes) acquire momentum and energy during the process of acceleration. The momentum and energy gained are subsequently relaxed by the charge carriers interacting with phonons and crystal defects. How effective such scattering centers are in returning the perturbed distribution towards equilibrium depends on the nature of the scatterer, the energy of the carrier, and on the temperature. I will consider here the most relevant scattering processes encountered in semiconductors and, in particular, those pertaining to thermoelectric materials. While I will not provide detailed calculations of the respective relaxation times, I will outline the steps to obtain the desired formulas, and what physical reasoning lies underneath, as well as present the final results.

The scattering processes can be categorized into two main areas: processes where charge carriers interact with lattice phonons, and processes where the carriers are scattered by lattice defects. Since the charge carriers in semiconductors often reside in more than one carrier pocket (energy valley), and because the number of symmetrically equivalent pockets N_v is highly beneficial to thermoelectricity, we consider both intravalley and intervalley scattering events. I will discuss the relevant scattering processes in turn.

4.6.1 SCATTERING OF CHARGE CARRIERS BY PHONONS

4.6.1.1 Acoustic Deformation Potential Scattering

As the temperature increases, the lattice atoms oscillate with the increasing amplitude about their equilibrium positions, periodically expanding and compressing the lattice. This gives rise to local density fluctuations that alter the periodic potential and cause small shifts in the energy of the conduction and/or valence band edges, δE_c and δE_v, respectively. Charge carriers are sensitive to such changes, as they impede their drift velocity acquired by the influence of the electric field or temperature gradient. While in a monoatomic solid the vibrations are strictly of acoustic nature, optical vibrations are usually also important, if not the dominant, scattering mechanism in systems with more than a single atom in a unit cell. The formalism describing the scattering of charge carriers by acoustic vibrations was developed by Bardeen and Shockley (1950) by invoking the concept of the *deformation potential scattering*.

Very briefly, as a consequence of local density fluctuations, a small fractional change in the unit cell volume, $\dfrac{\delta V}{V}$, gives rise to a proportionally small energy variation in the conduction band edge, δE_c (or the valence band edge, δE_v, in the case of a hole-dominated semiconductor),

$$\delta E_c = E_{ac}\Delta \frac{\delta V}{V}, \tag{4.24}$$

with the proportionality constant $E_{ac}\Delta$ called the acoustic deformation potential, where E_{ac} is the acoustic deformation constant. The local change in volume of the unit cell arises on account of the lattice displacement $\vec{u} = \vec{r}' - \vec{r}$, and mathematically is described as the divergence of this displacement, $\vec{\nabla} \cdot \vec{u}$. The lattice displacement contains the polarization vector of phonons, $\vec{e}_q$, and, applying the divergence, brings into play a factor $\vec{e}_q \cdot \vec{q}$, which is zero for *transverse phonon* modes. Thus, only *longitudinal phonons* contribute to deformation potential scattering. The lattice vibrations are described as phonons governed by the Bose-Einstein distribution

$$N_q = \left(e^{\frac{\hbar \omega_q}{k_B T}} - 1\right)^{-1}, \tag{4.25}$$

where N_q is the number of phonons with wavevector $\vec{q}$ and frequency ω_q at temperature T. Considering the scattering process as a transition of an electron from the state $\vec{k}$ to the state $\vec{k}'$, while absorbing

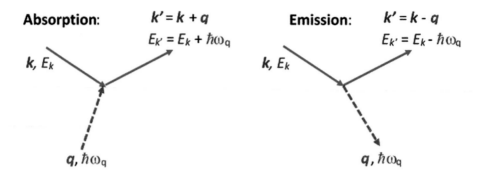

FIGURE 4.3 Scattering of an electron by absorbing or emitting a phonon.

or emitting a phonon of wavevector $\vec{q}$, conservation laws of energy and momentum must be satisfied, as depicted in Figure 4.3.

At high enough temperatures, $k_B T \gg \hbar\omega_q$, an equipartition approximation (mathematically Eq. 4.25 with $e^x \cong 1 + x$) can be used with the number of phonons becoming $N_q \approx \dfrac{k_B T}{\hbar\omega_q}$. The quantum-mechanical probability $S(k, k')$ that an electron in state k is scattered to state k' while absorbing or emitting a longitudinal phonon q is given by the Fermi golden rule,

$$S(k,k') = \frac{2\pi}{\hbar} \left| H_{k,k'} \right|^2 f(k) \left[1 - f(k') \right] \left\{ N_q \delta \left[E(k') - E(k) - \hbar\omega_q \right] \right.$$

$$\left. + \left(N_q + 1 \right) \delta \left[E(k') - E(k) + \hbar\omega_q \right] \right\}, \tag{4.26}$$

where $H_{k,k'}$ stands for the matrix element connecting electron states k and k', and f(k) and N_q are the Fermi-Dirac and Bose-Einstein distribution functions, respectively. The term $f(k)\left[1 - f(k')\right]$ implies that an electron comes from an occupied state k and the empty state k' can accept the electron. The Dirac δ function reflects the conservation of energy in the absorption and emission process, respectively. In the case of long wavelength phonons (small $\hbar\omega_q$), which are mostly considered in acoustic phonon deformation potential scattering, and at temperatures above the ambient where $N_q + 1 \cong N_q$, the two δ in Eq. 4.26 are the same. By integrating Eq. 4.26 over all possible final states k' an electron can go to, one finally arrives at the scattering rate, the inverse of which is the relaxation time for the process. For the acoustic deformation potential scattering, the relaxation time can be expressed as

$$\tau_{m,ac}(E) = \frac{\pi \hbar^4 c_\ell}{\sqrt{2} \, m_t^{*3/2} E_{ac}^2 k_B T} E^{-1/2} \equiv \tau_0 E^{-1/2} \tag{4.27}$$

Equation 4.27 contains the longitudinal elastic constant $c_\ell = \rho v_s^2$, where ρ is the mass density and v_s is the speed of sound. The subscript ℓ indicates that longitudinal phonons are involved. In the case of a cubic crystal, $c_\ell = c_{11}$ for strain in the $\langle 100 \rangle$ direction, $c_\ell = \dfrac{1}{2}(c_{11} + c_{12} + c_{44})$ for strain in the $\langle 110 \rangle$ direction, and $c_\ell = \dfrac{1}{3}(c_{11} + 2c_{12} + 4c_{44})$ for strain in the $\langle 111 \rangle$ direction, where c_{ij} are tensorial components of elastic strain constants. Typical values of the deformation potential are in the range of 5–25 eV.

Obviously, the main effect of a long-wavelength acoustic phonon is to relax electron momentum, hence the subscript m in $\tau_{m,ac}(E)$, while there is little energy transfer, i.e., the scattering process is nearly elastic.

4.6.1.2 Polar Optical Scattering

Compound semiconductors consist of dissimilar atoms that often have unequal valence, leading to a small charge transfer and thus a partly ionic character of bonding. The presence of opposite charges gives rise to an electric dipole, which is perturbed by lattice vibrations. The resulting electric field scatters charge carriers. If the lattice vibrations are associated with acoustic phonons, the scattering is referred to as *piezoelectric scattering*. Such scattering might make a contribution in pure compound semiconductors at low temperatures, but it is often far overwhelmed by impurity scattering. Thus, I will not consider it further.

More relevant is the so-called polar optical scattering, which is associated with the disturbance caused by a passage of an optic vibrational mode of phonons. Optic modes are displacements of dissimilar atoms in opposite directions in a unit cell (unlike acoustic modes where all atoms move in the same direction and cause dilation varying in space and time). Assuming the unit cell contains two atoms with opposite charges, their relative motion leads to non-zero polarization $\vec{P}$, while the center of mass of the unit cell is at rest, Figure 4.4. The electric field associated with the polarization field is the cause of electron scattering. Polar optical scattering is dominant at around room temperature for polar semiconductors, such as the III–V compound GaAs. The derivation of the scattering rate or its inverse, the relaxation time, is rather lengthy and the interested reader is referred to books by Lundstrom (2000) or by Yu and Cardona (2010). The scattering rate is given by

$$\frac{1}{\tau_{po}(E)} = \frac{e^{2}\omega_{LO}\left(\dfrac{\varepsilon_{0}}{\varepsilon_{\infty}}-1\right)}{2\pi\varepsilon_{0}\hbar\left(\dfrac{2E(k)}{m^{*}}\right)^{1/2}}\left[N_{LO}\sinh^{-1}\left(\frac{E(k)}{\hbar\omega_{LO}}\right)^{1/2}+\left(N_{LO}+1\right)\sinh^{-1}\left(\frac{E(k)}{\hbar\omega_{LO}}-1\right)^{1/2}\right]\quad (4.28)$$

with the understanding that the emission process (the second term in square brackets) is non-zero only for electrons with energy larger than the energy of the emitted longitudinal optic phonon. From Eq. 4.28. it follows that the relaxation time $\tau(E)\propto E^{1/2}$, i.e., the scattering parameter (an exponent in the above relation) $r = \frac{1}{2}$, has a positive value unlike the scattering parameter for acoustic deformation potential scattering. In Eq. 4.28, ε_{0} and $\varepsilon\infty$ are static and high frequency permitivities, ω_{LO} is the frequency of the longitudinal optic mode (it may be taken as the value at q = 0), and N_{LO} is the number of longitudinal optic phonons. Equation 4.28 is sometimes presented in an alternative form using the identity $\sinh^{-1}(x)\equiv \ln\left[x+\sqrt{1+x^{2}}\right]$. Unlike acoustic deformation potential scattering, the polar optical phonon scattering is inherently inelastic and anisotropic. Because of the inelastic nature of optic phonon scattering, optic phonons relax both the momentum and energy of electrons. Since ionicity is usually not conducive to good thermoelectric materials, polar optic phonon scattering is not often encountered in the analysis of transport properties of thermoelectrics.

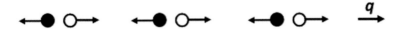

FIGURE 4.4 Longitudinal optic mode in a crystal with a unit cell containing two oppositely charged ions.

Moreover, the importance of the polar optical phonon scattering tends to decrease with increasing carrier concentration and rising temperature.

4.6.1.3   Nonpolar Optical Scattering

In acoustic deformation potential scattering, the change in the conduction or valence band edge was proportional to the divergence of the displacement, $\vec{\nabla} \cdot \vec{r}$. In nonpolar crystals with two different atoms, the motion of the two types of atoms is against each other, and the change in the energy of the band edge is proportional to the displacement $\delta \vec{r}$ with the proportionality constant being the optical deformation potential, E_{npo}. The relaxation time associated with charge carriers being scattered by nonpolar optical phonons is given by Hamaguchi (2010) as

$$\frac{1}{\tau_{npo}} = \frac{1}{2}\left(\frac{E_{npo}}{E_{ac}}\right)^2 \frac{\theta_D}{T}\left(e^{\theta_D/T}-1\right)^{-1}\left[\left(1+\frac{k_B\theta_D}{E}\right)^{1/2} + e^{\theta_D/T}\left(1-\frac{k_B\theta_D}{E}\right)^{1/2}\right]\frac{1}{\tau_{ac}} \qquad (4.29)$$

The second term in the square bracket vanishes when the carrier energy is less than $k_B\theta_D$.

4.6.2   IONIZED IMPURITY SCATTERING

Doping a semiconductor is essential to tune its carrier concentration to the optimal value to maximize the thermoelectric power factor and thus the figure of merit. For a dopant to fulfill its role, it must be ionized. Moreover, the material may also contain other charged defects, such as vacancies and interstitials. The presence of a Coulomb field associated with the ionized impurities and charged defects will clearly affect the passage of electrons and lead to their scattering.

There is a fundamental difference between interactions of charge carriers with phonons and with impurities. While electron-phonon processes are mainly inelastic (beside the noted electron-acoustic deformation potential scattering), electron scattering by charged impurities is considered purely elastic because the mass of an ion is some three orders of magnitude heavier than the mass of an electron, and the impurity ion is fixed in the crystal lattice. As with all scattering processes, the key is to determine the relevant transition probability $S(k, k')$. For parabolic bands, the probability for ionized impurity scattering has the form

$$S(k,k') = \frac{2\pi}{\hbar}\frac{N_{imp}e^2}{\Omega}\left|V(q)\right|^2 \delta\left(E'-E\right) \qquad (4.30)$$

where N_{imp} is the number of randomly located impurities, Ω is the crystal volume, $V(q)$ is the Fourier transform of the impurity potential $V(r)$, and the Dirac δ function indicates the elastic nature of the scattering process. The literature describes two main approaches to impurity scattering: the Conwell-Weisskopf (1950) treatment, and the Brooks-Herring treatment, Brooks (1955). They differ in a way how the divergence in zero angle scattering is treated when a potential of the ionized impurity is taken as the Coulomb potential,

$$V(r) = \frac{Ze}{4\pi\varepsilon_r\varepsilon_0\,r}, \qquad (4.31)$$

where Z is the charge of the impurity, ε_r is the relative permittivity, and ε_0 is the permittivity of free space. In the Conwell-Weisskopf scheme, the problem is solved by cutting off the impact parameter (perpendicular distance between the scattering center and the initial line of approach of an electron) at $d/2$, where $d = N_{imp}^{-1/3}$ is the average distance between impurities. Integrating Eq. 4.30 over all possible states k' the electron can attain, the scattering rate becomes

$$\frac{1}{\tau_{CW}} = \frac{N_{imp}Z^2e^4}{16\sqrt{2}\pi\varepsilon_r\varepsilon_0 m^{*1/2}} E^{-3/2} \ln\left(1 + \frac{16\pi^2\varepsilon_r\varepsilon_0 E^2}{Z^2e^4 N_{imp}^{2/3}}\right). \tag{4.32}$$

Neglecting the logarithmic term (slowly varying function), the Conwell-Weisskopf relaxation time for impurity scattering is proportional to $E^{3/2}$, i.e. the scattering parameter $r = 3/2$.

The Brooks-Herring treatment circumvents the problem of small angle scattering in the Coulomb potential by accounting for the screening of the ionic charge by the surrounding cloud of electrons. The point is that the conduction electrons do not see the Coulomb potential of Eq. 4.31 with its long $1/r$ tail, but a much shorter range screened Coulomb potential of the form

$$V(r) = \frac{Ze}{4\pi\varepsilon_r\varepsilon_0} \frac{e^{-\lambda r}}{r} \tag{4.33}$$

where the reciprocal value of the parameter λ is referred to as the screening length (or screening radius). The potential in Eq. 4.33 is also known as the Yukawa potential. In a simple Thomas-Fermi approximation appropriate for a degenerate electron gas that treats the electrons as free, the screening length becomes

$$\frac{1}{\lambda_{TF}} = \left(\frac{e^2 D(E_F)}{\varepsilon_0}\right)^{-1/2} \cong \frac{a_B^{1/2}}{2n^{1/6}}, \tag{4.34}$$

where a substitution for the density-of-states was made $(D(E_F) = \frac{3}{2}\frac{n}{E_F}$ with the Fermi energy given

by $E_F = \frac{\hbar^2}{2m_e}(3\pi^2 n)^{2/3})$ and the Bohr radius, $a_B = \frac{4\pi\hbar^2}{me^2}\varepsilon_0$, was introduced. In copper, the screen-

ing length is about 0.55 Å. As the carrier density decreases, the screening length becomes larger. The Fourier transform of the screened Coulomb potential is

$$V(q) = \frac{Ze}{\varepsilon_r\varepsilon_0\left(q^2 + \lambda_{TF}^2\right)} \tag{4.35}$$

Substituting this potential in Eq. 4.30, and integrating over all possible final states k', the Brooks-Herring relaxation rate becomes

$$\frac{1}{\tau_{BH}} = \frac{N_{imp}Z^2e^4}{16\pi\sqrt{2}\left(\varepsilon_r\varepsilon_0\right)^2 m^{*1/2}} E^{-3/2}\left[\ln\left(1 + \frac{4k^2}{q_D^2}\right) - \frac{4k^2}{q_D^2 + 4k^2}\right], \tag{4.36}$$

where $q_D^2 = \frac{ne^2}{\varepsilon_r\varepsilon_0 k_B T}$, with q_D being the Debye screening length.

The relaxation rates of Conwell and Weisskopf in Eq. 4.32 and of Brooks and Herring in Eq. 4.36 differ only in the form of the logarithmic term, but both show the relaxation times as proportional to $E^{3/2}$. In subsequent years, numerous corrections were proposed to both the Conwell-Weisskopf and Brooks-Herring formulas, but the main outcome, namely the $r = 3/2$ energy power law of the relaxation time, has not changed. Readers interested in various improvements to the above two treatments are referred to an excellent review article by Chattopadhyay and Queisser (1981).

While the ionized impurity scattering is important at all temperatures, it becomes the dominant scattering process at low temperatures, where the number of phonons rapidly decreases and so does their effectiveness to scatter charge carriers.

4.6.3 ALLOY SCATTERING

Many excellent thermoelectric materials are solid solutions. While the primary reason for working with solid solutions is to lower the lattice thermal conductivity, the random atomic compositional disorder inevitably leads to fluctuations in the crystal potential that affects the charge carrier transport and degrades the carrier mobility. Disorder scattering was originally considered by Nordheim (1931) and later developed by Harrison and Hauser (1976), who derived an expression for the relaxation time due to scattering in a random alloy of the form

$$\tau_{alloy} = \frac{8 N_A \hbar^4}{3\sqrt{2}\pi c_A \left(1 - c_A\right)\left(\Delta E_{aff}\right)^2 m^{*3/2}} E^{-1/2}. \tag{4.37}$$

Here, N_A is the number of atoms A with concentration c_A, and ΔE_{aff} is the electron affinity difference of the two components constituting the alloy. In general, it is difficult to detect the presence of alloy scattering as it is usually not the dominant scattering process at any temperature interval. At low temperatures, the dominant role is typically played by ionized impurity scattering, while at high temperatures the transport is governed by deformation potential scattering. The alloy scattering is often invoked to explain discrepancies observed in these two dominant processes when one studies solid solutions, but estimates by Harrison and Hauser place a 25% limit on the contribution of alloy scattering to the overall scattering events depending on the solid-solution and its composition. An extension of alloy scattering to non-parabolic bands (Kane model) was made by Auslender and Hava (1993).

4.6.4 INTERVALLEY SCATTERING

In semiconductors having N_v symmetry-related carrier pockets (valleys), there is a possibility that a charge carrier will be scattered to valleys located symmetrically at other points of the Brillouin zone. Multi-valley semiconductors are of special interest to thermoelectricity as the number of equivalent energy pockets N_v supports large Seebeck coefficients via the density-of-states effective mass enhanced by a factor $N_v^{2/3}$, while leaving the transport mass unaffected. In order to span the separation between the valleys, the phonons involved in intervalley scattering must have large wavevectors, i.e., they lie close to the Brillouin zone boundary. There, the acoustic and optical phonons have comparable energies and very little ω vs. q dependence. Therefore, the intervalley phonon is usually taken as having constant frequency, ω_{int}, lying between the optical and acoustic phonons at the zone boundary. Both acoustic and optical phonons may participate in intervalley scattering. Some texts distinguish between the so-called g- and f-intervalley scattering, the former meaning that a charge carrier is scattered to a valley on the opposite side of the same axis, while the latter indicates scattering to one of the other symmetrically related valleys. In principle, the necessary momentum for the intervalley scattering can also be provided by impurity scattering. However, such scattering is primarily limited to low temperatures and will not be considered here.

The scattering rate for intervalley scattering between symmetry equivalent carrier pockets is given by Balkanski and Wallis (2000) as

$$\frac{1}{\tau_{int}} = \frac{\left(N_v - 1\right) E_{ac}^2}{4\pi \rho \omega_{int}} \left(\frac{2m^*}{\hbar^2}\right)^{3/2} \left[\left(N_q + 1\right)\left(E_k - \hbar\omega_q\right)^{1/2} + N_q \left(E_k + \hbar\omega_q\right)^{1/2}\right] \tag{4.38}$$

with the first term (emission) being non-zero for $E_k > \hbar\omega_{int}$. Here, N_v is the number of equivalent energy valleys. As the relaxation rate τ_{int}^{-1} varies as $E^{1/2}$, it mimics the energy dependence of the acoustic deformation potential scattering. At low temperatures, the number of phonons N_q rapidly decreases, dramatically lowering the intervalley scattering rate. It should also be noted that the electron transitions between equivalent band extrema are governed by the selection rules. In some cases, such as transitions between the L-point valleys in lead chalcogenides, the transitions are forbidden, Ravich et al. (1970).

There are also intervalley scattering processes between nonequivalent valleys, in which the energy of the carrier changes. They could lead to a significant reduction of the carrier mobility when the carrier is scattered to a higher energy pocket, where the carriers have larger effective mass than in the initial valley. As they are specific to a particular band structure, I do not consider them here.

4.6.5 Averaging and the Combined Relaxation Time

In the expression for the carrier mobility, $\mu = e\langle\tau\rangle / m_i^*$, the relaxation time enters as an averaged quantity over the carrier energies. The proper form of averaging the momentum relaxation time over the distribution of electrons is achieved by taking, e.g., Ravich et al. (1970),

$$\langle\tau_m\rangle = \frac{\int_0^\infty \tau\left(-\partial f_0 / \partial E\right)k^3 dE}{\int_0^\infty \left(-\partial f_0 / \partial E\right)k^3 dE}, \tag{4.39}$$

where f_0 is the Fermi-Dirac distribution function, and k is the wavevector. Taking a simple band structure where $E = \dfrac{\hbar^2 k^2}{2m_e}$, and assuming the relaxation time is given by a power law function of energy $\tau(E) = \tau_0 E^r$. It is not too difficult to show, using integration by parts in the numerator of Eq. 4.39, that the average momentum relaxation time becomes

$$\langle\tau_m\rangle = \frac{2}{3}\tau_0\left(r + \frac{3}{2}\right)(k_B T)^r \frac{F_{r+1/2}(\eta)}{F_{1/2}(\eta)}, \tag{4.40}$$

where $\eta = \xi / k_B T$. In some texts, the relaxation time is taken as $\tau = \tau_0\left(E / k_B T\right)^r$, in which case the term $(k_B T)^r$ disappears from Eq. 4.40. As I have noted, the two most important elastic scattering processes are those with $r = -1/2$ (acoustic deformation potential scattering in Eq. 4.27), and $r = 3/2$ (ionized impurity scattering in Eq. 4.32). The relevant expression for a particular state of degeneracy is determined by the degenerate or classical form of the distribution function f_0 in Eq. 4.39, or equivalently, by appropriate approximations of the Fermi integrals in Eq. 4.40.

The individual scattering processes discussed above are usually not realized in isolation, but are accompanied by other forms of scattering, depending on the carrier concentration and temperature. Assuming the processes are independent, the overall relaxation time is obtained by the Matthiessen rule,

$$\frac{1}{\tau} = \sum_i \frac{1}{\tau_i}, \tag{4.41}$$

where τ_i is the relaxation time for the participating i-th scattering process.

4.7   FORMS OF THE CHARGE CARRIER MOBILITY

With the relaxation rates or relaxation times for various scattering processes of electrons given in Section 4.6, it is possible to write formulae for respective carrier mobilities for degenerate and nondegenerate carrier systems. Before doing so, it should be noted that by the carrier mobility one strictly understands the drift mobility μ_d (sometimes called the transport or conduction mobility), which is the group velocity a charge carrier acquires when acted upon by an electric field, thus

$$\mu_d = \frac{e}{m_t^*} \langle \tau \rangle \tag{4.42}$$

where e is the charge, m_t^* is the transport effective mass, and $\langle \tau \rangle$ is the averaged relaxation time. The drift mobility is rather difficult to measure, see, e.g., Haynes and Shockley (1951), or in a more modern form, e.g., Stassen et al. (2004). As a result, the drift mobility is essentially always replaced by the more readily available Hall mobility, μ_H, which measures the deflection of charge carriers due to the Lorentz force acting on an electron or hole in the presence of a transverse magnetic field. The Hall mobility is defined as

$$\mu_H = \frac{e}{m_t^*} \frac{\langle \tau \rangle^2}{\langle \tau \rangle}. \tag{4.43}$$

Strictly speaking, the drift and Hall mobilities are the same only when the relaxation time is independent of energy, or when the charge carriers of one particular energy contribute to the transport, as is the case of metals, where the carriers have the Fermi energy. In other situations, the two mobilities are not the same, but are related, Blatt (1957),

$$\mu_H = \frac{3\pi}{8} \mu_d \quad \text{for carriers scattered by phonons,} \tag{4.44}$$

$$\mu_H = \frac{315\pi}{512} \mu_d \quad \text{for carriers scattered by ionized impurities.} \tag{4.45}$$

In the next section, I will consider charge carrier mobilities for the two most important scattering processes encountered in the analysis of thermoelectric materials: the mobility limited by acoustic deformation potential scattering, and the mobility when ionized impurity scattering dominates.

4.7.1   MOBILITY OF ELECTRONS UNDER ACOUSTIC DEFORMATION POTENTIAL SCATTERING

With the expression for the acoustic deformation potential scattering in Eq. 4.27 substituted into Eq. 4.43 for the mobility, and using the averaged value of $\langle \tau_m \rangle$ in Eq. 4.40 with $r = -1/2$, we obtain for an arbitrary degenerate carrier system

$$\mu_{ac} = \frac{e}{m^*} \langle \tau_m \rangle = \frac{\sqrt{2}}{3} \frac{e}{k_B^{3/2}} \frac{\pi \hbar^4 c_\ell}{E_{ac}^2} m^{*-5/2} T^{-3/2} \frac{F_0(\eta)}{F_{1/2}(\eta)}. \tag{4.46}$$

The mobility of nondegenerate carriers under the acoustic deformation potential scattering is obtained by approximating the Fermi integrals by the appropriate Γ-functions

$$\mu_{ac}^{nd} = \frac{2(2\pi)^{1/2}}{3} \frac{e}{k_B^{3/2}} \frac{\hbar^4 c_\ell}{E_{ac}^2} (m^*)^{-5/2} T^{-3/2} \tag{4.47}$$

In general, acoustic deformation potential scattering dominates at and above room temperature. However, at temperatures significantly above 300 K, optical phonons may start to exert their influence and modify the above expression.

4.7.2 Mobility of Electrons under Ionized Impurity Scattering

The relevant relaxation time for the scattering of charge carriers by ionized impurity scattering is given by Eq. 4.36, i.e., the Brooks-Herring formula. Since the term in squared brackets has a rather weak logarithmic dependence, the averaging of the relaxation time is done by neglecting it. Applying Eq. 4.40 with $r = 3/2$, and substituting into Eq. 4.43, the mobility under impurity scattering for an arbitrary degeneracy becomes

$$\mu_{imp} = \frac{e}{m^*}\left\langle \tau_m^{BH}\right\rangle = \frac{32\pi\sqrt{2}\left(\varepsilon_r\varepsilon_0\right)^2}{N_{imp}m^{*1/2}Z^2e^3}\left(k_BT\right)^{3/2}\frac{F_2(\eta)}{F_{1/2}(\eta)} \tag{4.48}$$

When the carrier system is nondegenerate, we replace the Fermi integrals with the appropriate Γ-functions and obtain

$$\mu_{imp}^{nd} = \frac{128\left(2\pi\right)^{1/2}\left(\varepsilon_r\varepsilon_0\right)^2}{N_{imp}m^{*1/2}Z^2e^3}\left(k_BT\right)^{3/2}. \tag{4.49}$$

Ionized impurity scattering is typically the dominant process well below ambient temperature. However, in heavily doped semiconductors, its influence may extend to room temperature where it competes with the acoustic deformation potential scattering. The overall behavior of the mobility then follows from

$$\frac{1}{\mu_{total}} = \frac{1}{\mu_{ac}} + \frac{1}{\mu_{imp}} \tag{4.50}$$

4.8 ELECTRONIC TRANSPORT PROPERTIES OF SKUTTERUDITES

The focus of the discussion in the following sections will be on surveying the topic of electronic transport in skutterudites, starting with their binary forms, progressing to ternary systems, and, finally, finishing with filled skutterudites. The emphasis will be on skutterudites that have a tangible relevance to efficient energy conversion via thermoelectricity.

4.8.1 Electrical Conductivity of Skutterudites

4.8.1.1 Pure Binary Skutterudites

As detailed in Section 3.1, binary skutterudites, typified by $CoSb_3$, are small band gap semiconductors. Accordingly, their resistivity increases rapidly with decreasing temperature. An example of the behavior of binary skutterudites is illustrated in the electrical resistivity of a variety of nominally pure $CoSb_3$ samples shown in Figure 4.5. Both single crystals and polycrystals are intrinsically p-type semiconductors, and their room temperature carrier concentration spans the range between 6.9×10^{16} cm^{-3} and 3.1×10^{18} cm^{-3}. Although the resistivity rises sharply as the temperature decreases, the behavior cannot be approximated by a single exponential dependence on account of a contribution of the impurity bands formed by the presence of shallow acceptor levels. This is well documented by the trend in the Hall coefficient, shown in Figure 4.6. While all three sample (two single crystals, Mandrus et al. (1995) and Arushanov et al. (1997), as well as a rather pure polycrystalline specimen, Dyck et al. (2002)), show comparable R_H temperature dependences, the behavior

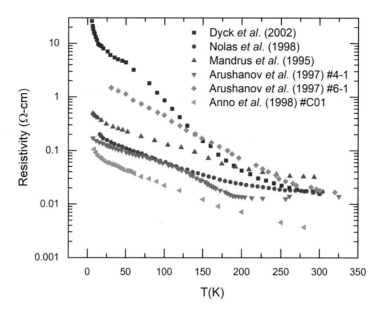

FIGURE 4.5 Electrical resistivity of a collection of pure CoSb$_3$ samples below ambient temperature. The form of the specimens and their room temperature carrier (hole) concentrations are as follows: Dyck et al. (2002), a polycrystal with 3.7×10^{17} cm^{-3}; Nolas et al. (1998), a polycrystal with 1.88×10^{18} cm^{-3}; Mandrus et al. (1995), a single crystal with 1.1×10^{17} cm^{-3}; Arushanov et al. (1997), their sample #4-1, a single crystal with 6.9×10^{16} cm^{-3}; Arushanov et al. (1997), their sample #6-1, a single crystal with 1.1×10^{17} cm^{-3}; Anno et al. (1998), their sample #C01, a polycrystal with 3.1×10^{18} cm^{-3} and average grain size of 3 μm.

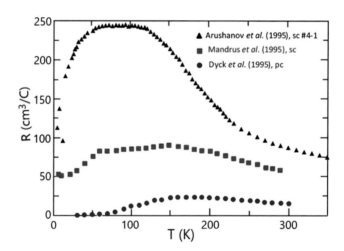

FIGURE 4.6 Temperature dependence of the Hall coefficient of CoSb$_3$ for a few of the samples the electrical resistivity of which is shown in Figure 4.5.

is particularly exemplified in the two single crystals. Roughly speaking, one can distinguish three regimes in the Hall data: a low temperature range up to about 50 K characterized by a sharply rising Hall coefficient, indicating the onset of impurity band conduction; the middle range between about 60 K and 120 K where the Hall coefficient is substantially constant and all impurity states are ionized; and a high temperature range where the decreasing Hall coefficient signals the onset of intrinsic conduction with the thermally excited carriers greatly exceeding the concentration of

TABLE 4.1

Parameters Used in the Modeling by Arushanov et al. (1997) of Single Crystal Samples of CoSb$_3$ (samples 4-1 and 6-1) and the Parameters Obtained by Dyck et al. (2002) for Their Polycrystalline Sample. Sample CoSb$_3$, sc is the Single Crystal Sample of Mandrus et al. (1995)

Sample	$N_{A,shallow}$ (10^{17} cm^{-3})	$N_{A,deep}$ (10^{17} cm^{-3})	N_d (10^{17} cm^{-3})	$N_d/N_{A,shallow}$ (%)	$E_{shallow}$ (meV)	E_{deep} (meV)
CoSb$_3$, 4-1	1.5	1.37	0.93	62	< 1	47
CoSb$_3$, 6-1	3.1	1.24	1.62	52	< 1	38
CoSb$_3$, sc	1.92	6.6	0.68	35	4	64
CoSb$_3$, pc	-	14	2.7	-	-	31.5

extrinsic carriers. Detailed modeling of the transport properties by Arushanov et al. (1997) assumed the presence of an acceptor impurity band (ionization energy $E_{shallow}$, carrier concentration $N_{A,shallow}$) with an additional deep acceptor level (ionization energy E_{deep}, carrier concentration $N_{A,deep}$), and compensating donors (concentration N_d) with all bands considered to be parabolic. With this same model they were also able to explain the data of Mandrus et al. (1995). The respective activation energies and the carrier concentrations are summarized in Table 4.1.

Although on the scale of the single crystals, the Hall data for the polycrystalline sample of Dyck et al. (2002) look much less eventful, on the much finer scale (see Figure 6 in the original paper) and plotted as R_H vs. 1000/T, the Hall coefficient shows a sharp and symmetrical peak near 170 K. Assuming again a parabolic band model, Dyck et al. (2002) was able to explain their transport data with just one rather than two acceptor levels (because the activation energies on either side of $R_{H,max}$ were nearly the same). The results are entered on the last row in Table 4.1. In fact, at temperatures below 30 K, the resistivity of the polycrystalline sample was best fitted by the variable range hopping model of Mott (1968) with the characteristic temperature of the system $T_0 = 584$ K.

More than a dozen years later, Kajikawa (2014a) reevaluated the above transport measurements by assuming that, rather than being parabolic, the valence band of CoSb$_3$ is nonparabolic of the form

$$\frac{\hbar^2 k^2}{2m_v^*} = E(1 + aE), \tag{4.51}$$

with the nonparabolicity parameter $a = (35 \text{ meV})^{-1}$, and the effective mass of holes at the Γ point of $m_v^* = 0.016 m_e$. The values of the fit parameters for the four samples in Table 4.1 indicated that, although the ionization energies have not changed much upon treating the valence band as nonparabolic, the carrier concentrations in the respective states changed significantly, in some instances by more than an order of magnitude. Thus, the influence of nonparabolicity of the valence band at the Γ-point was rather dramatic.

The temperature dependence of the carrier (hole) mobility of several pure CoSb$_3$ samples (the same samples as shown in Figure 4.5) is depicted in Figure 4.7. Less defected single crystals with mobilities of several thousand cm^{2}V^{-1}s^{-1} at room temperature are well documented. The mobility tends to increase as the temperature decreases down to about 200 K, but then it progressively decreases as the temperature falls further. Caillat et al. (1996a) reported acoustic phonon scattering as the dominant scattering mechanism around room temperature for their single crystal of CoSb$_3$. The same mechanism, down to about 200 K, was observed by Mandrus et al. (1995), and at lower temperatures was replaced by the ionized impurity scattering. In addition to the above two scattering processes, Arushanov et al. (1997) invoked the non-polar and polar optical scattering to explain the temperature dependence of mobility of their single crystals. The mobility of polycrystalline

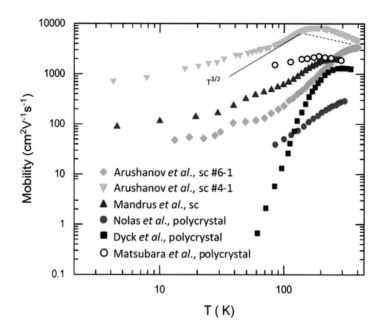

FIGURE 4.7 Temperature dependence of the carrier mobility (holes) in several pure $CoSb_3$ samples, the resistivity of which is shown in Figure 4.5. The solid line indicates the $T^{3/2}$ temperature dependence, and dashed line is the $T^{-1/2}$ dependence.

samples appears to decrease much faster with the decreasing temperature below about 200 K, Dyck et al. (2002), showing up to a T^5 power law of the mobility believed to be associated with the variable range hopping prevailing in the impurity band of $CoSb_3$ at low temperatures. In his assessment of the experimental transport data, Kajikawa (2014a) pointed out the importance of hopping conduction in the impurity band of $CoSb_3$. Vastly different mobilities in single crystals, as opposed to polycrystals, were explained as a consequence of the impurity band in single crystals arising from shallow impurities that he viewed as resonant acceptor levels, while the impurity band in polycrystals is formed from deep acceptor states. Kajikawa also drew attention to the deviation of the mobility at low temperatures from the $T^{3/2}$ dependence that would be expected for a parabolic valence band. In reality, the band is non-parabolic and, as such, the mobility follows a weaker T-linear variation.

Caillat et al. (1996a) measured room temperature carrier mobility in numerous p- and n-type single crystals of $CoSb_3$ as a function of their carrier concentration. The data, together with mobilities of other binary skutterudites, are shown in Figure 4.8 and indicate a more than order of magnitude higher mobility of holes than electrons. Moreover, it seems that p-type single crystals of $CoSb_3$ can be prepared with very low carrier concentrations of several 10^{16} cm^{-3}, while the lowest concentration of electrons in n-type $CoSb_3$ crystals is about 10^{18} cm^{-3}. In very pure p-type samples with concentrations below about 10^{18} cm^{-3}, the room temperature mobility stays essentially constant as a function of carrier concentration. At higher concentrations, the mobility decreases, following approximately a $p^{-1/3}$ power law dependence with the carrier concentration, expected for a linearly dispersing valence band.

The common structure and bonding of all binary skutterudites (disregarding the size of the band gap and unintentional impurities introduced during the synthesis) imply similar transport behavior. However, apart from $CoSb_3$, $IrSb_3$, and $RhSb_3$, detailed transport measurements have rarely been performed on other binary skutterudites.

The low temperature resistivity of $IrSb_3$ was measured by Slack and Tsoukala (1994), Nolas et al. (1996a), Tritt et al. (1996), and Suzuki et al. (2016). Above 100 K, the resistivity has a positive

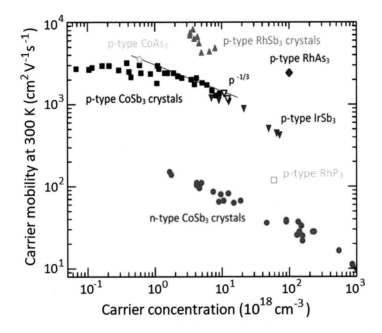

FIGURE 4.8 Room temperature carrier mobility as a function of carrier concentration. P-and n-type single crystals of CoSb$_3$ (black solid squares and red solid circles, respectively) are the data of Caillat et al. (1996a). P-type crystals of RhSb$_3$ (solid grey up-triangles) are from Caillat et al. (1996b). The figure also shows the carrier mobility for p-type RhAs$_3$ (red solid diamond) and p-type CoAs$_3$ (yellow open circle) from Caillat et al. (1996b); p-type RhP$_3$ (open light blue square) from Odile et al. (1978); and samples of p-type IrSb$_3$ (solid blue down-triangles), Caillat et al. (1995b), open blue down-triangle, Slack and Tsoukala (1994), and open red down-triangle, Nolas et al. (1996a).

temperature coefficient that switches over to a sharply rising resistivity as the temperature falls below 100 K. The mobility is high, on the order of 1000 cm^2 V^{-1}s^{-1} at room temperature, especially considering that the samples were polycrystalline with hole concentration on the order of 10^{19} cm^{-3}, see Figure 4.8. At higher temperatures, measurements by Caillat et al. (1992) on single crystals of IrSb$_3$ grown from Sb-rich melts indicated a rising resistivity, with the mobility decreasing to one-half of its room temperature value at 700 K. As shown in Figure 4.8, RhSb$_3$ is notable for its record-high room temperature hole mobility among skutterudites (and one of the highest mobility values of any bulk semiconductor) with the value of some 8000 cm^{2}V^{-1}s^{-1}, Caillat et al. (1996b). Binary skutterudites with the valence electron count deviating from 72 are metals, as documented by the NiP$_3$ structure that follows a distinctly metallic temperature dependence in its electrical resistivity from 1.8 K to ambient temperatures (Shirotani et al. 1993).

4.8.1.2 Intentionally Doped Binary Skutterudites

Within the rigid band picture, believed to be adequate for binary skutterudites, doping means a shift in the Fermi energy up or down depending on whether the dopants generate more electrons or more holes. In binary skutterudites, there are basically two ways of generating an excess of electrons or holes. One can substitute either on the site of the cation or replace some of the anion atoms. In Section 1.1.3, we have seen a broad range of substitutions one can make by replacing some fraction of Co by either Fe or Ni to drive the system p-type or n-type. One can achieve a similar outcome by substituting a fraction of a percent of the pnicogen atom (Column 15) with elements from Column 14 or Column 16. In either case, the charge carrier concentration is significantly enhanced, resulting in higher electrical conductivity. Of course, there is a price to pay in the form of somewhat reduced carrier mobility.

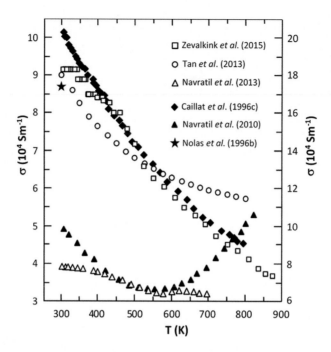

FIGURE 4.9 Electrical conductivity of several ternary skutterudites. Open symbols represent ternaries formed by isostructural substitutions on the pnicogen ring, and the left-hand scale applies to them. They are: open squares – Zevalkink et al. (2015) for $RhSn_{1.5}Te_{1.5}$; open circles – Tan et al. (2013) for $FeSb_{2.2}Te_{0.8}$; and open triangles – Navrátil et al. (2013) for $RuSb_2Te$. Solid symbols stand for isoelectronic substitutions on the cation sites, and the scale for their conductivity is on the right-hand side. Solid diamonds – Caillat et al. (1996c) for $Ru_{0.5}Pd_{0.5}Sb_3$; solid triangles – Navrátil et al. (2010) for $Fe_{0.5}Pd_{0.5}Sb_3$; and a solid star – room temperature value for $Ru_{0.5}Pd_{0.5}Sb_3$ measured by Nolas et al. (1996b). All structures were polycrystals.

4.8.1.3 Electrical Conductivity of Ternary Skutterudites

Ternary skutterudites have been explored primarily for their prospect of having much reduced thermal conductivity. While diminished values of κ have been realized, the ensuing disorder unfortunately degrades the carrier mobility too strongly, which results in too low values of the electrical conductivity. Drops in μ_H are especially noted when the isoelectronic substitutions are made on the pnicogen ring, as the disorder here affects the mobility to a far greater extent than when the ternary structure is made by substitutions on the cation site. Figure 4.9 shows electrical conductivities of several ternary skutterudites that have reasonably large values of σ. Please note that the samples formed by isoelectronic substitutions on the cation site (all drawn as solid symbols) have larger conductivity and their scale is on the right-hand side of the figure. As the Seebeck coefficients of ternary skutterudites are not much improved, if at all, the overall thermoelectric performance of ternary skutterudites is not promising for applications.

4.8.1.4 Electrical Conductivity of Filled Skutterudites

While the most beneficial effect of filling is the reduction of the lattice thermal conductivity, inserting foreign ions into the voids of the skutterudite structure also strongly impacts the electronic properties. Filling with electropositive fillers always leads to a substantial increase in the electron concentration that depends on the degree of filling and the charge state of the filler, and the structure turns into a heavily n-type doped semiconductor with a metallic form of the temperature-dependent resistivity. The vast majority of studies on filled skutterudites aiming to develop an efficient thermoelectric material was done on structures with the $[Co_4Sb_{12}]$, $[Fe_4Sb_{12}]$, $[Ni_4Sb_{12}]$, $[Co_{4-x}Fe_xSb_{12}]$, $[Co_{4-x}Ni_xSb_{12}]$, and $[Ni_{4-x}Fe_xSb_{12}]$ frameworks. Such structures offer not only the best thermoelectric

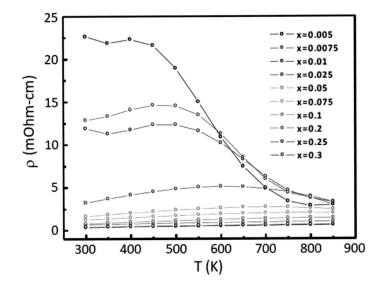

FIGURE 4.10 Temperature dependence of the electrical resistivity of $Yb_xCo_4Sb_{12}$ for the values of x indicated in the figure. Reproduced from Figure S4a of Y. Tang et al., *Nature Materials* 14, 1223 (2015). With permission from Springer Nature.

performance but also the least expensive forms of all skutterudites. The crossover from the intrinsically semiconducting nature of transport in pure $CoSb_3$ to a degenerate metal-like dependence of the resistivity upon gradually increasing the level of filling, common to all filled skutterudites, is exemplified in Figure 4.10 with Yb fillers. As I discuss later on when considering the Seebeck coefficient and the thermal conductivity, Yb is one of the most effective fillers on account of its small size and heavy mass. Moreover, the valence of Yb is intermediate between Yb^{2+} and Yb^{3+}, allowing for a higher occupancy of voids than is typical of trivalent rare earths.

4.8.2 SEEBECK COEFFICIENT OF SKUTTERUDITES

4.8.2.1 Seebeck Coefficient of Pure $CoSb_3$

Expressions for the Seebeck coefficient, assuming a single parabolic band, were derived in online appendix. The general formula of the Seebeck coefficient simplifies under the strongly degenerate and the non-degenerate cases into

$$S_{sd} = \frac{\pi^2}{3} \frac{k_B}{q} \frac{\left(r+\frac{3}{2}\right)\eta}{\left[\eta^2 + \left(r+\frac{1}{2}\right)\left(r+\frac{3}{2}\right)\frac{\pi^2}{6}\right]} \cong \frac{\pi^2}{3} \frac{k_B}{q} \frac{\left(r+\frac{3}{2}\right)}{\eta}$$

(4.52)

$$= \frac{2\pi^{2/3}}{3^{5/3}} \frac{k_B^2 T}{q} \frac{m^*}{\hbar^2} \frac{\left(r+\frac{3}{2}\right)}{n^{2/3}}$$

and

$$S_{nd} = \frac{k_B}{q}\left(-\eta + r + \frac{5}{2}\right),$$

(4.53)

respectively. Here, η is the reduced Fermi energy and r is the scattering parameter defined as an energy exponent in the relaxation time, $\tau(E) = \tau_0 E^r$. Often, the readers might encounter the so-called Mott formula,

$$S = \frac{\pi^2}{3} \frac{k_B}{q} k_B T \left[\frac{\partial \ln \sigma}{\partial E} \right]_{E=\xi} = \frac{\pi^2}{3} \frac{k_B}{q} k_B T \left[\left(\frac{\partial \ln n}{\partial E} \right)_{E=\xi} + \left(\frac{\partial \ln \mu}{\partial E} \right)_{E=\xi} \right], \tag{4.54}$$

which nicely expresses the dependence of the Seebeck coefficient on the carrier concentration and the carrier mobility. It should be kept in mind, however, that the expression was derived for metals or highly degenerate carrier systems, even though it is often used indiscriminately outside of this regime.

Taking into account the linearly dispersing valence band in $CoSb_3$, Singh and Pickett (1994) derived an expression for the Seebeck coefficient of the form

$$S = \frac{2\pi k_B^2 T}{3q\alpha} \left(\frac{\pi}{3n} \right)^{1/3} \tag{4.55}$$

where α is the slope of the linearly dispersing valence band (-3.10 eV Å for $CoSb_3$ and -3.45 eV Å for $IrSb_3$). Taking the carrier concentration at room temperature as between 10^{17} to 10^{18} cm^{-3} (pure, undoped $CoSb_3$), Eq. 4.55 predicts the Seebeck coefficient between 330 μV/K and 153 μV/K. Looking at Figure 4.11, this is, indeed, the range of most room temperature Seebeck coefficients

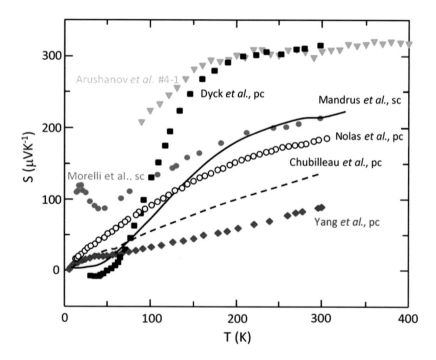

FIGURE 4.11 Temperature dependence of the Seebeck coefficient at low temperatures measured on a variety of $CoSb_3$ samples. The solid line represents the data of Mandrus et al. (1995) on the same single crystal used for the resistivity and the mobility shown in Figs. 4.5 and 4.7, respectively. The dashed curve stands for the data of Chubilleau et al. (2012a) obtained on a polycrystalline sample. Light blue triangles represent the Seebeck coefficient measured on a single crystal #4-1 by Arushanov et al. (1997), gray circles indicate the data on a single crystal collected by Morelli et al. (1995), red diamonds are measurements of Yang et al. (2000) on a polycrystalline sample, and open circles are the data of Nolas et al. (1998) on a polycrystalline specimen.

measured on a variety of pure $CoSb_3$ samples, including both single crystals and polycrystals. The Seebeck coefficient decreases with the decreasing temperature as expected based on Eq. 4.55. In some very pure crystals, the Seebeck coefficient develops a low temperature peak that might indicate a contribution from the phonon drag effect. In less pure, although undoped, polycrystals (typically structures where Co contains a Ni impurity, Matsubara et al. 1997), the Seebeck coefficient might start as slightly negative at low temperatures, but rapidly switches over to positive values as the temperature increases.

Measurements of the Seebeck coefficient of $CoSb_3$ at temperatures above 300 K have been numerous, aided by the fact that doping studies and explorations of composite structures have essentially always used $CoSb_3$ as the matrix. In Figure 4.12, I collect only a small fraction of such measurements to illustrate the high temperature trend in the Seebeck coefficient of $CoSb_3$.

Caillat et al. (1996a) measured the Seebeck coefficient on a number of p-type $CoSb_3$ single crystals spanning nearly two decades in the room temperature carrier concentration. The behavior of crystals with the lowest and highest carrier concentration bracketing the entire series is shown in Figure 4.12. As expected, the lowest carrier concentration corresponds to the highest magnitude of the Seebeck coefficient. Figure 4.12 also includes the Seebeck coefficients of four polycrystalline samples prepared by different synthesis routes. The trend in the Seebeck coefficient of all samples

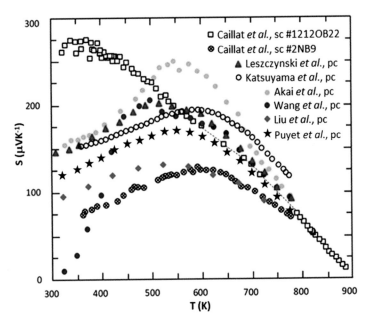

FIGURE 4.12 Seebeck coefficient of p-type $CoSb_3$ at high temperatures. A range of single crystals with various carrier concentrations measured by Caillat et al. (1996a) is bracketed by the sample #1212OB22 with the lowest carrier concentration of 1.2×10^{17} cm^{-3} (open square) and the sample #2NB9 with the highest carrier concentration of 9.45×10^{18} cm^{-3} (crossed circles and a dashed line). Crystals with various carrier concentrations were obtained by cutting samples along the length of a large single crystalline ingot. Polycrystals included in the figure were synthesized by different methods: the sample of Leszczynski et al. (2013) (red triangles) was prepared by the traditional melting, annealing, grinding and hot pressing (unfortunately, the value of the carrier concentration was not provided); the sample of Wang et al. (2009) (blue circles) was prepared by a solvothermal synthesis followed by melting and annealing, and had the room temperature carrier concentration of 8.33×10^{16} cm^{-3}, with negative Seebeck coefficient below 300 K; the sample of Liu et al. (2007a) (gray diamonds) was synthesized by ball milling with subsequent compaction by spark plasma sintering at 673 K; the sample of Akai et al. (1998) was made by a solid-state reaction followed by hot pressing; and so was the sample of Katsuyama et al. (1998). No carrier concentration was provided for the last three samples.

is clearly apparent. In the extrinsic regime of transport, the magnitude of the Seebeck coefficient increases, reaches a peak, the position of which depends on the carrier concentration, and then the Seebeck coefficient rapidly decreases as the intrinsic excitations set in. In some samples, particularly those prepared by the solvothermal method, there is an unintended n-type doping taking place during the synthesis process, and the Seebeck coefficient may start with negative values below room temperature. The same likely occurs in some arc-melted samples, where an excessive loss of Sb often happens, e.g., Kawaharada et al. (2001). Since $CoSb_3$ naturally grows as a p-type conductor, to make bona fide n-type $CoSb_3$ requires intentional n-type doping either on the cation or anion sites, the topic discussed in the next section.

4.8.2.2 Seebeck Coefficient of Doped $CoSb_3$

Doping studies of $CoSb_3$ were made already in the mid- to late 1950s by Dudkin and his collaborators (Dudkin and Abrikosov 1957, 1959), Zobrina and Dudkin 1959). By intentionally introducing 0.1 at% to 1 at% of Te or Pd to the starting melt, Caillat et al. (1996a) prepared n-type single crystals of $CoSb_3$ with room temperature carrier concentrations spanning the range between 4.5×10^{18} cm^{-3} and 1.38×10^{20} cm^{-3}. Seebeck coefficients of the two extreme carrier concentrations are replotted in Figure 4.13. In addition, the figure contains the data for two polycrystalline n-type samples (unintentionally doped) that, at elevated temperatures, either become p-type conductors, Li et al. (2005), or show a tendency to become p-type conductors had the temperature been increased well past 600 K (Wojciechowski et al. 2003).

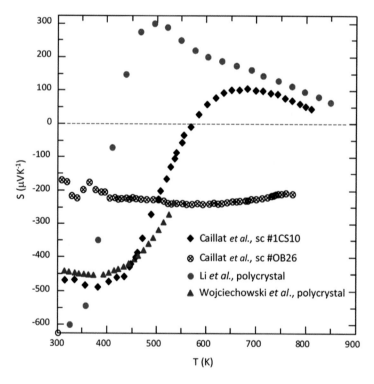

FIGURE 4.13 High temperature Seebeck coefficient of (very weakly doped) n-type skutterudites. The two single crystals of Caillat et al. (1996a) are samples with the lowest (4.5×10^{18} cm^{-3}) and the highest (1.38×10^{20} cm^{-3}) carrier concentration of several crystals investigated. They were intentionally doped with 0.08 at% Te and 1 at% of Pd, respectively. The polycrystalline sample of Li et al. (2005) has a room temperature concentration of electrons of about 5.3×10^{18} cm^{-3}, and the polycrystal of Wojciechowski et al. (2003) has a room temperature concentration of 3.0×10^{18} cm^{-3}. Both polycrystals were undoped. The dashed line divides n-type and p-type at zero Seebeck coefficient.

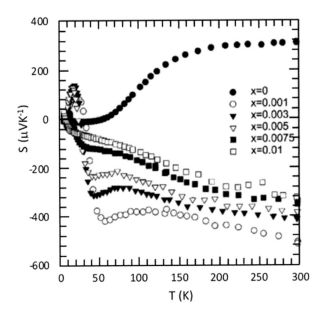

FIGURE 4.14   Seebeck coefficient of $Co_{1-x}Ni_xSb_3$ at low temperatures. Reproduced from J. S. Dyck et al., *Physical Review B* 65, 115204 (2002). With permission from the American Physical Society.

The p-type nature of transport in $CoSb_3$ is reinforced by substituting a fraction of Co by Fe, e.g., Katsuyama et al. (1998) and Yang et al. (2000), or by doping on the site of Sb with elements of Column 14, particularly with Sn, Koyanagi et al. (1996). On the other hand, n-type conduction is promoted by introducing Ni, Pd, or Pt on the site of Co (Matsubara et al. 1994, Tashiro et al. 1997, Anno et al. 1999, Dyck et al. 2002), or by doping with elements of Column 16, particularly Te, on the Sb sublattice (Nagamoto et al. 1998, Wojciechowski 2002, Wojciechowski et al. 2003, Li et al. 2005, Liu et al. 2007b, 2008). As an illustration, the data of Dyck et al. (2002) in Figure 4.14 document a very strong dependence of the low temperature carrier transport on a miniscule amount of Ni in the lattice. Undoubtedly, at high temperatures the Seebeck coefficient of these samples would tend towards positive values, but the data were not pursued beyond the ambient temperature.

4.8.2.3   Seebeck Coefficient of Other Binary Skutterudites

In comparison to studies of $CoSb_3$, other binary skutterudites have not been pursued much, except perhaps for $IrSb_3$. Significantly higher peritectic temperature of $IrSb_3$ (1414 K) compared to that of $CoSb_3$ (1146 K) provided for a possibility of extending the usability of $IrSb_3$ as a thermoelectric to higher temperatures. A collection of Seebeck coefficient measurements on $IrSb_3$ is presented in Figure 4.15. The data resemble those on $CoSb_3$, except that the peak in the Seebeck coefficient of $IrSb_3$ is at temperatures some 200 degrees higher, reflecting the higher room temperature carrier concentration on the order of 10^{19} cm^{-3}. Seebeck coefficient data of other binary skutterudites are rare and those available are presented in Figure 4.16.

Recently, a theoretical calculation of the Seebeck coefficient of binary skutterudites was performed by Saeed et al. (2019) using the Boltzmann transport equation. The electronic band structure parameters were calculated by DFT with the improved Tran-Blaha modified Becke-Johnson potential (see Chapter 3). The results of these calculations are shown in Figure 4.17. While the theory correctly predicts positive and rising Seebeck coefficients with the temperature, there are obviously great discrepancies as far as the trend in the behavior of the Seebeck coefficient at high temperatures is concerned. While all experimental studies indicate a distinct peak followed by a decreasing Seebeck coefficient at some temperature below 800 K, the theoretical curves keep rising at 800 K

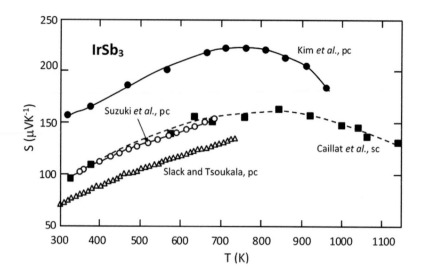

FIGURE 4.15 Seebeck coefficient of IrSb₃. The solid squares are the data of Caillat et al. (1992) obtained on a single crystal prepared by the vertical gradient freeze technique (carrier concentration at 300 K of about 2×10^{19} cm^{-3}), the open triangles are the results of Slack and Tsoukala (1994) on a polycrystalline specimen with the concentration of 1.1×10^{19} cm^{-3}, the solid circles are measurements by Kim et al. (2004) on a polycrystal (carrier concentration not provided), and the open circles stand for the data of Suzuki et al. (2016) obtained on a polycrystal having the carrier concentration of 8.7×10^{18} cm^{-3}.

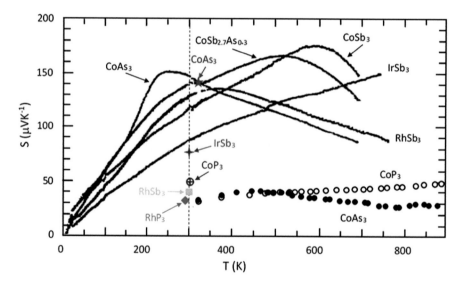

FIGURE 4.16 Seebeck coefficient of several binary skutterudites. Line traces are the data of Sharp et al. (1995) on polycrystalline samples with the room temperature carrier concentrations: 0.8×10^{18} cm^{-3} for CoAs₃, $(2–5) \times 10^{18}$ cm^{-3} for CoSb₃, $(3–10) \times 10^{18}$ cm^{-3} for IrSb₃. The above samples were cold-pressed with mass densities not higher than 70%. In fact, CoAs₃ had the density merely 50% of its theoretical value. The RhSb₃ sample was hot-pressed and had a density of 90%. The carrier concentration of this sample was 1×10^{19} cm^{-3}. The figure also presents Seebeck coefficients of CoP₃ (open circles) and CoAs₃ (solid circles) measured at elevated temperatures by Watcharapasorn et al. (2000). The carrier concentrations of CoP₃ and CoAs₃ were 2.69×10^{19} cm^{-3} and 5.7×10^{17} cm^{-3}, respectively. Included also are several near room temperature values of the Seebeck coefficient for CoAs₃ (red star) reported by Pleass and Heyding (1962); IrSb₃ (four-point star) measured by Tritt et al. (1996); CoP₃ (cross in circle) obtained by Ackermann and Wold (1977); RhP₃ (grey diamond) measured by Odile et al. (1978); and RhSb₃ (orange square) reported by Sirimart et al. (2019), the sample having the carrier concentration of 3.5×10^{18} cm^{-3}.

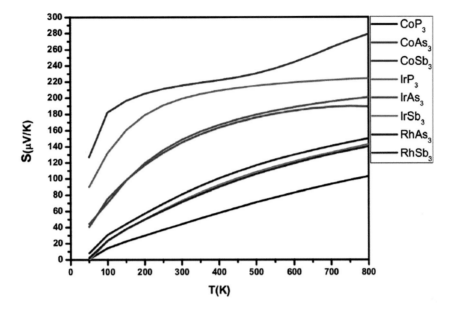

FIGURE 4.17 Theoretical Seebeck coefficients for binary skutterudites computed using the Boltzmann transport equation with the electronic energy band parameters obtained using an improved version of the Tran-Blaha modified Becke-Johnson potential. Reproduced from M. Saeed et al., *RSC Advances* 9, 24981 (2019). With permission from the Royal Society of Chemistry.

or, in a couple of cases, indicate saturation. The most glaring case is that of $CoSb_3$, which, according to Figure 4.17, shows a robustly increasing Seebeck coefficient at 800 K, while all experimental evidence (see Figure 4.12), documents a clear turnover already near 600 K. A similar situation is with the Seebeck coefficient of other binary skutterudites, including $IrSb_3$ that experimentally turns around 800 K. The only experimental data that depict a mildly increasing Seebeck coefficient at the high temperature end are measurements by Watcharapasorn et al. (2000) on CoP_3. It seems to me that the band gaps calculated by Saeed et al. (2019) are significantly overestimated, which shifts the onset of intrinsic excitations to much higher temperatures. However, because the authors did not provide information on the band gaps used in their calculations, I can only speculate.

4.8.2.4 Seebeck Coefficient of Composites Having the $CoSb_3$ Matrix

The primary reason behind the formation of composite structures with the $CoSb_3$ matrix is to enhance phonon scattering and thus reduce the thermal conductivity. Provided the electrical conductivity is degraded to a much lesser extent than the thermal conductivity, the approach is, indeed, effective. The Seebeck coefficient in such structures can be either increased or decreased, depending on whether the additives tend to increase the carrier effective mass or increase the carrier concentration, respectively. In any case, the changes in the Seebeck coefficient are typically much smaller than those in the electrical and thermal conductivities. However, depending on the nature of grain boundaries and the effectiveness of trap states at the interfaces, there is a possibility for an extra enhancement of the Seebeck coefficient via the process of energy filtering. By having a potential barrier at the interface that blocks the passage of low energy charge carriers, the average energy of the carriers with respect to the Fermi energy is effectively increased, leading to a larger magnitude of the Seebeck coefficient, e.g., Ravich (1995), Moyzhes and Nemchinsky (1998). Even though the electrical conductivity is a bit decreased, the overwhelming influence of the Seebeck coefficient results in an increased power factor. In Figures 4.18(a)–4.18(c), are presented three composite systems with the $CoSb_3$ matrix that show different trends in the behavior of the Seebeck coefficient,

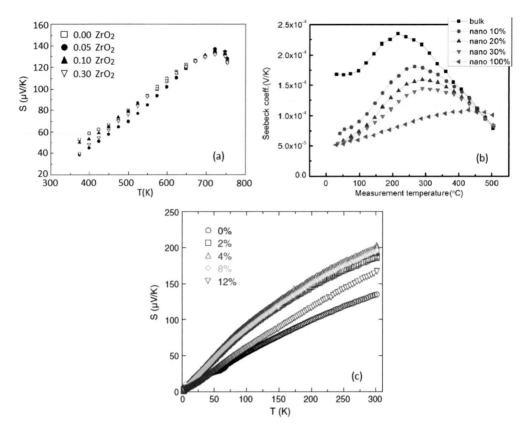

FIGURE 4.18 Seebeck coefficient of composite structures with the CoSb$_3$ matrix. (a) CoSb$_3$ with admixed 25 nm ZrO$_2$ powder, homogenized by a brief ball milling treatment and compacted by hot pressing at 853 K under 100 MPa applied for 30 min, He et al., *Nanotechnology* 18, 235602 (2007). (b) Composite formed by adding sub-30 nm CoSb$_3$ nanoparticles prepared by the polyol process (with excess of Sb to be p-type conducting) to micron-size CoSb$_3$ prepared by solid-state reaction, well mixed by brief ball milling, and compacted by hot pressing at 773 K for 120 min under 100 MPa, Yang et al., *Journal of Applied Physics* 106, 013705 (2009). (c) Nanosized ZnO powder purchased from Alfa Aesar added to micron-size CoSb$_3$, mixed in an ultrasonic bath of methanol/hexane (3:1) for 20 min and compacted by SPS at 873 K for 6 min under 50 MPa, Chubilleau et al., *Journal of Electronic Materials* 41, 1181 (2012b). With permission from IOP Publishing, Ltd., AIP Publishing, and Springer Nature, respectively.

depending on the role of additives in the structure. Figure 4.18(a) is the temperature dependence of a composite made by adding ZrO$_2$ powder of 25 nm size to a powder of CoSb$_3$ in a short ball milling process to achieve good dispersion, with subsequent compaction by hot pressing (He et al. 2007). The effect of ZrO$_2$ additions on the Seebeck coefficient is rather modest and notable only at lower temperatures. A strong suppression of the Seebeck coefficient depicted in Figure 4.18(b) was observed by Yang et al. (2009), where a nanopowder of CoSb$_3$ was admixed to a micron-size matrix powder of CoSb$_3$. In this case, it was believed that the degradation of the Seebeck coefficient was due to a significantly increased carrier concentration and electrical conductivity. Figure 4.18(c) shows a marked enhancement of the Seebeck coefficient when nano-sized powder (8–10 nm) of ZnO was mixed into a micron-sized powder of CoSb$_3$, Chubilleau et al. (2012b). Since the concentration of holes decreased by a factor of three in the 12 wt% of ZnO composite, the increased Seebeck coefficient is likely associated with the reduced carrier concentration. However, there is a possibility that some fraction of the enhancement might have resulted from carrier filtering at the grain boundaries. In all three cases, the overall effect was an enhanced thermoelectric figure of merit because the reduction of the thermal conductivity dominated over all other changes.

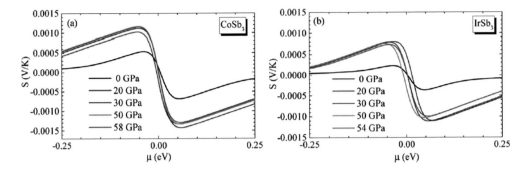

FIGURE 4.19　Calculated Seebeck coefficient of $CoSb_3$ and $IrSb_3$ as a function of Fermi energy for several pressures. Reproduced from the data of X. Yang et al., *Physical Chemistry Chemical Physics* 21, 851 (2019). With permission from the Royal Society of Chemistry.

4.8.2.5　Effect of Pressure on the Seebeck Coefficient

In Sections 1.1.4 and 3.1.1, the effect of pressure on the stability of the skutterudite structure and on the band structure was considered. As part of their study of the band structure of $CoSb_3$ and $IrSb_3$ under pressure, Yang et al. (2019) also estimated the effect of pressure on the Seebeck coefficient of the above two binary skutterudites. As the band gap increases up to about 30 GPa and the bands flatten considerably, see Figure 3.13, the larger effective mass resulted in higher Seebeck coefficients for both $CoSb_3$ and $IrSb_3$. Beyond about 30 GPa, the band gap starts to decrease gradually, and the Seebeck coefficient ceases to increase. From the data, it also follows that in both skutterudites, electron doping yields higher Seebeck coefficients than hole doping. The variation of the Seebeck coefficient as a function of either electron or hole doping, expressed through the Fermi level μ for several values of pressure, is illustrated in Figure 4.19. Unfortunately, no experimental studies have as yet verified the predictions.

4.8.2.6　Seebeck Coefficient as Input to Determine the Carrier Effective Mass

The effective mass of charge carriers is an important parameter that characterizes the nature of transport and the agility of a charge carrier to respond to applied fields and thermal gradients. Measurements of the Seebeck coefficient, in conjunction with the Hall coefficient, provide a convenient way to determine the carrier effective mass. From the general expression for the Seebeck coefficient reproduced here,

$$S = \frac{k_B}{q}\left[-\eta + \frac{\left(r+\dfrac{5}{2}\right)F_{r+3/2}(\eta)}{\left(r+\dfrac{3}{2}\right)F_{r+1/2}(\eta)}\right], \tag{4.56}$$

and assuming one knows the scattering parameter r (often, but not necessarily justifiably taken as the one applicable to acoustic phonon scattering), one can obtain the reduced Fermi energy η at each temperature at which the Seebeck coefficient is experimentally determined. Substituting thus obtained value of the reduced Fermi energy into the expression for the Hall carrier concentration, written here assuming holes as the major carrier,

$$p = 4\pi\left(\frac{2m^*k_BT}{h^2}\right)^{3/2}F_{1/2}(\eta), \tag{4.57}$$

the effective mass m^* can be readily determined. The transport coefficients used here were derived assuming a single parabolic band model. In the literature, the above approach is used quite liberally with no regard whether the band is parabolic or not, nor whether more than a single band contributes to the transport. In binary skutterudites, and in CoSb$_3$ in particular, the valence band is distinctly linearly dispersing, and taking it as parabolic might lead to a serious error in the estimate of the effective mass of holes. Moreover, at elevated temperatures other bands, apart from the Γ-point valence and conduction bands, may participate in the transport.

The initial assessment of the effective mass of holes in CoSb$_3$ crystals was made by Caillat et al. (1992) assuming acoustic phonon scattering, which yielded $m^* = 0.16\ m_e$. Mandrus et al. (1995), assuming that ionized impurity scattering dominates the transport below 300 K, estimated a more than an order of magnitude lighter effective mass of holes $m^* = 0.011\ m_e$. They pointed out that had they taken acoustic phonon scattering as the dominant process, their effective mass would have been much higher. In a subsequent paper, Caillat et al. (1996a) determined the variation of the hole and electron effective masses as a function of the carrier concentration, assuming that the valence band is described by a non-parabolic band model of Kane (1957). At low carrier concentration of 10^{17} cm^{-3}, the effective mass of holes was 0.071 m_e and rose roughly as $p^{1/3}$ with the carrier concentration, see Figure 4.20a. In contrast, the effective masses of n-type crystals were some two orders of magnitude larger and rose almost linearly on the log scale of the carrier concentration, as shown in Figure 4.20b. The figure also includes effective masses of several n-type doped CoSb$_3$ obtained by Anno et al. (1999), and a number of filled skutterudites with various filler species spanning from light fillers, such as Na and K to heavy rare earths Dy, Tb, Gd, Sm, and Nd. Interestingly, regardless of the filler, enhancements of the carrier effective mass are similar in all cases.

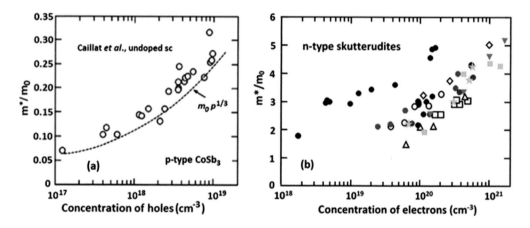

FIGURE 4.20 (a) Effective masses of holes versus the carrier concentration (from the Hall effect) measured by Caillat et al. (1996a) for nominally undoped p-type single crystals of CoSb$_3$. The data follow a $p^{1/3}$ dependence, consistent with the band structure model of Singh and Pickett (1994). (b) A collection of various intentionally n-type doped samples of CoSb$_3$. Solid black circles are the data of Caillat et al. (1996a), open symbols are measurements of Anno et al. (1999) for polycrystalline sample containing small quantities of Ni (diamonds), Pd (circles), Pt (up-triangles), and Pd+Pt (squares). Solid blue circles (Na$_y$Co$_4$Sb$_{12}$), orange stars (K$_y$Co$_4$Sb$_{12}$), light blue squares (Ba$_y$Co$_4$Sb$_{12}$), and down grey triangles (Eu$_y$Co$_4$Sb$_{12}$) are data of Pei et al. (2009). Solid red circles are measurements of Li et al. (2016) on CoSb$_3$ filled with heavy rare earths using a combination of melt spinning with high pressure sintering (at 0.5 GPa) that allowed for significantly enhanced filling fractions. In the ascending order of the carrier concentration, the solid red circles indicate Dy, Tb, Gd, Sm, and Nd. Regardless of the filler, the effective masses of n-type skutterudites show similar rising trends. Note more than one to two orders of magnitude larger effective masses of charge carriers in n-type skutterudites compared to p-type skutterudites. The panel (a) is redrawn from Caillat et al., *Journal of Applied Physics* 80, 4442 (1996). With permission from the American Institute of Physics.

Arushanov et al. (1997), who invoked additional bands (two acceptor bands plus one donor band) and assumed a variety of scattering processes, nevertheless used Eqs. 4.56 and 4.57 to estimate the effective mass of holes. The estimates turned out to be 0.067 m_e for acoustic phonon scattering and 0.036 m_e for polar optical scattering. Detailed analysis of the experimental transport parameters for all five single crystals reported by Caillat et al. (1996a) was carried out by Kajikawa (2014b). He assumed, apart from participation of the Γ-point valence and conduction bands, also a second valence band lying at point P about 0.4 eV below the Γ-point valence band, and a second conduction band at point H some 0.2 eV above the conduction band edge of the Γ-point conduction band. Moreover, he assigned nonparabolicity to the main (Γ-point) valence band of the form in Eq. 4.51 with the parameter $\alpha = 35$ meV. Fitting of the experimental transport data to all five crystals resulted in common values of effective masses for all four bands considered: 0.016 m_e for the main (Γ-point) valence band, 5.0 m_e for the second valence band, 7.0 m_e for the main (Γ-point) conduction band, and 14 m_e for the second conduction band. Clearly, while the exact values depend on the details of the bands and the predominant scattering mechanism, there is no doubt that the effective mass of holes in the linearly dispersing valence band of $CoSb_3$ and $IrSb_3$ is exceptionally small, supporting the record-high values of the hole mobility not only in single crystals but also in polycrystalline specimens. On the other hand, some two orders of magnitude larger effective mass of electrons bodes well for n-type skutterudite structures supporting unusually high Seebeck coefficients, in spite of large carrier concentrations typical in n-type skutterudites.

4.8.2.7 Seebek Coefficient of Filled Skutterudites

As we have seen in Chapter 3, due to strong hybridization between the states of the filler with p-states of the pnicogen atom and d-states of the transition metal, the electronic energy band structure is more complex than was the case of binary skutterudites. Nevertheless, because the density of states near the Fermi energy is not dramatically affected by filling, the multiple conduction bands with heavy electron masses maintain the magnitude of the Seebeck coefficient at high enough values, in spite of the much larger electron concentration. As documented in measurements by Cho et al. (2013), effective masses in p-type filled skutterudites also rise rapidly with the carrier concentration in a way similar to the trend seen in Figure 4.20(a) for p-type binary $CoSb_3$, except now the carrier concentrations are two orders of magnitude higher and the holes are some 20 times heavier.

Detailed theoretical assessment of electronic transport properties by Yang et al. (2009) indicated that to obtain the maximum power factor, the optimal carrier concentration in n-type skutterudites should be around 0.5 electrons per formula unit. In other words, filling with monovalent fillers, close to 50% of the voids should be filled, while for divalent fillers the requirement is to fill about 25% of the voids, and for trivalent fillers merely 17% void filling should lead to the optimal fraction. The above theoretical estimates are in good accord with the experimental findings, e.g., Morelli et al. (1997), Nolas et al. (2000b), Chen et al. (2001), Lamberton et al. (2002), Li et al. (2005), Puyet et al. (2005), Shi et al. (2008), and Pei et al. (2006, 2009). Consequently, from a purely electronic perspective, fillers having the same charge state should be equally effective to optimize the power factor at the same filling fraction. However, taking into account how effectively such fillers also degrade the thermal conductivity, Yb and Nd have been identified as perhaps the most promising filler species (Nolas et al. (2000b), Anno et al. 2000, Sales et al. 2000, Nagao et al. 2002, Zhao et al. 2006, Li et al. 2008, Zhang et al. 2013, and Shaheen et al. 2017). Of course, researchers have continued to look into the effect of other fillers, and an enormous number of publications covering every conceivable filler, including uranium (Arita et al. 2005), has accumulated over the years. The findings merely confirmed that it is difficult to surpass the performance achieved with Yb and Nd fillers.

The evolution of the Seebeck coefficient of Yb-filled $CoSb_3$ as the content of the filler gradually increases is illustrated in Figure 4.21. Increasing content of Yb results in both the enhanced electron concentration and an extension of the regime of extrinsic conduction to higher temperatures, while the magnitude of the Seebeck coefficient decreases. A collection of Seebeck coefficient measurements on $Yb_xCo_4Sb_{12}$ with the filling fraction x near its maximum value of $x \sim 0.3$ is depicted in

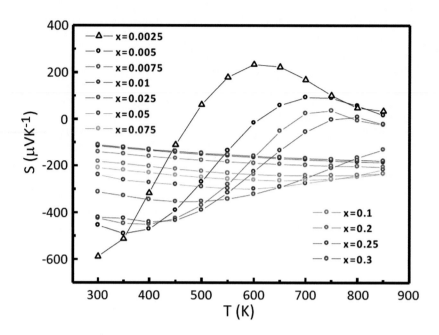

FIGURE 4.21 Temperature dependence of the Seebeck coefficient of $Yb_xCo_4Sb_{12}$ for various contents of Yb. Reproduced from Y. Tang et al., *Nature Materials* 14, 1223 (2015). With permission from Springer Nature.

Figure 4.22 by solid symbols and the dashed line, with the left-hand scale indicating the value. The same figure also shows by open symbols, and the right-hand scale being applicable, the Seebeck coefficient of two Nd-filled p-type structures having the $[Fe_2Co_2Sb_{12}]$ framework.

As discussed in Chapter 2, apart from divalent Eu and the intermediate valence of Yb, all other heavy and small rare earth elements past Gd cannot be trapped in the skutterudite cage under the usual synthesis conditions of ambient pressure. However, Nolas et al. (2000a) inserted Sn into the skutterudite voids under high pressure and Kihou et al. (2004) have shown that filled skutterudites with all heavy rare earth elements can be prepared using high pressure synthesis, described in Section 2.2.6. Unfortunately, no comprehensive transport data sets have been generated on such skutterudites. On the other hand, to prepare several skutterudites partially filled with heavy rare earth elements, Li et al. (2016) applied a melt-spinning technique that facilitates ultrafast cooling rates and homogeneous distribution of elements on the nanoscale. The obtained ribbons were powdered and then consolidated using high pressure sintering under 0.5 GPa. The subsequent transport measurements, including the Seebeck coefficient, are presented in Figure 4.23. The results clearly reflect the partial nature of filling achieved in the study. While near linear and overall negative Seebeck coefficients are obtained on Nd- and Sm-filled skutterudites with the filling fraction of $y = 0.15$, the Seebeck coefficient of Gd, Tb, and Dy turns around and decreases in magnitude on account of the merely 0.04, 0.03, and 0.02 filling fractions, respectively, attained in these three skutterudites. In the case of Tm, no measurable fraction of the rare earth was found in the voids, and the Seebeck coefficient actually becomes positive above about 600 K.

Praseodymium is an important filler as far as physical properties of filled skutterudites are concerned because it gives rise to a spectrum of fascinating superconducting, magnetic and heavy fermion characteristics in many skutterudite lattices. Here I note the effect of Pr on the Seebeck coefficient of $PrFe_4P_{12}$.

The structure of $PrFe_4P_{12}$ is typified by an antiferroquadrupolar (AFQ) ordering (Aoki et al. 2005), taking place at 6.5 K (previously believed to represent an antiferromagnetic ordering, Torikachvili et al. (1987)). Below this transition, the Seebeck coefficient rises to spectacular magnitudes, exceeding

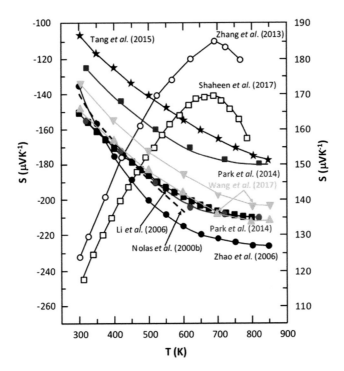

FIGURE 4.22 Seebeck coefficients of several Yb-filled skutterudites (solid symbols and the dashed line, left-hand scale) and two Nd-filled p-type skutterudites (open symbols and right-hand scale). The Seebeck coefficients shown are: solid black stars – $Yb_{0.3}Co_4Sb_{12}$ by Tang et al. (2015); solid blue squares - $Yb_{0.2}Co_4Sb_{12}$ by Park et al. (2014); solid red circles – $Yb_{0.3}Co_4Sb_{12}$ by Park et al. (2014); solid black squares – $Yb_{0.3}Co_4Sb_{12.3}$ by Li et al. (2008); solid black circles – $Yb_{0.15}Co_4Sb_{12.81}$ by Zhao et al. (2006); solid yellow up-triangles – $Yb_{0.2}Co_4Sb_{12}$ made by Wang et al. (2017) by ball milling; solid yellow down-triangles – $Yb_{0.2}Co_4Sb_{12}$ made by Wang et al. (2017) by long-term annealing; and the dashed line – $Yb_{0.19}Co_4Sb_{12}$ by Nolas et al. (2000b). The two Nd-filled p-type samples are: open circles – $Nd_{0.6}Fe_2Co_2Sb_{12}$ by Zhang et al. (2013), and open squares – $Nd_{0.9}Fe_2Co_2Sb_{12}$ by Shaheen et al. (2017).

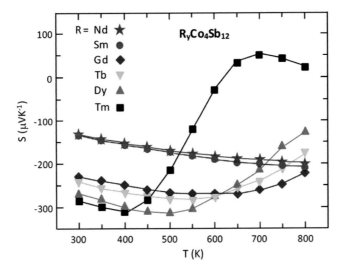

FIGURE 4.23 Seebeck coefficient of several partially filled skutterudites with heavy rare earth elements. Redrawn from the original data of Li et al., *Journal of Materials Chemistry C* 4, 4374 (2016). With permission from the Royal Society of Chemistry.

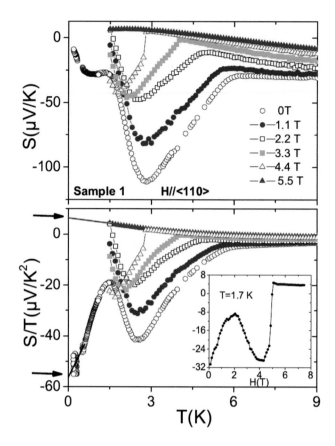

FIGURE 4.24 Seebeck coefficient of $PrFe_4P_{12}$ at low temperatures. Note a sharp decrease in the Seebeck coefficient below about 6 K due to the antiferroquadrupolar (AFQ) ordering, and exceptionally large magnitudes of the Seebeck coefficient near 3 K. The effect of a magnetic field is to suppress the ordering. Reproduced from Pourret et al., *Physical Review Letters* 96, 176402 (2006). With permission from the American Physical Society.

100 μVK^{-1} near 3 K, Sato et al. (2000) and Pourret et al. (2006) (see Figure 4.24). While the magnetic field strongly suppresses the Seebeck coefficient, the Nernst coefficient increases, and in the field of 4 T and temperatures around 1 K, the Nernst coefficient reaches values of 100 μVK^{-1}, making an Ettingshausen-type cooler, e.g., Goldsmid (1986), a tantalizing possibility.

REFERENCES

Ackermann, J. and A. Wold, *J. Phys. Chem. Solids* **33**, 1013 (1977).

Akai, K., H. Kurisu, T. Moriyama, S. Yamamoto, and M. Matsuura, *Proc. 17th Int. Conf. on Thermoelectrics,* IEEE Catalog Number 98TH8365, Piscataway, NJ, p. 105 (1998).

Anno, H., K. Hatada, H. Shimizu, and K. Matsubara, *J. Appl. Phys.* **83**, 5270 (1998).

Anno, H., K. Matsubara, Y. Notohara, T. Sakakibara, and H. Tashiro, *J. Appl. Phys.* **86**, 3780 (1999).

Anno, H., Y. Nagamoto, K. Ashida, E. Taniguchi, T. Koyanagi, and K. Matsubara, *Proc. 19th Inter. Conf. on Thermoelectrics*, ed. D. M. Rowe, Babrow Press, UK, p. 90 (2000).

Aoki, Y., H. Sugawara, H. Hisatomo, and H. Sato, *J. Phys. Soc. Jpn.* **74**, 209 (2005).

Arita, Y., T. Ogawa, H. Kobayashi, K. Iwasaki, T. Matsui, and T. Nagasaki, *J. Nucl. Mater.* **344**, 79 (2005).

Arushanov, E., K. Fess, W. Kaefer, C. Kloc, and E. Bucher, *Phys. Rev. B* **56**, 1911 (1997).

Auslender, M. and S. Hava, *Solid State Commun.* **87**, 335 (1993).

Balkanski, M. and R. F. Wallis, *Semiconductor Physics and Applications*, Oxford University Press (2000).

Bardeen, J. and W. Shockley, *Phys. Rev.* **80**, 72 (1950).

Blatt, F. J. Theory of mobility of electrons in solids, in *Solid State Physics*, ed. F. Seitz and D. Turnbull, Vol. 4, 199–366, Academic Press, NY (1957).

Bloch, F., Z. *Physik* **52**, 555 (1928).

Brooks, H., in *Adv. Electron. Electron Phys.* **7**, 85, Academic Press, NY (1955).

Caillat, T., A. Borshchevsky, and J.-P. Fleurial, *Proc. 11th Inter. Conf. on Thermoelectrics*, Copyright with The University of Texas at Arlington, pp. 98–101 (1992).

Caillat, T., A. Borshchevsky, and J.-P. Fleurial, *Proc. 13th Inter. Conf. on Thermoelectrics*, American Institute of Physics, NY, pp. 58–61 (1995).

Caillat, T., A. Borshchevsky, and J.-P. Fleurial, *J. Appl. Phys.* **80**, 4442 (1996a).

Caillat, T., J.-P. Fleurial, and A. Borshchevsky, *J. Crystal Growth* **166**, 722 (1996b).

Caillat, T., J. Kulleck, A. Borshchevsky, and J.-P. Fleurial, *J. Appl. Phys.* **79**, 8419 (1996c).

Chen, L. D., T. Kawahara, X. F. Tang, T. Goto, T. Hirai, J. S. Dyck, W. Chen, and C. Uher, J. Appl. Phys. **90**, 1864 (2001).

Chattipadhyay, D. and H. J. Queisser, *Rev. Mod. Phys.* **53**, 745 (1981).

Cho, J. Y., Z. X. Ye, M. M. Tessema, J. R. Salvador, R. A. Waldo, Jiong Yang, W. Q. Zhang, Jihui Yang, W. Cai, and H. Wang, *J. Appl. Phys.* **113**, 143708 (2013).

Chubilleau, C., B. Lenoir, A. Dauscher, and C. Godart, *Intermetallics* **22**, 47 (2012a).

Chubilleau, C., B. Lenoir, P. Masschelein, A. Dauscher, and C. Godart, *J. Electron. Mater.* **41**, 1181 (2012b).

Conwell, E. M. and V. F. Weisskopf, *Phys. Rev.* **77**, 388 (1950).

Drude, P., *Ann. Physik* **1**, 566 (1900).

Dudkin, L. D. and N. K. Abrikosov, *Zh. Neorg. Khimii*, **2**, 212 (1957).

Dudkin, L. D. and N. K. Abrikosov, *Sov. Phys.-Solid State*, **1**, 126 (1959).

Dyck, J. S., W. Chen, J. Yang, G. P. Meisner, and C. Uher, *Phys. Rev. B* **65**, 115204 (2002).

Goldsmid, H. J., in *Electronic Refrigeration*, Pion (1986).

Hamaguchi, C., in *Basic Semiconductor Physics*, Springer, Berlin (2010).

Harrison, J. W. and J. R. Hauser, *Phys. Rev. B* **13**, 5347 (1976).

Haynes, J. R. and W. Shockley, *Phys. Rev.* **81**, 835 (1951).

He, Z., C. Stiewe, D. Platzek, G. Karpinski, E. Müller, S. H. Li, M. Toprak, and M. Muhammed, *Nanotech.* **18**, 235602 (2007).

Kajikawa, Y., *J. Appl. Phys.* **116**, 153710 (2014a).

Kajikawa, Y., J. Appl. Phys. **115**, 203716 (2014b).

Kane, E. O., *J. Phys. Chem. Solids* **1**, 249 (1957).

Katsuyama, S., Y. Shichijo, M. Ito, K. Majima, and H. Nagai, *J. Appl. Phys.* **84**, 6708 (1998).

Kawaharada, Y., K. Kurosaki, M. Uno, and S. Yamanaka, *J. Alloys Compd.* **315**, 193 (2001).

Kihou, K., I. Shirotani, Y. Shimaya, C. Sekine, and T. Yagi, *Mater. Res. Bull.* **39**, 317 (2004).

Kim, S. W., Y. Kimura, and Y. Mishima, *Sci. Technol. of Adv. Mater.* **5**, 485 (2004).

Kolodziejczak, J., *Acta Phys. Pol.* 20, 289 (1961).

Koyanagi, T., T. Tsubouchi, M. Ohtani, K. Kishimoto, H. Anno, and K. Matsubara, *Proc. 15th Int. Conf. on Thermoelectrics*, IEEE Catalog Number 96TH8169, Piscataway, NJ, p. 107 (1996).

Kronig, R. L. and W. J. Penney, Proc. Roy. Soc. London, A**130**, 499 (1930).

Lamberton, Jr., G. A., S. Bhattacharya, R. T. Littleton IV, M. A. Kaeser, R. H. Tedstrom, T. M. Tritt, J. Yang, and G. S. Nolas, *Appl. Phys. Lett.* **80**, 598 (2002).

Leszczynski, J., V. Da Ros, B. Lenoir, A. Candolfi, P. Masschelein, J. Hejtmanek, K. Kutorasinski, J. Tobola, R. I. Smith, C. Stiewe, and E. Müller, *J. Phys. D: Appl. Phys.* **46**, 495106 (2013).

Li, H., X. F. Tang, Q. J. Zhang, and C. Uher, *Appl. Phys. Lett.* **93**, 252109 (2008).

Li, X. Y., L. D. Chen, J. F. Fan, W. B. Zhang, T. Kawahara, and T. Hirai, *J. Appl. Phys.* **98**, 083702 (2005).

Li, Y. L., P. F. Qiu, H. Z. Duan, J. Chen, G. J. Snyder, X. Shi, B. B. Iversen, and L. D. Chen, *J. Mater. Chem C* **4**, 4374 (2016).

Liu, W.-S., B.-P. Zhang, J.-F. Li, and L.-D. Zhao, *J. Phys. D: Appl. Phys.* **40**, 566 (2007a).

Liu, W.-S., B.-P. Zhang, J.-F. Li, H.-L. Zhang, and L.-D. Zhao, *J. Appl. Phys.* **102**, 103717 (2007b).

Liu, W.-S., B.-P. Zhang, L.-D. Zhao, and J.-F. Li, *Chem. Mater.* **20**, 7526 (2008).

Lundstrom, M., *Fundamentals of Carrier Transport*, 2nd ed., Cambridge University Press (2000).

Mandrus, D., A. Migliori, T. W. Darling, M. F. Hundley, E. J. Peterson, and J. D. Thompson, *Phys. Rev. B* **52**, 4926 (1995).

Matsubara, K., T. Iyanaga, T. Tsobouchi, K. kishimoto, and T. Koyanagi, *AIP Conf. Proc.* 316, 226 (1994).

Matsubara, K., T. Sakakibara, Y. Notohara, H. Anno, H. Shimizu, and T. Koyanagi, *Proc. 15th Int. Conf. on Thermoelectrics*, IEEE Catalog Number 96TH8169, Piscataway, NJ, p. 96 (1997).

Morelli, D. T., T. Caillat, J. P. Fleurial, J. Vandersande, B. Chen, and C. Uher, *Phys. Rev. B* **51**, 9622 (1995).

Morelli, D. T., G. P. Meisner, B. X. Chen, S. Q. Hu, and C. Uher, *Phys. Rev. B* **56**, 7376 (1997).

Mott, N. F., *J. Non-Cryst. Solids* **1**, 1 (1968).

Moyzhes, B. and V. Nemchinsky, *Appl. Phys. Lett.* **73**, 1895 (1998).

Nagamoto, Y., K. Tanaka, and T. Koyanagi, *Proc. 17th Int. Conf. on Thermoelectrics*, IEEE Catalog Number 98TH8365, Piscataway, NJ, p. 302 (1998).

Nagao, J., D. Nataraj, M. Ferhat, T. Uchida, S. Takeya, T. Ebinuma, H. Anno, K. Matsubara, E. Hatta, and K. Musaka, *J. Appl. Phys.* **92**, 4135 (2002).

Navrátil, J., F. Laufek, T. Plecháček, and J. Plášil, *J. Alloys Compd.* **493**, 50 (2010).

Navrátil, J., T. Plecháček, Č. Drašar, and F. Laufek, *J. Electron. Mater.* **42**, 1864 (2013).

Nolas, G. S., G. A. Slack, D. T. Morelli, T. M. Tritt, and A. C. Ehrlich, *J. Appl. Phys.* **79**, 4002 (1996a).

Nolas, G. S., V. G. Harris, T. M. Tritt, and G. A. Slack, *J. Appl. Phys.* **80**, 6304 (1996b).

Nolas, G. S., J. L. Cohn, and G. A. Slack, *Phys. Rev. B* **58**, 164 (1998).

Nolas, G. S., H. Takizawa, T. Endo, H. Sellinschegg, and D. C. Johnson, *Appl. Phys. Lett.* **77**, 52 (2000a).

Nolas, G. S., M. Kaeser, R. T. Littleton, and T. M. Tritt, *Appl. Phys. Lett.* **77**, 1855 (2000b).

Nordheim, L., Ann. Phys. (Leipzig) **9**, 607 (1931).

Odile, J. P., S. Soled, C. A. Castro, and A. Wald, *Inorg. Chem.* **17**, 283 (1978).

Park, K.-H., W.-S. Seo, D.-K. Shin, and I.-H. Kim, *J. Korean Phys. Soc.* **65**, 491 (2014).

Pei, Y. Z., L. D. Chen, W. Q. Zhang, X. Shi, S. Q. Bai, X. Y. Zhao, Z. G. Mei, and X. Y. Li, *Appl. Phys. Lett.* **89**, 221107 (2006).

Pei, Y. Z., J. Yang, L. D. Chen, W. Q. Zhang, J. R. Salvador, and J. H. Yang, *Appl. Phys. Lett.* **95**, 042101 (2009).

Pleass, C. M. and R. P. Heyding, *Can. J. Chem.* **40**, 590 (1962).

Pourret, A., K. Behnia, D. Kikuchi, Y. Aoki, H. Sugawara, and H. Sato, *Phys. Rev. Lett.* **96**, 176402 (2006).

Price, P. I., *Phil. Mag.* **46**, 1252 (1955).

Price, P. I., *Phys. Rev.* **102**, 1245 (1956).

Puyet, M., A. Dauscher, B. Lenoir, M. Dehmas, C. Stiewe, E. Müller, and J. Hejtmanek, J. Appl. Phys. **97**, 083712 (2005).

Ravich, Y. L., B. A. Efimova, and I. A. Smirnov, *Semiconducting Lead Chaclogenides*, Plenum Press, N.Y.-London (1970).

Ravich, Y. L., in *CRC Handbook of Thermoelectrics*, ed. D. M. Rowe, CRC Press, Boca Raton FL, Ch. 7, p. 67 (1995).

Saeed, M., B. Khan, I. Ahmad, A. S. Saleemi, N. Rehman, H. A. Rahnamaye Aliabad, and S. Uddin, *RSC Adv.* **9**, 24981 (2019).

Sales, B. C., B. C. Chakoumakos, and D. Mandrus, *Mater. Res. Soc. Symp. Proc.* **626**, Z7.1.1 (2000).

Sato, H., Y. Abe, H. Okada, T. D. Matsuda, K. Abe, H. Sugawara, and Y. Aoki, *Phys. Rev. B* **62**, 15125 (2000).

Shaheen, N., X. C. Shen, M. S. Javed, H. Zhan, L. J. Guo, R. Alsharafi, T. Y. Huang, X. Lu, G. Y. Wang, and X. Y. Zhou, *J. Electron. Mater.* **46**, 2958 (2017).

Sharp, J. W., E. C. Jones, R. K. Williams, P. M. Martin, and B. C. Sales, *J. Appl. Phys.* **78**, 1013 (1995).

Shi, X., H. Kong, C.-P. Li, C. Uher, J. Yang, J. R. Salvador, H. Wang, L. D. Chen, and W. Q. Zhang, *Appl. Phys. Lett.* **92**, 182101 (2008).

Shirotani, I., E. Takahashi, N. Mukai, K. Nozawa, M. Kinoshita, T. Yagi, K. Suzuki, T. Enoki, and S. Hino, *Jap. J. Appl. Phys.* 32, Suppl. 32–3, 294 (1993).

Singh, D. J. and W. E. Pickett, *Phys. Rev. B* **50**, 11235 (1994).

Sirimart, J., J.-I. Hayashi, Y. Kawamura, K. Kihou, H. Nishiate, C.-H. Lee, and C. Sekine, *Jap. J. Appl. Phys.* **58**, 081006 (2019).

Slack, G. A. and V. G. Tsoukala, *J. Appl. Phys.* **76**, 1665 (1994).

Sommerfeld, A., *Z. Physik* **47**, 43 (1928).

Stassen, A. F., R. W. I. De Boer, N. N. Iosad, and A. F. Morpungo, *Appl. Phys. Lett.* **85**, 3899 (2004).

Suzuki, T., A. Kikkawa, Y. Tokura, and Y. Taguchi, *Phys. Rev. B* **93**, 155101 (2016).

Tan, G. J., W. Liu, H. Chi, X. Su, S. Wang, Y. G. Yan, X. F. Tang, W. Wong-Ng, and C. Uher, *Acta Mater.* **61**, 7693 (2013).

Tang, Y., Z. M. Gibbs, L. A. Agapito, G. D. Li, H.-S. Kim, M. B. Nardelli, S. Curtarolo, and G. J. Snyder, *Nature Mater.* **14**, 1223 (2015).

Tashiro, H., Y. Notohara, T. Sakakibara, H. Anno, and K. Matsubara, *Proc. 16th Int. Conf. on Thermoelectrics*, IEEE Catalog Number 97TH8291, Piscataway, NJ, p. 326 (1997).

Thomson, J. J., *Phil Mag.* **44**, 293 (1897).

Torikachvili, M. S., J. W. Chen, Y. Dalichaouch, R. P. Guertin, M. W. McElfresh, C. Rossel, M. B. Maple, and G. P. Meisner, *Phys. Rev. B* **36**, 8660 (1987).

Tritt, T. M., G. S. Nolas, G. A. Slack, A. C. Ehrlich, D. J. Gillespie, and J. L. Cohn, *J. Appl. Phys.* **79**, 8412 (1996).

Wang, L., K. C. Cai, Y. Y. Wang, H. Li, and H. F. Wang, *Appl. Phys. A* **97**, 841 (2009).

Wang, Y. M., J. Mao, Q. Jie, B. H. Ge, and Z. F. Ren, *Appl. Phys. Lett.* **110**, 163901 (2017).

Watcharapasorn, A., R. C. DeMattei, and R. S. Feigelson, T. Caillat, A. Borshchevsky, G. J. Snyder, and J.-P. Fleurial, *Mater. Res. Soc. Symp. Proc.*, Vol. 626, Z.1.4.1 (2000).

Wojciechowski, K. T., *Mater. Res. Bull.* **37**, 2023 (2002).

Wojciechowski, K. T., J. Tobola, and J. Leszczynski, *J. Alloys Compd.* **361**, 19 (2003).

Yang, J., G. P. Meisner, D. T. Morelli, and C. Uher, *Phys. Rev. B* **63**, 014410 (2000)

Yang, J., Y. Xi, W. Zhang, L. D. Chen, and J. Yang, *J. Electron. Mater.* **38**, 1397 (2009).

Yang, L., H. H. Hng, D. Li, Q. Y. Yan, J. Ma, T. J. Zhu, X. B. Zhao, and H. Huang, *J. Appl. Phys.* **106**, 013705 (2009).

Yang, X. X., Z. H. Dai, Y. C. Zhao, W. C. Niu, J. N. Liu, and S. Meng, *Phys. Chem. Chem. Phys.* **21**, 851 (2019).

Yu, P. Y. and M. Cardona, *Fundamentals of Semiconductors*, 4th ed., Springer (2010).

Zawadzki, W., *Adv. Phys.* **23**, 435 (1974).

Zevalkink, A., K. Star, U. Aydemir, G. J. Snyder, J.-P. Fleurial, S. Bux, T. Vo, and P. von Allmen, *J. Appl. Phys.* **118**, 035107 (2015).

Zhang, L., F. F. Duan, X. D. Li, X. L. Yan, W. T. Hu, L. M. Wang, Z. G. Liu, Y. J. Tian, and B. Xu, *J. Appl. Phys.* **114**, 083715 (2013).

Zhao, X. Y., X. Shi, L. D. Chen, W. Q. Zhang, S. Q. Bai, Y. Z. Pei, X. Y. Li, and T. Goto, *Appl. Phys. Lett.* **89**, 092121 (2006).

Zobrina, B. N. and L. D. Dudkin, *Sov. Phys.-Solid State* **1**, 1668 (1959).

5 Thermal Transport Properties of Skutterudites

The important role the thermal conductivity plays in determining whether a given material might or might not be a prospective thermoelectric necessitates a thorough examination of phonon vibrations, dispersion of phonon modes, and the phonon density of states (PDOS). Given that binary skutterudites have outstanding electronic properties but are handicapped by a large thermal conductivity, it is critical to impede the propagation of phonons in the skutterudite structure while minimally perturbing their electronic states. Actually, an early interest in the vibrational properties of skutterudites predates the interest driven by thermoelectricity and goes back to the 1980s, when the far-infrared spectra of binary skutterudites CoP_3, $CoAs_3$, and $CoSb_3$ and their solid solutions were studied by Lutz and Kliche (1981, 1982) and Kliche and Bauhofer (1988). Following the promulgation of the Phonon-Glass-Electron-Crystal (PGEC) paradigm by Slack (1995), and experimental verification that filling voids of the structure suppresses the thermal conductivity, Morelli and Meisner (1995), the study of vibrational properties of filled skutterudites has dramatically intensified.

To illustrate the essence of lattice vibrations and to introduce the terminology used in this section, I will first introduce the concept of normal phonon modes, and then describe two simplest cases of one-dimensional chains to illuminate how acoustic and optic modes of vibrations arise. With the background set, the lattice dynamics and thermal conductivity of various forms of skutterudites will be discussed.

5.1 NORMAL PHONON MODES

A crystal consists of atoms bound together by forces between the nearest neighbors. Let us assume there are N such atoms and their equilibrium positions are expressed *via* vectors $r_i^{(0)}$ with the Cartesian coordinates $x_{i,1}^{(0)}$, $x_{i,2}^{(0)}$, and $x_{i,3}^{(0)}$. Thermal motion displaces the atoms from their equilibrium positions by distances designated by vectors u_i to new positions r_i; we thus have

$$u_i \equiv r_i - r_i^{(0)}. \qquad (5.1)$$

The vibrating atom has a kinetic energy $\frac{1}{2}m_i \dot{u}_i^2$, and thus the kinetic energy of all vibrating atoms is

$$K = \frac{1}{2}\sum_{i=1}^{N} m_i \dot{u}_i^2, \qquad (5.2)$$

where $\dot{u}_i$ is the velocity of the i-th atom, and m_i is its mass. The potential energy of the crystal of N atoms is $V(r_1, r_2, \ldots, r_N)$. Since the equilibrium position corresponds to a minimal value of the potential energy, and since we assume small deviations from the equilibrium position, the potential energy can be expanded (by a Taylor series) around the equilibrium position, i.e.,

$$V(r_1, r_2, \ldots, r_N) = V_0 + \sum_{i=1}^{N}\left(\frac{\partial V}{\partial r_i}\right)_0 u_i + \frac{1}{2}\sum_{i,j}\left(\frac{\partial^2 V}{\partial u_i \partial u_j}\right)_0 u_i u_j + \ldots, \qquad (5.3)$$

169

where all derivatives are taken at the equilibrium position. In Eq. 5.3, the term V_0 (potential energy at equilibrium) is constant, and the linear term is zero because the force on an atom at equilibrium must be zero. The term of significance is the quadratic (harmonic) term. Designating the second-order partial derivative by $G_{i,j}$, the potential energy becomes

$$V(r_1, r_2, \ldots, r_N) = V_0 + \frac{1}{2} \sum_{i,j} G_{i,j}\, u_i u_j. \tag{5.4}$$

Combining Eqs. 5.2 and 5.4, the Hamiltonian (total energy) of the system is

$$H = V_0 + \frac{1}{2} \sum_{i=1}^{N} m_i \dot{u}_i^2 + \frac{1}{2} \sum_{i,j} G_{i,j} u_i u_j. \tag{5.5}$$

While the kinetic energy in Eq. 5.5 is simple, as each term refers to a single atom, the potential energy term is more complicated, as each term involves coordinates of two different atoms i and j, reflecting interactions between a given atom i and all other atoms j. However, with a transformation of the form

$$u_i = \sum_{s=1}^{3N} A_{i\alpha,s} q_s, \tag{5.6}$$

where $A_{i\alpha,s}$ are appropriately chosen coefficients with the subscript α indicating one of the three Cartesian coordinates, the Hamiltonian in Eq. 5.5 can written as

$$H = V_0 + \frac{1}{2} \sum_{s=1}^{3N} \left(\dot{q}_s^2 + \omega_s^2 q_s^2 \right). \tag{5.7}$$

The term $\dfrac{1}{2} \displaystyle\sum_{s=1}^{3N} \left(\dot{q}_s^2 + \omega_s^2 q_s^2 \right)$ *de facto* represents 3N independent one-dimensional harmonic oscilla-

tors, each having an angular frequency ω_s, otherwise known as the normal mode frequency or normal phonon mode.

The next two subsections will illustrate the concept of normal phonon modes on two simple examples.

5.1.1   NORMAL MODES OF A MONATOMIC LINEAR CHAIN

A section of the one-dimensional chain of N identical atoms of mass m held together by harmonic springs with the spring constant k that act only between adjacent atoms is illustrated in Figure 5.1. In equilibrium, the spacing of atoms is a. At some particular time t, the small displacements of atoms from equilibrium are $u_1, u_2, \ldots u_{n-1}, u_n, u_{n+1}, \ldots u_N$, as indicated in the figure.

The force on the atom at position na, which is displaced from equilibrium by u_n, comes from its neighbor on the left (displaced by u_{n-1}) and the neighbor on the right, displaced by u_{n+1}. The force on the atom at na is thus $F_n = -k(u_n - u_{n-1}) - k(u_n - u_{n+1}) = -k(2u_n - u_{n-1} - u_{n+1})$. By Newton's second law, this must equal $m\ddot{u}_n$, hence

$$m\ddot{u}_n = -k(2u_n - u_{n-1} - u_{n+1}). \tag{5.8}$$

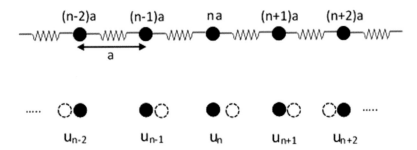

FIGURE 5.1 A linear chain of identical atoms interacting between the nearest neighbors only. The lower sketch indicates displacements of atoms from their equilibrium positions.

Similar equations apply for any other atom, except for the one at the beginning and one at the end of the chain. Usually, these two atoms are treated using the Born-von Karman periodic boundary conditions, which means that the beginning and the end of the chain are joined by a spring with the same spring constant as connects all other atoms. Since we have thus made a circle of atoms, the labels n and $n+N$ obviously refer to the same atom, hence

$$u_n(t) = u_{N+n}(t). \tag{5.9}$$

In analogy to vibrations of a string, we anticipate a solution in the form of traveling waves, i.e., we take the displacement as

$$u_n(t) = Ae^{i(qna-\omega t)}, \tag{5.10}$$

where q is a wave vector of the wave, $q = 2\pi/\lambda$, (λ being the wavelength), and ω its angular frequency. Substituting Eq. 5.10 into Eq. 5.8 leads to

$$-m\omega^2 = -k\left(2 - e^{iqa} - e^{-iqa}\right) = -4k\sin^2\frac{qa}{2}. \tag{5.11}$$

Isolating the frequency, we arrive at

$$\omega = \left(\frac{4k}{m}\right)^{1/2}\left|\sin\frac{qa}{2}\right| = \omega_{max}\left|\sin\frac{qa}{2}\right| \equiv \omega_{max}\left|\sin\frac{\pi a}{\lambda}\right|, \tag{5.12}$$

where the maximum angular frequency is

$$\omega_{max} = 2\pi f_{max} = \left(\frac{4k}{m}\right)^{1/2}. \tag{5.13}$$

The solution, i.e., the angular frequency as a function of the wave vector, is depicted in Figure 5.2, and represents a traveling wave on the chain. Substituting Eq. 5.9 into Eq. 5.10, we arrive at $e^{i(qna-\omega t)} = e^{i(q[n+N]a-\omega t)}$, from which

$$q = \frac{2\pi}{Na}j = \frac{2\pi}{L}j, \tag{5.14}$$

where Na is the length of the chain L and j is an integer. Equation 5.14 stipulates the number of allowed vibrational modes. They must fall within the first Brillouin zone, i.e., they must

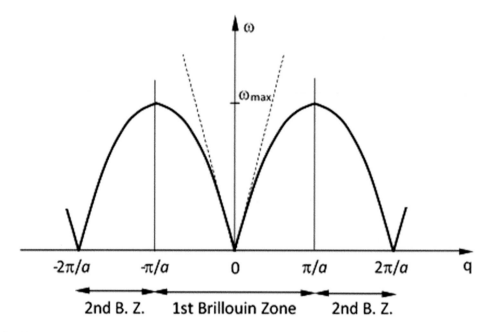

FIGURE 5.2 Dispersion curve, i.e., the dependence of the angular frequency ω on the wave vector q in a monatomic linear chain connected by harmonic springs of the spring constant k, with the interaction restricted to the nearest neighbors only. The first Brillouin zone and the two sections of the second Brillouin zone are shown. Note that the frequency is linear for small values of q, and the group velocity $v_g = \partial\omega/\partial q$ has its largest value. At the Brillouin zone boundary, $\partial\omega/\partial q$ is zero.

satisfy $-\dfrac{\pi}{a} \leq q \leq \dfrac{\pi}{a}$. Substituting Eq. 5.8 into the inequality, a restriction on the integer values j follows,

$$-\frac{N}{2} \leq j \leq \frac{N}{2}. \tag{5.15}$$

Equation 5.15 states that the values of j are restricted to $j = \pm 1, \pm 2, \pm 3, \ldots \pm \dfrac{N}{2}$. Note, $j = 0$ means that all atoms are at rest and there are no vibrations, hence we do not consider $j = 0$. The total number of distinct vibration modes is thus equal to N, the number of atoms in the chain. One may ask how densely distributed the modes are. From Eq. 5.14, we have

$$dj = \frac{Na}{2\pi} dq = \frac{L}{2\pi} dq. \tag{5.16}$$

The branch depicted in Figure 5.2 is called the acoustic branch. It is characterized by the frequency becoming zero as the wave vector vanishes. In monatomic chains, acoustic vibrations are the only type of vibrations. It is also interesting to note that the monatomic chain cannot support waves with arbitrarily small wavelengths. The restriction comes from the fact that $\sin(qa/2)$ has a maximum value of unity. Thus, the wavelength $\lambda = \dfrac{2\pi}{q} \geq \dfrac{2\pi}{\pi/a} = 2a$, i.e., there cannot be a wave on the mona-

tomic chain with wavelength shorter than twice the distance between the atoms.

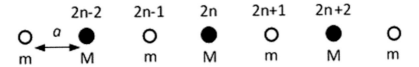

FIGURE 5.3 Diatomic chain consisting of atoms of mass M alternating with atoms of mass m, all atoms at the same equilibrium spacing a, and interacting between the nearest neighbors only.

5.1.2 Normal Modes of a Diatomic Linear Chain

If there are two or more kinds of atoms on the chain, the situation is a bit more complicated. Let us consider what happens when two atoms with different mass alternate on a linear chain. The simplest case is when the nearest neighbor separations are the same and the interaction again takes place only between the immediate neighbors. For instance, we can take every even number atom as having mass M and every odd number atom with a smaller mass m (Figure 5.3).

Proceeding in exactly the same way as in the case of a monatomic chain, we write the Newton's second law equations

$$M\ddot{u}_{2n} = -k\left(2u_{2n} - u_{2n-1} - u_{2n+1}\right),\tag{5.17a}$$

$$m\ddot{u}_{2n+1} = -k\left(2u_{2n+1} - u_{2n} - u_{2n+2}\right),\tag{5.17b}$$

and assume two traveling waves, one for the even-numbered atoms of mass M and one for the odd-numbered atoms of mass m,

$$u_{2n}(t) = Ae^{i(2nqa-\omega t)},\tag{5.18a}$$

$$u_{2n+1}(t) = Be^{i((2n+1)qa-\omega t)}.\tag{5.18b}$$

Substituting these wave forms into Newton equations, one obtains, after simple manipulation,

$$\left(M\omega^2 - 2k\right)A + 2Bk\cos qa = 0,\tag{5.19a}$$

$$\left(m\omega^2 - 2k\right)B + 2Ak\cos qa = 0.\tag{5.19b}$$

This set of equations will have a nonzero solution when its determinant is equal to zero, i.e.,

$$\begin{vmatrix} M\omega^2 - 2k & 2k\cos qa \\ 2k\cos qa & m\omega^2 - 2k \end{vmatrix} = 0.\tag{5.20}$$

By setting $x = \omega^2$ and solving a quadratic equation, the resulting frequencies become

$$\omega^2 = \left(\frac{1}{M} + \frac{1}{m}\right)k \pm k\left[\left(\frac{1}{M} + \frac{1}{m}\right)^2 - \frac{4\sin^2 qa}{Mm}\right]^{\frac{1}{2}}.\tag{5.21}$$

Since vibrations are real, there are two values of ω for any given q. The upper branch of vibrations goes with the + sign in Eq. 5.21, and is called the optic branch (because the frequency falls in the infrared range),

$$\omega_+^2 = k\left(\frac{1}{M} + \frac{1}{m}\right) + k\left[\left(\frac{1}{M} + \frac{1}{m}\right)^2 - \frac{4\sin^2 qa}{Mm}\right]^{\frac{1}{2}}.\tag{5.22}$$

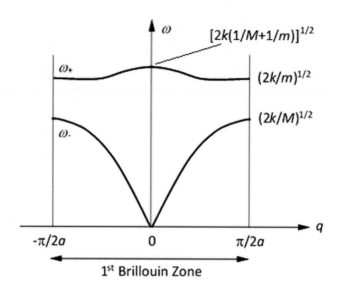

FIGURE 5.4 Dispersion of a diatomic chain with masses M > m, including the higher lying optic branch ω_+ and the lower lying acoustic branch ω_-.

The lower branch of vibrations, called the acoustic branch, follows by taking a negative sign,

$$\omega_-^2 = k\left(\frac{1}{M}+\frac{1}{m}\right) - k\left[\left(\frac{1}{M}+\frac{1}{m}\right)^2 - \frac{4\sin^2 qa}{Mm}\right]^{1/2}. \tag{5.23}$$

The two branches are shown in Figure 5.4. The $\sin^2 qa$ term makes the frequency a periodic function of the wave vector q. The maximum value of $\sin^2 qa$ is unity, which leads to $q = \pm\dfrac{\pi}{2a}$. The first Brillouin zone is thus delineated by values of q given by $-\dfrac{\pi}{2a} \leq q \leq \dfrac{\pi}{2a}$.

The important outcome of the presence of a second type of atoms is the appearance of the higher frequency optic branch of vibrations. Anytime more than a single type of atoms is present in the lattice or when the unit cell contains more than one atom, optic branches will form.

Let us first look at frequencies at the zone center, $q = 0$. Equations 5.22 and 5.23 lead to

$$\omega_+ = \left[2k\left(\frac{1}{M}+\frac{1}{m}\right)\right]^{1/2}, \tag{5.24}$$

and

$$\omega_- = 0. \tag{5.25}$$

At the zone boundaries, $q = \pm\dfrac{\pi}{2a}$, and substituting this into ω_+ and ω_-, leads to

$$\omega_+ = \left(\frac{2k}{m}\right)^{1/2}, \tag{5.26}$$

$$\omega_- = \left(\frac{2k}{M}\right)^{\frac{1}{2}}. \tag{5.27}$$

Both values are indicated in Figure 5.4. Obviously, in a diatomic linear chain with masses that are the same, $m = M$, the branches ω_+ and ω_- will have the same value at the zone boundary.

It is instructive to see how the two vibrating lattices move relative to each other. From Eq. 5.19a there follows a relation between the amplitudes of vibrations of atoms A and atoms B:

$$\frac{A}{B} = \frac{2kcosqa}{2k - M\omega^2}. \tag{5.28}$$

At the zone center, $q = 0$, and Eq. 5.28 simplifies into

$$\frac{A}{B} = \frac{2k}{2k - M\omega^2}. \tag{5.29}$$

For the acoustic branch, $\omega_-(q = 0) = 0$, and $A = B$. In other words, at the zone center, all atoms of the acoustic branch move in the same direction and have the same amplitude.

For the optic branch at the zone center, $\omega_+ = \left[2k\left(\frac{1}{M} + \frac{1}{m}\right)\right]^{\frac{1}{2}}$, and substituting into Eq. 5.29

leads to $AM = -mB$. Hence, in the optic branch at the zone center, atoms M move in opposite direction to atoms m, but in such a way that the center of mass remains fixed. For any other wave vector within the first Brillouin zone, the phase relation between atoms M and atoms m can be obtained from Eq. 5.28 and the appropriate values of frequencies ω_+ and ω_- given by Eqs. 5.22 and 5.23.

For any set of displacements $\{u_{na}\}$, the total energy of the system will be $E\{u_{na}\}$. The force on the j-th atom is then $F_j = -\partial E\{u_{na}\}/\partial u_j$. It is a standard practice to expand the energy by applying Taylor's expansion and write

$$E\{u_{na}\} = E(0) + \frac{1}{2}\sum_{ij}\left(\frac{\partial^2 E}{\partial u_i \partial u_j}\right)_0 u_i u_j + \frac{1}{3!}\sum_{ijk}\left(\frac{\partial^3 E}{\partial u_i \partial u_j \partial u_k}\right)_0 u_i u_j u_k + \cdots. \tag{5.30}$$

By keeping only the second term in Eq. 5.30, a harmonic approximation, appropriate for small enough displacements, is obtained. The first term does not influence vibrational properties (it is relevant only for the cohesive energy of a solid), and the term where the energy is linear in displacement is missing in Eq. 5.30 because at equilibrium $\partial E/\partial u_j = 0$ for any j. Thus, within the harmonic approximation,

$$E_{\{u_{na}\}} = \frac{1}{2}\sum_{ij}\left(\frac{\partial^2 E}{\partial u_i \partial u_j}\right)_0 u_i u_j \equiv \frac{1}{2}\sum_{ij}L_{ij} u_i u_j, \tag{5.31}$$

where L_{ij} is a force constant defined by

$$L_{ij} = \left(\frac{\partial^2 E}{\partial u_i \partial u_j}\right)_0. \tag{5.32}$$

The force constants relate the force acting on a particular atom as a result of its displacement, and their determination is the key step in developing the lattice dynamics of a system.

Developing the lattice dynamics of a three-dimensional crystal would proceed in the same way as illustrated for a one-dimensional chain of atoms, except that the notation becomes quite messy as now one has to deal with vibrations in three dimensions in a crystal that may contain a basis of

n_b atoms in a unit cell, Moreover, a relation between the direction of propagation q (q now being a vector) and the polarization vector must be considered. As q spans the first Brillouin zone, at every value of q there will be $3n_b$ normal modes of vibrations, three of them being acoustic, characterized by $\omega(q \to 0) \to 0$, and $3n_b - 3$ being optic modes. Each acoustic and optic mode consists of one branch polarized parallel to q (called the longitudinal branch) and two branches polarized perpendicular to the direction of q (called the transverse branches).

To start with, one determines the force constants acting on atoms of the basis as they are displaced by a small amount from their equilibrium position, taking into account the symmetry of the crystal lattice. The next step is to solve a set of coupled differential equations of the type given in Eq. 5.19 for all atoms forming the basis of the unit cell.

It should be pointed out that in a three dimensional space, the first Brillouin zone (or simply the Brillouin zone) is a geometrical construct in a reciprocal space corresponding to a particular real space crystal structure, i.e., a real space Bravais lattice. In the language of solid-state physics, it is the Wigner-Seitz primitive cell of the reciprocal lattice. It is obtained as a result of bisecting the nearest neighbor reciprocal lattice vectors by a plane, and taking the smallest enclosed volume. The Brillouin zone of a skutterudite structure is depicted in Figure 5.5, including some high symmetry points.

The last thing to note is an expression for the phonon density of states, $g(\omega)$, defined so that $g(\omega)$ $d\omega$ stands for the number of modes contained in a small frequency interval between ω and $\omega + d\omega$, normalized per total volume V of the crystal. Since each allowed wavevector q occupies a volume $(2\pi)^3/V$ of the Brillouin zone, then a volume element $\Delta\Omega_q$ of the Brillouin zone will contain the number of allowed wavevectors given by $\Delta n_q = \dfrac{\Delta\Omega_q}{\left(2\pi\right)^3/V} = \dfrac{V\Delta\Omega_q}{\left(2\pi\right)^3}$. As the volume in the Brillouin zone we take the volume enclosed between surfaces on which the angular frequency has constant values ω and $\omega + d\omega$, shown in Figure 5.6. As an element of this volume, we take a small cylinder having base dS_ω and height $\Delta q \perp$ and integrate over the surface S_ω. Since an increment of the angular frequency $\Delta\omega(q)$ can be written as $\Delta\omega(q) = \nabla_q\omega(q)\Delta q_\perp$, the infinitesimal element of the volume contained between the surfaces S_ω and $S_{\omega+\Delta\omega}$ is $\Delta\Omega_q = dS_\omega dq_\perp = \dfrac{dS_\omega}{\left|\nabla_q\omega(q)\right|}\Delta\omega(q)$. Dividing the number of allowed wavevectors Δn_q by $\Delta\omega(q)$ and taking the limit of this expression for $\Delta\omega(q) \to 0$, the phonon density of states becomes

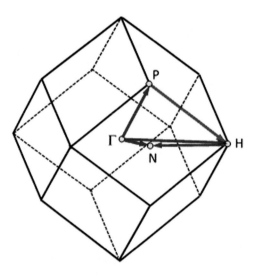

FIGURE 5.5 Brillouin zone of a skutterudite indicating the zone center Γ and some high symmetry points and directions usually marked in the dispersion relations.

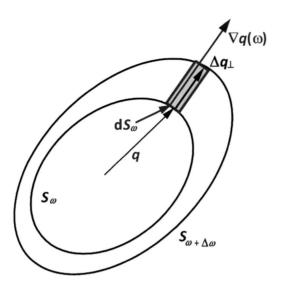

FIGURE 5.6 A schematic drawing illustrating the calculation of the density of states.

$$g(\omega) = \frac{1}{(2\pi)^3} \int_{S_\omega} \frac{dS_\omega}{\left|\nabla_q \omega(q)\right|}. \tag{5.33}$$

If the basis motif consists of s atoms, we have to add their contributions and write the phonon density of states as

$$g(\omega) = \sum_s \frac{1}{(2\pi)^3} \int \frac{dS_\omega}{\left|\nabla_q \omega_s(q)\right|}. \tag{5.34}$$

Obviously, high density of states occurs where $\left|\nabla_q \omega(q)\right|$ is small, i.e., where the dispersion curve is rather flat. At frequencies where the dispersion has a horizontal tangent, the derivative of the density of states with respect to frequency has a singularity, referred to as van Hove singularity. With the above as a background, we now consider lattice modes in skutterudites.

5.2 LATTICE MODES AND THE PHONON DENSITY OF STATES IN SKUTTERUDITES

5.2.1 LATTICE MODES IN BINARY SKUTTERUDITES

According to group theory, the crystal structure of skutterudites with the symmetry $Im\overline{3} - T_h^5$ features the total of 45 zone center vibrations, which can be decomposed into the irreducible representation

$$\Gamma = 2A_g + 2E_g + 4F_g + 2A_u + 2E_u + 7F_u. \tag{5.35}$$

Thus, there are 19 distinct zone center phonon modes. Of these, the seven F_u modes are infrared-active and include vibrations of metal ions, while eight modes

$$\Gamma_R = 2A_g + 2E_g + 4F_g, \tag{5.36}$$

are first-order Raman modes, exclusively related to vibrations of the pnicogen atoms. A_g modes are singly degenerate, E_g modes are doubly degenerate, and F_g modes are triply degenerate.

Frequencies of the infrared modes were determined experimentally from far-infrared spectra of $CoSb_3$ by Lutz and Kliche (1982). By fitting the seven frequencies with six bond-stretching force constants (interactions of the nearest-neighbor Co–Sb, the nearest-neighbor Co–Co, two nearest-neighbor Sb–Sb within the ring, and two nearest-neighbor Sb–Sb between the rings), they developed what is known as the Lutz and Kliche model. Feldman and Singh (1996) used DFT-LDA calculations to determine frequencies and eigenvectors of A_g and A_u zone center phonons. The A_g modes reflect changes in the positional parameters y and z, while the symmetry of the structure is maintained, often referred to as the breathing modes of the pnicogen rings. There are two such modes because the sides of the pnicogen ring d_1 and d_2 are not equal. These Raman-type frequencies turned out to be 150 cm^{-1} and 178 cm^{-1} (1 cm^{-1} = 1.24 × 10^{-4} eV), having the polarization vectors very nearly parallel to the sides of the Sb$_4$ rings. The higher frequency here corresponds to modulations of the shorter d_1 edge of the ring. The two A_u modes have frequencies of 110 cm^{-1} and 241 cm^{-1}, with the lower frequency corresponding to in-phase motion of nearest-neighbor Sb and Co atoms, while the higher frequency is primarily due to their out-of-phase motion. The above four frequencies, together with the Lutz-Kliche force-constant model, served Feldman and Singh as input to compute frequencies of all zone center phonon modes, presented in Table 5.1 as the LK model. Moreover, Feldman and Singh improved on the LK model by going beyond the central forces and included two Sb–Co–Sb bond angle force constants to account for the two distinct bond angles of 109.2° and 107.8°. They considered a single additional bond-stretching force constant between the second- and third-neighbor Co and Sb atoms, the distance between which is shorter than the distance between Co–Co.

Depending on the bond angle terms, Feldman and Singh computed four variants of the model, all giving similar results. The final set of zone center phonon frequencies of CoSb$_3$ reported by Singh

TABLE 5.1

Frequencies of Zone Center Phonon Modes of CoSb$_3$ in Units of cm^{-1}

Mode	LK	Singh (1996)	Theory FS LDA (1996)	R LDA (2008)	G LDA (2007)	LK (1982)	Experiment N (1999)	P (2012)
A_g	162	149	150	153	141		135	149
	183	177	178	183	169		186	183
A_u	69	109	110	115	116			
	250	242	241	258	236			
F_u	79	78		80	80	78		
	119	120		123	117	120		
	143	145		146	136	144		
	174	176		178	167	174		
	242	242		253	232	247		
	258	260		263	245	257		
	277	275		281	258	275		
E_g	140	141		134	122			133
	194	181		190	175			
F_g	71	84		85	87			
	103	96		110	105			
	162	158		158	143			163
	188	176		190	173			176
E_u	95	139		134	133			
	273	262		274	246			

LK stands for the force constant model of Lutz and Kliche (1982), FS LDA are LDA-computed frequencies by Feldman and Singh (1996), R LDA are frequencies computed by Rotter et al. (2008), G LDA are frequencies reported by Ghosez and Veithen (2007). LK (1982) stands for experimental frequencies obtained by Lutz and Kliche (1982) using far-infrared spectra of CoSb$_3$, N (1999) are experimental frequencies reported by Nolas et al. (1999), and P (2012) are experimental frequencies obtained by Peng et al. (2012a).

et al. (1996) is entered in Table 5.1. The table also includes the already noted four frequencies of A_g and A_u modes computed by Feldman and Singh using DFT-LDA calculations, DFT-LDA zone center frequencies calculated by Ghosez and Veithen (2007), and LDA-computed zone center frequencies obtained from inelastic X-ray scattering measurements by Rotter et al. (2008) on single crystals of $CoSb_3$ that took into account bond-angle force constants. On the experimental side, the table lists frequencies from far-infrared measurements by Lutz and Kliche (1982), frequencies of the A_g modes from Raman measurements by Nolas et al. (1999), and several Raman frequencies measured by Peng et al. (2012a), There is excellent agreement among all low frequency modes. At high frequencies, there is some discord, partly due to difficulties in assigning certain modes. The phonon density-of-states (PDOS) $g(\omega)$ obtained by Feldman and Singh (1996) by sampling 256 points in the irreducible element of the Brillouin zone is depicted in Figure 5.7. There are several important aspects of the PDOS. Starting at low frequencies, the parabolic low-frequency range, dominated by acoustic phonons, is interrupted at about 70 cm⁻¹ as the low-lying optic branches start to make a significant contribution. These evolve from the substantially rigid librational Sb_4 ring character to the twisting motions of the rings, then to the rigid translational motion of Sb_4 and, finally, to the bond-stretching motion within the rings. Interestingly, there is a gap in the spectrum of vibrating Sb atoms just above 100 cm⁻¹. The essentially Sb-derived modes extend to near 200 cm⁻¹, and the dominant transition metal modes are well separated from the Sb modes by a gap and exert their influence above 230 cm⁻¹. The importance of the low-frequency part of the spectrum to the heat transport is made even more apparent by considering the PDOS weighted by the squared group velocity, $\left\langle v_g^2 \right\rangle g(\omega)$.

This parameter, multiplied by the relaxation time τ, is called the phonon transport function and enters in the Boltzmann theory of thermal conductivity as an indicator of modes that dominate heat transport. The phonon transport function for $CoSb_3$ computed by Singh et al. (1999) is shown in Figure 5.8. It is clear that the phonon modes of the transition metal are substantially irrelevant and that the heat transport in $CoSb_3$ (and other binary skutterudites) is due to acoustic and, to a lesser extent, optic modes originating from vibrations of Sb atoms.

Phonon dispersion in $CoSb_3$, including element-specific PDOS, computed by Rotter et al. (2008) from first principles, is shown in Figure 5.9. The results verify previous conclusions that phonon

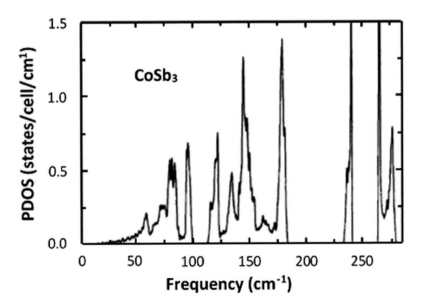

FIGURE 5.7 Phonon density of states of $CoSb_3$. Adapted from J. L. Feldman and D. J. Singh, *Physical Review B* 53, 6273 (1996). With permission from the American Physical Society.

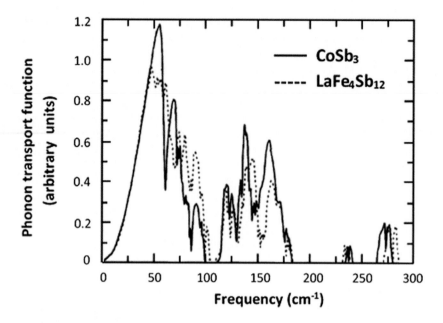

FIGURE 5.8 Phonon transport function, $\langle v_g^2 \rangle g(\omega)\tau$, for CoSb$_3$ (solid curve) and LaFe$_4$Sb$_{12}$ (dashed curve), assuming constant relaxation time τ. Adapted from D. J. Singh et al., *Materials Research Society Symposia Proceedings* 545, 3 (1999). With permission from the Materials Research Society.

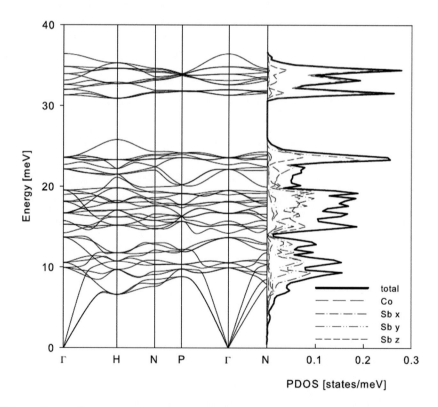

FIGURE 5.9 Phonon dispersion and element-specific phonon density of states in CoSb$_3$ computed from first principles using LDA. Reproduced from M. Rotter et al., *Physical Review B* 77, 144301 (2008). With permission from the American Physical Society.

modes below 25 meV (about 200 cm^{-1}) are mostly due to vibrations of Sb atoms and are well separated from high frequency modes dominated by Co. Thus, to lower thermal conductivity of binary skutterudites, it is imperative to scatter pnicogen-related phonons!

There are a number of ways how to scatter phonons of pnicogen atoms. It is possible to introduce impurity atoms via doping, bring into play alloy scattering by forming solid solutions with other binary skutterudites where solubility permits it, increase boundary scattering by making composite skutterudite structures, or perhaps even synthesize the entire sample with nanometer-scale grains. All these approaches have been tried and can typically reduce the thermal conductivity of binary skutterudites at room temperature by a factor of two or so. However, that is not enough to make skutterudites competitive with the best classical thermoelectrics, such as PbTe and Bi$_2$Te$_3$, the room temperature thermal conductivity of which is 2.2 Wm^{-1}K^{-1} and 1.6 Wm^{-1}K^{-1}, respectively. Major reductions in the thermal conductivity of skutterudites had to wait until void filling was verified as an effective approach.

5.2.2 LATTICE MODES IN TERNARY SKUTTERUDITES

Given how sensitive the lattice dynamics are to the disorder on the pnicogen rings, ternary skutterudites formed by substitution of Columns 14 and 16 atoms on the rings are expected to have significantly reduced thermal conductivity. Filling ternary skutterudites should further reduce their ability to conduct heat by bringing into play low frequency optic phonons that effectively scatter heat conducting acoustic modes of the structure. The effect of filling ternary skutterudites by alkaline earth fillers was explored by DFT calculations by Bang et al. (2016). While the two heavier alkaline earth fillers, Sr and Ba, have their lowest optic modes in a narrow frequency range 63–67 cm^{-1}, Ca has its optic modes at a significantly lower frequency of about 49 cm^{-1}, resulting in more effective scattering of acoustic phonons. The likely reason is a significantly smaller ionic radius of Ca^{2+} (100 pm) in comparison to that of Sr^{2+} (132 pm) and Ba^{2+} (135 pm), which results in much weaker bonding of Ca with the framework atoms and a more rattling character of its motion in the oversized cage. As shown in calculations by Bang et al., the smaller mass of Ca cannot be the reason as the smaller mass ions have their optic modes shifted upward rather than downward in frequency. Another notable feature in the dispersion of ternary skutterudites upon filling is the reduced slope of transverse acoustic modes, which implies lower group velocity for heat conducting phonons and thus further reduced thermal conductivity.

5.2.3 LATTICE MODES AND THE DENSITY OF STATES OF FILLED SKUTTERUDITES

Motivated by the PGEC concept, a truly impressive reduction of the thermal conductivity of skutterudites was achieved by filling the voids of the structure, first documented by Morelli and Meisner (1995), with numerous other studies following in short order (Sales et al. 1996, Nolas et al. 1996a,b, Tritt et al. 1996, Fleurial et al. 1997, Morelli et al. 1997, Chen et al. 1997, Nolas et al. 1997, Sales et al. 1997, Uher et al. 1998, Sellinschegg et al. 1998a, Nolas et al. 1998, Meisner et al. 1998). The PGEC paradigm rests on the premise that a filler ion is only weakly bonded in the oversized void and thus executes large, incoherent, substantially anharmonic Einstein-like vibrations, often called 'rattling'. It is the interference of this rattling motion of the filler with the normal phonon modes that was postulated as responsible for a dramatic degradation of the thermal conductivity. Moreover, since the electronic transport is facilitated by the framework atoms, which are little perturbed by the presence of the fillers, the carrier mobility is not greatly affected by filling. Slack's PGEC paradigm was instrumental in moving forward the development of novel thermoelectric materials, especially those having some kind of cage structures, like skutterudites and clathrates. Apart from the reduced values of the thermal conductivity, the early experimental support for the PGEC concept in filled skutterdites came from the much larger atomic displacement parameters (ADP) of the filler species compared to the ADPs of the framework atoms, measured by Sales et al. (1997) and shown in Figure 1.26, and low temperature specific heat studies combined with measurements of the elastic constants and inelastic neutron scattering by Keppens et al. (1998). The focus here was on a comparison of the low temperature specific heat of $La_{0.9}Fe_3CoSb_{12}$ with $CoSb_3$, and the difference in the inelastic neutron scattering (INS) spectra of $LaFe_4Sb_{12}$ and $CeFe_4Sb_{12}$, i.e., the compounds that differed either in the transition metal or in the filler. The results suggested the presence of two low-energy localized modes of La with the Einstein temperatures $\theta_{E1} = 70$ K and $\theta_{E2} = 200$ K.

The appearance of two independent vibration modes of La was certainly puzzling and led to a detailed investigation of the lattice dynamics of filled skutterudites by Feldman et al. (2000). Having previously developed the lattice dynamics model for $CoSb_3$, Feldman and colleagues added a limited number of coupling constants obtained from first principles calculations of atomic forces arising from small atomic displacements (within harmonic range) of a filler from its ideal position at the 2a site of the skutterudite structure along various directions of the crystal. Thus generated model they fitted to their DFT–LDA-based calculations. The important step in the process was to determine the total energy as a function of displacement of a rare-earth filler ion in the bcc skutterudite lattice. The functional form turned out to be a nearly parabolic (harmonic) with only a minor quartic (anharmonic) correction. For the two filler species considered, Ce and La, the 'bare' frequency (to be understood as the frequency computed with all other atoms of the structure frozen in their positions) came to 68 cm^{-1} and 74 cm^{-1}, respectively. To appreciate how little the vibrations of the filler deviated from the harmonic motion, the authors noted that even at 1000 K, the frequency shift would be no more than a few cm^{-1}. The major outcome of the calculations was a realization that the presence of a filler gives rise to its rather strong coupling with the neighboring Sb atoms, which significantly modified the Sb-weighted phonon DOS in the 80 cm^{-1} to 100 cm^{-1} range, Figure 5.10 and also Figure 5.8. The effect of the coupling is a notable upshift of frequencies of Sb-derived modes while the vibration frequency of the filler (La) is pushed down from its 'bare' value of 74 cm^{-1} to around 50 cm^{-1}. Moreover, the filler also softens the bonds within the Sb_4 rings by some 30%.

Given that the filler's vibrations have a substantially harmonic nature, that the void filling significantly modifies the dispersion of heat-carrying Sb vibrations, softens bonding on the Sb_4 rectangle, and that often incomplete filling introduces disorder among force constants within the structure, it

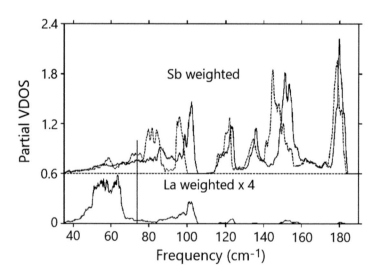

FIGURE 5.10 Sb (top) and La (bottom) weighted PDOS for $LaFe_4Sb_{12}$. The dashed line indicates the Sb weighted PDOS in $CoSb_3$, which is similar to that of $FeSb_3$ within the same dynamical model. Note that the La weighted PDOS is magnified by a factor of four, and the 'bare' La frequency is denoted by the vertical line. Reproduced from J. L. Feldman et al., *Physical Review B* 61, R9209 (2000). With permission from the American Physical Society.

seemed unlikely that the large anharmonic excursions of the filler implied by PGEC in the early experiments are the dominant mechanism of thermal conductivity reduction.

Feldman et al. noted that the observation of two vibration modes of La in the measurements of Keppens et al. is a consequence of the assumed transferability of the interatomic forces between unfilled and filled skutterudite structures, something they showed was not realistic. The closely lying 'bare' frequency of La and a significant number of Sb vibrations become spread apart by the La–Sb coupling, with the La mode shifted down to about 50 cm^{-1} while the Sb-derived modes are upshifted closer to 100 cm^{-1}. Such La–Sb hybridization gave rise to two extra contributions to the specific heat and a two-peak structure in the INS data, creating an impression that La vibrates in two distinct modes.

To avoid uncertainties arising from comparing skutterudites with different transition metals, Hermann et al. (2003) made a similar set of experiments as Keppens et al., this time using $Tl_{0.8}Co_4Sb_{11}Sn$ and $CoSb_3$. Thallium was chosen based on its scattering cross section being ~ 95% coherent, and thus any independent features in the scattering angle should reflect incoherent motion of Tl. In order to increase the content of Tl to $y = 0.8$, it was necessary to charge compensate the structure by replacing one atom of Sb with Sn. Unlike the results of Keppens et al., only a single Einstein mode of Tl with θ_E ~ 53 K was extracted from the specific heat data. The Tl mode was manifested by a well-developed single peak at energy ~ 5 meV (57 K) in the plot of the PDOS (corrected for the Debye parabolic contribution) as a function of energy a neutron gained in its inelastic scattering process (see Figure 5.11).

In subsequent publications, Feldman and colleagues made improvements to their lattice dynamics model by testing and retesting it against measurements of the Raman spectra, Feldman et al. (2003), INS studies of their own, Feldman et al. (2006), as well as INS studies by other researchers, Viennois et al. (2005), and against nuclear inelastic scattering data obtained by Long et al. (2005). The simulations took into account improvements in the algorithms that allowed to consider larger and larger supercell sizes so that force constants for pairs of atoms at larger distances could be examined, something that was not possible in earlier studies where such force constants were either altogether neglected or simply guessed. This resulted in re-evaluation of the strength of some of the force constants (Feldman et al. 2014). The most notable in this respect is the cubic interaction between the nearest neighbor La-Fe atoms, which turned out to be dramatically smaller (essentially negligible) when a large supercell was used compared to previous estimates based on a smaller supercell.

Regarding the eight symmetry-allowed Raman modes in the skutterudite structure, Figure 5.12 depicts intensities of the experimental Raman peaks measured by Feldman et al. (2003) on $La_{0.75}Fe_3CoSb_{12}$ in both parallel (VV) and crossed (HV) light polarizations. Table 5.2 collects the

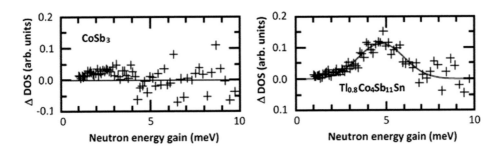

FIGURE 5.11 Experimental phonon density of states obtained from inelastic neutron scattering after subtracting the Debye parabolic dependence. The left panel are data for $CoSb_3$, the right panel are data for $Tl_{0.8}Co_4Sb_{11}Sn$. Note a well-defined excitation peak near 5 meV in the spectrum of $Tl_{0.8}Co_4Sb_{11}Sn$ reflecting incoherent vibrations of Tl fillers. Adapted from R. P. Hermann et al., *Physical Review Letters* 90, 135505 (2003). With permission from the American Physical Society.

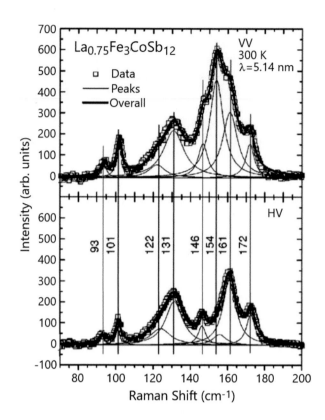

FIGURE 5.12 Experimental Raman spectra for $La_{0.75}Fe_3CoSb_{12}$. VV indicates parallel polarization between incident and scattered light, while HV designates cross-polarized incident and scattered light. The thin lines are eight Lorentzian functions fitted to the data (open squares). Reproduced from J. L. Feldman et al., *Physical Review B* 68, 094301 (2003). With permission from the American Physical Society.

TABLE 5.2

Zone Center Frequencies and Relative Intensities of Raman Modes in Parallel (VV) and Perpendicular (HV) Polarizations of Incident and Scattered Light

Mode	Theory			Experiment			
	Frequency	I (VV)	I (HV)	Frequency	I (VV)	I (HV)	Width
F_g	95.1	0.085	0.066	93.8	0.053	0.037	3.5
F_g	101.1	0.092	0.071	101.8	0.143	0.081	3.4
E_g	133.5	0.377	0.270	122.0	0.154	0.239	10.6
F_g	136.9	0.711	0.552	130.8	0.701	0.540	13.2
A_g	148.2	0.281	-	146.5	0.248	0.110	6.3
A_g	155.9	1.000	-	153.9	1.000	0.106	9.0
E_g	157.6	0.784	0.562	161.2	0.714	0.625	9.6
F_g	164.3	0.013	0.010	172.0	0.218	0.199	5.6

Theoretical intensities are computed based on the bond polarizability model. Experimental frequencies and intensities are from Lorentzian fits of the experimental data. Note in particular an excellent agreement (the difference less than 1%) between A_g experimental and computed modes, with the former identified by the HV-VV comparison. The group theoretical symmetries of the other Raman modes have not been confirmed experimentally because measurements were not performed on a single crystal. However, the placement of the A_g modes with respect to the other modes seems to be consistent with the theoretical symmetry ordering. All frequencies in the table are in units of cm^{-1}, including the width of the Raman peak. The table is constructed based on the data of Feldman et al. (2003). With permission from the American Physical Society.

zone center frequencies and intensities of computed Raman modes for $LaFe_4Sb_{12}$ as well as the parameters of Lorentzian fits to the experimental data of $La_{0.75}Fe_3CoSb_{12}$. Particularly relevant is the experimental identification of the two A_g symmetry modes at 146 cm^{-1} and 154 cm^{-1}, which show strong polarization dependence and are identified by their greatest difference in the Raman intensity between the top and bottom panels in Figure 5.12.

Overall, the agreement between the computed frequencies and the experimental Raman frequencies is excellent, attesting to the reliability of the model. While a comparison of the experimental data obtained on $La_{0.75}Fe_3CoSb_{12}$ with the theoretically generated Raman data on $LaFe_4Sb_{12}$, i.e., two different compounds being involved, might be questioned as not comparing apples with apples, one should not forget that the Raman modes are generated exclusively by vibrations on Sb sites, which have nominally the same environment in the two compounds. Nevertheless, there is a possibility that, due to the different filling content and the distribution of La in the voids of the two compounds, the La–Sb coupling could be somewhat altered and cause a small shift in the Raman frequencies.

I have already described some inelastic neutron scattering measurements made on filled skutterudites, and numerous others were made on a variety of filled skutterudites. Among those using polycrystalline forms of the structure are measurements by Viennois et al. (2005) [$LaFe_4Sb_{12}$ and $CeFe_4Sb_{12}$], Dimitrov et al. (2010) [$Yb_{0.2}Co_4Sb_{12}$], Möchel et al. (2011a) [$YbFe_4Sb_{12}$], Peng et al. (2012b) [$In_xYb_yCo_4Sb_{12}$], and Koza et al. (2008, 2010, 2011, 2013, 2014) [$LaFe_4Sb_{12}$ and $CeFe_4Sb_{12}$; $(Ca,Sr,Ba,Yb)Fe_4Sb_{12}$; LaT_4X_{12}, T = Fe, Ru, Os, X = As, Sb; $YbFe_4Sb_{12}$]. Single crystal specimens were used in the studies by Iwasa et al. (2006) [$PrOs_4Sb_{12}$] and Koza et al. (2015) [$LaFe_4Sb_{12}$].

Lattice dynamics of filled skutterudites has also been explored with inelastic X-ray scattering (IXS) making use of high brilliance, hard X-ray synchrotron radiation sources. The chief difference between INS and IXS is the target species being probed, namely the nuclear motion *versus* the coherent motion of electrons around atoms, respectively. The major advantage of IXS is its decoupling of energy and momentum transfer in the scattering process with the measured energy transfer of 1–200 meV compared to the incident X-ray energies of ~ 20 keV. Other advantages are nearly no intrinsic background, limited multiple scattering as X-rays are readily attenuated, and perhaps the most appealing aspect is the ability to probe very small samples (microgram quantities) with incident beams focused down to 10–100 μm. In contrast, the advantages of INS is tunability of the energy of incident neutrons to the energy of phonons, high energy resolution down to the sub-meV range, and the shorter 'tails' of Lorenzians fitted to the data, which allows to resolve weaker phonon modes in the proximity of strong modes. The readers interested in a more detailed account of attributes of INS and IXS are referred to an excellent article by Baron (2009). The IXS investigations of lattice dynamics of skutterudites included polycrystalline samples, Tsutsui et al. (2007, 2008a, 2009) [$SmRu_4P_{12}$, $LaOs_4Sb_{12}$, $SmOs_4Sb_{12}$] as well as single crystals, Tsutsui et al. (2010, 2012) [ROs_4Sb_{12}, R = La, Ce, Pr, Nd, Sm, Eu]. An example of the IXS data is shown in Figure 5.13, depicting the phonon mode associated with the motion of fillers in the icosahedral Sb cage as well as the acoustic phonon mode lying nearby.

I also wish to note nuclear resonant inelastic scattering (NRIS) as an additional probe of lattice dynamics, which, unlike the above two techniques, can directly access the element-specific phonon density of states, even in a complex multi-element lattice. This is important, because INS and IXS lack the element of specificity when it comes to assigning features in the spectrum to vibration modes of specific constituent elements. The power of specificity in NRIS arises from the resonant nature of the scattering process (a typical bandwidth of nuclear resonances is in the range of neV to μeV), which shares similarities with Mössbauer spectroscopy. However, the specificity comes with a constraint, namely the element must have a Mössbauer-active nuclide. In filled skutterudites of the form RFe_4Sb_{12}, this is no problem as both iron and antimony have [57]Fe and [121]Sb isotopes, respectively. Moreover, selecting europium as a filler, as done in the first NRIS study with skutterudites by Long et al. (2005), one has also available a [151]Eu isotope for an independent assessment of the filler's vibrations. Typically, the NRIS data are of high quality, and a large nuclear cross-section allows

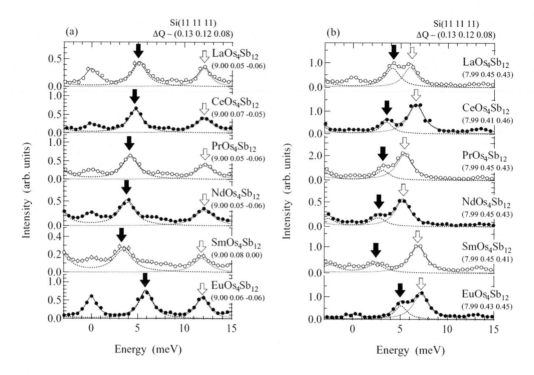

FIGURE 5.13 Inelastic X-ray scattering spectra of filled skutterudites ROs_4Sb_{12} (R: rare earth element) at 300 K. Closed (open) arrows show excitations of the filler guest mode (acoustic mode). (a) Spectra of ROs_4Sb_{12} compounds were measured around Q = (9.00 0.00 0.00). (b) Spectra of ROs_4Sb_{12} compounds were measured around Q = (8.00 0.45 0.45). Reproduced from S. Tsutsui et al., *Physical Review B* 86, 195115 (2012). With permission from the American Physical Society.

access to small samples. Figure 5.14(a) presents the [151]Eu-specific PDOS in $EuFe_4Sb_{12}$ obtained by NRIS at 25 K, with Figure 5.14(b) displaying the [57]Fe PDOS in $EuFe_4Sb_{12}$ obtained at 295 K. In each case, the respective partial PDOS predicted by Feldman et al. (2003) for $LaFe_4Sb_{12}$ is shown for comparison. Unfortunately, it was not possible to extract the [151]Eu partial PDOS at 295 K from the measurements due to very intense multiphonon scattering contributions at zero energy that overwhelmed the spectrum. The partial PDOS data for Eu at 25 K reveal a clear Einstein-like mode at 7.3 meV (58.9 cm^{-1}), which is smaller than the 'bare' Einstein energy of 9 meV for La, but consistent with a somewhat heavier Eu. In addition, there is a hint of small peaks near 12 meV and 17.8 meV, again, at slightly lower frequencies compared to small peaks computed for La that were previously identified as a result of hybridization between filler and antimony vibrations. The experimental partial PDOS for Fe in $EuFe_4Sb_{12}$ is in excellent agreement with the calculated partial PDOS of Fe in $LaFe_4Sb_{12}$, and the authors claimed no coupling between the modes of Eu and Fe. NRIS measurements have subsequently found much appeal and several studies of filled skutterudites have been pursued, Tsutsui et al. (2006a,b, 2008b) [$SmFe_4P_{12}$ and $SmRu_4P_{12}$ using here [149]Sm resonance], Wille et al. (2007) [$EuFe_4Sb_{12}$].

First principle calculations of the lattice dynamics of filled skutterudites continued to generate much interest as they attempted to provide physical interpretation of fascinating experimental data generated during the first decade of the new millennium. Following in the spirit of Feldman and his colleagues to gain insight into the nature of filler vibrations and how they relate to the vibrations of the framework, several complementary *ab initio* studies of the lattice dynamics have been carried out. Hasegawa et al. (2008) calculated La vibrations in nine La-filled skutterudites LaT_4X_{12} (T = Fe, Ru, Os; X = P, As, Sb) and pointed out that La vibrations distort the icosahedral cage of pnicogen

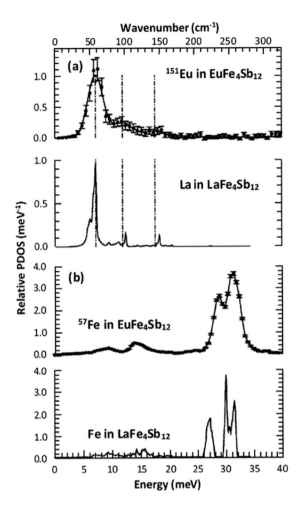

FIGURE 5.14 (a) The europium partial PDOS in $EuFe_4Sb_{12}$ obtained at 25 K by ^{151}Eu nuclear resonant inelastic scattering. The solid line is a fit with three Gaussian peaks with energies of 7.3 meV, 12.0 meV, and 17.8 meV and linewidths of 3.2 meV, 4.4 meV, and 4.4 meV, respectively. For comparison, the curve underneath is the La projected PDOS in $LaFe_4Sb_{12}$ computed by Feldman et al. (2003). (b) The ^{57}Fe partial PDOS obtained at 295 K by ^{57}Fe nuclear inelastic scattering. For comparison, the curve above is the Fe projected PDOS in $LaFe_4Sb_{12}$ computed by Feldman et al. (2003). Adapted from G. J. Long et al., *Physical Review B* 71, 140302(R) (2005). With permission from the American Physical Society.

atoms. In turn, the distortion decreases the energy of La vibrations, with the effect increasing as one goes from phosphide skutterudites to antimonide skutterudites. The distortion of the pnicogen cage upon filling was also the main topic in the theoretical work by Wee et al. (2010). In this case, the authors considered a rather large Ba filler in a void of $CoSb_3$, and compared the results with Na, K, Ca, and Sr fillers. By treating the presence of the filler as a perturbation of the $CoSb_3$ lattice, they proceeded to explore various filling levels, including those well beyond what $CoSb_3$ can normally accommodate. Their *ab initio* calculations indicated an important aspect, namely that the strain arising from filling is highly localized, affecting only the nearest neighboring pairs of Sb atoms. The relative softness of Sb–Sb bonds on the Sb_4 ring absorbs the strain and blocks it from spreading beyond the nearest Sb atoms. The situation is pictured in a sketch in Figure 5.15. The pressure exerted by the filler on the Sb_4 ring will tend to shorten the longitudinal bond d_2 and lengthen the transverse bond d_1 (still maintaining $d_2 > d_1$ as in $CoSb_3$, see also Figure 1.4). The vibration

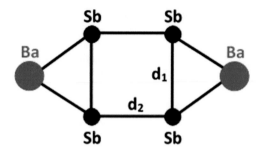

FIGURE 5.15 A sketch of the effect of a filler (Ba) on the Sb-Sb bonds d_1 and d_2 within the Sb_4 ring. The pressure of the filler tends to shorten d_2 and lengthen d_1. The softness of bonds within the ring accommodates the strain and prevents it from spreading beyond the nearest Sb-Sb atoms to the filler.

spectrum was computed by DFT within the LDA approximation, and returned substantially similar outcome as the previous PDOS calculations on filled skutterudites. The authors drew attention to the similarity of the lowest optic modes in $CoSb_3$ and in the filled structures, shown in Figure 5.16.

In $CoSb_3$, the lowest optic mode (F_u) is at 80 cm^{-1} and the next lowest mode (F_g) is at 83 cm^{-1}, in good agreement with the frequencies collected in Table 5.1. The modes are, of course, associated with vibrations of Sb, and are considered as translation (F_u) and in-plane rotation (F_g) of rings, the respective atomic motions sketched in Figure 5.17 in the left panels. In filled structures, here represented by Ca fillers, the lowest optic modes are associated with the motion of the fillers and, as we have seen in many examples in preceding paragraphs, are usually taken as localized and decoupled from other vibrations. However, looking at the panels on the right-hand side in Figure 5.17, the motion of Sb atoms have features very similar to motions of Sb atoms in $CoSb_3$. This similarity between the lowest optic modes in $CoSb_3$ and in the filled structures prompted the authors to view the filler-dominated optic modes as originating from the lowest optic modes of Sb in $CoSb_3$, rather

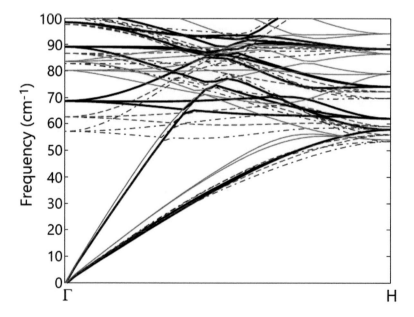

FIGURE 5.16 Computed phonon dispersion curves of $BaCo_4Sb_{12}$ (thick black curves), $SrCo_4Sb_{12}$ (dashed curves), $CaCo_4Sb_{12}$ (dash dot curves), and $CoSb_3$ (thin solid curves) along the Δ symmetry line from Γ to H. Reproduced from D. Wee et al., *Physical Review B* 81, 045204 (2010). With permission from the American Physical Society.

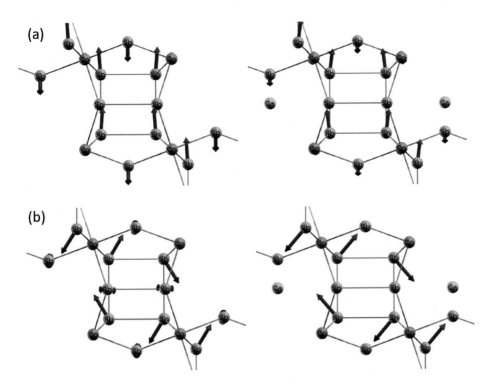

FIGURE 5.17 (a) Atomic motions corresponding to the lowest optic mode (F_u) at Γ in CoSb$_3$ (left panel) and in CaCo$_4$Sb$_{12}$ (right panel). Sb atoms are colored in yellow, Co atoms in green, and Ca atoms in blue. (b) Atomic motions corresponding to the second lowest optic mode (F_g). Again, the left panel is for CoSb$_3$, the right panel for CaCo$_4$Sb$_{12}$. The same color scheme used as in (a). Adapted from D. Wee et al., *Physical Review B* 81, 045204 (2010). With permission from the American Physical Society.

than arising solely from filler vibrations. In other words, there is considerable coupling (hybridization) between the vibrations of the filler and vibrations of the nearby Sb$_4$ rings. Such hybridization is apparently reflected in a much larger separation of frequencies of the two lowest optic modes F_u and F_g, which in CoSb$_3$ is 3 cm^{-1}, while upon filling with Ba or Ca increases to $90 - 71 = 19$ cm^{-1} and $84 - 55 = 29$ cm^{-1}, respectively.

As already pointed out, filler species typically have much larger atomic displacement parameters than the atoms of the framework. What gives the filler freedom to make large excursions from its equilibrium position is its loose bonding and a large cage size.

Of all skutterudite voids, the largest one belongs to the [Os$_4$Sb$_{12}$] framework. It was thus tempting to study rattling motion of rare earth fillers in such an oversized cage. The task was taken on by Yamaura and Hiroi (2011) who used high-quality single crystals of ROs$_4$Sb$_{12}$ (R = La, Ce, Pr, Nd, and Sm) and made precise X-ray analyses to map the relation between the crystal structure and rattling vibrations of rare earth filler R. The results indicated that the ADP of the filler R is a factor of 5 to 10 times larger than the ADP of Os and Sb atoms, in other words, the rare earth ions as a group, indeed, vibrate with exceptionally large amplitudes as suggested in previous studies by Evers et al. (1995) [NdOs$_4$Sb$_{12}$], by Kaneko et al. (2006 [PrOs$_4$Sb$_{12}$], and by Tsubota et al. (2008) [SmOs$_4$Sb$_{12}$]. In discussions of vibrations of the filler, it is instructive to have in mind the geometry of the skutterudite cage shown in Figure 5.18. The extent of the filler's motion from the center of the void is limited by the distance it can move before it encounters atoms of the framework. The shortest distance, often called the shortest guest free space, is toward the nearest Sb atom, $d_{gfs}(R\text{-}Sb)$.

The longest guest free space is in the [111] direction perpendicular to a blue triangle in Figure 5.18, designated as $d_{gfs}[111]$. Atomic displacements u_0 and Einstein temperatures θ_E of the fillers as a

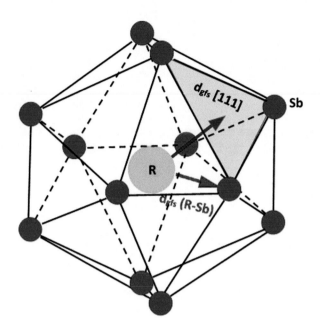

FIGURE 5.18 Icosahedral cage surrounding a filler atom R. Two distances limiting the motion of the filler are shown: the shortest one is toward the nearest Sb atom, designated as the shortest guest free space d_{gfs} (R–Sb), and shown with red arrow; the other one is the longest guest free space in the [111] direction shown by a blue arrow pointing toward the light blue triangle formed by three Sb atoms.

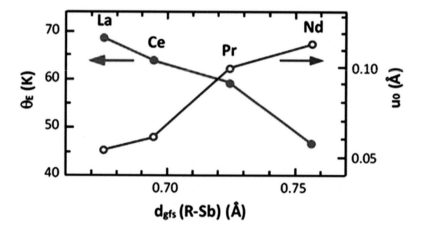

FIGURE 5.19 Einstein temperature θ_E and the atomic displacement u_0 as a function of the guest free space d_{gsf} (R–Sb) for ROs$_4$Sb$_{12}$, R = La, Ce, Pr, and Nd. Drawn from the data of J. Yamaura and Z. Hiroi, *Journal of the Physical Society of Japan* 80, 054601 (2011). With permission from the Physical Society of Japan.

function of the guest free space d_{gfs}(R-Sb) are shown in Figure 5.19. The Einstein temperature is calculated from the isotropic displacement parameter $U_{iso}(R)$ of each filler (obtained from the refinement of the X-ray data) according to

$$U_{iso}(R) = u_0^2 + \frac{\hbar^2}{mk_B\theta_E^2}T,\tag{5.37}$$

where u_0 is the temperature-independent static displacement parameter (if any), and m is the atomic mass of the filler R. From the X-ray experiments, the authors also extracted electron density maps for the rare earth fillers, which showed anisotropic broadening at high temperatures, especially for the small rare earth ions. Moreover, since there were no hints of any off-center motion of the fillers, the authors surmised that the exceptionally large atomic displacements of the small filler ions must be related to their large anharmonic vibrations. The lack of any off-center motion of Pr is particularly relevant as it counters an interpretation of ultrasonic measurements by Goto et al. (2004), which claimed splitting of the Pr site into four off-center positions with Pr vibrating between the four sites. Unlike in certain clathrates, where the off-center vibrations have been substantiated, e.g., Sales et al. (2001), the occasionally invoked off-center motion of fillers in skutterudites has never been verified. Theoretical treatment of phonon spectra, including the mean-square displacement $<u^2>$ within the anharmonic lattice model was considered by Yamakage and Kuramoto (2009) for a simple cubic lattice with each filler surrounded by eight cage atoms. Although the lattice is not the same as that of a skutterudite, the temperature dependence of the Einstein frequency and of $<u^2>$ reproduced faithfully the vibrational parameters of Pr in $PrOs_4Sb_{12}$. While all branches of phonons acquired temperature dependence, the effect depends on the wave vector, with the largest temperature dependence for phonons at the boundary of the Brillouin zone.

The discussion of the lattice dynamics of filled skutterudites contained considerable details. This was done intentionally to provide the readers with a broad spectrum of experimental and theoretical studies in order to appreciate fully the controversy presented in the next few paragraphs.

5.3   CHALLENGES TO THE PGEC CONCEPT IN SKUTTERUDITES

The original PGEC concept assumed that the filler vibrates essentially independently and incoherently with respect to the atoms of the framework and, as such, can be described by an Einstein temperature θ_E. The much reduced lattice thermal conductivity of skutterudites upon filling is then a consequence of normal phonon modes of the framework being resonantly scattered by vibrations of the fillers. The PGEC paradigm provided an intuitive and useful viewpoint that successfully guided the development of filled skutterudites (as well as clathrates) as efficient thermoelectric materials. The viewpoint was also consistent with experimental data showing that double and triple filling reduces the thermal conductivity even more, simply by introducing additional localized Einstein modes that resonantly scatter a broader range of normal phonon modes. Moreover, as we have seen, every experimental technique attempting to interrogate phonons in filled skutterudites documented the presence of a localized low frequency mode of the filler species, the observation further reinforced by theoretical modeling. However, as pointed out already early on by Feldman et al. (2000), based on their INS measurements, the vibrations of the filler (La), rather than being independent, appeared strongly hybridized with certain Sb branches, and thus the simplest picture of rattling is not applicable. The hybridization gave rise to a significant lowering of the 'bare' filler frequency seen in many experiments as well as in rigorously computed lattice dynamics, e.g., Ghosez and Veithen (2007). Furthermore, as seen in INS measurements on $CeRu_4Sb_{12}$ by Lee et al. (2006), vibrations of Ce were identified as the coherent optical phonon branch that undergoes avoided crossing with the transverse acoustic mode, rather than being an Einstein mode, and thus could not act as a scattering center, although the filler ion might somewhat reduce the acoustic phonon lifetime at the avoided crossing, Christensen et al. (2008). Consequently, the PGEC paradigm, in spite of its considerable appeal, started to be challenged as one of its foundation stones – independence of filler vibrations – seemed to be undermined, opening the door for alternative explanations of thermal conductivity reduction, such as anharmonicity. Although the well in which the filler vibrates is substantially harmonic, it has a small quartic anharmonic contribution, which could be reinforced in coupling with framework vibrations. Yet another alternative to resonant scattering of normal phonon modes by the Einstein modes of the filler is a possibility of opening additional scattering channels for three-body Umklapp processes, particularly when the low-lying filler mode actually intersects the acoustic phonon branch, Lee et al. (2006).

What looked like the final death blow to the PGEC concept in filled skutterudites were high resolution time-of-flight neutron scattering measurements by Koza et al. (2008) and their subsequent investigations (Koza et al. 2010, 2013, 2014, 2015). The key argument the authors waged concerned the low resolution nature of previous INS experiments that could not capture the microscopic mechanism underpinning the reduction of the thermal conductivity because they lacked the ability of simultaneously probing the phonon energy $\hbar\omega_q$ and phonon momentum $\hbar q$. Consequently, such experiments could not make any conclusions regarding the pivotal issues of phase coherence, collectivity of motion, and the filler and framework dynamics. With their experimental setup (claiming more than an order of magnitude higher energy resolution), they were able to map the vibrational dynamics over wide ranges of energy and momentum. However, because the authors used polycrystalline rather than single crystal samples, the inelastic response could be monitored only in terms of the modulus of the wave vector |$\mathbf{q}$| rather than distinct phonon eigenvectors. Supporting their experimental work with a newly developed *ab initio* theoretical treatment, called powder-averaged lattice dynamics (PALD) calculations, they addressed three pivotal issues a true 'rattler' should possess: (1) with increasing (decreasing) temperature, the vibrating filler explores the larger (smaller) geometrical limits of the cage and its frequency should increase (decrease); (2) the filler's vibrational energy should be independent of the wave-vector transfer, and its group velocity should be zero everywhere within the Brillouin zone; and (3) the filler's motion should completely lack phase coherence, i.e., the structure factor (intensity of rattling) should follow $S(\mathbf{q},\omega = \text{const}) \propto q^2$. Finding no such characteristics in the vibrations of the filler, the authors concluded that 'a freely rattling guest in a host cage is not applicable in the case of filled $LaFe_4Sb_{12}$-type skutterudites', and, instead, the heat transport is hindered by Umklapp scattering. In studies that followed, Koza and colleagues explored the lattice dynamics of a variety of fillers, including alkaline earths Ca, Sr, Ba (as well as Yb) (Koza et al. 2010), various La-filled skutterudites having different transition metals (Fe, Ru, Os) and pnicogens (As,Sb) (Koza et al. 2013), $Yb_{1-x}Fe_4Sb_{12}$ (Koza et al. 2014), and single crystal $LaFe_4Sb_{12}$ (Koza et al. 2015).

The denial of the relevance of the PGEC concept in filled skutterudites on microscopic grounds has turned out to be highly controversial and not universally accepted, especially in the community trying to develop skutterudites into efficient thermoelectrics. Until this day, many publications present the thermal conductivity of skutterudites as if the PGEC paradigm was valid or at least relevant. I shall return to this point when discussing models of the thermal conductivity and the most recent experimental measurements.

5.4   GOLDSTONE MODES IN CERTAIN SKUTTERUDITES

It was shown (see Chapters 1 and 2) that certain species, such as Sn, can be filled into skutterudite voids with the aid of high pressure. Fu et al. (2018) considered theoretically what Sn-filling the $[Fe_4Sb_{12}]$ framework might do to the thermal conductivity and arrived at a remarkably low thermal conductivity that they explained in terms of Goldstone-like modes associated with the collective motion of Sn with its neighboring Sb atoms. In general, by Goldstone modes one understands bosons that spontaneously break continuous symmetries. In the case of $SnFe_4Sb_{12}$, the Goldstone modes form as a consequence of the unusual off-center location of Sn in the voids, which leads to rather strong threefold covalent bonds with Sb, disturbing in the process the Sb_4 ring structure. Such a structural modification gives rise to an unusual lattice dynamics with Sn moving perpendicular to the off-centering direction in a three-dimensional potential best described as having the shape of a Mexican hat. The Goldstone-like modes result in exceptionally low-lying optic modes that interfere with the acoustic phonons of the framework, resulting in a very low lattice thermal conductivity, Figure 5.20. However, $SnFe_4Sb_{12}$ would be of no interest as a thermoelectric material on its own due to very limited charge transfer between the filler and the framework and, therefore, far too low carrier density. As noted by the authors, one might do better by co-filling the framework with Sn and rare earths, such as La, known to transfer readily its three valence electrons to the framework.

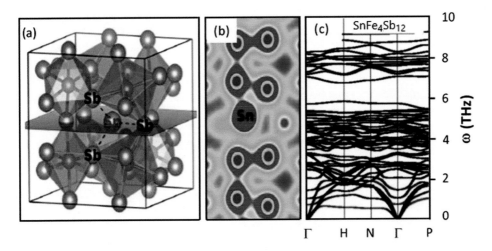

FIGURE 5.20   (a) Structure of $SnFe_4Sb_{12}$ with the atom of Sn off-center in the (001) plane (blue) and bonding with three Sb atoms. (b) Charge density in the (001) plane showing the disturbed Sb_4 rings and the Sn filler. (c) Calculated phonon dispersion of $SnFe_4Sb_{12}$ with the red traces indicating the low-lying Goldstone-like modes. Adapted from Y. H. Fu et al., *Physical Review B* 97, 024301 (2018). With permission from the American Physical Society.

DFT calculations indicated that, although the double-well potential is weakened as La substitutes for Sn, tin maintains its off-center position and the Goldstone-like modes keep generating low-lying optic modes with even lower frequency of 0.52 THz (versus 0.79 THz for fully filled $SnFe_4Sb_{12}$) for a skutterudite with the composition $Sn_{0.5}La_{0.5}Fe_4Sb_{12}$.

5.5   LATTICE DYNAMICS IN $FeSb_3$

In Chapters 1 and 2, it was mentioned that $FeSb_3$ is a metastable skutterudite that cannot form in bulk but can be stabilized as a thin film by MBE deposition or by modulated elemental reaction method. The structure has drawn considerable interest as it should reveal unique vibrational properties on account of its metastable nature. Softening of the phonon modes in $FeSb_3$ in comparison to $CoSb_3$ was documented in lattice dynamics investigations by Möchel et al. (2011b). In cases where sufficiently large bulk samples are available, the technique of choice in the study of lattice dynamics is neutron inelastic scattering. However, thin films of $FeSb_3$ could not provide a large enough sample volume. To avoid prohibitively long beam times, the authors made use of much improved resolution of ~ 1.3 meV achieved in nuclear resonant scattering experiments using ^{121}Sb isotope, Wille et al. (2007). This opened the road for element-specific studies of the density of phonon states in systems such as $FeSb_3$ films. Softer bonding of Sb in $FeSb_3$ than in $CoSb_3$ leads to smaller group velocities, which, in turn, degrade the thermal conductivity, particularly when most of the heat is carried by vibrations of Sb atoms. Interestingly, while Sb bonding became weaker, the Fe sublattice turned harder than the Co sublattice. The authors also observed that, upon filling the $[Fe_4Sb_{12}]$ framework, the low energy optic phonon modes shifted to higher energies.

Particularly illuminative study of the lattice dynamics of $FeSb_3$ is the work by Fu et al. (2016), in which the authors used DFT-based computations (the VASP code) in conjunction with the iterative solutions of the linearized Boltzmann transport equation (via the SHENGBTE package, Li et al. (2014)) and showed that the thermal conductivity of unfilled $FeSb_3$ is exceptionally low. With the computed room temperature value of 1.14 $Wm^{-1}K^{-1}$, it is an order of magnitude smaller than the value of 11.6 $Wm^{-1}K^{-1}$ computed for $CoSb_3$ using the same approach, and even lower than the thermal conductivity of most filled skutterudites, except perhaps for $YbFe_4Sb_{12}$, discussed later. The

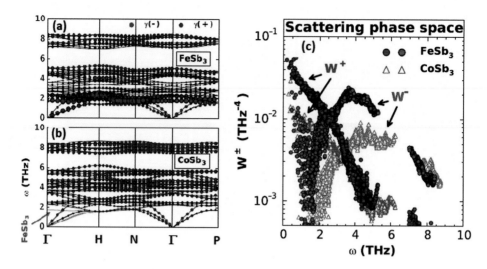

FIGURE 5.21 Calculated phonon dispersion curves of (a) $FeSb_3$ and (b) $CoSb_3$. The size of the circles indicates the magnitude of Grüneisen parameter γ for each mode. Positive γ is shown in blue, negative γ in red. For comparison, the upper red curve in (b) is the lowest optic mode of $FeSb_3$ and the other red curve in (b) is the acoustic mode of $FeSb_3$. (c) Scattering phase space $W^{\pm}$ as a function of frequency in $FeSb_3$ (filled circles) and in $CoSb_3$ (open triangles). The scattering phase space represents the available three-phonon scattering channels among all modes. The two different colors stand for three-phonon processes of absorption W^+, where two phonons merge into one, and emission W^-, where one phonon splits into two. In the low-frequency range below 3.5 THz, both W^+ and W^- of $FeSb_3$ are several times larger than those of $CoSb_3$, the primary reason for an order of magnitude lower thermal conductivity of $FeSb_3$ compared to $CoSb_3$. Adapted from Y. Fu et al., *Physical Review B* 94, 075122 (2016). With permission from the American Physical Society.

reason for such low thermal conductivity is not only the general softening of Sb bonds and associated lower frequency of optic phonon modes (see Figure 5.21a and 5.21b) but primarily the enhanced overlap of these low-lying optic modes with the acoustic modes that carry bulk of the heat. The overlap opens additional anharmonic scattering channels for phonons by increasing the phase space for three-phonon anharmonic scattering events, as documented by a much larger scattering phase space in $FeSb_3$ compared to that in $CoSb_3$ (Figure 5.21c). The authors also computed the cumulative thermal conductivity of $FeSb_3$ as a function of the phonon mean free path at 300 K, which served a notice that in order to avoid surface and boundary scattering and attain the intrinsic bulk thermal conductivity, the film thickness should be in the 1–1.5 μm range. Of course, to obtain such data from measurements of thermal conductivity on thin films, the thermal shunt via the substrate must first be eliminated by either freely suspending the film or by using techniques, such as thermoreflectivity (Paddock and Eesley 1986, Schmidt et al. 2009, and Braun et al. 2019), or 3ω method (Cahill 1990), that effectively avoid the presence of the substrate.

5.6 PHONON DISPERSION IN Yb-FILLED SKUTTERUDITES

In a worldwide search during the late 1990s and early 2000s for a filler that can impede heat transport in skutterudites to the greatest extent and make them a competitive novel thermoelectric, Yb emerged as one of the most effective. Partially filled $Yb_xCo_4Sb_{12}$ and essentially fully filled $YbFe_4Sb_{12}$ attained room temperature thermal conductivities several times lower than the thermal conductivity of $CoSb_3$, yet maintained excellent power factors (Dilley et al. 2000a,b, Nolas et al. 2000a, Anno and Matsubara 2000). As major advantages of Yb were viewed its small size and heavy mass augmented by the intermediate valence (Dilley et al. 1998, Leithe-Jasper et al. 1999), that allowed this late lanthanide to establish bonding even in the large icosahedral cage of antimonide skutterudites. Moreover, Yb displayed a large atomic displacement parameter (Dilley et al. 2000a). Since then, filling skutterudites with Yb and exploring their physical properties, in particular the heat transport, turned out to be important steps in the development of thermoelectric skutterudites. Numerous studies with Yb single-filled structures based on $[Co_4Sb_{12}]$ (Anno et al. 2002, Nolas and Fowler 2005, Li et al. 2007, 2008, 2009, Geng et al. 2007, Salvador et al. 2009, J. Yang et al. 2009, Mi et al. 2011, Liu et al. 2012, Dong et al. 2013, Park et al. 2014, Dahal et al. 2014, Li et al. 2015, Wang et al. 2016, 2017, Bashir et al. 2018b, Ryll et al. 2018), and $[Fe_4Sb_{12}]$ frameworks (Kuznetsov and Rowe 2000, Bauer et al. 2000, Qiu et al. 2011, Cho et al. 2012, 2013), sometimes charge-compensated on the transition metal or on the pnicogen sites (Yang et al. 2001, Yang et al. 2003, Mori et al. 2005, Zhou et al. 2011, Kaltzoglou et al. 2012, Ballikaya et al. 2013, Dong et al. 2014, Thompson et al. 2015, Li et al. 2019), double-filled skutterudites with Yb as one of the fillers (Guo et al. 2007, Uher et al. 2008, Shi et al. 2008, Peng et al. 2009, Liu et al. 2011, Salvador et al. 2010, 2013a, Xu et al. 2013, Zhou et al. 2013, Joo et al. 2015, Hobbis et al. 2017, Choi et al. 2018), triple-filled skutterudites with impressive thermoelectric performance (Shi et al. 2009, Ballikaya et al. 2012, Guo et al. 2013, Salvador et al. 2013b, Rogl et al. 2015a, Matsubara and Asahi 2016), and composite skutterudites with the $Yb_yCo_4Sb_{12}$ matrix (Zhao et al. 2006, Ding et al. 2012, Fu et al. 2015), document the interest this filler has generated. In the best performing triple-filled skutterudites containing Yb, the room temperature lattice thermal conductivity has been reduced to a level of 1 $Wm^{-1}K^{-1}$ or below (Shi et al. 2011a).

Notwithstanding the above obvious phenomenological attributes of Yb, questions remained how exactly the presence of Yb in the voids causes such an impressive reduction in the thermal conductivity, i.e., how it modifies the phonon spectrum of $[Co_4Sb_{12}]$ and $[Fe_4Sb_{12}]$. The microscopic origin of the degradation of heat transport of skutterudites by Yb was revealed by detailed calculations by Li and Mingo (2015), who combined a DFT-based approach using the VASP package to obtain the vibration frequencies, phonon velocities and phonon transition probabilities with iteratively solved Boltzmann transport equation (BTE) using the SHENGBTE algorithm developed by Li et al. (2014). A comparison of phonon density of states of $YbFe_4Sb_{12}$ with another quite low thermal conductivity skutterudite, $BaFe_4Sb_{12}$, Figure 5.22, indicates a dramatic lowering of the optic phonon frequencies of the former and distinctly flat modes in the range $5 < \omega < 7.2$ rad/ps (0.8 THz < f < 1.15 THz) highlighted by a shadowed band in Figure 5.22a, which are absent in the phonon density-of-states of $BaFe_4Sb_{12}$ in Figure 5.22b. The significantly different dispersion in $YbFe_4Sb_{12}$ compared to other filled skutterudites is a consequence of an order of magnitude higher scattering rates of phonons, particularly at the range of frequencies between 4 rad/ps and 25 rad/ps, as shown in Figure 5.23. The

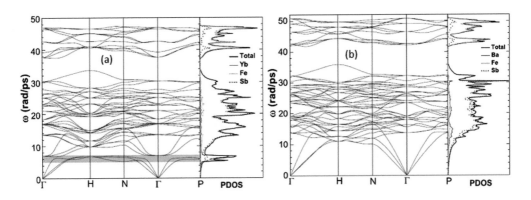

FIGURE 5.22 Phonon dispersion and phonon density-of-states with contributions from each element of (a) $YbFe_4Sb_{12}$ and of (b) $BaFe_4Sb_{12}$. The shadowed band in (a) indicates the range of flat modes of $YbFe_4Sb_{12}$. Adapted from W. Li and N. Mingo, *Physical Review B* 91, 144304 (2015). With permission from the American Physical Society.

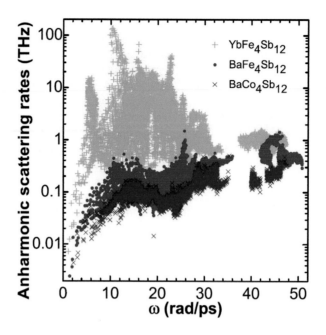

FIGURE 5.23 Anharmonic scattering rates of $YbFe_4Sb_{12}$ (yellow) compared to the scattering rates of $BaFe_4Sb_{12}$ (red) and $BaCo_4Sb_{12}$ (blue). Reproduced from W. Li and N. Mingo, *Physical Review B* 91, 144304 (2015). With permission from the American Physical Society.

exceptionally low thermal conductivity of $YbFe_4Sb_{12}$ is thus chiefly due to a much shorter lifetime of phonons undergoing three-phonon anharmonic processes. In fact, Li and Mingo have shown that phonons with frequencies in the range $5 < \omega < 7.2$ rad/ps (the region of flat bands in Figure 5.22a) essentially do not contribute to a cumulative thermal conductivity on account of their very high scattering rates and small group velocity associated with the flat modes. Since the authors have also shown that the anharmonic third-order interatomic force constants do not differ significantly between $YbFe_4Sb_{12}$ and $BaFe_4Sb_{12}$, the reason for the vastly different scattering rates must be a dramatically increased number of scattering channels for phonons. The direct measure of allowed scattering processes of phonons is the weighted phase space W that sums all frequency-containing factors that enter the expression for three-phonon transition probabilities for phonon absorption W^+ and phonon emission W^-,

$$W_\lambda^+ = \frac{1}{2N} \sum_{\lambda' p''} \left\{ 2\left(f_{\lambda'} - f_{\lambda''}\right) \right\} \frac{\delta\left(\omega_\lambda + \omega_{\lambda'} - \omega_{\lambda''}\right)}{\omega_\lambda \omega_{\lambda'} \omega_{\lambda''}}, \tag{5.38}$$

$$W_\lambda^- = \frac{1}{2N} \sum_{\lambda' p''} \left\{ f_{\lambda'} + f_{\lambda''} + 1 \right\} \frac{\delta\left(\omega_\lambda - \omega_{\lambda'} - \omega_{\lambda''}\right)}{\omega_\lambda \omega_{\lambda'} \omega_{\lambda''}}, \tag{5.39}$$

where N is the number of uniformly spaced q points in the Brillouin zone. The weighted phase space for both phonon absorption and emission in $YbFe_4Sb_{12}$ as well as in $BaFe_4Sb_{12}$ and $BaCo_4Sb_{12}$ is shown in Figure 5.24. Obviously, both W^+ and W^- for $YbFe_4Sb_{12}$ are much larger (note a logarithmic scale) below 25 rad/ps (below 4 THz) than the respective (similar) weighted phase space values of $BaFe_4Sb_{12}$ and $BaCo_4Sb_{12}$. Moreover, as the frequency approaches the flat mode region near $\omega_{flat} \approx 6$ rad/ps, the emission W^- value jumps and reaches nearly an order of magnitude larger value at its peak at 12 rad/ps. This peak in W^- and a much larger W^+ at low frequencies are distinct features of $YbFe_4Sb_{12}$ and the consequence of its flat modes. Since the three-phonon processes

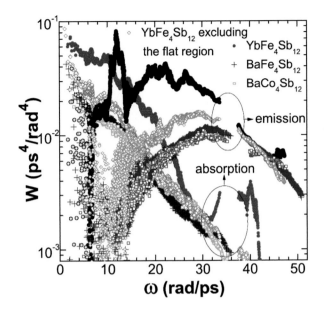

FIGURE 5.24 Weighted phase space at room temperature plotted as a function of frequency for both absorption and emission phonon processes in YbFe$_4$Sb$_{12}$ (red and black solid circles, respectively), in BaFe$_4$Sb$_{12}$ (crosses), in BaCo$_4$Sb$_{12}$ (open circles), and in YbFe$_4$Sb$_{12}$ where flat modes were excluded (open diamonds). Note a rapid, nearly an order of magnitude increased W^- in YbFe$_4$Sb$_{12}$ when the frequency of phonons is in the region of flat phonon modes. Reproduced from W. Li and N. Mingo, *Physical Review B* 91, 144304 (2015). With permission from the American Physical Society.

must conserve both energy and momentum, in the case of phonon emission where a phonon mode $(\boldsymbol{q}_0, \omega_0)$ decays into two phonons $(\boldsymbol{q}_1, \omega_1)$ and $(\boldsymbol{q}_2, \omega_2)$, the existence of flat modes in the frequency range of YbFe$_4$Sb$_{12}$ between about 5 rad/ps and 7.2 rad/ps enables one to find for any $(\boldsymbol{q}_1, \omega_0 - \omega_{flat})$ a partner mode $(\boldsymbol{q}_0 - \boldsymbol{q}_1, \omega_{flat})$ in the flat region of the phonon spectrum that facilitates the process. Consequently, for frequencies $\omega_0 > \omega_{flat}$, the number of emission channels has increased compared to structures with no flat phonon modes, as documented in Figure 5.24 by the values of W for YbFe$_4$Sb$_{12}$ when its flat modes were removed. The lesson learned is that the presence of flat phonon modes opens extra scattering channels for phonon emission processes and this dramatically decreases thermal conductivity.

5.7 THEORETICAL FOUNDATIONS OF THE THERMAL CONDUCTIVITY

Thermal conductivity is an important transport parameter for any solid and in thermoelectric materials it is one of three variables (electrical conductivity and the Seebeck coefficient being the other two) that determines the thermoelectric performance *via* the thermoelectric figure of merit. Moreover, because the lattice part of the thermal conductivity typically accounts for more than half of the overall heat flow in thermoelectric materials and is the only variable that can be altered without affecting other transport properties, it has become an important focus of studies aiming to improve the thermoelectric performance. In fact, major advances in thermoelectric materials made during the past two dozen years have been made by devising strategies how to minimize the lattice thermal conductivity. In the following sections, I will first provide theoretical foundations of the thermal conductivity and then discuss key studies relevant to the heat transport in the skutterudite structures.

With a few exceptions, thermoelectric materials are semiconductors and thus both lattice vibrations (phonons in a quantized form) and charge carriers participate in the flow of heat. The two systems are considered as independent entities and are described by their respective unperturbed wave functions. Any kind of interaction between the two is entered subsequently as transition between the unperturbed states. With this assumption, the overall thermal conductivity is then obtained by adding the two independent contributions, one due to lattice phonons κ_p and one due to charge carriers κ_e,

$$\kappa = \kappa_e + \kappa_p. \tag{5.40}$$

Thermal conductivity relates the amount of heat flux $\vec{Q}$ (the rate of heat that flows across a unit cross sectional area perpendicular to the direction of the flow) to the applied thermal gradient $\vec{\nabla} T$ as

$$\kappa = -\frac{\vec{Q}}{\vec{\nabla} T}. \tag{5.41}$$

The negative sign indicates that the heat always flows down the thermal gradient, i.e., from the hotter toward the colder region. The expression in Eq. 5.41 is the well-known isotropic form of the Fourier law. For structures that do not have cubic symmetry, the direction of the heat flow need not be in the same direction as the thermal gradient, and the thermal conductivity is a tensor,

$$\kappa_{ij} = -\frac{\vec{Q}_i}{\partial T / \partial x_j}. \tag{5.42}$$

We do not have to worry about this complication because skutterudites are cubic systems. In SI units, the thermal conductivity is measured and expressed in Wm^{-1}K^{-1}. If, for some unlikely reason, one would like to express thermal conductivity in other units, the conversion is

$$1\,\mathrm{Wm^{-1}K^{-1}} = 10^{-2}\,\mathrm{Wcm^{-1}K^{-1}} = 2.389 \times 10^{-3}\,\mathrm{cal\,s^{-1}cm^{-1}K^{-1}}$$

$$= 0.578\,\mathrm{BTU\,ft^{-1}h^{-1}degF^{-1}}.$$

The simplest expression for the thermal conductivity comes from the kinetic theory of gases, which expresses the heat flux in terms of a number of particles n per unit volume (particles can be charge carriers or phonons), each with the heat capacity c that all travel under the thermal gradient with velocity $\vec{v}$. On average, the particles travel a distance called the mean free path $\ell = v\tau$ before they are scattered. Here, τ is the time between the collisions, also referred to as the relaxation time. Hence,

$$\vec{Q} = -nc\tau \langle \vec{v}\vec{v} \rangle \vec{\nabla} T = -\frac{1}{3} ncv^2 \vec{\nabla} T, \tag{5.43}$$

from which follows the well-known expression

$$\kappa = \frac{1}{3} nc\tau v^2 = \frac{1}{3} Cv\ell. \tag{5.44}$$

Here, $C = nc$ is the heat capacity per unit volume. In a solid, the entries in Eq. 5.44 are functions of frequency ω, and thus the spectral thermal conductivity, i.e., the conductivity associated with a particular mode of frequency ω is

$$\kappa(\omega) = \frac{1}{3} C(\omega)v(\omega)\ell(\omega). \tag{5.45}$$

Summing (integrating) over all phonon modes of the structure, the thermal conductivity becomes

$$\kappa = \frac{1}{3}\int_0^\infty C(\omega)v(\omega)\ell(\omega) = \frac{1}{3}\int_0^\infty C(\omega)v^2(\omega)\tau(\omega), \tag{5.46}$$

where in the last equality $\ell = v\tau$ was substituted.

5.7.1 ELECTRONIC PART OF THE THERMAL CONDUCTIVITY

An expression for the electronic thermal conductivity in its general form in terms of the Fermi integrals, applicable for any degree of degeneracy, is

$$\kappa_e = \frac{1}{3\pi^2 mT}\left(\frac{2m}{\hbar^2}\right)^{3/2}(k_B T)^{r+7/2}\tau_0\left\{\left(r+\frac{7}{2}\right)F_{r+\frac{5}{2}}(\eta) - \frac{\left(r+\frac{5}{2}\right)^2 F_{r+\frac{3}{2}}^2(\eta)}{\left(r+\frac{3}{2}\right)F_{r+\frac{1}{2}}(\eta)}\right\}. \tag{5.47}$$

It has been noted since the middle of the 19th century that metals, which conduct electricity well happen to be also very good conductors of heat. Empirical formulation of this interdependence was provided by Wiedemann and Franz (1853) after whom the effect is known as the Wiedemann-Franz law. Forming the ratio $\kappa_e/(\sigma T)$ and designating it as the Lorenz number L, Lorenz (1881), we obtain

$$L = \frac{\kappa_e}{\sigma T} = \left(\frac{k_B}{Q}\right)^2\left\{\frac{\left(r+\frac{7}{2}\right)F_{r+\frac{5}{2}}(\eta)}{\left(r+\frac{3}{2}\right)F_{r+\frac{1}{2}}(\eta)} - \left[\frac{\left(r+\frac{5}{2}\right)F_{r+\frac{3}{2}}(\eta)}{\left(r+\frac{3}{2}\right)F_{r+\frac{1}{2}}(\eta)}\right]^2\right\}. \tag{5.48}$$

The Lorenz ratio makes a tacit but important assumption that the electrical and thermal scattering processes are governed by the same energy-dependent relaxation time $\tau(E) = \tau_0 E^r$, where τ_0 is independent of energy of charge carriers but depends on temperature. The assumption is valid, provided all scattering processes are of elastic nature, i.e., at high temperatures and at sufficiently low temperatures. At intermediate temperatures the relaxation time for carrier scattering and the relaxation time for thermal transport may differ, affecting the behavior of the Lorenz function.

In metals, representing the case of *high degeneracy*, the Fermi integrals are approximated by the expression

$$F_n(\eta) \approx \frac{\eta^{n+1}}{n+1} + \eta\eta^{n-1}\frac{\pi^2}{6} + n(n-1)(n-2)\eta^{n-3}\frac{7\pi^4}{360} + \dots. \tag{5.49}$$

After tedious but not difficult manipulations, one arrives at a well-known result

$$L = \frac{\kappa_e}{\sigma T} = \frac{\pi^2}{3}\frac{k_B^2}{q^2} = 2.44\times10^{-8}\,V^2K^{-2} \equiv L_0. \tag{5.50}$$

Since thermoelectrics are semiconductors, albeit some of them with a fairly high degeneracy, a more appropriate form of the electronic thermal conductivity and the Lorenz number is obtained in the *nondegenerate approximation*, which consists of replacing the Fermi integral with a Γ-function,

$$F_n(\eta) \equiv \int_0^\infty \frac{\varepsilon^n}{e^{\varepsilon-\eta}}d\varepsilon \;\rightarrow\; e^\eta\int_0^\infty \varepsilon^n e^{-\varepsilon}d\varepsilon \equiv e^\eta\Gamma(n+1). \tag{5.51}$$

Values of the Γ function frequently needed are $\Gamma(n+1) = n\Gamma(n)$, $\Gamma(1/2) = \pi^{1/2}$, and $\Gamma(1) = \Gamma(2) = 1$. The electronic thermal conductivity in this nondegenerate approximation becomes

$$\kappa_e = \frac{1}{3\pi^2 mT} \left(\frac{2m}{\hbar^2} \right)^{3/2} \tau_0 e^\eta \left(r + \frac{5}{2} \right) \Gamma \left(r + \frac{5}{2} \right), \tag{5.52}$$

with the Lorenz number equal

$$L \equiv \frac{\kappa_e}{\sigma T} = \frac{k_B^2}{q^2} \left(r + \frac{5}{2} \right). \tag{5.53}$$

Here q is the charge of a carrier, positive for holes and negative for electrons.

5.7.2   Lattice Thermal Conductivity

In Section 5.1.2, we have seen that vibrations of a crystal consist of three acoustic branches (one longitudinal and two transverse) and the remaining $3n_b - 3$ modes (n_b being the number of atoms of the basis) are optic modes. By and large, the heat is carried by acoustic phonons because their group velocity ($\partial \omega_q / \partial q$) is many times larger than the very small group velocity of optic modes, the result of the flat profile of their dispersion curves. However, it does not mean that we can entirely forget about optic phonons. Because they interact with acoustic phonons, they may redistribute momenta of the latter and thus affect the heat flow. In general, transport of heat is one of the more complicated phenomena because, unlike the flow of electricity or water, which can be contained within the wire or inside a pipe, heat is very difficult to contain as it can dissipate not just by conduction but also by convection and radiation.

There are three prime approaches to calculate the thermal conductivity of a solid. The first, and by far the most popular, is the time honored Boltzmann Transport Equation (BTE) with the relaxation time ansatz. With the development of effective algorithms and the availability of ever-increasing computing power, it became feasible during the past 15–20 years to compute the thermal conductivity from first principles. The approach branched into two areas: the direct *ab initio* calculations, and the use of Molecular Dynamics (MD) simulations. Unlike the BTE computations that depend on input from experimental data, the first principles computations are independent of any experimental input and have a predictive power. I will outline the key features of all three methods in turn.

5.7.2.1   Boltzmann Transport Equation for Heat

The Boltzmann transport equation traces the evolution of particles perturbed from equilibrium by external stimuli, such as the electric field or thermal gradient. In the case of phonons, while the BTE is formally similar, it is the thermal gradient rather than the electric field, which drives phonons out of equilibrium. Although the BTE has classical (or at most semiclassical) roots, it is still very popular, surprisingly effective, intuitive, relatively easily comprehended, and most importantly, it lends itself naturally to the analysis of the experimental transport data.

Since we consider lattice vibrations and not charge carriers, the Fermi-Dirac distribution functions of electrons must be replaced by the Bose-Einstein distribution function relevant to boson-like phonons. The average number of phonons with wavevector q in equilibrium, designated as $N^0(q)$, is given by the distribution function,

$$N^0(q) = \frac{1}{e^{\left(\hbar \omega_q / k_B T \right)} - 1}, \tag{5.54}$$

with the usual meaning of the appearing parameters. Please note that at high temperatures taken as $\hbar\omega_q/k_B T \ll 1$, Eq. 5.54 simplifies to $N^0(q) = k_B T / \hbar\omega_q$. Since the energy of the phonon is $\hbar\omega_q$, the average energy of a normal mode at high temperatures is $k_B T$.

The BTE appropriate for phonons has the form

$$\left(\frac{dN(q)}{dt}\right)_{coll} = \frac{dN(q)}{dt} + \frac{d\vec{r}}{dt}\nabla_r N(q). \tag{5.55}$$

Assuming for simplicity that the temperature gradient is applied along the z-axis, dT/dz, the perturbed phonons will move along z-direction, and we rewrite Eq. 5.55 for this one-dimensional situation. At a steady state, $dN(q)/dt = 0$, and Eq. 5.55 becomes

$$\left(\frac{dN(q)}{dt}\right)_{coll} = \frac{dz}{dt}\frac{\partial N(q)}{\partial z} = v_z \frac{\partial N(q)}{\partial T}\frac{dT}{dz} = v_z \frac{\partial N^0(q)}{\partial T}\frac{dT}{dz}. \tag{5.56}$$

In Eq. 5.56, the z-component velocity of drifting phonons $v_z = dz/dt$, and the partial derivative of the distribution function was expanded. Further simplification is made by replacing $\partial N(q)/\partial T$ by the derivative of the equilibrium value of the distribution function $\partial N^0(q)/\partial T$. Justification for the step is the fact that a modest temperature gradient perturbs the distribution only mildly, in which case the temperature derivatives of the equilibrium and the perturbed distributions are essentially the same. The three-dimensional version of Eq. 5.56 is

$$\left(\frac{dN(q)}{dt}\right)_{coll} = \vec{v}\cdot\vec{\nabla}T\frac{\partial N^0(q)}{\partial T}. \tag{5.57}$$

The perturbed system with redistributed positions of the particles attempts to come back to equilibrium by undergoing collisions. The way the particles get back to equilibrium is usually modeled by introducing the relaxation time τ, the rate of which is assumed proportional to how far the system was perturbed from equilibrium, i.e.,

$$\left(\frac{dN(q)}{dt}\right)_{coll} = -\frac{N(q)-N^0(q)}{\tau}. \tag{5.58}$$

Rewriting Eq. 5.58, and substituting for the collision term from Eq. 5.56, the BTE becomes

$$N(q)-N^0(q) = -v_z\tau\frac{dN^0(q)}{dT}\frac{dT}{dz} = -\frac{\hbar\omega}{k_B T^2}v_z\tau\frac{dT}{dz}\frac{\exp\left(\dfrac{\hbar\omega}{k_B T}\right)}{\left[\exp\left(\dfrac{\hbar\omega}{k_B T}\right)-1\right]^2}, \tag{5.59a}$$

or, in three dimensions,

$$N(q)-N^0(q) = -\frac{\hbar\omega}{k_B T^2}\tau\left(\vec{v}\vec{\nabla}T\right)\frac{\exp\left(\dfrac{\hbar\omega}{k_B T}\right)}{\left[\exp\left(\dfrac{\hbar\omega}{k_B T}\right)-1\right]^2}. \tag{5.59b}$$

The heat flux $Q(z,t)$ is the energy content of mode $\hbar\omega_q$, multiplied by the velocity with which energy propagates (group velocity) and by the population of phonons $N(q)$ with wavevector q, and summing over all modes and all polarizations (not shown)

$$Q(z,t) = \sum_q \hbar \omega_q v_z N(q). \tag{5.60}$$

Substituting for $N(q)$ from Eq. 5.59a and realizing that no heat flows when phonons are at equilibrium (because $\omega(-q) = \omega(q)$, $v(-q) = -v(q)$, and $N^0(-q) = N^0(q)$), the heat flow becomes

$$Q(z,t) = -\sum_q \hbar \omega_q v_z^2 \tau \frac{dN^0(q)}{dT} \frac{dT}{dz}. \tag{5.61}$$

According to Eq. 5.41, the thermal conductivity is then

$$\kappa = -\frac{Q(z,t)}{dT/dz} = \frac{1}{3} \sum_q \hbar \omega_q v^2 \tau \frac{dN^0(q)}{dT}, \tag{5.62}$$

where v_z^2 was replaced with $v^2/3$.

Converting summation into an integral, we must introduce the density of phonon states $g(\omega)d\omega$.

$$\kappa = \frac{1}{3} \int_0^{\omega_{max}} \hbar \omega v^2 \tau \frac{dN^0(q)}{dT} g(\omega) d\omega. \tag{5.63}$$

Assuming the Debye model, i.e., taking the dispersion $\omega(q) = vq$ for all modes and assuming velocities of all three acoustic branches being the same, Eq. 5.34 for the density of states simplifies, because $|\nabla_q \omega(q)| = v$. The integral over the surface S_ω (now a sphere) is evidently $4\pi q^2 dq$, and the density of q-values is $g(q)dq = 3p^2 dp/2\pi^2$, as each q-vector has three polarizations. A straightforward conversion to the density of phonon states $g(\omega)d\omega$ yields

$$g(\omega)d\omega = \frac{3\omega^2}{2\pi^2 v^3} d\omega. \tag{5.64}$$

Thus, in the Debye model, the phonon density of states increases quadratically with the angular frequency. However, the frequency cannot reach arbitrarily large values as it is limited by a requirement that the number of allowed vibrations must equal $3N$, where N is the number of atoms in a crystal. Thus,

$$3N = \int_0^{\omega_D} g(\omega)d\omega = \frac{\omega_D^3}{2\pi^2 v^3}, \tag{5.65}$$

where ω_D is the Debye frequency equal

$$\omega_D = v\left(6\pi^2 N\right)^{1/3}. \tag{5.66}$$

Substituting Eq. 5.64 into Eq. 5.62, differentiating the phonon distribution function in Eq. 5.53 with respect to temperature, and introducing a new variable $x = \hbar\omega/k_B T$, the thermal conductivity becomes

$$\kappa = \frac{k_B}{2\pi^2 v}\left(\frac{k_B}{\hbar}\right)^3 T^3 \int_0^{\theta_D/T} \tau(x) \frac{x^4 e^x}{\left(e^x - 1\right)^2} dx. \tag{5.67}$$

The upper limit of the integral contains the Debye temperature θ_D defined as

$$\theta_D = \frac{\hbar}{k_B}\omega_D. \tag{5.68}$$

Given that the differential specific heat of phonons within the Debye approximation is given by

$$C(x)\,dx = \frac{3k_B}{2\pi^2 v^3}\left(\frac{k_B}{\hbar}\right)^3 T^3 \frac{x^4 e^x}{(e^x - 1)^2}\,dx \tag{5.69}$$

the expression for the thermal conductivity in Eq. 5.67 attains the form

$$\kappa = \frac{1}{3}\int_0^{\theta_D/T} C(x)\tau(x)v^2 dx = \frac{1}{3}\int_0^{\theta_D/T} C(x)v\ell(x)dx, \tag{5.70}$$

equivalent to Eq. 5.46 derived from the kinetic theory.

To proceed further, we need to consider contributions of the terms in Eq. 5.70. The velocity of phonon propagation (group velocity) v stays substantially constant. The specific heat at high temperatures (above the Debye temperature) is also constant, given by the Dulong-Petit value. At very low temperatures, the specific heat attains the T^3 power law dependence and will rapidly drive the thermal conductivity toward zero. The term that is most important over a broad range of temperatures is the phonon relaxation time $\tau(x)$ or, alternatively, the mean free path of phonons $\ell(x)$. These two respective parameters are determined by scattering processes phonons participate at various temperatures, the topic of the following sections.

5.8 SCATTERING PROCESSES OF PHONONS

There are basically two types of scattering events phonons participate in: intrinsic scattering processes, whereby phonons interact among themselves; and extrinsic scattering processes, where phonons are scattered by structural imperfections spanning from atomic scale defects (dopants, vacancies, interstitials, isotope), to line defects (dislocations), to grain boundaries and, in very pure crystals, even the boundaries of the sample. The task is to capture the various scattering processes phonons undergo and properly evaluate them. Provided the scattering events are independent, they can then be subsumed in the overall scattering rate of phonons by writing

$$\tau^{-1}(x) = \sum_i \tau_i^{-1}(x) \tag{5.71}$$

Let us consider the most important scattering evens that limit the lifetime of phonons.

5.8.1 INTRINSIC PHONON SCATTERING PROCESSES

In perfect crystals (no defects), waves propagate independently of other waves, do not interact, and do not decay. Harmonic waves are thus stationary waves with zero probability of changing from one to another. Equivalently, we may say that a phonon has an infinite lifetime. In real crystals, the lattice modes are no longer true normal modes, and they can exchange energy. Given enough time, the modes can attain equilibrium. Classically, it means that on average each mode will have energy $k_B T$, while quantum mechanically we would say that the average number of phonons in a mode of frequency ω will be given by Eq. 5.54. Phonon interaction is essentially driven by the anharmonicity of the lattice. The implication is that instead of truncating the energy expansion in Eq. 5.31 at the harmonic term, we must admit higher order terms, most notably the cubic term, which allows for processes, such as a single phonon decaying into two phonons, and an inverse process of two phonons combining to form a third phonon. In principle, including also the quartic term in the

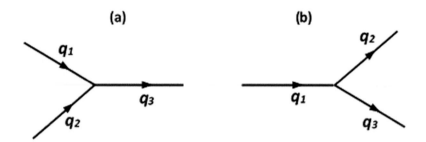

FIGURE 5.25   Three-phonon processes depicting (a) phonons with wavevectors q_1 and q_2 combining to form a third phonon with wavevector q_3, and (b) a phonon q_1 decaying into phonons q_2 and q_3.

expansion, four-phonon processes could participate, but the probability of such events is considerably smaller. Hence, only three-phonon processes will be considered.

Typical three-phonon processes are depicted in Figure 5.25.

Since a phonon has energy $\hbar\omega$ and momentum $\hbar\vec{q}$, the processes in Figure 5.25 can be represented as satisfying the conservation of energy and momentum,

$$\omega_1 + \omega_2 = \omega_3 \quad \omega_1 = \omega_2 + \omega_3 \tag{5.72}$$

$$\vec{q}_1 + \vec{q}_2 = \vec{q}_3 \quad \vec{q}_1 = \vec{q}_2 + \vec{q}_3 \tag{5.73}$$

Processes in Figure 5.26 and Eqs. 5.72 and 5.73 describe what is referred to as *normal* phonon processes. However, a complication arises when two vectors with large momenta combine so that the resultant vector lands outside of the Brillouin zone. An example in the case of a two-dimensional reciprocal lattice is shown in Figure 5.26. In the left-hand panel, representing the relation $\vec{q}_3 = \vec{q}_1 + \vec{q}_2$, is depicted a normal three-phonon process, in which the overall momentum of phonons does not change, as some phonons lose their momentum by transferring it to other phonons of the system. In the right-hand panel, depicting an umklapp (flip-over) phonon process, we see something entirely different. The addition of vectors $\vec{q}_1$ and $\vec{q}_2$ places the resultant vector $\vec{q}_3$ into the next zone and the reciprocal vector $\vec{g}$ of magnitude $2\pi/a$ is used to bring the resultant vector back into the first Brillouin zone with the wavevector $\vec{q}_3$. The relation between the wavevectors is thus

$$\vec{q}_3 = \vec{q}_1 + \vec{q}_2 + \vec{g}. \tag{5.74}$$

Obviously, the original direction of vectors $\vec{q}_1$ and $\vec{q}_2$, i.e., their momentum, is now drastically altered in this resistive process.

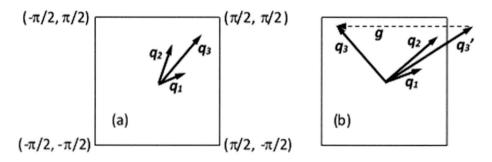

FIGURE 5.26   (a) Normal and (b) Umklapp three-phonon process illustrated in two-dimensional representation. The reciprocal lattice vector $\vec{g}$ has magnitude $2\pi/a$.

Since normal phonon processes do not themselves cause thermal resistance, one might ask why to bother about them at all when discussing the thermal conductivity. The answer is simple, normal phonons do matter. They matter because they redistribute energy and momentum between the modes. As we shall see shortly, nearly all resistive processes of phonons are frequency dependent. Consequently, there are certain frequency ranges where phonon modes are barely or not at all affected by scattering, and if not for normal scattering processes, such modes would run away with the heat flow. In normal processes, these weakly affected phonon modes interact with other phonon modes and transfer their excess energy and momentum to modes that do undergo resistive scattering events, thus preventing attaining very high (infinite) thermal conductivity.

5.8.1.1 Model of Callaway

The role of normal phonon processes in the thermal conductivity was originally considered by Klemens (1951, 1955a, 1958) and, based on his ideas, a particularly successful treatment was developed by Callaway (1959). His model is used in the vast majority of analyses and fits to the experimental thermal conductivity data.

Since the normal phonon processes are distinctly different from resistive umklapp processes in that they are not able to bring the system to equilibrium, their relaxation time cannot be considered on the same footing with relaxation times of resistive processes. If we simply add the inverse relaxation time applicable for normal phonon scattering to Eq. 6.71, we would be underestimating the thermal conductivity and we would have to compensate for it somehow. How such compensation should look like is at the heart of the Callaway model.

The key point is that, as the thermal gradient pumps momentum to the phonon system, normal phonon processes cannot bring phonons to equilibrium even when the source of the thermal gradient is withdrawn. Rather, normal processes shift the distribution to a 'displaced' distribution given by

$$N(\vec{u}) = \frac{1}{\exp\left(\hbar(\omega - \vec{u}\vec{q})/k_B T\right) - 1}, \tag{5.75}$$

where $\vec{u}$ is a constant velocity vector in the direction of the temperature gradient, corresponding to the total momentum of the drifting gas of phonons. While $N(\vec{u})$ is not an equilibrium phonon distribution, it would be one in a coordinate system moving along with the velocity $\vec{u}$. For small values of $\vec{u}$ (not very large thermal gradients), the distribution $N(\vec{u})$ can be approximated in terms of the equilibrium distribution as

$$N(\vec{u}) \approx N^0(q) + \vec{u}.\vec{q}\frac{dN^0(q)}{d\omega} = N^0(q) - \frac{\hbar \vec{u}.\vec{q}}{k_B T}\frac{\exp\left(\hbar\omega/k_B T\right)}{\left[\exp\left(\hbar\omega/k_B T\right) - 1\right]^2} \tag{5.76}$$

We note that the deviation $N(\vec{u}) - N^0(q)$ in Eq. 5.76 is formally the same as the deviation from equilibrium in Eq. 5.59b, provided we take the velocity vector $\vec{u}$ equal to

$$\vec{u} = v^2\tau \vec{\nabla}T / T. \tag{5.77}$$

Thus, with assigning the relaxation time τ_N to normal processes and the relaxation time τ_R to resistive processes that restore the equilibrium state, the total rate of change of the distribution of phonons can be written as

$$\frac{N(q) - N^0(q)}{\tau_R} + \frac{N(q) - N(\vec{u})}{\tau_N} = -\left(\vec{v}.\vec{\nabla}T\right)\frac{\partial N^0(q)}{\partial T}. \tag{5.78}$$

Callaway then introduced a combined relaxation time τ_c by

$$\tau_c^{-1} = \tau_R^{-1} + \tau_N^{-1}, \tag{5.79}$$

and wrote the overall thermal conductivity in terms of two components, $\kappa = \kappa_1 + \kappa_2$, where

$$\kappa_1 = \frac{k_B}{2\pi^2 v}\left(\frac{k_B}{\hbar}\right)^3 T^3 \int_0^{\theta_D/T} \tau_c \frac{x^4 e^x}{\left(e^x - 1\right)^2}\, dx, \tag{5.80}$$

$$\kappa_2 = \frac{k_B}{2\pi^2 v}\left(\frac{k_B}{\hbar}\right)^3 T^3 \frac{\left(\displaystyle\int_0^{\theta_D/T} \frac{\tau_c}{\tau_N}\, \frac{x^4 e^x}{\left(e^x - 1\right)^2}\, dx\right)^2}{\displaystyle\int_0^{\theta_D/T} \frac{\tau_c}{\tau_N \tau_R}\, \frac{x^4 e^x}{\left(e^x - 1\right)^2}\, dx}. \tag{5.81}$$

In the first term of the thermal conductivity κ_1, the reader will note that Callaway uses the relaxation time τ_c, which subsumes the relaxation time for normal phonon scattering as if it was a resistive process. As I have already pointed out, this underestimates the value of the thermal conductivity. Callaway then uses the second term κ_2 to correct for it by bringing back some of the thermal conductivity not captured in the term κ_1.

Let us look how well the Callaway model fares in extreme cases of highly imperfect samples and exceptionally pure and defect-free crystals:

A. In the case of highly imperfect specimens, resistive phonon processes are very frequent and thus $\tau_R \ll \tau_N$, leading to $\tau_c \approx \tau_R$ by Eq. 5.79. To get an idea about the relative contribution of terms κ_1 and κ_2, assume that all relaxation times are frequency independent. Then, as follows from Eqs. 5.80 and 5.81, $\dfrac{\kappa_1}{\kappa_2} \approx \dfrac{\tau_N}{\tau_R} \gg 1$ and, hence, $\kappa_1 \gg \kappa_2$. In other words, we do not introduce a significant error in this case by relying on the term κ_1 only and neglecting completely the role of normal processes.

B. If the crystal is exceptionally pure and free of defects, the normal phonon processes vastly dominate, and $\tau_N \ll \tau_R$. In fact, we can take $\tau_R \to \infty$, and this leads to $\tau_c = \tau_N$. The denominator of κ_2 in Eq. 5.81 is then zero and κ_2 approaches infinity, as one would expect when only normal phonon processes are present.

Thus, the extreme situations are well reproduced by the Callaway model. In general, the correction term κ_2 is rather small. Since thermoelectric materials are, by their design, rather highly doped and often include submicron-size grains, resistive processes dominate and Eq. 5.80, i.e., the term κ_1, is a very good approximation of the overall thermal conductivity.

The next step is to evaluate the frequency and temperature dependence of the relaxation time for various phonon scattering processes. I first consider intrinsic phonon processes and then discuss extrinsic phonon scattering.

5.8.1.2   Temperature and Frequency Dependence of Intrinsic Phonon Processes

Starting with *normal phonon-phonon* scattering, the relaxation rate is typically presented in the form

$$\tau_N^{-1} = A\omega^a T^b, \tag{5.82}$$

with B, a, and b constants determined from the best fits to the experimental data. Morelli et al. (2002) quoted exponents $a = 1$ and $b = 4$ for several group IV semiconductors and $a = 2$ and $b = 3$

for III–V semiconductors. Occasionally, different relaxation rates are assigned to normal processes involving longitudinal and transverse branches of phonons, Herring (1954).

The relaxation rate for *umklapp processes* was first considered by Peierls (1929) who proposed to explain the nearly 20-year-old data of Eucken (1911) that indicated the thermal conductivity of insulators at elevated temperatures being proportional to $1/T$. The expression Peierls came up with was of the form

$$\tau_U^{-1} = T^n e^{-\theta_D/mT},\tag{5.83}$$

with n and m of about unity. There is really nothing one could call the exact form of the phonon umklapp scattering rate and many variants of Eq. 5.83 have been used. I include several of them in Table 5.3. The one form that is very popular and has often been relied on is the expression of Slack and Galginaitis (1964)

$$\tau_u^{-1} = A\omega^2 \frac{T}{\theta_D} e^{-\theta_D/3T} \approx \frac{\hbar\gamma^2}{M_a v^2} \frac{T}{\theta_D} e^{-\theta_D/3T},\tag{5.84}$$

who used for the prefactor A a combination of parameters developed by Leibfried and Schlömann (1954) for their intrinsic thermal conductivity. Regardless of the exact values of prefactors and exponents, the umklapp scattering rate of phonons must satisfy the following asymptotic behavior observed in experiments: (1) at high temperatures, understood to be temperatures above the Debye temperature θ_D (in practice usually above room temperature), the relaxation rate ought to follow the $1/T$ dependence. The expected high temperature $1/T$ behavior follows from a small value of $\hbar\omega/k_BT$, which reduces Eq. 5.54 to $N^0(q) \approx k_BT/\hbar\omega$; (2) in very pure crystals, roughly between $1/30 < T/\theta_D < 1/10$, the thermal conductivity should be exponentially rising, i.e., the relaxation rate should attain the $\exp(-\theta_D/T)$ dependence.

In reality, it is often found that the thermal conductivity above room temperature diminishes faster than the $1/T$ dependence. The discrepancy has nothing to do with not capturing adequately the complexity of three-phonon umklapp processes, but rather with a simple fact that theoretical expressions assume specific heat (and other variables, including the Grüneisen parameter) at constant volume, while the experiments are carried out at constant pressure. As shown by Slack (1972), it is the volume changes with temperature that produce such departures from the $1/T$ behavior.

TABLE 5.3

Various Forms of Intrinsic Phonon Relaxation Rates in Three-Body Processes

	Longitudinal	Transverse	Reference
N-process	$\tau_{NL}^{-1} = A_L\omega^2 T^3$ low T	$\tau_{NT}^{-1} = A_T\omega T^4$ low T	Herring (1954)
N-process	$\tau_{NL}^{-1} = A'_L\omega^2 T$ high T	$\tau_{NT}^{-1} = A'_T\omega T$ high T	Herring (1954)
U-process	$\tau_U^{-1} = B_P T^n e^{-\theta_D/mT}$		Peierls (1929)
U-process	$\tau_U^{-1} = B_K\omega^2 T^3 e^{-\theta_D/\alpha T}$		Klemens (1951)
U-process	$\tau_U^{-1} = B_K\omega T^3 e^{-\theta_D/\alpha T}$		Klemens (1958)
U-process	$\tau_U^{-1} = B_K\omega^2 T$ high T		Klemens (1958)
U-process	$\tau_U^{-1} = B_S\omega^2 \dfrac{T}{\theta_D} e^{-\theta_D/3T}$		Slack & Galginaitis (1964)

At low temperatures, the umklapp scattering rate proportional to $\exp(-\theta_D/T)$ reflects 'freezing out' of three-phonon umklapp processes as the magnitude of the phonon wavevectors becomes small and the resultant wavevector falls short of the Brillouin zone boundary. Thermal conductivity thus rises exponentially with decreasing temperature, and is prevented to blow to infinity by phonons undergoing scattering on boundaries of a crystal, which is a resistive process.

5.8.2 Temperature and Frequency Dependence of Extrinsic Phonon Processes

There are several kinds of extrinsic phonon scattering processes and they are conveniently classified by the physical size of the scatterer. Starting with the largest defects, they are crystal boundaries and diameters of grains in polycrystalline specimens. The next category are extended defects that include dislocations. Then come point defects, which include solute atoms, vacancies, interstitials, and even isotopes mixtures. Finally, important phonon scatterers are charge carriers. I will briefly discuss them in turn.

5.8.2.1 Boundary Scattering

As alluded to in the previous section, the exponentially rising thermal conductivity at low temperatures, as the umklapp phonon processes are gradually being 'frozen out', is prevented from growing without bounds by the mean free path of phonons eventually approaching the geometrical extent of a crystal. The issue of boundary scattering was first raised by Casimir (1938), with the relaxation rate independent of both temperature and frequency and taken as

$$\tau_B^{-1} = \frac{v}{L}, \tag{5.85}$$

where L is the smallest crystal dimension or, in the case of polycrystalline samples, the average diameter of grains. How effective boundary scattering is as a resistive process depends on the degree of roughness at boundary surfaces. In case the boundary is parallel to the thermal gradient and perfectly smooth (specular phonon reflection), the boundary scattering would lose its resistive nature. The effect of boundary scattering is easily documented by a considerably larger thermal conductivity of a single crystal compared to a polycrystalline sample having the exactly same chemical composition. While the influence of boundary scattering is usually most notable at low temperatures near the peak in the thermal conductivity, the effect persists even at ambient temperatures, as documented by Savvides and Goldsmid (1973) who observed a drop from 8.2 $Wm^{-1}K^{-1}$ in undoped $Si_{70}Ge_{30}$ down to 4.3 $Wm^{-1}K^{-1}$ when the grain size was reduced to 2 μm in the polycrystalline sintered sample. Substantial reduction in the thermal conductivity as the grain size decreased from 10 μm to 1 μm was also observed in measurements of half-Heusler alloys by Bhattacharya et al. (2008). The effectiveness of boundary scattering in impeding heat flow is the reason why nanostructuring has become such an important tool in trying to develop thermoelectrics with superior performance. Although the population of low frequency phonons is rather small, they tend to have large mean free path and, as such, their contribution to the thermal conductivity cannot be neglected. The only way to mitigate their effect on the thermal conductivity is to make grains sufficiently small, ≤ 1 μm, i.e., on the nanometer scale.

5.8.2.2 Scattering of Phonons by Dislocations

As pointed out by Nabarro (1951), scattering of sound waves by dislocations has two aspects: the effect of the core of the dislocation (basically a change in the local density within the core radius r) and the effect of strain field that has a rather long reach. In the case of lattice waves (phonons), the effect of the core was treated by Klemens (1955b) and yielded the relaxation rate

$$\tau_{core}^{-1} \propto N_D \frac{r^4}{v^2} \omega^3, \tag{5.86}$$

where N_D is the number of dislocations per unit area, and r is the radius of the core. Strain field with its rather long reach depends on the type of a dislocation, and the subsequent refinements of the Klemens' work by Parrott and Stuckes (1975) gave the relaxation rates for the strain field scattering due to screw and edge dislocations of the form

$$\tau_{SD}^{-1} = \frac{2^{3/2} b^2 \gamma^2 N_D}{27 \left(3^{1/2} + 2^{1/2}\right)} \omega \tag{5.87}$$

and

$$\tau_{ED}^{-1} = \frac{2^{3/2} b^2 \gamma^2 N_D}{27 \left(3^{1/2} + 2^{1/2}\right)} \omega \left\{ \frac{1}{2} + \frac{\beta^2}{24} \left[1 + 2^{1/2} \left(\frac{v_\ell}{v_t} \right)^2 \right]^2 \right\}, \tag{5.88}$$

with $\beta = (1 - 2v_P)/(1 - v_P)$, where v_P stands for Poisson's ratio. Other parameters are the Burgers vector b, and the Grüneisen constant γ. Calculations by Carruthers (1961) give somewhat higher scattering rate but the same frequency dependence. In any case, it is often found, e.g., Sproull et al. (1959), that the experimental thermal resistance is several times (or even an order of magnitude) larger than Eqs. 5.87 and 5.88 predict.

5.8.2.3 Scattering of Phonons by Point Defects

Similar to the case of dislocation scattering, phonon scattering by point defects has two contributions; one associated with the local mass defect ΔM as the substituting and the original atom have different masses (in the case of vacancy the mass defect is 100%), and the other contribution originates from the strain caused by different sizes or bonding of the solute and solvent atoms. The point defect scattering also arises even in otherwise perfect lattice when different naturally occurring isotopes are present in the sample. The anharmonicity of the lattice assures that phonons 'feel' such local variations, resulting in the thermal resistance. Since the phonon wavelength is typically much larger than the extent of the point defect, the Rayleigh-type scattering applies and the relaxation rate for point defect scattering is proportional to the fourth power of frequency, i.e., $\tau_{PD}^{-1} \propto \omega^4$. Thus, in a regime where point defects are important, high frequency phonons do not contribute to the thermal conductivity as they are highly preferentially scattered.

In general, the relaxation rate for point defect scattering is written as

$$\tau_{PD}^{-1} = \frac{\Omega}{4\pi v_s^3} \omega^4 \Gamma, \tag{5.89}$$

where Ω is the volume per atom, v_s is the average sound velocity, and Γ is the disorder scattering parameter that includes contributions from the mass Γ_M and strain Γ_S fluctuations, i.e.,

$$\Gamma = \Gamma_M + \Gamma_S. \tag{5.90}$$

The original derivation of the mass fluctuation parameter was given by Klemens (1955b). Later modifications were made by Slack (1957) and Abeles (1963). The generalized formula developed by Yang et al. (2004a) is used here.

Assuming the composition of a structure is specified by $A_{1c_1} A_{2c_2} A_{3c_3} \ldots A_{nc_n}$ with individual crystallographic sublattices A_i having site degeneracies c_i, the mass fluctuation parameter Γ_M becomes

$$\Gamma_M = \frac{\sum_{i=1}^{n} c_i \left(\dfrac{\overline{M_i}}{\overline{\overline{M}}}\right)^2}{\sum_{i=1}^{n} c_i}, \tag{5.91}$$

with the mass fluctuation parameter for the i-th sublattice given by

$$\Gamma_M^i = \sum_k f_i^k \left(1 - \frac{M_i^k}{\overline{M_i}}\right)^2, \tag{5.92}$$

and the average atomic mass of the compound $\overline{\overline{M}}$ of

$$\overline{\overline{M}} = \frac{\sum_{i=1}^{n} c_i \overline{M_i}}{\sum_{i=1}^{n} c_i}. \tag{5.93}$$

In the case of solid solutions where two different atoms can occupy each of the i-th sublattices, i.e., $k = 1, 2$, with the masses of the two atoms being M_i^1 and M_i^2 and their fractional concentrations f_i^1 and f_i^2 such that $f_i^1 + f_i^2 = 1$ give $\overline{M_i} = f_i^1 M_i^1 + f_i^2 M_i^2$, then Eq. 5.91 becomes

$$\Gamma_M = \frac{\sum_{i=1}^{n} c_i \left(\dfrac{\overline{M_i}}{\overline{\overline{M}}}\right)^2 f_i^1 f_i^2 \left(\dfrac{M_i^1 - M_i^2}{\overline{M_i}}\right)^2}{\sum_{i=1}^{n} c_i}. \tag{5.94}$$

The strain fluctuation parameter is arrived at similarly, except mass differences are replaced with differences in radii of the impurity and host atoms, and an adjustable parameter ε_i is introduced for the i-th sublattice, which is a function of the Grüneisen constant γ characterizing anharmonicity of the lattice. We thus have an equation analogous to Eq. 5.94,

$$\Gamma_S = \frac{\sum_{i=1}^{n} c_i \left(\dfrac{\overline{M_i}}{\overline{\overline{M}}}\right)^2 f_i^1 f_i^2 \varepsilon_i \left(\dfrac{r_i^1 - r_i^2}{\overline{r_i}}\right)^2}{\sum_{i=1}^{n} c_i}. \tag{5.95}$$

Let us illustrate how the mass fluctuation parameter is obtained for a filled skutterudite of the form $R_y T_4 X_{12}$. Obviously, we now have $n = 3$ with $A_1 = R$, $A_2 = T$, and $A_3 = X$, with degeneracies $c_1 = 1$, $c_2 = 4$, and $c_3 = 12$. Since there is no mass fluctuation on the transition metal site T and the pnicogen site X, we have $\overline{M_2} = M_T$ and $\overline{M_3} = M_X$, where M_T and M_X are the atomic masses of the transition metal T and pnicogen atom X, respectively. Then, clearly, $\Gamma_M^2 = \Gamma_M^3 = 0$. At the filler site A_1 with the degeneracy $c_1 = 1$ and occupancy y we have $\overline{M_1} = f_1^R M_1^R + f_1^{\text{void}} M_1^{\text{void}} = y M_R + (1 - y)0 = y M_R$.

Hence, $\Gamma_M^1 = f_1^R \left(1 - \dfrac{M_1^R}{\overline{M_1}}\right)^2 + f_1^{\text{void}} \left(1 - \dfrac{M_1^{\text{void}}}{\overline{M_1}}\right)^2 = y \left(1 - \dfrac{M_R}{y M_R}\right)^2 - (1 - y) \left(1 - \dfrac{0}{y M_R}\right)^2 = \dfrac{1 - y}{y}$. The

average mass of the partially filled skutterudite is then, from Eq. 5.93,

$$\overline{\overline{M}} = \frac{\overline{M}_1 + 4\overline{M}_2 + 12\overline{M}_3}{1 + 4 + 12} = \frac{yM_R + 4M_T + 12M_X}{17}.$$ Finally, the mass fluctuation parameter is

$$\Gamma_M = \frac{\left(\dfrac{\overline{M}_1}{\overline{\overline{M}}}\right)^2 \Gamma_M^1}{17} = \left(\frac{17M_R}{yM_R + 4M_T + 12M_X}\right)^2 y(1-y).$$

In the case of alloy scattering, which is an intense form of point defect scattering in solid solutions, one can calculate the thermal conductivity of a solid solution κ_{alloy}, based on the experimental value of the thermal conductivity of pure compound κ_{pure}, by the theoretical approach developed by Callaway and von Baeyer (1960). At high temperatures, the thermal conductivity of the solid solution κ_{alloy} with respect to a pure compound having thermal conductivity κ_{pure} is given by

$$\frac{\kappa_{alloy}}{\kappa_{pure}} = \frac{\tan^{-1}(u)}{u}, \tag{5.96}$$

with

$$u = \left(3G\Gamma\kappa_{pure}\right)^{1/2}, \tag{5.97}$$

where

$$G = \frac{\pi^2 \theta_D \delta^3}{3hv_s^2}. \tag{5.98}$$

Here, θ_D is the Debye temperature, δ^3 is the average volume per atom of the crystal, v_s is the average sound velocity, h is the Planck constant, u is a scattering scaling parameter of the alloy, and Γ is the scattering parameter in Eq. 5.90.

5.8.2.4 Scattering of Phonons by Charge Carriers

Electron-phonon scattering is yet another resistive process that can reduce thermal conductivity. To be effective, it obviously requires a rather high concentration of electrons, which is found in metals. In semiconductors, the contribution of this scattering mechanism as a limiting agent on the thermal conductivity is usually neglected. However, since the carrier concentration in filled skutterudites is significant, being typically in the 10^{20} to high 10^{21} cm^{-3} range, phonons might find electrons as limiting their mean free path.

Of course, electron-phonon scattering is essentially an absorption of a phonon of wavevector $\vec{q}$ by an electron of wavevector $\vec{k}_1$, in the process of which the electron attains a wavevector $\vec{k}_2$. There is also a reverse process, whereby an electron $\vec{k}_1$ emits a phonon $\vec{q}$ and alters is momentum to a state with wavevector $\vec{k}_2$. In other words, for a phonon being scattered by an electron, restrictions on the energy and momentum must be satisfied,

$$E\left(\vec{k}_1\right) \pm E\left(\vec{q}\right) = E\left(\vec{k}_2\right) \tag{5.99}$$

$$\vec{k}_1 \pm \vec{q} = \vec{k}_2 + \vec{g}, \tag{5.100}$$

where $\vec{g}$ is a reciprocal lattice vector, which is zero for the normal scattering process and nonzero for the umklapp process, in a similar way as in the case of phonon-phonon scattering. In the case of absorption, Eq. 5.99 can be written as

$$\hbar\omega_q = E\left(\vec{k_1}\right) - E\left(\vec{k_2}\right). \tag{5.101}$$

Since the maximum energy of a phonon is $\hbar\omega_D$ and it is much smaller than the Fermi energy, the electron does not lose nor gain much energy in the process. We can then write

$$\hbar\omega_q = E\left(\vec{k_1}\right) - E\left(\vec{k_2}\right) \approx \left(\frac{\partial E(k)}{\partial k}\right)_{E_F} \cdot \vec{q} = \hbar\vec{v}_F \cdot \vec{q}. \tag{5.102}$$

Equation 5.102 can be written as

$$\frac{\omega_q}{q} = \frac{\vec{v}_F \cdot \vec{q}}{q}, \tag{5.103}$$

and states that for the scattering to take place the electron velocity in the direction of phonon $\vec{q}$ must be equal to the phase velocity of the phonon.

Formal derivation of the relaxation rate for phonon-electron scattering was done by Ziman (1956) and, including a correction, Ziman (1957), has a somewhat unwieldy form

$$\tau_{pe}^{-1} = \frac{C^2 m^{*3} v}{4\pi\hbar^4 \rho} \frac{k_B T}{\frac{1}{2} m^* v^2}$$

$$\left\{ \frac{\hbar\omega}{k_B T} - \ln \frac{1 + \exp\left[\left(\frac{1}{2} m^* v^2 - E_F\right)/k_B T + \hbar^2\omega^2/8m^* v^2 k_B T + \hbar\omega/2k_B T\right]}{1 + \exp\left[\left(\frac{1}{2} m^* v^2 - E_F\right)/k_B T + \hbar^2\omega^2/8m^* v^2 k_B T - \hbar\omega/2k_B T\right]} \right\}, \tag{5.104}$$

where C is the phonon-electron interaction constant (deformation potential) and ρ is the mass density. A more appealing is the form of the relaxation rate derived by Pippard (1955) when explaining ultrasonic attenuation in metals,

$$\tau_{pe}^{-1} = \frac{4nm^* v_e \ell_e}{15\rho v_s^2} \omega^2. \tag{5.105}$$

Here, n is the concentration of electrons, m^*, v_e, and ℓ_e their effective mass, velocity and mean free path, respectively, ρ is the mass density, and v_s is the sound velocity. The expression is valid for phonon wavevectors satisfying $q\ell_e \ll 1$, i.e., the wavelength of phonons ($\lambda_p = 2\pi/q$) much longer than the electron mean free path. The above quadratic frequency dependence of the relaxation rate was used by Yang et al. (2002) to explain the temperature dependence of the thermal conductivity of $Co_{1-x}Ni_x Sb_3$ in the range between 5 K and 30 K, where the thermal conductivity of samples with the Ni concentration $x \geq 0.003$, corresponding to electron concentrations above 10^{19} cm^{-3}, invariably attained a linear T-dependence. Shi et al. (2011b) reported a somewhat stronger dependence on the carrier concentration n in their measurements on $Mo_3 Sb_{7-x} Te_x$, yielding the exponent 4/3.

5.8.2.5 Resonant Scattering of Phonons

Fillers occupy site 2a (Wickers notation) in the skutterudite lattice, which is surrounded by a large structural void. Consequently, the filler species are weakly bonded to pnicogen atoms forming the icosahedral void (cage), and have considerably larger atomic displacement parameters than the atoms of the framework. Slack suggested that the large amplitude vibrations of the filler are essentially independent of vibrations of the framework atoms, and coined the term 'rattling' that was supposed to indicate strong scattering of normal phonon modes by the fillers, leading to a dramatic reduction of the thermal conductivity to a level typical of glassy structures, hence the PGEC concept. Although widely embraced by the thermoelectric community, the PGEC paradigm has been criticized, see discussion in Section 5.3, based on the fact the motion of the filler is coherent rather than incoherent with the motion of the framework atoms. While the controversy persists, the relaxation rate describing the resonant scattering of lattice phonons by the rattling fillers that successfully accounted for the suppression of the thermal conductivity in filled skutterudites will be considered and included here.

The rattling mode is viewed as an Einstein oscillator. Taking its angular frequency simply as $\omega_E = (k/M)^{1/2}$, where k is the force constant reflecting the strength of bonding between the filler and pnicogen atoms, and M being the rattler's mass, the rattling frequencies fall in the range of acoustic phonon modes. As discussed in Section 5.2.3, experimental evidence for rattling modes of fillers comes from a broad spectrum of experiments. As far as the thermal conductivity is concerned, resonant scattering is modelled by a relaxation rate of the form

$$\tau_{res}^{-1} = \frac{C\omega^2}{\left(\omega_0^2 - \omega^2\right)^2}, \tag{5.106}$$

where C is a proportionality factor related to the concentration of oscillators, and ω_0 is the resonance frequency. The resonance term was originally introduced by Pohl (1962) and was used with success by Cohn et al. (1999) in modeling the influence of rattlers in clathrates. In skutterudites, resonant scattering was used to model the thermal conductivity in a number of studies, among them by Chen et al. (2001) to fit the thermal conductivity of Ni-doped $Ba_{0.3}Co_4Sb_{12}$, by Yang et al. (2003) to explain the effect of alloying Sn on the Sb site of Yb-filled skutterudites, by Nolas et al. (2006) in a series of La- and Yb-filled skutterudites, and by Wang et al. (2009) to explain time-resolved reflectivity and thermal conductivity measurements on misch-metal-filled skutterudites. The resonance scattering term was also indispensable in very recent *ab initio* molecular dynamics calculations by Wang et al. (2018), and the discussion of the chemical bond hierarchy in thermoelectric materials by Yang et al. (2019).

5.9   MOLECULAR DYNAMICS SIMULATIONS OF THERMAL CONDUCTIVITY

With the improved algorithms and faster computers, *ab initio* calculations of thermal conductivity, including the molecular dynamics *Green-Kubo formalism*, have become feasible during the past 20 years or so.

The Green-Kubo method is an equilibrium Molecular Dynamics (MD) approach that relates the heat flux Q to the thermal conductivity via the fluctuation-dissipation theorem. The essence of the Green-Kubo calculations is an evaluation of the heat flux autocorrelation function,

$$\kappa = \frac{V}{3k_B T^2} \int_0^\infty \left\langle Q(t).Q(0) \right\rangle dt, \tag{5.107}$$

where $Q(t)$ is the instantaneous heat flux at time t and V is the volume of the system. The bracket indicates average over time, and a factor of three accounts for averaging over three dimensions. The thermal conductivity thus relates to the time needed for fluctuations in the heat flux to dissipate. Although the upper limit of the integral is infinity, the actual integration limit is a relaxation time beyond which the integrand attains vanishingly small value. In the actual MD simulation the time is discretized into finite time-steps Δt, and Eq. 5.107 becomes

$$\kappa = \frac{V\Delta t}{3k_B T^2} \sum_{m=1}^N (N-m) \sum_{n=1}^{N-m} Q(m+n)Q(n), \tag{5.108}$$

with N being the total number of time-steps, each of duration Δt, $Q(m + n)$ the instantaneous heat flux at time step $m + n$, and $Q(n)$ the instantaneous heat flux at time step n. The instantaneous heat flux is obtained from the energy of each atom in the simulation,

$$\vec{Q} = \frac{d}{dt}\frac{1}{V}\sum_{i=1}^{N}\vec{r_i}\varepsilon_i, \tag{5.109}$$

where $\vec{r_i}$ is the position vector of atom i, and the total energy of atom i is $\varepsilon_i = \frac{1}{2}m_i v_i^2 + \frac{1}{2}\sum_{j=1}^{N}u_{ij}\left(r_{ij}\right)$.

The sum is over all atoms of the system. Assuming central forces, carrying out the time derivative, and neglecting the first term $\frac{1}{V}\sum_{i=1}^{N}\vec{v_i}\varepsilon_i$, which represents convection, the heat flux becomes

$$\vec{Q} = \frac{1}{2V}\sum_{i=1}^{N}\sum_{j\neq i}^{N}\left(\vec{v_i} + \vec{v_j}\right).\vec{F_{ij}}\vec{r_{ij}^0}, \tag{5.110}$$

with the equilibrium relative position between atoms i and j of $\vec{r_{ij}^0} = \vec{r_i^0} - \vec{r_j^0}$, and the force on atom i by atom j of F_{ij}. As the size of the system increases, the calculated thermal conductivity should approach the experimental value. An interested reader can find useful information about applications of the Green-Kubo method to computations of the thermal conductivity in, e.g., McGaughey and Kaviany (2006) and Stackhouse and Stixrude (2010).

An example of the Green-Kubo MD simulations of the thermal conductivity of skutterudites is presented in Figure 5.27 by the work by Bernstein et al. (2010). Using central forces only and making a further simplification by assuming that the anharmonic parameters are the same for the filled and unfilled forms of the structure, they used the force parameters of Lutz and Kliche (1982)

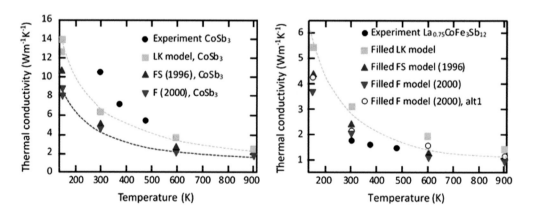

FIGURE 5.27 Thermal conductivity calculated by the Green-Kubo method for (a) $CoSb_3$ and (b) $La_{0.75}CoFe_3Sb_{12}$. Force parameters used in the calculations are those of Lutz and Kliche (1982) designated by orange squares, Feldman and Singh (1996) marked as blue up-triangles, Feldman et al. (2000) indicated by red down-triangles, and open circles represent Feldman et al. (2000) model, designated as Filled F model (2000) alt1 where Fe-La cubic interaction was omitted. Solid circles in (a) are partial data measured for $CoSb_3$ by Katsuyama et al. (1998a), while solid circles in (b) are partial data measured for $La_{0.75}CoFe_3Sb_{12}$ by Yang et al. (2004b). The curves are guides to the eye proportional to $1/T$, expected from the Boltzmann transport theory. Adapted and redrawn from N. Bernstein et al., *Physical Review B* 81, 134301 (2010). With permission from the American Physical Society.

and of Feldman and Singh (1996) for CoSb$_3$ and the force parameters of Feldman et al. (2000) for La$_{0.75}$CoFe$_3$Sb$_{12}$ (the models were discussed in Section 5.2.1 and presented in Table 5.1) and compared the calculated results with the experimental temperature dependence of the thermal conductivity. The Green-Kubo computations clearly captured the decrease in the thermal conductivity upon filling, but the computed decrease at 300 K is only a factor of 2, compared to the experimental decrease of about a factor of 5. The computed results turned out to be little dependent on the model used to generate the force parameters.

Interestingly, the computed thermal conductivity matches better the experimental data in the case of the filled skutterudite, even though computations assume a fully filled structure while the experimental sample had the filling fraction of only 75%. The authors surmised that while the overriding anharmonicity of the filled skutterudite was captured well, the same degree of anharmonicity used in computations of the thermal conductivity of CoSb$_3$ was simply too much and resulted in far too great suppression of the conductivity.

At about the same time, Huang and Kaviany (2010) performed Green-Kubo MD simulations for the thermal conductivity of CoSb$_3$ and Ba$_y$Co$_4$Sb$_{12}$, the choice of Ba based on the documented high filling fraction of $y = 0.44$, W. Chen et al. (2001). Using first the DFT and the response-function theory (RFT), they computed the phonon density of states and the dispersion relations for both structures, and then carried out Green-Kubo MD simulations of the thermal conductivity by changing the potential in the model of Feldman et al. (2006) to that of the Morse form,

$$\varphi = \varphi_0 \left\{ \left[1 - \exp\left(-a\left(r - r_0\right)\right) \right]^2 \right\}, \tag{5.111}$$

where φ_0 is the depth of the potential energy minimum and r_0 is the equilibrium bond length. Surprisingly, the thermal conductivity of the filled skutterudite at 300 K (10 Wm^{-1}K^{-1}) turned out to be considerably larger than the thermal conductivity of CoSb$_3$, with the latter yielding too low value of 6.3 Wm^{-1}K^{-1} in comparison to the experimental data. Such significantly different outcomes between the results of Bernstein et al. (2010) and Huang and Kaviany (2010), raise concerns whether the parameterized potentials necessary for the Green-Kubo MD simulations of the thermal conductivity can yield reliable results for a structure as complex as skutterudites.

5.10   *AB INITIO* CALCULATIONS OF THE THERMAL CONDUCTIVITY

Similar to MD-based simulations, *ab initio* computations of the thermal conductivity have been enabled by the increasing computing power and have become an important approach in the evaluation of *intrinsic* thermal conductivity of solids. In the case of skutterudites, it has provided an essential window into the atomistic origin of the reduction in the thermal conductivity upon filling.

Ab initio calculations combine first-principles treatment of harmonic and anharmonic interatomic force constants (IFC) with the Boltzmann transport equation. The starting point of computations is Eq. 5.57, where the collision term is equal to the drift term of phonons driven out of equilibrium by the thermal gradient. In Section 5.7.2.1, the collision term was substituted with the relaxation time ansatz, i.e., Eq. 5.58. Here, in *ab initio* computations, the collision term, i.e., the relaxation time, is actually computed. The only input needed are harmonic and anharmonic interatomic force constants that are determined using density functional perturbation theory. Thus, the initial equation is, Broido et al. (2007),

$$\vec{v}_\lambda \cdot \vec{\nabla} T \frac{\partial N^0\left(q\right)}{\partial T} = \sum_{\lambda'}\sum_{\lambda''} \left[W^+_{\lambda\lambda'\lambda''}\left(\Psi_{\lambda''} - \Psi_{\lambda'} - \Psi_\lambda\right) + \frac{1}{2} W^-_{\lambda\lambda'\lambda''}\left(\Psi_{\lambda''} + \Psi_{\lambda'} - \Psi_\lambda\right) \right], \tag{5.112}$$

where the right hand side is the collision term for three-phonon processes. Here, λ is used as a short hand for a mode with wave vector $\vec{q}$ and branch index j, $W^{\pm}_{\lambda\lambda'\lambda''}$ are three-phonon scattering rates for absorption and emission of phonons, ψ_λ is the deviation function defined as $\psi_\lambda = N_{1\lambda} / \left[N^0_\lambda \left(N^0_\lambda + 1 \right) \right]$, where $N_{1\lambda} = N_\lambda - N^0_\lambda$, i.e., a nonequilibrium part producing thermal current, and N^0_λ is the equilibrium phonon distribution function. The summation in Eq. 5.112 is over all energy and momentum conserving normal and umklapp three-phonon processes. The required three-phonon scattering rates are computed from Fermi's golden rule and are

$$W^+_{\lambda\lambda'\lambda''} = \frac{\hbar\pi}{4N} \frac{\left(N^0_\lambda + 1\right)\left(N^0_{\lambda'} + \frac{1}{2} + \frac{1}{2}\right)N^0_{\lambda''}}{\omega_\lambda \omega_{\lambda'} \omega_{\lambda''}} \left|V_+\left(\lambda, \lambda', \lambda''\right)\right|^2 \delta\left(\omega_\lambda + \omega_{\lambda'} - \omega_{\lambda''}\right) \tag{5.113}$$

$$W^-_{\lambda\lambda'\lambda''} = \frac{\hbar\pi}{4N} \frac{\left(N^0_\lambda + 1\right)\left(N^0_{\lambda'} + \frac{1}{2} - \frac{1}{2}\right)N^0_{\lambda''}}{\omega_\lambda \omega_{\lambda'} \omega_{\lambda''}} \left|V_-\left(\lambda, \lambda', \lambda''\right)\right|^2 \delta\left(\omega_\lambda - \omega_{\lambda'} - \omega_{\lambda''}\right), \tag{5.114}$$

where N is the number of unit cells, the δ-function guarantees energy conservation, and the three-phonon scattering matrix elements $V_\pm\left(\lambda, \lambda', \lambda''\right) \equiv V\left(j, -\vec{q}; j', \mp\vec{q}; j'', \vec{q}''\right)$ measure the strength of the scattering processes and depend on the third-order interatomic force constants $\Phi_{\alpha\beta\gamma}\left(0k, \ell'k', \ell''k''\right)$ as

$$V\left(j, \vec{q}; j', \vec{q}'; j'', \vec{q}''\right) = \sum_k \sum_{\ell'k'} \sum_{\ell''k''} \sum_{\alpha\beta\gamma} \Phi_{\alpha\beta\gamma}\left(0k, \ell'k', \ell''k''\right) e^{i\vec{q}'\cdot R_{\ell'}} e^{i\vec{q}''\cdot R_{\ell''}}$$

$$\times \frac{e^j_{\alpha k}\left(\vec{q}\right) e^{j'}_{\beta k'}\left(\vec{q}'\right) e^{j''}_{\gamma k''}\left(\vec{q}''\right)}{\sqrt{M_k M_{k'} M_{k''}}}, \tag{5.115}$$

where R_ℓ is a lattice vector, M_k is the mass of the k-th atom, and functions $e^j_{\alpha k}\left(\vec{q}\right)$ are phonon wave vectors.

One proceeds by solving Eq. 5.112 iteratively until a nonequilibrium distribution function ψ_λ is obtained, i.e., using a criterion that if after n iterative steps the function is ψ_λ, after $n+1$ steps it has not changed. The relaxation time τ_λ is directly proportional to ψ_λ, and the thermal conductivity follows from

$$\kappa \equiv \kappa_{\alpha\alpha} = \frac{1}{NV} \sum_\lambda \frac{\partial N^0_\lambda}{\partial T} \left(\hbar\omega_\lambda\right) v^\alpha_\lambda v^\alpha_\lambda \tau_\lambda. \tag{5.116}$$

Here, N is the number of uniformly spaced $\vec{q}$ vectors in the Brillouin zone, V is the volume of the unit cell, v^α_λ is the velocity of mode λ in the direction α, and N^0_λ is the equilibrium phonon distribution function.

As already noted, the all-important third-order interatomic force constants $\Phi_{\alpha\beta\gamma}\left(0k, \ell'k', \ell''k''\right)$ are obtained by the density functional perturbation theory. How to do it, and summation and symmetry rules to obey are discussed, among others, by Li et al. (2012).

Ab initio computations of thermal conductivity have been used on many systems and returned values in reasonable agreement with experiment, particularly at temperatures around the ambient. Even if some discrepancies exist, the important point regarding first principles approach is its atomistic nature that allows probing harmonic and anharmonic interactions and reveals their influence on the thermal transport.

5.11   BIPOLAR HEAT TRANSPORT

For a semiconductor exposed to elevated temperatures, sooner or later comes a point where the thermal energy $k_B T$ is comparable to its band gap. When this happens, a large number of electrons and holes will be generated. If a thermal gradient is applied to the sample, electrons and holes will travel down the thermal gradient and carry the rather large energy gained at the hot end of the specimen where electrons were excited across the band gap. This energy is transported along the length of the sample and released as the pairs of electrons and holes reach the cold end of the sample and recombine. The energy (heat) transported by this process is known as the bipolar thermal conductivity and comes on top of the heat carried by electrons and holes individually. In fact, depending on the temperature, the amount of heat transported via the bipolar conduction can be many times larger than the heat associated with individual contributions of electrons and holes.

The formal derivation of the bipolar thermal conductivity contribution is presented in Section 4.3 in the context of considering transport in a semiconductor consisting of a pocket of electrons and a pocket of holes. Consequently, only the final result is presented here:

$$\kappa = \kappa_{1,q} + \kappa_{2,q} + \frac{\sigma_1 \sigma_2}{\sigma_1 + \sigma_2} \left(S_2 - S_1 \right)^2 T. \tag{5.117}$$

The important point is that, in addition to the ordinary thermal conductivity of electrons and holes, there is a term proportional to the temperature and the square of the difference between the Seebeck coefficients of the two types of carriers, which is known as *the bipolar term* (Price 1955). Note that if the two types of carriers have the same sign, the bipolar term is very small. However, in the presence of both electrons (negative Seebeck coefficient) and holes (positive Seebeck coefficient), the bipolar term can be very large and dominate the heat transport of a semiconductor at high temperatures. One often encounters the bipolar conduction when studying transport properties of thermoelectric materials at elevated temperatures. The most vivid manifestations of its presence are the rising thermal conductivity that takes over the otherwise umklapp process-limited $1/T$ temperature dependence, and frequently even more spectacular turnover in the Seebeck coefficient as the presence of both electrons and holes is highly detrimental to the magnitude of the Seebeck effect. In fact, the bipolar thermal conductivity seriously limits the high temperature range of operation of many thermoelectric materials.

5.12   MINIMUM THERMAL CONDUCTIVITY

In many technological areas it is of interest to know how low a thermal conductivity can be attained in a given material. In particular, this is important in thermal management and very relevant in thermoelectricity, where a high conversion efficiency necessitates thermoelectric materials to have very low thermal conductivity. Traditionally, estimates of the minimum thermal conductivity were developed based on quantized lattice vibrations (phonons), where the mean free path of phonons ℓ_p was curtailed by some physically reasonable criterion, such as the interatomic spacing. An early interest in how small a phonon mean free path can be dates back to the work of Kittel (1949), who analyzed thermal conductivity of glasses in comparison to the crystalline forms of matter and arrived at $\ell_p \sim$ unit cell (7 Å). Recasting the problem in terms of the Young's modulus Y and the material's density ρ, by taking the mean acoustic velocity $v_{am} = A \left(Y / \rho \right)^{1/2}$, with A having a value 0.87 ± 0.02, Clarke (2003) expressed the minimum thermal conductivity in the form

$$\kappa_{min} \approx 0.87 k_B N_A^{\frac{2}{3}} \frac{\rho^{\frac{1}{6}} Y^{\frac{1}{2}}}{\bar{M}^{\frac{2}{3}}}, \tag{5.118}$$

where N_A is the Avogardo's number and $\bar{M}$ is the mean atomic mass.

The forms of minimum thermal conductivity that have often been used are due to Slack (1979) and Cahill and Pohl (1989). Slack's concept of minimum thermal conductivity is based on modeling lattice vibrations as waves and, as such, their mean free path cannot be shorter than one wavelength. Cahill and Pohl developed an alternative approach based on the Einstein (1911) model of thermal conductivity. Einstein assumed a collection of harmonic oscillators, each vibrating with the same frequency and, in order to facilitate transport of thermal energy, the coupling between oscillators was provided by harmonic forces between the first, second, and third neighbors. Although the Einstein model of thermal conductivity failed rather miserably in explaining heat transport in crystalline medium (due to the lack of coherence between the motions of neighboring atoms), it fared much better in describing thermal conductivity of amorphous solids. Anyway, Cahill and Pohl modified the Einstein model by making oscillators larger than just single atoms by dividing the sample into segments of size of one half of the wavelength and assumed that such regions oscillate with the angular frequency $\omega = 2\pi v/\lambda$. Since each oscillating region had a lifetime of one half of the vibration period, i.e, $\tau = \pi/\omega$, the thermal conductivity was then a result of the random walk between the oscillating regions, and turned out to be given by

$$\kappa_{\min} = \left(\frac{\pi}{6}\right)^{1/3} k_B n^{2/3} \sum_{i=1}^{3} v_i \left(\frac{T}{\theta_{Di}}\right)^2 \int_{0}^{\theta_{Di}/T} \frac{x^3 e^x}{\left(e^x - 1\right)^2} dx, \tag{5.119}$$

where the sum is over three acoustic modes (one longitudinal and two transverse) with speeds of sound v_i, the Debye temperatures for each mode θ_{Di}, and n the number density of atoms. At high temperatures, the spectral heat capacity $C(\omega)$ is k_B, and the minimal thermal conductivity becomes

$$k_{\min}^{HT} = 1.21 n^{2/3} k_B \frac{1}{3} \left(v_L + 2v_T\right). \tag{5.120}$$

The difference between Slack's minimum thermal conductivity and that of Cahill and Pohl is a factor of 2, as Slack effectively took the shortest mean free path of phonons $\ell_p = \lambda$, while Cahill and Pohl set it at $\ell_p = \lambda/2$. Both approaches are useful in judging how far the samples' thermal conductivity is from their minimal values and whether it makes sense to expand further efforts in attempting to try to bring the conductivity closer to its minimum theoretical value.

Occasionally, thermal conductivity at high temperatures falls below the minimal values predicted by Eq. 5.119. This prompted Agne et al. (2018) to consider an alternative form of energy propagation based on diffusion. The thermal conductivity is then written as

$$\kappa = \int_{0}^{\infty} g(\omega)C(\omega)D(\omega)d\omega, \tag{5.121}$$

where instead of the usual term $v\ell_p$ now enters the diffusivity $D(\omega)$. The diffusion mediated thermal transport was developed in a series of papers by Allen and Feldman (1989), Feldman et al. (1993), and Allen et al. (1999), and considers thermal transport as the harmonic coupling between nonpropagating, nonlocalized atomic vibrations. The quantized entities are called diffusons and their propagation mode is random walk.

In a random walk, after taking N discrete steps, each step of length a, the net distance traveled is $d = a\sqrt{N}$. The diffusivity in random walk is defined as $D_{RW} = d^2/t$, where t is the elapsed time. The number of steps per unit time, N/t, is viewed as the frequency of attempts to transfer energy. The success of the transfer is specified by the probability P, the number between 0 and 1. Agne et al. assumed that an oscillator makes two attempts during one period to transfer the energy, i.e.,

$\frac{N}{t} = \left(\frac{2\omega}{2\pi}\right)P$. If there are n atoms in a volume, the spacing between the oscillators is $n^{-1/3}$ and this

is then the length of each step a. The maximum diffusivity of diffusons ($P = 1$) is thus

$$D_{\text{diff}}(\omega) = \frac{1}{3}\frac{n^{-2/3}}{\pi}\omega. \qquad (5.122)$$

The factor of one-third comes from the three-dimensional nature of the problem.

The diffusive thermal conductivity is defined as $\kappa_{\text{diff}} = \int_0^\infty g(\omega)C(\omega)D_{\text{diff}}(\omega)d\omega$, where $g(\omega)$ is

the density of vibrational states, $C(\omega) = \frac{\partial}{\partial T}\left(\frac{\hbar\omega}{\exp(\hbar\omega/k_BT)-1}\right)$ is the usual spectral heat capacity,

and D_{diff} is the diffusivity of diffusons. At high temperatures, the spectral heat capacity is a constant k_B, and thus, upon substituting Eq. 5.122, the diffusive thermal conductivity becomes

$$\kappa_{\text{diff}} = k_B \int_0^\infty g(\omega)\left(\frac{1}{3}\frac{n^{-\frac{2}{3}}}{\pi}\omega\right)d\omega = \frac{k_B}{3\pi}\int_0^\infty g(\omega)\omega\, d\omega$$

$$\qquad (5.123)$$

$$= \frac{k_B n^{1/3}}{\pi}\frac{\int_0^\infty g(\omega)\omega\, d\omega}{\int_0^\infty g(\omega)d\omega} = \frac{k_B n^{1/3}}{\pi}\omega_{av}.$$

Equation 5.123 indicates that a measure of the minimum thermal conductivity in this diffusive heat transport is the average frequency of oscillators ω_{av}, which could be obtained from inelastic neutron scattering. Short of that, the authors have shown that the usual Debye temperature $\theta_D = \hbar\omega_D/k_B = \hbar\left(6\pi^2 n\right)^{\frac{1}{3}}/k_B$ can serve very well as a metric to estimate ω_{av}. Writing for the average speed of sound $v_s = \frac{1}{3}\left(v_L + 2v_T\right)$, Agne et al. compared the average sound velocities, the number densities n, and values of $\theta_D k_B$ for some 24 compounds for which ω_{av} was known and found out that the average energy $\hbar\omega_{av}$ can be expressed as

$$\hbar\omega_{av} \approx 0.61 k_B\theta_D. \qquad (5.124)$$

Consequently, the minimum diffusive thermal conductivity in Eq. 5.123 becomes

$$\kappa_{\text{diff}} = \frac{k_B n^{\frac{1}{3}}}{\pi}\omega_{av} \approx \frac{k_B n^{\frac{1}{3}}}{\pi}\frac{0.61 k_B\theta_D}{\hbar} = \frac{k_B n^{\frac{1}{3}}}{\pi}\frac{0.61 k_B}{\hbar}\frac{\hbar v_s\left(6\pi^2 n\right)^{\frac{1}{3}}}{k_B}$$

$$\qquad (5.125)$$

$$= 0.76 n^{2/3}k_B\frac{1}{3}\left(v_L + 2v_T\right).$$

A direct comparison of Eqs. 5.120 and 5.125 indicates that the diffusive thermal conductivity criterion yields about 37% lower value for the minimum thermal conductivity than does the formula of Cahill and Pohl.

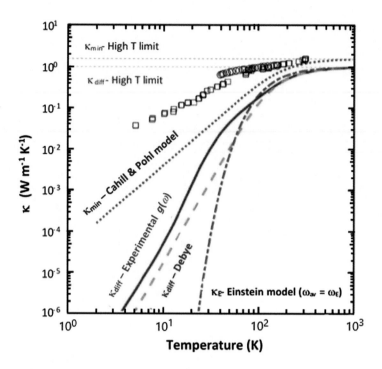

FIGURE 5.28 Temperature dependence of the minimum thermal conductivity for the Cahill & Pohl model, diffusive models of Agne et al. using experimental density of states and truncated Debye model, and the Einstein model. The dashed blue and red lines indicate the limiting high temperature value of the Cahill & Pohl and Agne et al. models, respectively. Open circles and open squares are experimental data for amorphous Si measured by Cahill et al. (1989) and Zink et al. (2006), respectively. Adapted and redrawn from M. T. Agne et al., *Energy & Environmental Science* 11, 609 (2018). With permission from the Royal Society of Chemistry.

If one desires the value of the minimum thermal conductivity at temperatures below the Debye temperature, it is obtained by inserting in Eq. 5.121 the temperature dependent specific heat,

$$C(\omega) = k_B (\hbar\omega / k_B T)^2 \exp\left\{(\hbar\omega / k_B T) / \left[\exp(\hbar\omega / k_B T) - 1\right]^2\right\}.$$ Equation 5.123 then acquires a tem-

perature dependent average frequency $\omega_{av}(T)$. Figure 5.28 shows the temperature dependent minimum thermal conductivity for the model of Cahiil and Pohl, for the diffusive treatment of Agne et al., where either the experimental density of states $g(\omega)$ or the truncated Debye model (the maximum energy taken not as $k_B\theta_D$ but instead as $0.95\ k_B\theta_D$) is used, and for the Einstein model, i.e., taking $\omega_{av} = \omega_E$. The data points indicated in the figure are measurements for amorphous Si by Cahill et al. (1989) [open circles] and by Zink et al. (2006) [open squares].

With this background, let us discuss thermal conductivity of skutterudites.

5.13 THERMAL CONDUCTIVITY OF SKUTTERUDITES

5.13.1 THERMAL CONDUCTIVITY OF BINARY SKUTTERUDITES

Over the years, there have been numerous measurements of the thermal conductivity of binary skutterudite not just because of the interest in their heat transport ability but also because they were usually used as a yardstick against which the filled skutterudites have been assessed. In pure binary skutterudites, on account of their low carrier concentration, the heat transport is carried essentially by phonons. Even in heavily doped $CoSb_3$, the carrier contribution to heat transport does not exceed more than 20% of the total thermal conductivity. Thus, there is a good chance (at least in theory) to bring the thermal conductivity down from high values ~ 10 Wm⁻¹K⁻¹ that preclude binary skutterudites to be effective thermoelectrics. Moreover, considering that the minimum thermal conductivity estimated for $CoSb_3$ based on the Cahill and Pohl (1989) criterion is only about 0.3 $Wm^{-1}K^{-1}$, and a slightly less stringent criterion of Slack (1979) gives about 0.4 $Wm^{-1}K^{-1}$, there are ample opportunities to devise effective strategies to degrade the heat transport in skutterudites.

Room temperature values of the thermal conductivity of binary skutterudites are collected in Table 5.4. Clearly, the overwhelming number of entries for $CoSb_3$ indicates where the interest has been the greatest.

TABLE 5.4

A Collection of Room Temperature Thermal Conductivity Values for Binary Skutterudites

Skutterudite	$\kappa(Wm^{-1}K^{-1})$	Form of the Material	Reference
$CoSb_3$	5.2	Polycrystal	Dudkin and Abrikosov (1959)
	5.5	Cold-pressed, 70% density	Sharp et al. (1995)
	8.4	Hot-pressed, 90% density	Sharp et al. (1995)
	10.5–12	Single crystal	Morelli et al. (1995)
	10.5–11	Single crystal	Caillat et al. (1995a)
	~10	Single crystal	Caillat et al. (1996a)
	4.8	Submicron grain size	Nakagawa et al. (1996)
	10.5	Hot-pressed, annealed	Nakagawa et al. (1996)
	11.8	Hot-pressed, 98% density	Fleurial et al. (1996)
	11.5	Hot-pressed, > 95% density	Uher et al. (1997)
	10.5	Hot-pressed, 97% density	Katsuyama et al. (1998b)
	8.4	Hot-pressed	Nolas et al. (1998)
	9.3	Hot-pressed	Stokes et al. (1999)
	9.2	Hot-pressed, >95% density	Sales et al. (2000)
	10.6	Hot-pressed, > 95% density	Yang et al. (2000)
	10	Hot-pressed	Nagao et al. (2000)
	8.4	SPS	Chen et al. (2001b)
	9.7	SPS	Shi et al. (2002)
	8.9	Polycrystal, direct synthesis	Tobola et al. (2003)
	9.3	Hot-pressed	Wojciechowski et al. (2003)
	10.6	SPS	Yu et al. (2003)
	9.4	SPS	Shi et al. (2004)
	8.6	Hot-pressed	Puyet et al. (2004)
	10.8	SPS	Kitagawa et al. (2005)
	9.25	SPS	Arita et al. (2005)
	8.7	Hot-pressed	Da Ros et al. (2006)
	11.1	Ind. melted, quenched, annealed	Mallik et al. (2006)
	11.0	Ind. melted, quenched, annealed	Mallik et al. (2007)
	8.6	SPS, 90–96% density	Zhang et al. (2008)
	11.0	Ind. Melted, annealed	Jung et al. (2008)
$IrSb_3$	8–9.5	Polycrystal, 85% density	Caillat et al. (1995b)
	17.7	Hot-pressed	Slack and Tsoukala (1994)
	16	Hot-pressed, 98% density	Slack and Tsoukala (1994)
	20	Hot-pressed	Tritt et al. 1996)
	12	Cold-pressed, 70% density	Sharp et al. (1995)
$RhSb_3$	13	Hot-pressed	Caillat et al. (1995c)
	10.7	Hot-pressed, 90% density	Sharp et al. (1995)
$CoAs_3$	14.5	Polycrystal	Caillat, private commun.
CoP_3	18.5	Hot-pressed, 87% density	Watcharapasorn et al. (1999)

Inspecting the data for $CoSb_3$, one notes that the values of the room temperature thermal conductivity cluster near 5.5 $Wm^{-1}K^{-1}$ and near 10.5 Wm^{-1} K^{-1}. The lower value is associated with poorly compacted or small grain polycrystalline samples that are typically of no more than 70% theoretical density. The higher values of thermal conductivity reflect well compacted (usually hot-pressed or SPSed) polycrystalline samples with larger grains and single crystal specimens, all with densities well in excess of 90%.

The thermal conductivity of single crystals of $CoSb_3$ at low temperatures measured by Morelli et al. (1995) is shown in Figure 5.29a. Also included are two curves that represent fits to the data assuming the Debye model with umklapp and boundary scattering of phonons (curve A) and the fit that includes scattering by vacancies with the density 1×10^{18} cm^{-3} (curve B). The charge carrier contribution to the total thermal conductivity is less than 10% for all samples and temperatures. The very steep rise with decreasing temperature is characteristic of the dominance of phonon-phonon umklapp scattering. The different peak heights near 12 K are most likely due to the presence of vacancies on the pnicogen sites. For comparison, Figure 5.29a also includes thermal conductivity of a well-compacted polycrystalline $CoSb_3$, Uher et al. (1997), where the effect of boundary scattering is obvious from a strong reduction of the peak height (and a shift of the peak to higher temperatures), while there is little difference in room temperature values of the thermal conductivity. The temperature dependence of the thermal conductivity of a single crystal of $CoSb_3$ at temperatures above the ambient measured by Caillat et al. (1996a) and of a polycrystal of $CoSb_3$ measured by Katsuyama et al. (1998b) is displayed in Figure 5.29b. The conductivity decreases from its room temperature value of about 10.5 Wm^{-1} K^{-1} and reaches a minimum near 700 K in the case of the single crystal and between 600 and 650 K in the case of the polycrystal. At still higher temperatures, intrinsic excitations start to set in and the bipolar thermal conductivity contribution gives rise to an increasing thermal conductivity. This general trend has also been confirmed in measurements with arsenide skutterudites, Caillat et al. (1995c), and the decreasing thermal conductivity with the increasing temperature up to about 750 K was reported for $IrSb_3$ by Slack and Tsoukala (1994).

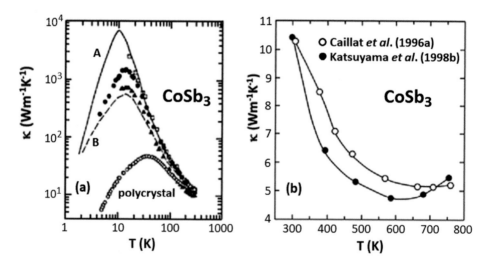

FIGURE 5.29 (a) Thermal conductivity of $CoSb_3$ single crystals below 300 K measure by Morelli et al. (1995). Curve A is a fit to the Debye model assuming Umklapp and boundary scattering of phonons; curve B includes the effect of 1×10^{18} cm^{-3} vacancies. Open circles represent thermal conductivity of a polycrystalline $CoSb_3$. Adapted from C. Uher, *Semiconductors and Semimetals*, Vol. 69, ed. T. M. Tritt, Academic Press, p.139 (2001). (b) Thermal conductivity above room temperature of a single crystal of $CoSb_3$ (open circles) from measurements of Caillat et al. (1996a) and of a polycrystalline sample of $CoSb_3$ (solid circles) from measurements of Katsuyama et al. (1998b). With permission from Elsevier and the American Institute of Physics, respectively.

Although the data on arsenide and phosphide skutterudites are sparse, it seems that the thermal conductivity increases on going from antimonides to arsenides to phosphides, the trend being consistent with the general rule that heavier compounds have lower thermal conductivity. However, the rule is clearly broken within the antimonide series, where $IrSb_3$ has higher thermal conductivity than $RhSb_3$, which in turn has higher thermal conductivity than $CoSb_3$, as documented in measurements by Sharp et al. (1995). The reason why such a trend is observed in antimonide skutterudites was explored by Li and Mingo (2014a) in their first principles study of the thermal conductivity of $CoSb_3$ and $IrSb_3$. Computing first phonon dispersion for $CoSb_3$ and $IrSb_3$ (see Figures 5.30a and 5.30b), the authors then used the general methodology outlined in Section 5.8.4 and calculated the respective thermal conductivities. In spite of Ir having larger atomic mass than Co, it tends to increase the thermal conductivity in antimonide skutterudites because the ensuing smaller thermal displacement parameter results in weaker anharmonic scattering, which the computations indicate is the dominant factor influencing the thermal conductivity. Inspecting Figures5.30a and 5.30b, the most notable difference in the phonon dispersion of the two binary skutterudites is a clear gap in the frequency spectrum of $CoSb_3$ and the lack of it in $IrSb_3$. The reason for it is the change of mass ratio of the constituent atoms. As the red lines in Figure 5.30a indicate, by replacing the mass of Ir by that of Co, the gap is recovered. Moreover, such an artificial change in $IrSb_3$ brings the acoustic modes to a near coincidence with the acoustic modes of $CoSb_3$, documenting that changes in the low frequency phonon spectrum are solely due to the mass increasing from Co to Ir. Because on going from $CoSb_3$ to $IrSb_3$ the mass of the unit cell changes from 1696.85 amu to 2230 amu, i.e., by a factor of 1.31, the acoustic frequencies change by $(1.31)^{1/2} = 1.15$ when the mass of Ir is replaced by the mass of Co, in accord with the changes depicted in Figure 5.30a. At higher frequencies, somewhat stronger harmonic IFCs of $IrSb_3$ contribute to an increase of optic mode frequencies by about a factor of 1.1. From the ω^2 dependence of the phonon density of states, the Debye temperatures of $CoSb_3$ and $IrSb_3$ are 342 K and 303 K, respectively.

As Figure 5.31 indicates, the computed thermal conductivities of both $CoSb_3$ and $IrSb_3$ are in very good agreement with the respective experimental values. Any overestimation of the theoretical thermal conductivities in comparison to the experimental values are most likely the result of the computations being restricted to intrinsic phonon scattering processes only and not including scattering of phonons by defects. Overall, the thermal conductivity of $IrSb_3$ is about 35% larger than the thermal conductivity of $CoSb_3$. Since at ambient temperatures the heat capacities of $CoSb_3$ and $IrSb_3$ ought to approach the Dulong-Petit value, and because $1/V$ and ν_λ are smaller in $IrSb_3$ than in $CoSb_3$, the increased thermal conductivity in $IrSb_3$ must have the origin in lower scattering rates of

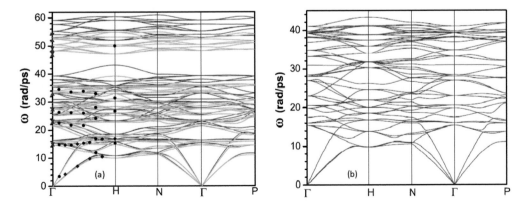

FIGURE 5.30   (a) Computed phonon dispersions for $CoSb_3$ (green lines) and for $IrSb_3$ with the mass of Ir being replaced by that of Co (red lines). (b) Calculated phonon dispersion for $IrSb_3$. Adapted from W. Li and N. Mingo, *Physical Review B* 90, 094302 (2014). With permission from the American Physical Society.

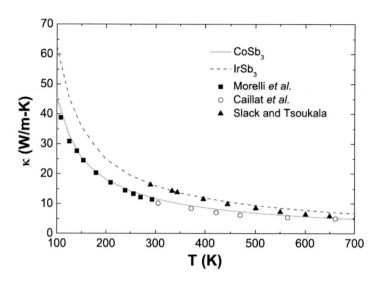

FIGURE 5.31 Calculated lattice thermal conductivity of CoSb₃ (solid line) and IrSb₃ (dashed line). Experimental data for CoSb₃ are those of Morelli et al. (1995) (solid squares) and Caillat et al. (1996a) (open circles). Experimental data for IrSb₃ are due to Slack and Tsoukala (1994). Adapted from W. Li and N. Mingo, *Physical Review B* 90, 094302 (2014). With permission from the American Physical Society.

phonons in this skutterudite. To check this point, the authors first looked if the third-order IFCs of IrSb₃ are smaller than those of CoSb₃. Finding that this is not the case, the attention shifted toward the role of the weighted phase space parameters W^+ and W^- defined in Eqs. 5.113 and 5.114, and written this time in the form

$$W_\lambda^\pm = \sum_{\lambda'p''} \left\{ \begin{array}{c} N_{\lambda'} - N_{\lambda''} \\ N_{\lambda'} + N_{\lambda''} + 1 \end{array} \right\} \frac{\delta\left(\omega_\lambda \pm \omega_{\lambda'} - \omega_{\lambda''}\right)}{\omega_\lambda \omega_{\lambda'} \omega_{\lambda''}}. \tag{5.126}$$

Here, the + sign stands for absorption and − sign for emission processes. These weighted phase space parameters are exceptionally sensitive to small changes in the phonon spectrum. If the phonon spectrum is altered (scaled) by a factor c, the parameters $W^\pm$ scale by a factor c^{-5}! Since on average, the phonon spectrum of IrSb₃ is suppressed compared to that of CoSb₃, the weighted phase space parameters $W^\pm$ are expected to increase in IrSb₃, as is, indeed, the case, Figure 5.32.

Hence, even the weighted phase space parameters on their own suggest that IrSb₃ should have higher scattering rates than CoSb₃, and thus smaller thermal conductivity, contrary to experiments. So, what causes the thermal conductivity of IrSb₃ to be higher than that of CoSb₃?

An inspection of the three-phonon matrix elements $V_\pm(\lambda, \lambda'\lambda'')$ in Eq. 5.115 reveals a dependence on the atomic mass, with heavier atoms having reduced matrix elements. Since the unit cell mass of IrSb₃ is a factor of 1.314 larger than that of CoSb₃, the scattering rates of IrSb₃ on average should decrease by a factor of $1.314^3 = 2.27$. Although the changes in the third-order IFCs and the phase space parameters $W^\pm$ go against the effect of the atomic mass, the influence of the latter dominates, and is the chief reason for the thermal conductivity of IrSb₃ being larger than that of the lighter CoSb₃.

Thus, the *ab initio* calculations confirmed the surprising trend within the binary antimonide skutterudites, where the heavier metal atom results in an increased rather than decreased thermal conductivity. Moreover, the calculations also offered an explanation why this happens by analyzing in detail the role of atomic masses, different phonon velocities, the degree of anharmonicity, and the weighted phase space in the scattering rates.

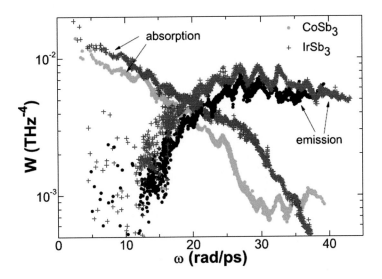

FIGURE 5.32 Weighted phase space W$^+$ (absorption) and W$^-$ (emission) given in Eq. 5.126, plotted as a function of angular frequency for CoSb$_3$ (circles) and IrSb$_3$ (crosses). Reproduced from W. Li and N. Mingo, *Physical Review B* **90**, 094302 (2014). With permission from the American Physical Society.

5.13.1.1 Effect of Grain Size on the Thermal Conductivity of Binary Skutterudites

Different synthesis routes to prepare binary skutterudites result in a spectrum of grain sizes spanning from tens of micrometers down to tens of nanometers. It is no surprise then that the effect of grain boundaries on the thermal conductivity was explored in numerous studies with the hope of minimizing the thermal conductivity. In one of the first reports, Nakagawa et al. (1997) used extensive ball milling to reduce the grain size of the synthesized CoSb$_3$. The powder was subsequently hot-pressed and annealed for various times to obtain samples with a range of grain sizes from submicron to 40 µm. Thermal conductivity, measured as a function of the average grain size at 293 K and 673 K decreased significantly as the grain size decreased. In fact, in subsequent studies of CoSb$_3$ by Bertini et al. (2003a), the authors have demonstrated nearly an order of magnitude reduction in the thermal conductivity (about 1.6 Wm^{-1}K^{-1} at 300 K) when the grain size was reduced down to 140 nm. The authors analyzed their data using the model of Nan and Birringer (1998), where the ratio of the bulk thermal conductivity κ_0 to the thermal conductivity κ of a structure with the grain size d is written as

$$\frac{\kappa_0}{\kappa} = 1 + \frac{2R_k\kappa_0}{d}. \tag{5.127}$$

Here, R_k is the Kapitza resistance, which describes the thermal boundary resistance at the interfaces of two materials or grains of the same material in terms of the macroscopic effective medium approaches, see, e.g., Nan (1993). Instead of the Kapitza resistance, one often uses the Kapitza length, defined as $L_K = R_K \kappa_0$, as a characteristic length to describe the grain size influence on the thermal conductivity. The Kapitza length in the above experiment turned out to be 72 nm.

Comparable reductions in the thermal conductivity of CoSb$_3$ having grain size of a couple of hundreds of nm was also observed by Yu et al. (2003), by Lu et al. (2006), and by Liu et al. (2007a). The last-mentioned authors controlled the grain size of their CoSb$_3$ samples prepared by ball-milling + SPS by adjusting the SPS temperature between 400°C and 600°C while the same 50 MPa pressure was applied for 5 minutes. By lowering the SPS temperature from 600°C to 400°C, the average grain size was reduced from 300 nm down to 100 nm as the rapid grain growth was dramatically slowed down. As a consequence, the room temperature thermal conductivity of 4.29 Wm^{-1}K^{-1} (SPS

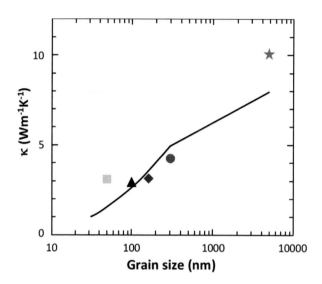

FIGURE 5.33 Thermal conductivity of CoSb$_3$ at room temperature as a function of the grain size. Samples were prepared by ball milling followed by spark plasma sintering (SPS). By varying the SPS temperature, grain size is controlled. Experimental data are indicated by symbols while the solid curve is a theoretically predicted dependence. The star symbol indicates the thermal conductivity of a bulk sample with the average grain size of 5 μm. Redrawn from W.-S. Liu et al., *Journal of Physics D: Applied Physics* 40, 566 (2007a). With permission from the Institute of Physics.

at 600°C) was reduced to 2.9 Wm^{-1}K^{-1} (SPS at 400°C), the data are shown in Figure 5.33. It is important to add that the sample density has remained identical at about 98% and had no effect on the value of the thermal conductivity.

Although these reductions in the thermal conductivity upon reducing the grain size are impressive, a word of caution is in order. Since the Seebeck coefficient is usually substantially independent of the grain size, one always has to make sure that the submicron grains do not degrade the carrier mobility to an even greater extent than they reduce the thermal conductivity, otherwise, there would be no benefit to the thermoelectric performance.

5.13.1.2 Effect of Doping on the Thermal Conductivity of Binary Skutterudites

Low concentrations of impurities act as point defects and strongly scatter phonons. This so-called Rayleigh scattering is strongly frequency dependent, $\tau_{PD}^{-1} \sim \omega^4$, (see Section 5.8.2.3), and acts as a low-pass filter, scattering most effectively high frequency phonons. The strength of the scattering depends on the mass difference between the impurity and the host atom, and on the elastic strain generated by the different ionic radii.

A strong influence of doping on the thermal conductivity of binary skutterudites was first noted by Dudkin and Abrikosov (1957) who observed a 50% reduction at 300 K upon the presence of 10 at% Ni, or 10 at% Fe, or 0.25 at% of Te in CoSb$_3$. Although left without a comment, these are large effects on the thermal conductivity, particularly in the case of Fe-substituted Co where the mass and size differences are not more than 5%. Numerous subsequent studies with doping agents on the cation site have also reported large reductions in the thermal conductivity regardless of whether CoSb$_3$ was driven p-type with Fe substituting for Co or n-type with Ni replacing some Co. The effect of Ni on the thermal conductivity of CoSb$_3$ was studied, for instance, by Caillat et al. (1996a), Uher et al. (2000), Bertini et al. (2003b), Kitagawa et al. (2005), and Kim et al. (2006), while Fe-doped CoSb$_3$ was investigated by Katsuyama et al. (1998b), Stokes et al. (1999), Yang et al. (2000, 2002), and Park et al. (2006). A direct comparison of the effect of Fe and Ni on the thermal conductivity was provided in measurements by Zhang et al. (2008). Extensive investigations of CoSb$_3$ doped

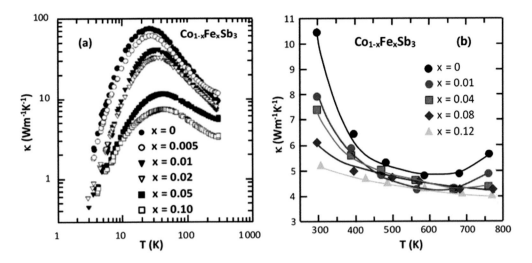

FIGURE 5.34 (a) Temperature-dependent low temperature thermal conductivity of $Co_{1-x}Fe_xSb_3$. Redrawn from J. Yang et al., *Physical Review B* 63, 014410 (2000). (b) Temperature dependence of $Co_{1-x}Fe_xSb_3$ above room temperature for samples with comparable Fe content as in (a). Adapted and redrawn from S. Katsuyama et al., *Journal of Applied Physics* 84, 6708 (1998b). With permission from the American Physical Society and the American Institute of Physics, respectively.

with Ni, Pd, Pt, and PdPt were undertaken by Tashiro et al. (1997) and Anno et al. (1999a,b). The effect of Fe on the thermal conductivity of $CoSb_3$ at both low and high temperatures is illustrated in Figures 5.34a and 5.34b.

A somewhat peculiar finding was made by He et al. (2008), who prepared $Co_{1-x}Ni_xSb_{3+y}$ with $x = 0.1$ and 0.2 and $y = 0$, and 0.05 by ball milling followed by the direct current-induced hot pressing, and finally annealing the samples to volatize the excess Sb in structures with $y = 0.05$. The volatized excess of Sb left behind pores with the diameter between 1 and 1.5 µm, reducing the density of such samples to 92% of the theoretical value. Comparing the thermal conductivity of samples having the same Ni content but with and without pores, the authors noted about a 1 $Wm^{-1}K^{-1}$ reduction in the thermal conductivity of samples containing pores (from about 4.1 $Wm^{-1}K^{-1}$ at 300 K for $Co_{0.8}Ni_{0.2}Sb_3$ with no pores down to about 3.15 $Wm^{-1}K^{-1}$ at 300 K for $Co_{0.8}Ni_{0.2}Sb_3$ containing pores). That in itself is not surprising. However, the fact that the pores apparently significantly enhanced the Seebeck coefficient while had surprisingly no effect on the electrical conductivity, resulted in a dramatically increased $ZT \sim 0.7$ at 773 K for $Co_{0.8}Ni_{0.2}Sb_3$ with pores compared to the value of ZT of less than 0.2 at the same temperature recorded for the fully dense $Co_{0.8}Ni_{0.2}Sb_3$ with no pores.

In general, the level of doping one can attain on the anion site, i.e., on the ring structure of skutterudites, is considerably smaller than on the cation site. Already early on, Zobrina and Dudkin (1960) reported that $CoSb_3$ can accommodate no more than about 0.25 at% of Te, a rather low solubility limit. However, as the dopant here creates disorder at a location that governs the low frequency vibration spectrum that determines the thermal conductivity, one may expect a significant effect on the magnitude of the lattice thermal conductivity in spite of a rather low doping level. For instance, Caillat et al. (1996a) reported a nearly halved thermal conductivity of $CoSb_3$ single crystals when Te concentrations between 0.1 at% and 0.2 at% were dissolved in the crystal. Of course, depending on the type of the dopant, one may also observe a significantly altered electronic part of the thermal conductivity when the dopant's charge state differs from that of Sb. However, because the electronic part of the thermal conductivity is a small fraction of the overall conductivity, the increased carrier concentration is likely to have a stronger effect via degrading the lattice thermal conductivity on account of the enhanced electron-phonon scattering than via increasing the thermal conductivity by a larger electronic term.

There are numerous reports on the effect of doping on the pnicogen rings, and the consequent degradation of the thermal conductivity caused mostly by Te, occasionally by Se, and sometimes compensating with the Column 14 (group IVB) elements Si, Ge, and Sn to enhance the solubility of Te. Before discussing the relevant studies, I wish to alert the reader to a possible confusion that might arise due to inconsistent specification of the amount of dopant replacing Sb. Most of the authors designate the content of Te as $CoSb_{3-x}Te_x$, some use $Co(Sb_{1-x}Te_x)_3$, and one can even find $Co_4Sb_{12-x}Te_x$, a rather improper form as the structure is a binary skutterudite and not filled skutterudite. To avoid the confusion and make a comparison meaningful, I will refer to the content x corresponding to the designation $CoSb_{3-x}Te_x$, or state the amount of substituted Te as a percentage.

The original solubility limit of 0.25 at% of Te quoted by Zobrina and Dudkin (1960) is far too low in view of the studies by Nagamoto et al. (1998) who claimed 5.6 at% solubility of Te, the work of Wojciechowski (2002) reporting 1.5 at% solubility, 2 at% Te solubility observed by Wojciechowski et al. (2003), 4.6 at% solubility measured by Li et al. (2005), and 5 at% solubility found by Liu et al. (2007b). Most of the values are based on a break in the linearly dependent lattice parameter as a function of the content of Te. The presence of Te on the pnicogen rings at its maximum solubility can degrade the thermal conductivity of $CoSb_3$ to half of its value, especially when the average grain size of the structure is on the order of 100 nm. Since the electrical conductivity benefits from the increased concentration of electrons, while the Seebeck coefficient is still robust, the ZT can attain quite high values of 0.72 at 850 K in $CoSb_{2.875}Te_{0.125}$, Li et al. (2005), and 0.93 at 820 K in $CoSb_{2.85}Te_{0.15}$, Liu et al. (2007b). Even higher value of $ZT = 1.1$ at 823 K was achieved by Liu et al. (2008) when the Te content was raised to $x = 0.20$ by codoping with Sn, which partly electronically compensated the structure and perhaps even relaxed the strain due to the different size of Sb (1.53 Å) and Te (1.42 Å), as Sn is larger (1.72 Å). In addition to the degradation of the thermal conductivity, Te doping benefits thermoelectricity also via raising the temperature where the intrinsic processes start to limit the Seebeck coefficient. In all cases, the ionization activity of Te as a dopant is limited and is typically no more than 30–40%.

Apart from Te, Se doping was explored by Wojciechowski et al. (2003), but had a lower solubility of about 1.3 at%, and considerably lower degree of ionization of only about 2.5%. Temperature dependence of the thermal conductivity of several Te-doped samples of $CoSb_3$ prepared by the same technique (mechanical alloying plus SPS at 500°C for 5 min under 50 MPa) is shown in Figure 5.35. All samples in the figure had a very fine grain structure between 140 nm and 160 nm that was responsible for the dramatically reduced thermal conductivity, with the undoped $CoSb_3$ sample having the conductivity at 300 K reduced by a factor of 3 compared to a typical $CoSb_3$ with the grain size on the scale of tens of microns. A further reduction of the thermal conductivity due to the increased Te doping is notable. To enhance the doping limit of Te, Liu et al. (2008) codoped the skutterudite with Sn. Because of charge compensation, the carrier concentrations (electrons) in $CoSb_{2.75}Te_{0.20}Sn_{0.05}$ and $CoSb_{2.85}Te_{0.15}$ are similar. Having comparable carrier densities and grain sizes, the significantly reduced thermal conductivity of $CoSb_{2.75}Te_{0.20}Sn_{0.05}$ with respect to $CoSb_{2.85}Te_{0.15}$ must come from an additional point defect scattering of phonons on Sn.

Direct p-type doping on the pnicogen rings by elements of Column 14 (group IVB) were tried only sporadically, Koyanagi et al. (1996), and suggested that the solubility limit of Ge and Sn is about 1.4 at% and 2.4 at%, respectively, with Pb not substituting at all. It is interesting to note that the lattice constant of $CoSb_3$ increases when a smaller Te atom is substituting for a larger Sb atom, while a larger Sn atom seems to decrease the lattice constant upon substituting for Sb. Obviously, it is not just the relative size of atoms that matters, but perhaps also alterations in bonding caused by such substitutions.

The above doping studies were done with species that have a different valence state with respect to the element they substitute for. In spite of impressive reductions in the thermal conductivity, and some surprisingly large ZT values, there is always a danger that the ionized impurity scattering will strongly suppress the charge carrier mobility. To avoid this possibility, one may try to reduce the lattice thermal conductivity by doping with isoelectronic impurities. The impurity in this case causes

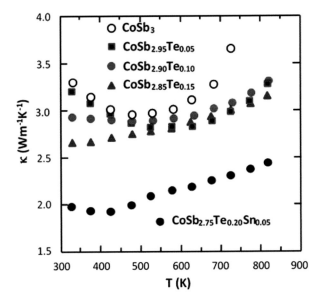

FIGURE 5.35 Thermal conductivity of $CoSb_{3-x}Te_x$, $x = 0, 0.05, 0.10$, and 0.15 prepared by the same mechanical alloying followed by spark plasma sintering under 50 MPa pressure at 500°C for 5 min and measured by Liu et al. (2007b). Also included are measurements by Liu et al. (2008) on a sample prepared by the same technique but where the solution limit of Te was enhanced to 0.20 by codoping with Sn. All samples had a very fine average grain structure between 140 nm and 160 nm, which is responsible for the much-reduced thermal conductivity in comparison to samples with the grain size on the scale of tens of microns. Additional progressive reduction in the thermal conductivity due to Te doping at 300 K is noted as the content of Te increases. Because of charge compensation, the carrier concentration in $CoSb_{2.75}Te_{0.20}Sn_{0.05}$ is similar to that in $CoSb_{2.85}Te_{0.15}$. Having comparable carrier density and grain size, the much-reduced thermal conductivity of $CoSb_{2.75}Te_{0.20}Sn_{0.05}$ with respect to the conductivity of $CoSb_{2.85}Te_{0.15}$ is then primarily the result of an additional point defect scattering on Sn. With permission from the American Institute of Physics.

the Rayleigh-like scattering of phonons, but its electrically neutral state is far less damaging to the mobility of charge carriers. An example is $CoSb_3$ doped isoelectronically on both the cation and anion sites to form $Co_{0.97}Ir_{0.03}Sb_{2.85}As_{0.15}$, which resulted in the thermal conductivity being reduced down to 2.4 $Wm^{-1}K^{-1}$ at 300 K (Sharp et al. 1995).

5.13.1.3 Thermal Conductivity of Solid Solutions of Binary Skutterudites

Solid solutions are archetypal examples of point defect scattering where both mass and strain fluctuations tend to effectively scatter high-frequency phonons. The formation of solid solutions from isostructural compounds is the time-honored approach of reducing lattice thermal conductivity and has been used in the development of virtually all state-of-the-art thermoelectrics. A solid solution will be of benefit to thermoelectricity provided the ratio of its mobility to its lattice thermal conductivity is larger than for the individual compounds forming the solid solution.

The formation of solid solutions of binary skutterudites and the ranges of their solubility were discussed in Section 2.1.3.1. The effectiveness of solid solutions in reducing the lattice thermal conductivity was first documented by Slack and Tsoukala (1994) in their study of $Ir_{0.5}Rh_{0.5}Sb_3$, which had the thermal conductivity of 9 $Wm^{-1}K^{-1}$ compared to the conductivity of $IrSb_3$ of some 16 $Wm^{-1}K^{-1}$. Although the reduction is significant, the two end members unfortunately have far too high thermal conductivity and the resulting solid solution does not suppress the thermal conductivity enough to make it of interest for thermoelectricity. A better outcome is possible with other combinations of binary and ternary skutterudites, as reported by Borshchevsky et al. (1996). For instance, solid solutions of $(CoSb_3)_{0.96}$–$(IrSb_3)_{0.04}$, $(CoSb_3)_{0.90}$ –$(IrSb_3)_{0.10}$, $(CoSb_3)_{0.88}$–$(IrSb_3)_{0.12}$,

$(CoAs_3)_{0.98}–(CoSb_3)_{0.02}$, $(CoSb_3)_{0.90}–(IrAs_3)_{0.10}$, $(CoSb_3)_{0.79}–(Fe_{0.5}Ni_{0.5}Sb_3)_{0.21}$, and $(Ru_{0.5}Pd_{0.5}Sb_3)_{0.5}–$ $(Fe_{0.5}Ni_{0.5}Sb_3)_{0.5}$ have room temperature thermal conductivity of 4.5 Wm^{-1} K^{-1}, 3.2 Wm^{-1} K^{-1}, 2.9 Wm^{-1}K^{-1}, 5.7 Wm^{-1}K^{-1}, 2.5 Wm^{-1}K^{-1}, and 3.5 Wm^{-1}K^{-1}, respectively.

Based on the experimental data of binary skutterudites, the authors used Eqs. 5.96–5.98 to predict the thermal conductivity of several other solid solutions. Of these estimates, the lowest value of the thermal conductivity of 0.8 Wm^{-1}K^{-1} was predicted for a solid solution of $CoSb_3–IrP_3$. There are no reports of anyone trying to confirm this exceptionally low value. Nevertheless, just as in other thermoelectric materials, solid solutions of skutterudites are certainly one of the available means of reducing the thermal conductivity.

5.13.1.4 Thermal Conductivity of Composite Structures Based on Binary Skutterudites

Composite structures and their intrinsic or extrinsic origin are described in detail in Section 1.4. Indeed, the primary driving force to explore the properties of composites were attempts to lower the thermal conductivity while making a limited damage to the carrier mobility. Again, the bulk of the studies were carried out with various forms of $CoSb_3$-based skutterudites, hoping to achieve uniformly dispersed secondary phases in the $CoSb_3$ matrix by either precipitation or by intentionally admixing foreign phases.

In the case of intrinsically formed composites, the typical approach is to precipitate a $FeSb_2$ or $NiSb_2$ (may change into NiSb plus Sb by decomposition) phase by starting with $Co_{1-x}Fe_xSb_3$ or $Co_{1-x}Ni_xSb_3$, where x exceeds the solubility limit. While a modest overstoichiometry will lower the thermal conductivity, excessive overstoichiometry will lead to a large volume fraction of the secondary phase, which, if its thermal conductivity exceeds the conductivity of the matrix, will result in an increased thermal conductivity. The case in point was a study with the highly overstoichiometric $Co_{1-x}Ni_xSb_3$ carried out by Katsuyama et al. (2003), where the precipitated $NiSb_2$ decomposed into a high heat-conducting NiSb and Sb.

Extrinsically formed nanocomposites by uniformly dispersing (typically by ball milling) a second component in the $CoSb_3$ matrix offer a much greater pallet of materials. Of the many examples in the literature, I illustrate the effect of three different additives in $CoSb_3$ that are representatives of the spectrum of extrinsic composites prepared with binary skutterudites.

Insulating oxides have been a very popular choice to add to the $CoSb_3$ matrix. An example of a typical suppression of the thermal conductivity attained with ZrO_2 nanoparticles in the $CoSb_3$ matrix measured by He et al. (2007) is shown in Figure 5.36a. Other interesting additives were various forms of carbon. Figure 5.36b shows the effect of micron-size conducting fullerene (C_{60}) particles dispersed in $CoSb_3$ by Shi et al. (2004) on the thermal conductivity of $CoSb_3$. The various curves represent composites with up to 6.54 mass% of dispersed C_{60}.

The last distinct class of composites I wish to mention are 'homo'-composite structures, where nanoparticles of $CoSb_3$ are dispersed in micron-sized powders of the $CoSb_3$ matrix. Mi et al. (2007) synthesized nanoparticles via a solvothermal method and mixed the nanopowder in wt% with micron-size $CoSb_3$ prepared by vacuum melting and annealing. The composites were compacted by hot pressing under 60 MPa at around 600°C for 60–90 min. Relative densities of the samples were around 90%. The range of homo-composites spanned from no nanoparticles to 100% nanoparticles, and their room temperature thermal conductivity covered 2–5 Wm^{-1}K^{-1}, with the thermal conductivity decreasing as the content of the nanophase increased. In a similar study a couple of years later by L. Yang et al. (2009), the nanoparticles were prepared by the modified polyol process, mixed with coarse-grained $CoSb_3$, and hot pressed under 100 MPa at 500°C for 120 min. Following hot pressing, the particle size grew from about 30 nm to 50 nm, and the denser composites had a somewhat higher thermal conductivity as depicted in Figure 5.36c. As the data indicate, all three types of additives impede the heat transport quite effectively. In all cases, the effect of additives to the $CoSb_3$ matrix can be estimated using the effective conductivity model proposed by Muta et al.

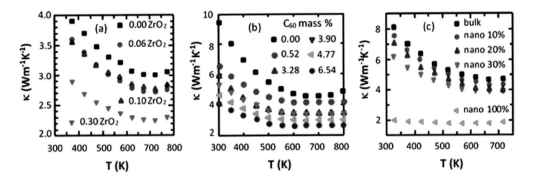

FIGURE 5.36 Temperature-dependent thermal conductivity of CoSb$_3$ with various amounts of additives. (a) Insulating ZrO$_2$ nanopowder with particle size of 25 nm dispersed by ball milling and sintered at 853 K for 30 min under 100 MPa. The composite's content is specified as (1-x)CoSb$_{3-x}$ZrO$_2$ with x up to 0.30. Drawn from the data by Z. He et al., *Nanotechnology* 18, 235602 (2007). (b) Fullerene with up to 6.54 mass % added to the starting powders of Co and Sb by Shi et al. (2004), reacted at 943 K under Ar for 150 hr, ground into fine powders, and sintered by SPS at 848 K for 15 min. Drawn from the data by X. Shi et al., *Applied Physics Letters* 84, 2301 (2004). (c) Nanosized CoSb$_3$ particles with various wt % were added to micron-size powders of CoSb$_3$ and hot-pressed at 500°C for 120 min under 100 MPa. Drawn from the data by L. Yang et al., *Journal of Applied Physics* 106, 013705 (2009). With permission from the Institute of Physics and the American Institute of Physics, respectively.

(2003), where the thermal conductivity of the composite κ_c is related to the thermal conductivity of the matrix κ_0 by

$$\kappa_c \approx \kappa_0 \left(1 - \frac{3f}{2}\right),\qquad(5.128)$$

where f is the volume fraction of the dispersed phase.

5.13.2 THERMAL CONDUCTIVITY OF TERNARY SKUTTERUDITES

As discussed in Section 1.2, ternary skutterudites are man-made variants of the binary skutterudite structure, where either the transition metal M or the pnicogen atom X are replaced by an equal number of atoms from columns immediately to the left and right of the substituted element. The strict adherence to the Zintl concept maintains the valence electron count at 72. No surprise that ternary skutterudites show strong phonon scattering and, in general, have a significantly reduced thermal conductivity in comparison to their parent binary compounds. Unfortunately, such atomic replacements also lead to a large degradation in the carrier mobility, and the thermoelectric properties of ternary skutterudites are not very promising.

One of the first studies of thermal conductivity of ternary skutterudites targeted Ru$_{0.5}$Pd$_{0.5}$Sb$_3$, the structure derived from RhSb$_3$. Caillat et al. (1995d) and Fleurial et al. (1995) reported on extremely low values of thermal conductivity of 0.7–0.9 Wm^{-1}K^{-1} over a broad range of temperatures from 300 K up to 700 K, representing some 12–16 times smaller conductivity than for the parent RhSb$_3$. Subsequent reports from the same group, Caillat et al. (1996a) and Fleurial et al. (1997), revised the value upward to 2.5–3 Wm^{-1}K^{-1}, and comparable magnitude of the thermal conductivity was also reported by Nolas et al. (1996b). Subtracting the electronic thermal conductivity contribution amounting to about one-third of the total thermal conductivity at ambient temperature, the lattice thermal conductivity at 300 K was in the range 1.6–2 Wm^{-1}K^{-1}, more than a factor of 5 reduction in comparison to RhSb$_3$. Such a low value was deemed to be beyond the effect of mass and strain fluctuations and an additional scattering mechanism had to be acting.

Caillat et al. (1996b) noted that their Ru$_{0.5}$Pd$_{0.5}$Sb$_3$ samples were Pd deficient, which could suggest that Ru possesses a mixed valence state with the presence of both Ru^{2+} and Ru^{4+}. The issue was explored in detail by Nolas et al. (1996b), who confirmed the presence of both Ru^{2+} and Ru^{4+} by XANES (x-ray absorption near-edge structure) measurements, which returned nearly equal contents of Ru^{2+} and Ru^{4+} (48% vs. 52%). Thus, on top of the strong point-defect scattering of phonons, a proposal was made that, in the presence of both Ru^{2+} and Ru^{4+} charge states, phonons may induce a transition of a pair of $4d$ electrons from the Ru^{2+} ion ($4d^6 \rightarrow 4d^4$) to the neighboring Ru^{4+} ion ($4d^4 \rightarrow 4d^6$). The strength of this scattering depends on the concentration of both Ru^{2+} and Ru^{4+}, and since the respective concentrations are nearly equal, the scattering should be substantial. Slack et al. (1996) and Nolas et al. (1996b) calculated the thermal resistivity associated with this process and arrived at values as high as 1.4 and 1.18 mKW^{-1}, respectively.

Early measurements of the thermal conductivity of ternary skutterudites formed by replacing Sb with equal amounts of elements from Columns 14 and 16 were carried out by Nolas et al. (2003). Compared to CoSb$_3$, the thermal conductivity of CoGe$_{1.5}$Se$_{1.5}$ at 300 K was reduced by about a factor of 2. The carrier mode, n- or p-type, depends rather sensitively on the ratio of Ge/Se. Similar degradation of the heat transport in ternary skutterudites was obtained by the same group (Nolas et al. 2006), using a minute amount of Ni on the Co sites. In its effect on the thermal conductivity, the latter study resembled the effect of Ni on the conductivity of binary CoSb$_3$ (Dyck et al. 2002), except that the concentration of Ni that could be accommodated in the ternary skutterudite was much smaller (no more than $x = 0.05$ in the formula Co$_{4-x}$Ni$_{ix}$Ge$_6$Se$_6$) than in CoSb$_3$. Subsequently, thermal conductivity measurements on ternary skutterudites with comparable room temperature values were carried out by Navrátil et al. (2010) on CoSn$_{1.5}$Se$_{1.5}$, and by Zevalkink et al. (2015) on CoSn$_{1.5}$Te$_{1.5}$, RhSn$_{1.5}$Te$_{1.5}$, and IrSn$_{1.5}$Te$_{1.5}$. Te-containing ternaries indicated somewhat lower values of the thermal conductivity than the ternary samples having Se. Selenium containing ternary skutterudites are prone to have a slight deficiency of Se a result of the much higher vapor pressure and consequent Se loss during the synthesis. Rather low values of the room temperature thermal conductivity of ternary skutterudites (all below 2 Wm^{-1}K^{-1}) of compositions CoGe$_{1.5}$S$_{1.5}$, CoGe$_{1.5}$Te$_{1.5}$, and CoSn$_{1.5}$Te$_{1.5}$ were measured by Vaqueiro and Sobany (2008). In particular, CoSn$_{1.5}$Te$_{1.5}$, the same composition as measured by Zevalkink et al. that extrapolated to a room temperature value of about 2.8 Wm^{-1}K^{-1}, was reported by Vaqueiro and Sobany to be only about 0.75 Wm^{-1}K^{-1}, and substantially temperature independent. Since no information was provided how the pellets were compacted and no density was given, it is likely that such exceptionally low values of the thermal conductivity were due purely to very low densities of the samples.

A great sensitivity of the thermal conductivity to a slightly off-stoichiometric composition is documented by thermal conductivity measurements of ternary skutterudite FeSb$_{2+x}$Te$_{1-x}$ ($x = 0$–0.25) by Tan et al. (2013), shown in Figure 5.37. This isoelectronic p-type analog of CoSb$_3$ is stable, unlike FeSb$_3$, and is one of the few p-type unfilled skutterudites with a reasonably high figure of merit (0.65 at 800 K). The samples were prepared by melting followed by annealing with the final compaction done by SPS under 40 MPa for 5 min at around 820 K. The typical density was above 97% and the grain size was in the range 2–5 µm. In spite of the much larger grain size, this ternary system showed very low values of the thermal conductivity. XPS measurements indicated that Fe in FeSb$_2$Te appears in a mixed valence state as both Fe^{2+} and Fe^{3+} in roughly equal amounts. Electron transfer between the two valence states of equal abundance gives rise to phonon scattering that, in addition to the disturbance on the pnicogen rings, leads to a strongly reduced thermal conductivity. Extensive fitting of the thermal conductivity using the Callaway model below 300 K indicated that, by far, the greatest effect on the thermal conductivity is due to point defect scattering, particularly as x deviated more and more from the stoichiometric value of $x = 0$.

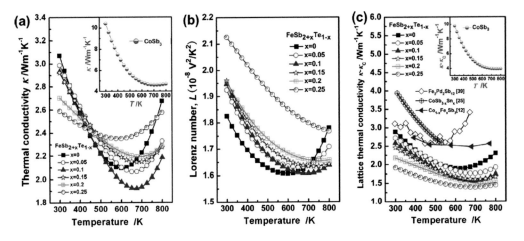

FIGURE 5.37 Thermal conductivity of FeSb$_{2+x}$Te$_{1-x}$ (x = 0 – 0.25) samples and of CoSb$_3$ (insets) reported by Tan et al., *Sci. Adv. Mater.* 5, 12 (2013). (a) Total thermal conductivity, (b) calculated Lorenz number, and (c) the lattice thermal conductivity. For comparison, lattice thermal conductivities of p-type Fe$_2$Pd$_2$Sb$_{12}$ from Navratil et al. (2010), of CoSb$_{3-x}$Sn$_x$ from Tobola et al. (1999), and of Co$_{1-x}$Fe$_x$Sb$_3$ from Katsuyama et al. (1998b) are included. Reproduced from G. Tan et al., *Acta Materialia* 61, 7693 (2013). With permission from Elsevier.

5.13.3 THERMAL CONDUCTIVITY OF FILLED SKUTTERUDITES

An impressive reduction in the thermal conductivity observed by Morelli and Meisner (1995) upon filling the skutterudite structure shortly after Slack (1995) suggested that fillers are likely to resonantly scatter normal phonon modes, has launched an intensive worldwide effort to develop skutterudites into useful thermoelectric materials. The structure used by Morelli and Meisner in their experiments was polycrystalline CeFe$_4$Sb$_{12}$, which attained nearly an order of magnitude lower thermal conductivity compared to CoSb$_3$, (1 Wm^{-1}K^{-1} compared to about 10 Wm^{-1}K^{-1}). Similarly large degradations of the heat transport were seen soon after also in various filled IrSb$_3$ (Nolas et al. 1996a and Tritt et al. 1996), in partially Ce-filled CoSb$_3$ (Uher et al. 1997), and in partially La-filled CoSb$_3$ (Sales et al. 1997 and Nolas et al. 1998). Subsequent efforts (in the late 1990s and the early 2000s) focused on identifying the most effective filler species that suppress the thermal conductivity to the greatest extent yet leave the electronic properties only modestly affected. Over the years, virtually the entire periodic table of naturally occurring elements was explored as fillers in the skutterudite voids, including even uranium, Arita et al. (2005), and the list of thermal conductivity studies counts nearly a hundred publications. Many of them have merely documented that a particular element can fill the skutterudite void and, otherwise, were inconsequential. On the other hand, several studies contributed greatly toward revealing fascinating physical properties of filled skutterudites, shed light on the unique role the fillers play in the lattice dynamics, and helped to identify fillers or filler groups that are most beneficial to the thermoelectric performance. Rare earth fillers in certain skutterudite frameworks have also given rise to a plethora of interesting and even exotic superconducting and magnetic properties; see, e.g., Sato et al. (2009).

As far as heat transport is concerned, all experimental data concur that, whatever the filling agent, the thermal conductivity of filled skutterudites is significantly lower that their unfilled parent structures. Most of the arguments why this is so have centered on the rattling nature of the filler as a disruptor of the normal phonon modes within the spirit of the PGEC concept. However, *ab initio* calculations, which now could be done with a fair degree of confidence, started to reveal a microscopic view of phonon dynamics and pointed out the underlying physical processes that are likely the cause of the dramatic reduction of the thermal conductivity in filled skutterudites. One of the first *ab initio* studies dedicated to skutterudites was a comparison of heat transport in BaCo$_4$Sb$_{12}$ with its parent $CoSb_3$ by Li and Mingo (2014b). Because the calculations assumed a fully filled structure, the Ba filler was chosen as it can fill a relatively large fraction of voids. The calculated values of the thermal conductivity turned out to be 11.5 $Wm^{-1}K^{-1}$ for $CoSb_3$, within the range of experimental values in Table 5.4, and 6.1 $Wm^{-1}K^{-1}$ for the Ba-filled skutterudite, i.e., a reduction by a factor of about 2. Previous measurements by Chen et al. (2001a), indicated that the thermal conductivity of $Ba_yCo_4Sb_{12}$ reaches its minimum value of 2.5 $Wm^{-1}K^{-1}$ at $y = 0.38$. Thus, the calculated value of 6.1 $Wm^{-1}K^{-1}$ might look too large compared to the experimental results. However, two things have to be kept in perspective: one does not know what might be the experimental value for the fully filled $BaCo_4Sb_{12}$, and perhaps even more important, the calculated thermal conductivity assumed only intrinsic phonon scattering processes, while point defect scattering is known to be very strong in partially filled skutterudites, such as $Ba_yCo_4Sb_{12}$, Nolas et al. (1998), Meisner et al. (1998), Wang et al. (2009), and Kim et al. (2010).

Li and Mingo explained the computed reduction in the thermal conductivity of $BaCo_4Sb_{12}$ with respect to the conductivity of $CoSb_3$ as arising chiefly from reduced phonon lifetimes limited by the anharmonic scattering that takes place across the entire phonon spectrum. While there is also a role played by a reduced group velocity in the filled structure, it is the increased scattering rate (roughly by a factor of 1.7) that dominates the suppression of the thermal conductivity, Figure 5.38. The result contrasts the decreased phonon scattering in $BaCo_4Sb_{12}$ reported by Huang and Kaviany (2010). Moreover, the findings fundamentally contradict the idea of a rattling model, where a significant reduction in the scattering rates would be expected only in a narrow range of frequencies near the characteristic rattler's frequency (Einstein frequency). The question is where the anharmonicity comes from. To shed the light on the problem, Li and Mingo recalculated the scattering rates by excluding anharmonicity between the Ba atom and the framework's neighbors. The outcome has not changed dramatically, the scattering rates in $BaCo_4Sb_{12}$ remained significantly larger than in $CoSb_3$, indicating that the filler ion is not an efficient scatterer of phonons. If not the filler, the enhanced scattering rates in filled skutterudites must originate from atoms of the host framework,

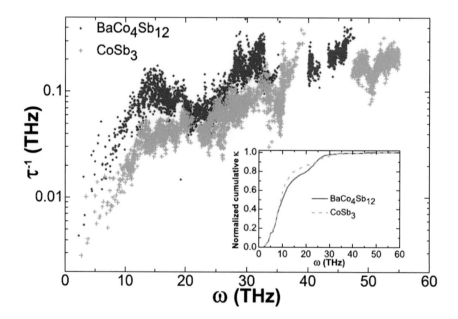

FIGURE 5.38 Anharmonic scattering rates as a function of angular frequency in $CoSb_3$ (crosses) and in $BaCo_4Sb_{12}$ (red circles). The inset shown the normalized cumulative thermal conductivity of both $CoSb_3$ and $BaCo_4Sb_{12}$. Reproduced from W. Li and N. Mingo, *Physical Review B* **89**, 184304 (2014b). With permission from the American Physical Society.

but how? Do the third-order, i.e., anharmonic, IFCs change? To isolate this issue, the authors computed the scattering rates in $BaCo_4Sb_{12}$ with the third-order IFCs of $CoSb_3$. Again, the scattering rates remained high and very similar to the rates computed with the IFCs of $BaCo_4Sb_{12}$. With this outcome, the attention then shifted to changes in the second-order (harmonic) IFCs, which govern vibration frequencies that enter in Eqs. 5.113 and 5.114 for the transition probabilities of absorption (+) and emission (−) processes. The frequency dependence of the respective transition probabilities for $CoSb_3$, $BaCo_4Sb_{12}$, and also for $LaCo_4Sb_{12}$, are shown in Figure 5.39. Clearly, both W^+ and W^- are significantly larger in the filled skutterudite structures than in $CoSb_3$. Li and Mingo argued that the enhanced scattering rates in the filled skutterudites are a consequence of the depressed phonon spectrum as the filler ion modifies the phonon dispersion. This leads to enhanced transition probabilities for phonon scattering in filled skutterudites, which, coupled with a somewhat reduced group velocity, entirely neglects any resonant scattering of phonons by the rattling fillers.

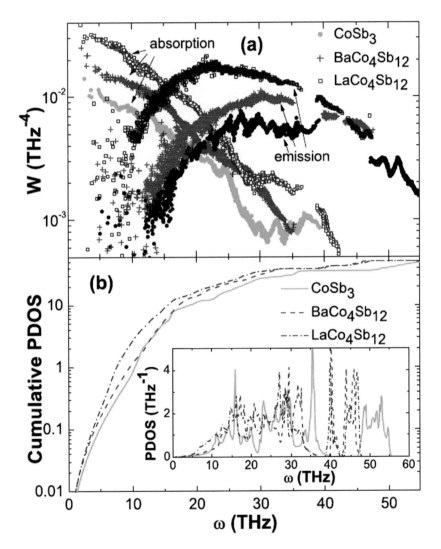

FIGURE 5.39 (a) Computed transition probabilities W^+ (absorption) and W^- (emission) for $CoSb_3$ (circles), $BaCo_4Sb_{12}$ (crosses), and $LaCo_4Sb_{12}$ (squares). (b) Cumulative phonon density of states (PDOS) for $CoSb_3$, $BaCo_4Sb_{12}$, and $LaCo_4Sb_{12}$. The inset shows individual frequency-dependent PDOSs. Reproduced from W. Li and N. Mingo, *Physical Review B* 89, 184304 (2014). With permission from the American Physical Society.

Obviously, it is not possible to discuss all publications on heat transport in skutterudites that have accumulated over the years. Rather, I have selected those that I view as the most insightful and important and apologize to all whose work I have not included in the discussion in the sections that follow.

5.13.3.1 Thermal Conductivity of Single-Filled $R_yCo_4Sb_{12}$ and RFe_4Sb_{12}

Of the myriad of filler species successfully filled into the skutterudite voids (either under the ambient pressure or with the aid of high pressure synthesis), the early thermal conductivity studies were carried out mostly with Ce (Morelli and Meisner 1995, Morelli et al. 1995, Sales et al. 1996, Uher et al. 1997, Morelli et al. 1997, Chen et al. (1997), and Meisner et al. 1998) and with La (Sales et al. 1997, Nolas et al. 1998, and Keppens et al. 1998). However, the chief drawback of the above two rare earth fillers is a very limited fraction of voids that can be filled, merely about 20% in the case of La and less than 10% in the case of Ce. Although even minute amounts of the fillers had a dramatic effect on the thermal conductivity, as documented by measurements of Uher et al. (1997) in Figure 5.40, the very restricted filling fraction limited the overall assessment of the fillers on the thermal conductivity. Subsequent measurements with rare earths up to Gd revealed the advantages of having smaller and heavier fillers as they decreased the thermal conductivity to a greater extent. Particularly effective in this regard was found Yb, which is positioned well beyond Gd in the periodic table but can be synthesized because of its mixed valence (both Yb^{2+} and Yb^{3+}) (Dilley et al. 2000a, Nolas et al. 2000a, and Anno and Matsubara 2000). In this class of rare earth fillers with less than trivalent valence, and thus a greater void occupancy, also belongs Eu, which is substantially divalent and was used in a few studies of heat transport and thermoelectric properties (Berger et al. 2001, Lamberton et al. 2002, Grytsiv et al. 2002, and Qiu et al. 2011). Grytsiv et al. reported the lowest room temperature lattice thermal conductivity in $Eu_{0.83}Fe_4Sb_{12}$ of 2.83 $Wm^{-1}K^{-1}$, which was further reduced in $Eu_{0.63}Co_{1.5}Fe_{2.5}Sb_{12}$ down to a very low value of 1.1 $Wm^{-1}K^{-1}$. Later, with the development of the high-pressure high-temperature synthesis (see Section 2.2.6), all rare earth elements were successfully inserted into the skutterudite voids. In this context, I note that studies by Giri et al. (2002) with Ce and Sm fillers in all three (Fe, Ru, Os) phosphide skutterudite frameworks indicated the decreasing trend in the room temperature thermal conductivity on going from Fe to Ru to Os, with Sm fillers reducing the thermal conductivity by nearly a factor of two more than

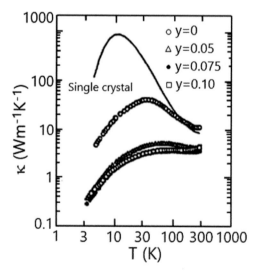

FIGURE 5.40 Temperature dependence of thermal conductivity of $Ce_yCo_4Sb_{12}$. The solid line represents the data for a single crystal of $CoSb_3$ from Morelli et al. (1995). Reproduced from C. Uher et al. *Mater. Res. Soc. Symp. Proc.* 478, 315 (1997). With permission from the Materials Research Society.

Ce fillers. This is not surprising as the heavier metals on the framework as well as heavier fillers should reduce the thermal conductivity more effectively. However, as already pointed out in the case of binary antimonide skutterudites, the trend is reversed and holds also for their filled forms. In the meantime, very early in the new millennium, alkaline-earth elements became of considerable interest because their divalent ions offered a much higher fraction of voids to be filled, particularly in the case of the large Ba^{2+} ions (Chen et al. 2000a,b, 2001a). At the same time, Sales et al. (2000) was able to fill the $[Co_4Sb_{12}]$ framework with a modest fraction (22 at%) of monovalent Tl, and other monovalent species, particularly alkali metal–filled skutterudites, synthesized by Leithe-Jasper et al. (2003) (fully-filled $NaFe_4Sb_{12}$) and Pei et al. (2009) (partially-filled $Na_yCo_4Sb_{12}$ with a high 50% filling fraction). Although the monovalent fillers have not been able to reduce the thermal conductivity to the same extent as some rare earth fillers, they demonstrated good electronic properties and several of them reached ZT greater than the unity. With the aid of either a modulated elemental reaction synthesis (Sellinschegg et al. 1998b), or high-pressure high-temperature synthesis (8 GPa, 550°C) (Takizawa et al. 1999 and Nolas et al. 2000b), it was possible to fill fully the voids of $CoSb_3$ with elemental Sn. The filling was documented by a very large ADP of Sn, the n-type nature of conduction (the structure would turn p-type if Sn instead substituted for some Sb), and a dramatically reduced lattice thermal conductivity to levels well below 2 $Wm^{-1}K^{-1}$ at 300 K. In view of these reports, it was surprising to find an article, Mallik et al. (2007), claiming that Sn can fill $CoSb_3$ voids under an ambient pressure synthesis. Moreover, the presumed $SnCo_4Sb_{12}$ displayed p-type conduction in both the Seebeck and Hall coefficients! Since no ADP parameters were provided to document the weak bonding of Sn in the void and no other supporting evidence for Sn being situated in the voids was offered, it is more likely that Sn actually substituted for Sb on the ring structure (where, indeed, it would drive the structure p-type) rather than filled the voids.

In the next couple of sections, I focus on the thermal conductivity of the two exceptionally effective single fillers, Yb and In, both generating much interest theoretically and experimentally.

5.13.3.2 Thermal Conductivity of Yb-Filled Frameworks $[Co_4Sb_{12}]$ and $[Fe_4Sb_{12}]$

Even though it is the second-last rare earth element, because of its partly divalent nature, it was possible to synthesize the Yb-filled skutterudite under ambient pressure (Dilley et al. 1998 and Leithe-Jasper et al. 1999). Early transport measurements (Dilley et al. 2000a,b, Nolas et al. 2000a, Anno and Matsubara 2000, Anno et al. 2000, and Yang et al. 2001), indicated that Yb reduces the thermal conductivity particularly strongly, partly due to its somewhat larger filling fraction, estimated in the uncompensated $[Co_4Sb_{12}]$ framework as $y = 0.25$–0.29 (see Table 1.4), partly due to the lowest vibrational frequency of $\omega_0 = 40$ cm^{-1} of any filler (Yang et al. 2007), and partly due its small size compared to the size of the void (large ADP parameter). This drew much attention and resulted in numerous subsequent explorations of the effect of Yb on the heat transport in both single-filled and multi-filled skutterudites.

As the measurements by Mi et al. (2011) presented in Figure 5.41 show, the greatest reduction in the thermal conductivity for an equal nominal content of the filler (at $y = 0.1$) is achieved with Nd, followed by Yb and Ce. However, even though the nominal filling fractions are identical, it does not mean that the actual void occupancy is the same for all rare-earth fillers. In fact, the occupancy decreases from La toward Sm, and for Yb and Eu it stays high only because of their partly divalent nature. The opportunity to maintain the occupancy of Yb at much higher levels than is possible with Nd, coupled with the high carrier mobility in Yb-filled skutterudites, is what makes Yb the most favored filler from the perspective of thermoelectricity. Detailed Rietveld analysis of the synchrotron radiation XRD data provided ADP parameters for all structures (see Table 5.5), and a clear correlation between ADPs and the thermal conductivity, originally formulated by Sales et al. (1999), was confirmed.

The filling fraction limit of Yb in the $[Co_4Sb_{12}]$ framework has changed dramatically since the early days when it was set at $y = 0.19$ in samples prepared by reaction of powders and long-term annealing (Bauer et al. 2000), or structures prepared by melting and annealing (Nolas et al. 2000a).

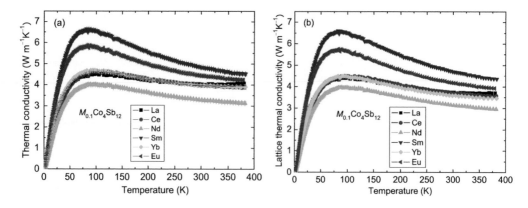

FIGURE 5.41 (a) Temperature dependence of the total thermal conductivity of $R_{0.1}Co_4Sb_{12}$, R = La, Ce, Nd, Sm, Yb, and Eu. (b) The lattice thermal conductivity obtained from the total thermal conductivity after subtracting the electronic term using the Wiedemann-Franz relation and the Lorenz number 2.0×10^{-8} V^2K^{-2}. The data are not corrected for any potential radiation loss, but its effect should be comparable for all samples as the sample sizes are similar and the same technique was used in all measurements. Reproduced from J.-L. Mi et al., *Physical Review B* 84, 064114 (2011). With permission from the American Physical Society.

TABLE 5.5

Selected Structural and Transport Parameters of $R_{0.1}Co_4Sb_{12}$ Obtained from the Measurements by Mi et al. (2011)

Property	$La_{0.1}Co_4Sb_{12}$	$Ce_{0.1}Co_4Sb_{12}$	$Nd_{0.1}Co_4Sb_{12}$	$Sm_{0.1}Co_4Sb_{12}$	$Yb_{0.1}Co_4Sb_{12}$	$Eu_{0.1}Co_4Sb_{12}$
Void occ.(%)	4.9	4.7	3.2	2.6	7.6	7.9
ADP guest (Å^2)	0.018	0.018	0.035	0.015	0.020	0.034
Ion. radius (Å)	1.36	1.34	1.27	1.24	1.02	1.17
θ_E (K) (ADP)	79	60	54	76	60	68
θ_D (K) (ADP)	268	267	267	265	274	266
θ_D (K) (C_p)	280	283	294	292	281	291
n_{exp} (10^{20} cm^{-3})	2.3	1.9	0.9	1.3	2.0	4.9
μ ($cm^2V^{-1}s^{-1}$)	14.2	14.9	16.1	8.6	17.2	6.4
S (μVK^{-1})	−152	−168	−204	−152	−177	−116
κ ($Wm^{-1}K^{-1}$)	4.02	3.97	3.28	4.84	3.97	4.44

With permission from the American Physical Society.

Shortly after, the FFL of Yb was increased to $y = 0.25$ by Anno and Matsubara (2000) in samples synthesized by SPS, and then increased to $y = 0.29$ by Li et al. (2008), who used a melt-spinning method characterized by ultra-fast solidification of the ejected melt. A further major increase in FFL to $y = 0.4$ was obtained by Xia et al. (2012) upon using very high temperature annealing (800°C for 7 days). Moreover, ball milling followed by hot pressing resulted in the Yb occupancy near 50%, as reported by J. Yang et al. (2009) and Dahal et al. (2014), the latter authors opting to make first an ingot by melting and quenching and then pulverizing it by ball milling. The recently used ball milling method by Ryll et al. (2018) confirmed the high filling fraction of Yb, this time reaching $y = 0.44$. Thus, the FFL of Yb in $[Co_4Sb_{12}]$ apparently depends strongly on the synthesis route used to prepare samples. Nevertheless, opinions still diverge regarding what is the actual filling limit of Yb in the voids of the $[Co_4Sb_{12}]$ framework, and many researchers believe that filling above $y = 0.29$ inevitably leads to precipitates of the metallic $YbSb_2$ phase. By detailed X-ray, structural, and quantitative compositional analysis, augmented by the composite theory by Bergman and Levy (1991), Wang et al. (2016) have shown that $Yb_yCo_4Sb_{12}$ with $y > 0.3$ is not a single-phase structure but, rather, a composite skutterudite with the $Yb_{0.29}Co_4Sb_{12}$ matrix and $YbSb_2$ inclusions. The $YbSb_2$ phase has a rather high room temperature thermal conductivity of 15 $Wm^{-1}K^{-1}$ and would account for increased lattice thermal conductivities seen in skutterudites filled with large nominal Yb filling fractions. As discussed in Section 1.3.1.1, the theoretical assessment of the Co–Sb–Yb phase diagram by Tang et al. (2015) also revealed a strong filling dependence on the exact starting composition and the annealing temperature. However, in no case the filling fraction of Yb in $CoSb_3$ could exceed $y = 0.44$ for the Co-rich starting composition, and a much smaller $y = 0.26$ fraction was viewed as a limiting occupancy in Sb-rich starting composition.

The effect of Yb filling on the room temperature lattice thermal conductivity of $Yb_yCo_4Sb_{12}$ is shown in Figure 5.42. As already pointed out in the case of filling with Ce, the thermal conductivity decreases initially very fast with a small content of Yb in the voids but tends to level off when the Yb filling reaches 20% and, in some cases, actually starts to increase when y exceeds about 0.30. The exact cause of the rising lattice thermal conductivity observed above a certain level of Yb filling is not clear but could be associated with the presence of a highly conductive secondary $YbSb_2$ phase that likely segregates at high filling fractions but remain undetected by XRD, or the skutterudite lattice might be distorted. In this context, it is interesting to note that, by using a highly sensitive spherical aberration-corrected electron microscopy Wang et al. (2017) observed the lattice distortion (unequal lattice parameter in directions a and b) in their Yb-filled cobalt antimonide with the nominal content $y = 0.20$ when the sample was annealed at a high temperature of 1023 K for 5 min. Such distortion was absent in the nominally same but ball-milled structure. Be as it may, the data suggest that the smallest thermal conductivity is obtained by the time the Yb filling fraction reaches $y = 0.3$. The high Yb content is also detrimental to the electronic properties, and the thermoelectric

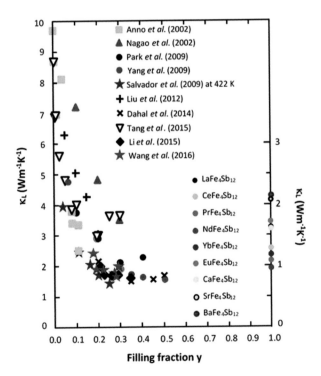

FIGURE 5.42 Lattice thermal conductivity of $Yb_yCo_4Sb_{12}$ at room temperature (left-hand scale) for various filling fractions y. The right-hand scale and the symbols indicate the lattice thermal conductivity at room temperature of fully filled RFe_4Sb_{12} from Qiu et al. (2011). Note a different scale on the right- and left-hand sides.

performance suffers. If, for some reason, higher filling levels of Yb were desired, one can always counter-dope on the pnicogen rings with Sn (Dilley et al. 2000b, Yang et al. 2001, 2003, or with Ge, Mori et al. 2005), or substitute some Fe on the Co site (Chen et al. 2016), the approach discussed in more detail in Section 5.13.3.4.

Differences in the magnitude of the lattice thermal conductivity between various sets of the data at comparable levels of filling in Figure 5.42 are caused primarily by the use of different synthesis methods, which result in different grain size, microstructure, pores, and lattice strain. The effect of different synthesis routes on the thermal conductivity is well documented, e.g., by Li et al. (2008) and Wang et al. (2017). It was also pointed out by Geng et al. (2007) that carefully controlled melt temperature and composition essentially eliminates the formation of pores in the skutterudite during the synthesis process of melting and annealing.

Recently, a significant extra reduction in the lattice thermal conductivity of $Yb_yCo_4Sb_{12}$ was obtained by Meng et al. (2017), who used a liquid phase compaction method to synthesize Yb-filled skutterudite with a large excess of Sb to form a eutectic composition. The technique is similar to the one used by Kim et al. (2015) with the excess of Te in $Bi_{0.5}Sb_{1.5}Te$, where the liquid phase was expelled, leaving behind low angle grain boundaries decorated with dense arrays of dislocations. Apart from an enhanced Seebeck coefficient claimed to benefit from energy filtering, and a small penalty paid for only a modestly affected carrier mobility, the dislocation arrays served as strong scattering centers for phonons and brought down the lattice thermal conductivity at room temperature very close to 2 $Wm^{-1}K^{-1}$ and, above 700 K, reaching values below 1 $Wm^{-1}K^{-1}$. However, the lattice thermal conductivity was surprisingly little dependent on the Yb filling fraction.

Yb has also been a favored co-filler in double and multifilled skutterudites, particularly in partnership with In (Peng et al. 2012b, Wei et al. 2011, Hobbis et al. 2017, and Lee et al. 2018). The topic of multifilling and its effect on the thermal conductivity is covered in Section 5.13.3.6.

In Figure 5.42, using the right-hand scale, are also plotted room temperature values for the fully-filled RFe_4Sb_{12} skutterudites reported by Qiu et al. (2011). Indeed, the lowest lattice thermal conductivity was attained for Nd, followed by Pr and Yb fillers.

As we have discussed, low thermal conductivity of Yb-filled skutterudites is generally ascribed to the small and heavy Yb ion and its partly divalent character. But what is the microscopic origin of the effectiveness of Yb to reduce the lattice thermal conductivity? The answer was provided by detailed *ab initio* calculations of the heat transport in $YbFe_4Sb_{12}$ by Li and Mingo (2015). The computational approach is the same as described in Section 5.9.1 and relies on the concept of the weighted phase space W defined in Eq. 5.126. To understand why Yb is able to reduce the lattice thermal conductivity to a considerably greater degree than other fillers, it is useful to compare the phonon dispersion and PDOS of $YbFe_4Sb_{12}$ with that of $BaFe_4Sb_{12}$, where Ba is a typical divalent filler. Referring to Figure 5.22 in Section 5.6, the most glaring difference between phonon dispersions in $YbFe_4Sb_{12}$ and $BaFe_4Sb_{12}$ is the presence of flat phonon modes in the frequency range $5 < \omega < 7.2$ rad/ps of the former. As the PDOS documents, these flat modes are exclusively tied to the vibrations of Yb and are missing in $BaFe_4Sb_{12}$. However, because of the distinct avoided crossing of the flat modes with the acoustic phonons, i.e., the flat modes 'sense' and are influenced by the vibrations of the framework, the flat modes should not be viewed as isolated in the spirit of PGEC. Nevertheless, the flat modes are the primary reason for the exceptionally low lattice thermal conductivity in $YbFe_4Sb_{12}$. A comparison of the calculated lattice thermal conductivities in $YbFe_4Sb_{12}$ with the calculated conductivities in $BaFe_4Sb_{12}$, and also in $BaCo_4Sb_{12}$, is shown in Figure 5.43a. The calculated lattice thermal conductivity of $YbFe_4Sb_{12}$ is exceptionally low. In fact, it is far too low in comparison to the experimental data, shown by solid squares in the figure and the data point in Figure 5.42. Such an outcome is surprising as the calculations are made for an ideal lattice with only intrinsic phonon scattering, and it is difficult to comprehend how a real sample with the additional defect scattering processes could have nearly five times higher lattice thermal conductivity! There is not as much difference between the conductivities of $BaFe_4Sb_{12}$ and $BaCo_4Sb_{12}$ (of course, in reality, 100% filling in the $[Co_4Sb_{12}]$ framework by Ba is not possible). Moreover, the experimental

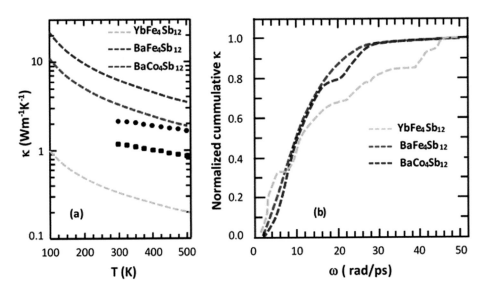

FIGURE 5.43 (a) Calculated lattice thermal conductivity of $YbFe_4Sb_{12}$, $BaFe_4Sb_{12}$, and $BaCo_4Sb_{12}$. Experimental lattice thermal conductivity for $YbFe_4Sb_{12}$ and for $BaFe_4Sb_{12}$, measured by Qui et al. (2011), is shown as solid squares and solid circles, respectively. (b) Normalized cumulative thermal conductivity for the same three skutterudites. The effect of flat Yb modes is reflected by essentially no contribution to the cumulative conductivity of $YbFe_4Sb_{12}$ in the frequency range between 5 rad/ps and 7.2 rad/ps. No such gap is seen in either of the Ba-filled skutterudites. Adapted and redrawn from W. Li and N. Mingo, *Physical Review B* 91, 144304 (2015). With permission from the American Physical Society.

value of the lattice thermal conductivity of $BaFe_4Sb_{12}$, shown by solid circles in Figure 5.43a, is reasonably close to the theoretical value, at least at high temperatures. The effect of the Yb flat modes is also revealed by virtually no contribution to the normalized cumulative thermal conductivity in the range between 5 rad/ps and 7.2 rad/ps, as shown in Figure 5.43b. The reason for the ultra-low thermal conductivity of $YbFe_4Sb_{12}$ is its order of magnitude larger scattering rates, shown in Figure 5.23, compared with the scattering rates in $BaFe_4Sb_{12}$ and $BaCo_4Sb_{12}$. By artificially replacing the anharmonic force constants of $YbFe_4Sb_{12}$ with those of $BaFe_4Sb_{12}$ and, seeing no significant change, the *ab initio* treatment documented a very similar strength of the anharmonic forces in both skutterudites and eliminated them as the source of the order of magnitude larger scattering rates in $YbFe_4Sb_{12}$ compared to those in $BaFe_4Sb_{12}$. The difference in the scattering rates then must arise from a greatly enhanced number of allowed scattering processes of phonons in $YbFe_4Sb_{12}$ and this, as we have already seen, is conveniently measured by the weighted phase space W, depicted in Figure 5.23 in Section 5.6. While the emission (W^+) and absorption (W^-) values for $BaFe_4Sb_{12}$ and $BaCo_4Sb_{12}$ are quite similar in both shape and magnitude, the trend and magnitude in $YbFe_4Sb_{12}$ are entirely different. In particular, the emission curve (black circles) jumps as the frequency enters the range of flat modes (around 6 rad/ps) and develops a sharp peak at 12 rad/ps that is totally missing in the Ba-filled skutterudites. The absorption curve (red circles) of $YbFe_4Sb_{12}$, too, lies significantly higher than in the other two skutterudites, although no dramatic peak is notable in the range of flat modes. As a reminder, the allowed three-phonon processes are restricted to those that simultaneously conserve both the energy and momentum. Thus, in the emission process, a mode (q_0, ω_0) decays into (q_1, ω_1) and (q_2, ω_2), and for any $(q_1, \omega_0 - \omega_{flat})$ there are plenty of modes $(q_0 - q_1, \omega_{flat})$ in the flat mode range enabling the scattering process. Consequently, for ω_0 larger than ω_{flat}, the number of emission channels is vastly increased in comparison to a situation where the flat modes are missing. This is documented by plots of $W^{\pm}$ for $YbFe_4Sb_{12}$ where the flat modes were excluded (diamonds). The artificial exclusion of the flat modes results in a factor of 3 increase in the thermal conductivity. The remaining contributing factor to the low thermal conductivity of $YbFe_4Sb_{12}$ comes from the depressed phonon dispersion in Figure 5.22a.

Although the *ab initio* calculations offer a detailed microscopic view of scattering in a structure and can discern important processes that limit heat transport, the computed magnitude of the lattice thermal conductivity is troublesome, at least in the case of $YbFe_4Sb_{12}$, given its factor of five underestimated value with respect to the numerous experimental data.

5.13.3.3 Thermal Conductivity of In-Filled Frameworks [Co₄Sb₁₂] and [Fe₄Sb₁₂]

As discussed in Section 1.3.1.2, indium fails the filling criterion in Eq. 1.18 and should not be able to enter structural voids of $CoSb_3$. Yet, there are numerous reports of successful void occupancy by In, e.g., Akai et al. (1997), He et al. (2006), Jung et al. (2008, 2010a,b), Wang et al. (2009), Mallik et al. (2008, 2009), Leszczynski et al. (2013), Visnow et al. (2015), Deng et al. (2017). Moreover, many In-filled skutterudites show quite low lattice thermal conductivity, and, as we shall see in Section 5.13.3.6, In is a favorite partner in multifilled skutterudites displaying high thermoelectric performance (Zhao et al. 2007, 2009, Peng et al. 2008, 2009, Li et al. 2009, Tang et al. 2011, Harnwunggmoung et al. 2012, and Bashir et al. 2018a.

As far as In filling fraction is concerned, the reader will find an extensive discussion in Section 1.3.1.3, where it was shown that In is an amphoteric impurity in skutterudites, which means that is can act as a donor when filling voids and also as an acceptor when it substitutes on the pnicogen site. In the former case, it is a monovalent In^{+1} ion, while in the latter case it is a divalent In^{2-} acceptor. Theoretical filling estimates, Grytsiv et al. (2013) and Tang et al. (2014), depend on the exact Co/Sb ratio in the starting material, and give the highest In content of $y = 0.22$ and $y = 0.27$, respectively. Such filling fraction limits are compatible with the experimentally determined 22% void occupancy (He et al. 2006 and Mallik et al. 2009), and 26% occupancy measured by Leszczynski et al. (2013). Again, the filling limit can be extended by replacing a fraction of Co by Fe, e.g., Park et al. (2010) and Kim et al. (2017).

A collection of lattice thermal conductivities of $In_yCo_4Sb_{12}$ is displayed in Figure 5.44. The data include measurements on samples synthesized by various techniques, among them a combination

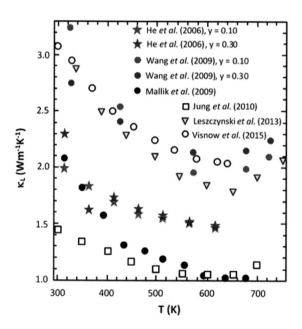

FIGURE 5.44 Temperature dependence of the lattice thermal conductivity of $In_yCo_4Sb_{12}$ with indium filling fractions $y \geq 0.1$.

of the solvothermal preparation with melting (Wang et al. 2009). Two sets of measurements, in He et al. (2006) and Wang et al. (2009), present data on samples prepared by the identical synthesis process but with different content of In. Where available, the lattice thermal conductivity was obtained from the original measurements. In several reports, only the total thermal conductivity was presented, in which case I estimated the lattice conductivity by subtracting from the total measured thermal conductivity the electronic term, obtained from the original electrical conductivity by using the Lorenz number 2×10^{-8} V^2K^{-2} deemed appropriate for a degenerate semiconductor. As follows from the figure, small concentrations of In rapidly degrade the lattice conductivity. As the In content approaches $y \sim 0.20$, the rate of decrease flattens, and above $y = 0.30$, one often finds the lattice thermal conductivity to increase, undoubtedly on account of the increasing concentration of the high heat conducting InSb phase precipitating as the filling fraction of In is exceeded. In fact, even structures with a modest void filling of $y \sim 0.18$, which appear perfectly single phase by XRD, have been shown to possess traces of InSb when examined by high resolution synchrotron X-ray diffraction (Visnow et al. 2015). While all data sets are in a general agreement as far as the trend in the temperature dependence is concerned, the exact values of the lattice thermal conductivity reflect different synthesis processes and densities, unique microstructures, and unequal strength of point defect scattering as the samples have different impurity levels.

5.13.3.4 Thermal Conductivity of Single-Filled Charge Compensated Skutterudites

We have seen in Chapter 1 that the filling fraction limit in the $[Co_4Sb_{12}]$ framework is very limited, and it is difficult to achieve the optimal level of doping to maximize the power factor. Nevertheless, the effect on the thermal conductivity of even a minute amount of the filler in the voids is quite dramatic, as documented in the case of Ce- and Yb-filled skutterudites in Figures 5.40 and 5.42.

To enhance the filling fraction and to explore its effect on the thermal conductivity, there are two main approaches one can try. In the first one, a small fraction of Sb on the pnicogen ring is substituted by a Column 14 element, either Sn or Ge, $[Co_4Sb_{12-x}(Sn/Ge)_x]$, which partly charge compensates for the filler and increases its filling fraction. Nolas et al. (1998) used Sn up to $x = 2.58$ and managed to increase the filling fraction of La in $CoSb_3$ to $y = 0.9$, while Dilley et al. (2000b) partly replaced Sb with $x = 2$ of Sn to enhance filling of Yb to $y = 0.5$. Yang et al. (2003) measured and fit two series of Yb-filled $[Co_4Sb_{12}]$ frameworks, one with $y = 0.19$, where a small amount of Sn substituted for Sb, and one with $y = 0.5$, using Sn substitutions in the range $x = 0.5$ to 0.9 in a formula $Yb_yCo_4Sb_{12-x}Sn_x$. Lamberton et al. (2002) substituted Ge with up to $x = 0.50$ to increase filling by Eu. In their measurements of $La_yCo_4Sb_{12-x}Sn_x$, Nolas et al. (1998) noted that partial filling of the voids degrades the thermal conductivity to greater extent than full filling, as shown in Figure 5.45. Moreover, partially filled skutterudites tend to have a higher carrier mobility than the completely filled structure. Although Sn disrupts the order on the pnicogen ring, its presence seems to be less damaging to the carrier mobility than the presence of Fe in the $[Co_{4-x}Fe_xSb_{12}]$ framework, discussed in the next paragraph.

An alternative, and often more frequently used approach to enhance the filling fraction, is to replace some Co with Fe, i.e., to form filled skutterudites of the form $R_yCo_{4-x}Fe_xSb_{12}$. The framework $[Co_{4-x}Fe_xSb_{12}]$ served as the matrix in numerous studies of thermal conductivity of filled skutterudites, among them Fleurial et al. (1996, 1997), Chen et al. (1997), Sales et al. (1997), Uher et al. (1998), Meisner et al. (1998), Arita et al. (2005), Lu et al. (2005), Tang et al. (2001, 2005, 2006), Narazu et al. (2007), Liu et al. (2007), Park et al. (2010), Leszczynski et al. (2010), Uher et al. (2010), Zhou et al. (2011), Park et al. (2012), Thompson et al. (2015), Chen et al. (2016), Matsubara and Asahi (2016), Kim et al. (2017), and Li et al. (2019). As an illustration of the effect of the $[Co_{4-x}Fe_xSb_{12}]$ framework on the thermal conductivity of a single-filled skutterudite, I have chosen the $Yb_{0.6}Co_{4-x}Fe_xSb_{12}$ system studied by Chen et al. (2016). Subtracting the electronic thermal conductivity contribution using the Wiedemann-Franz law with the Lorenz number taken as 2.0×10^{-8} V^2K^{-2} from the total thermal conductivity, the lattice part is plotted in Figure 5.46. The nominal and actual composition of the samples, together with the carrier

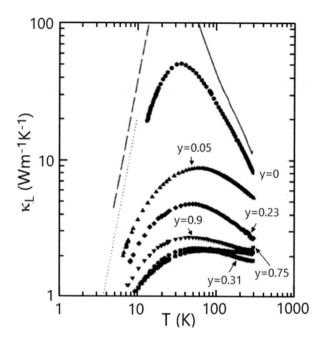

FIGURE 5.45 Lattice thermal conductivity of $La_yCo_4Sb_{12-x}Sn_x$. The filling fraction of La is indicated. The solid line indicates the thermal conductivity of the $CoSb_3$ single crystal. The dotted and dashed lines are calculated values of κ_L for 4 μm and 7 μm grain sizes, respectively. Adapted and redrawn from G. S. Nolas et al., *Physical Review B* 58, 164 (1998). With permission from the American Physical Society.

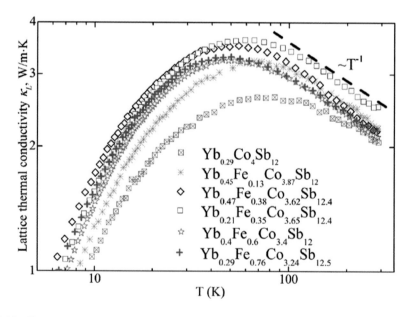

FIGURE 5.46 Temperature dependence of the lattice thermal conductivity of $Yb_yCo_{4-x}Fe_xSb_{12}$ below 300 K. Adapted and redrawn from Y. Chen et al., *Journal of Applied Physics* 120, 235105 (2016). With permission from the American Institute of Physics.

TABLE 5.6

Composition and Some Transport Parameters of $Yb_yCo_{4-x}Fe_xSb_{12}$ at 300 K

Nominal Content	Actual Content	n (10^{21} cm^{-3})	ρ (mΩ.cm)	κ(Wm^{-1}K^{-1})	S(μVK^{-1})
$Yb_{0.6}Co_4Sb_{12}$	$Yb_{0.29}Co_4Sb_{12}$	0.6	0.5	3.6	−64
$Yb_{0.6}Fe_{0.2}Co_{3.8}Sb_{12}$	$Yb_{0.45}Fe_{0.13}Co_{3.87}Sb_{12.2}$	3.0	0.42	3.89	−74.4
$Yb_{0.6}Fe_{0.5}Co_{3.5}Sb_{12}$	$Yb_{0.47}Fe_{0.38}Co_{3.62}Sb_{12.4}$	2.2	1.47	2.66	−115.9
$Yb_{0.6}Fe_{0.8}Co_{3.2}Sb_{12}$	$Yb_{0.21}Fe_{0.35}Co_{3.65}Sb_{12.4}$	0.32	3.2	2.7	−101.3
$Yb_{0.6}Fe_{0.9}Co_{3.1}Sb_{12}$	$Yb_{0.40}Fe_{0.60}Co_{3.4}Sb_{12}$	0.020	4.7	2.27	62.2
$Yb_{0.6}FeCo_3Sb_{12}$	$Yb_{0.29}Fe_{0.76}Co_{3.24}Sb_{12.5}$	0.034	2.6	2.46	71.1

Data are from tables of Chen et al. (2016). With permission from the American Institute of Physics.

density, electrical resistivity, and Seebeck coefficient are presented in Table 5.6. The reader will note that, although the nominal Yb content was fixed at $y = 0.6$, the actual content varied over a rather large range. At low Fe content, the structures are n-type semiconductors, while the last two entries with a significant amount of Fe are p-type skutterudites. As originally proposed by Morelli et al. (1995, 1997), the boundary delineating p- and n-type conduction (assuming perfect crystal chemistry) in $R_yFe_{4-x}Co_xSb_{12}$ is (p,n) = 4 − x − 3y, taking the filler as trivalent. For the currently more frequently used designation of the structure as $R_yCo_{4-x}Fe_xSb_{12}$, the formula becomes (p,n) = 4 − (4 − x) − 3y = x − 3y. Assuming that Yb is substantially divalent, then the boundary between p and n conduction in $Yb_yCo_{4-x}Fe_xSb_{12}$ should be given by (p,n) = x − 2y, with the positive value for p-type and the negative value for n-type. The data in Table 5.6 conform to the above rule, except for the second last entry, where the value of (p,n) = −0.2 is two small and the structure is close to the p-n crossover.

As already pointed out, anytime Fe substitutes for some fraction of Co in the skutterudite's framework, the thermal conductivity of the structure is greatly depressed. Given that the mass and ionic radius of Fe and Co are nearly the same, there should be only a very small reduction in the thermal conductivity, based on the mass- and strain-fluctuation scattering. To clarify this puzzling situation, Meisner et al. (1998) investigated thermal conductivity of a series of 'optimally' filled $Ce_yCo_{4-x}Fe_xSb_{12}$ compounds, where the term 'optimally' refers to the content of Ce equal to the maximum filling consistent with the solid solubility limit of Ce for a given Co concentration, as illustrated in Figure 1.29. The authors concluded that the fractionally filled skutterudites can be viewed as solid solutions of fully filled $CeFe_4Sb_{12}$ and completely empty $\square Co_4Sb_{12}$ skutterudites, where $\square$ refers to a void in the skutterudite structure. Taking this position, it is easy to explain the reduction in the thermal conductivity upon the presence of Fe in the structure. It is not the mass and strain difference between Fe and Co, but the mass difference between a site occupied by Ce and a vacancy $\square$, which, of course, amounts to a mass difference of 100%! The same argument would presumably apply to any filler partially occupying voids in the $[Co_{4-x}Fe_xSb_{12}]$ framework or alternative frameworks based on $[Co_{4-x}Ni_xSb_{12}]$ (Wei et al. 2010), and $[Fe_{4-x}Ni_xSb_{12}]$ (Choi et al. 2014), the latter two frameworks used only sporadically in measurements of the thermal conductivity. However, beyond the solid solution concept above, it is undisputable that the presence of Fe in the framework always leads to a low thermal conductivity. This puzzle might possibly be resolved by postulating that a phonon is scattered by a rapid dynamic exchange of electrons, a process proposed by Slack (1962) to operate in Fe_3O_4. In that case, an incoming phonon is absorbed by a Fe^{2+} ion, transferring in turn one of its electrons to a neighboring Co^{3+}. The electron is immediately transferred back, returning now the Fe^{3+} ion back to Fe^{2+}, and re-emitting (incoherently) a phonon. From the perspective of the phonon, this is an additional phonon scattering process, which could lead to a degradation of the thermal conductivity. Since each unit cell of volume a_0^3 contains eight metal atom sites, the concentrations of Co^{3+} and Fe^{2+} ions are obviously $c(Co^{3+}) = 8(4 - y)/a_0^3$ and $c(Fe^{2+}) = 8y/a_0^3$ (Li et al. 2019). Hence, the trend in the thermal conductivity due to this dynamic exchange of electrons can be described as $\kappa \sim \left[c\left(Co^{3+}\right) \times c\left(Fe^{2+}\right)\right]^{-1} = a_0^6 \left[64(4-y)y\right]^{-1}$. The expression has its maximum value when $y = 2$, and thus the effect of dynamic exchange of electrons should reduce the thermal conductivity particularly effectively when one half of Co atoms are substituted by Fe. An interplay of the Fe content and Yb filling on the thermal conductivity of $Yb_yCo_{4-x}Fe_xSb_{12}$ in measurements of Li et al. (2019) is shown in Figure 5.47.

Of course, one can combine both charge compensation approaches, substituting on the pnicogen rings and using the $[Co_{4-x}Fe_xSb_{12}]$ framework, as was done by Zhang et al. (2013) in their measurements of the thermal conductivity of $Nd_{0.6}Fe_2Co_2Sb_{12-x}Ge_x$, with $0 \le x \le 1$. Interestingly, the actual Nd content decreased significantly with the increasing Ge doping, and this contributed to the trend in the thermal conductivity shown in Figure 5.48a, where the smallest values of the conductivity were obtained for the nominal $x = 0.15$ and 0.3, while samples with the higher nominal doping levels of Ge had thermal conductivity twice as large. While the actual Ge content increased with the nominal values of x, it, too, fell far below the intended doping level of Ge, and the Fe/Co ratio shifted in favor of Co. The low thermal conductivity of these Nd-filled skutterudites also benefited from the presence of nanoparticles (40 nm) embedded on the surface of the matrix, shown in Figure 5.48b, and the nominal $Nd_{0.6}Fe_2Co_2Sb_{11.7}Ge_{0.3}$ sample attained one of the highest $ZT \sim 1.1$ for a single-filled skutterudite.

I also include in this section the thermal conductivity of $S_yCo_{4-x}Ni_xSb_{12}$, a skutterudite filled with an electronegative sulfur filler, the filling aided by replacing some Co with Ni. The measurements were done by Li et al. (2019) for the filling and substitution ranges $0 \le y \le 0.35$ and $0.25 \le x \le 1.5$. The lattice thermal conductivity is depicted in Figure 5.49. While Ni reduces the thermal conductivity of $CoSb_3$, a further significant reduction arises due to the presence of the filler, and the lattice thermal conductivity around 700–800 K approaches values of 1 $Wm^{-1}K^{-1}$. By the way, the large content of Ni, which would otherwise be well beyond its solubility in $CoSb_3$, is enabled by the presence of sulfur in the voids.

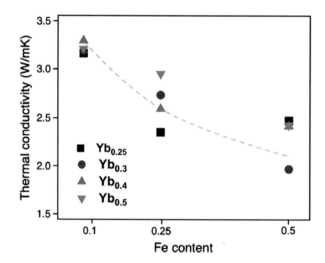

FIGURE 5.47 Total thermal conductivity of $Yb_yCo_{4-x}Fe_xSb_{12}$ as a function of the Fe content x for various Yb filling fractions indicated. Reproduced from W. Li et al., *Chemistry of Materials* 31, 862 (2019). With permission from the American Chemical Society.

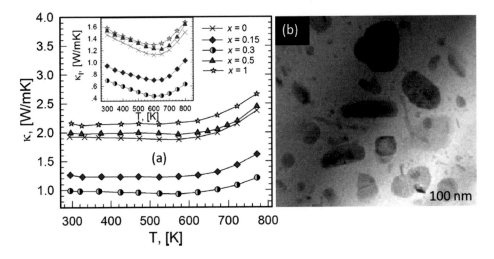

FIGURE 5.48 (a) Thermal conductivity of nominal $Nd_{0.6}Fe_2Co_2Sb_{12-x}Ge_x$, $0 \leq x \leq 1$. The inset shows the lattice thermal conductivity of the same samples. (b) Nanoprecipitates embedded in the skutterudite matrix of the $Nd_{0.6}Fe_2Co_2Sb_{11.7}Ge_{0.3}$ sample. Adapted and reproduced from L. Zhang et al., *Journal of Applied Physics* 114, 083715 (2013). With permission from the American Institute of Physics.

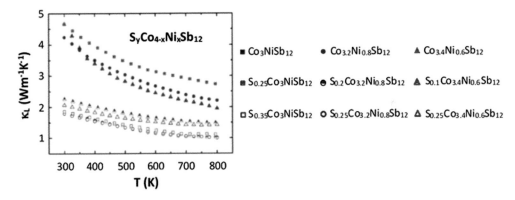

FIGURE 5.49 Lattice thermal conductivity of $S_yCo_{4-x}Ni_xSb_{12}$, filled with an electronegative sulphur filler and Ni substituting for a fraction of Co. Adapted from J. L. Li et al., *Journal of Materials Chemistry C* 7, 8079 (2019). With permission from the Royal Society of Chemistry.

5.13.3.5 Thermal Conductivity of Composite Skutterudites with a Filled Matrix

Introducing secondary phases into the skutterudite matrix, either by intrinsic or extrinsic means as discussed in Section 1.4, has shown to be an effective approach to lower the lattice thermal conductivity. The emphasis here is on strengthening point defect scattering of phonons, hoping that the mean free path of phonons will be reduced considerably more than any possible degradation in the mean free path of the charge carriers. Occasionally, the nanometer-scale inclusions that tend to locate at the grain boundaries may also give rise to potential barriers of the right size that can filter out low energy carriers, increasing in the process the Seebeck coefficient.

Numerous studies, accounted for in Section 1.4, have been carried out with a large number of different nanoinclusions in either the matrix of $CoSb_3$ or various matrices of filled skutterudites. Typically, concentrations of nanoinclusions that effectively impair the heat transport are a few wt%. At much higher contents of the secondary phase the benefits are usually lost, either because the

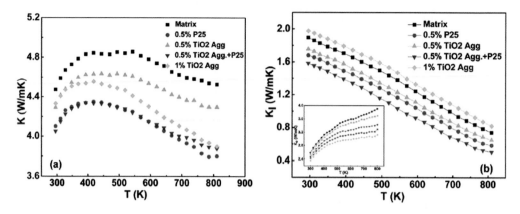

FIGURE 5.50 Temperature dependence of composite samples consisting of a matrix of $Ba_{0.3}Co_4Sb_{12}$ and TiO_2 nanoparticles. The nanoparticles are of two kinds: (i) small nanoaggregates of TiO_2 prepared by a carbon sphere-templated method that produces particles of about 12 nm diameter, which tend to aggregate into larger, about 830 nm beads, designated as TiO_2 aggr., and (ii) commercially available TiO_2 nanoparticles (Degussa Aeroxide P25) of 25 nm diameter, which consist of about 30% rutile and 70% anatase. Both kinds of additives were uniformly dispersed by ball milling in the powder of $Ba_{0.3}Co_4Sb_{12}$ prepared by a solid-state reaction. Samples were compacted by hot pressing and had the density better than 95% of the theoretical density. (a) The total thermal conductivity and (b) the lattice thermal conductivity, with the inset showing the corresponding electronic thermal conductivity. Reproduced from X. Y. Zhou et al., *Journal of Materials Chemistry A* 2, 20629 (2014). With permission from the Royal Society of Chemistry.

secondary phase has a higher thermal conductivity than the matrix, or because the degradation of the electrical conductivity starts to negate the benefit of lowering the lattice thermal conductivity. An example of the impact of the admixed nanoinclusions in the skutterudite matrix on the heat transport is shown in Figure 5.50 (Zhou et al. 2014). In this case, to the powdered $Ba_{0.3}Co_4Sb_{12}$ matrix was admixed, via brief ball milling, a nanometer-size powder of TiO_2 and the mixture was compacted to a better than 95% theoretical density by hot pressing. Two kinds of TiO_2 powder were used. The first kind was prepared by a carbon sphere-templated method and consisted of fine, 12 nm diameter nanoparticles that tended to aggregate into roughly 830 nm beads. In Figure 5.50, this kind of TiO_2 is designated as TiO_2 aggr. and represents the pure anatase phase. The second form of TiO_2 was the commercially obtained powder with particles in the range 20–40 nm (designated as P25) and consisting of roughly 30% rutile and 70% anatase phases. The degree of suppression of the lattice thermal conductivity in the presence of the TiO_2 nanoinclusions is similar at all temperatures and amounts to some 32% reduction at 813 K for the most affected composite that was loaded with 0.5 wt% of TiO_2 aggr. and 0.5 wt% of P25. The reader will note that when the content of the TiO_2 aggregate particles increased to 1 wt%, the thermal conductivity exceeded that of the $Ba_{0.3}Co_4Sb_{12}$ matrix. Since the authors used the fully degenerate value of the Lorenz number (instead of a likely more appropriate value of 2.0×10^{-8} V^2K^{-2}) in their estimates of the lattice thermal conductivity based on the Wiedemann-Franz law, the values of the lattice thermal conductivity in Figure 5.50 are likely underestimated by about 20%.

In any kind of comparison of the thermal conductivity of composite structures with the thermal conductivity of the matrix it is essential to make sure that the porosity of the samples is similar. Moreover, both the matrix and the composites should be compacted (sintered) at the same temperature in order to make the comparison meaningful.

5.13.3.6 Thermal Conductivity of Multiple-Filled Skutterudites

Although the idea of filling the voids of the skutterudite structure with more than a single filler was put forward earlier, e.g., Nolas et al. (1998), Chen et al. (2001), Bérardan et al. (2003, 2005), and

Lu et al. (2005), visualization of a wide range of tabulated resonant frequencies one can attain with various fillers (Yang et al. 2007), provided an extra impetus for exploring multi-filling as means of achieving very low lattice thermal conductivities. Double-filling and even triple-filling have become the norm in attempts to impede heat transport in skutterudites, and it did not take long for the exciting results to appear in the literature. Particularly eye catching were low values of the thermal conductivity achieved in triple-filled $Ba_{0.08}La_{0.05}Yb_{0.04}Co_4Sb_{12}$, that led to exceptional $ZT = 1.7$ at 850 K (Shi et al. 2011a) and thermal conductivities of $(Sr_xBa_xYb_{1-2x})_yCo_4Sb_{12}$ measured by Rogl et al. (2011). In Figure 5.51 are collected lattice thermal conductivities of several n-type multifilled skutterudites that attained exceptionally high values of the figure of merit exceeding 1.3 at their maximum.

Although it is instructive to show the lattice thermal conductivity and its temperature dependence because of its importance to the overall thermoelectric performance, let me emphatically note that such data should be treated with the utmost caution. Beyond the fact that the measurements of

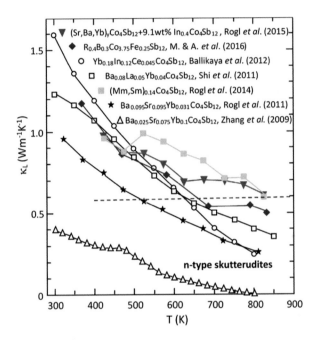

FIGURE 5.51 Lattice thermal conductivity of several n-type multifilled skutterudites with high thermoelectric figure of merit. The values of κ_L in measurements of Shi et al. (2011a) and Matsubara and Asahi (2016) were obtained with the Lorenz number of 2.0×10^{-8} V^{2}K^{-2}. In measurements of Rogl et al. (2015a), the Lorenz number was semi-experimentally established from the Seebeck coefficient by obtaining the Fermi energy ξ and assuming acoustic phonon scattering, i.e., $r = -1/2$. Thus, determined Lorenz numbers varied between about 1.7×10^{-8} V^{2}K^{-2} at 300 K and 1.62×10^{-8} V^{2}K^{-2} at 700 K. In all other measurements the Lorenz number was taken as 2.44×10^{-8} V^{2}K^{-2}, resulting in an underestimation of the lattice thermal conductivity. Solid blue diamonds represent the data of Matsubara and Asahi (2016) for an n-type skutterudite filled with seven elements, $R_{0.4} = Ba_{0.1}Yb_{0.2}Al_{0.1}$ and $B_{0.3} = La_{0.05}Eu_{0.05}Ga_{0.1}In_{0.1}$, in addition to the framework elements Co, Fe, and Sb. The solid red down-triangles are data of Rogl et al. (2015a) obtained on a high-energy ball milled composite consisting of $(Sr_{0.33}Ba_{0.33}Yb_{0.33})_{0.35}Co_4Sb_{12}$ and 9.1 wt% of $In_{0.4}Co_4Sb_{12}$, in other words, this was a quadruple-filled skutterudite, which attained the highest value of $ZT = 1.8$ at 823 K among all n-type skutterudites without the benefit of severe plastic deformation. Orange full squares represent the lattice thermal conductivity of commercially available skutterudite powder (Treibacher Industrie AG, Austria) of $(Mm,Sm)_{0.14}Co_4Sb_{12}$ measured by Rogl et al. (2014). The powder was sieved to particles of less than 50 μm, high-energy ball milled, and consolidated by hot pressing. While the total filling fraction was $y = 0.14$, the relative amount of the mischmetal and Sm was not provided. The blue dashed line indicates the minimum lattice thermal conductivity estimated by Rogl et al. (2017) based on the model of Cahill and Pohl (1989).

the thermal conductivity at elevated temperatures are rarely performed with the accuracy better than 10%, the chief problem with the lattice thermal conductivity data is how they are obtained. Being an indirectly determined transport parameter, the lattice thermal conductivity relies on an accurate assessment of the Lorenz number, which is the key parameter to determine the electronic part of the thermal conductivity from the experimentally established electrical conductivity using the Wiedemann-Franz law. In turn, the electronic part of the thermal conductivity is subtracted from the total thermal conductivity to arrive at the lattice term. The crux of the problem is the Lorenz number used, which in many studies is taken at its fully degenerate value of 2.44×10^{-8} V^2K^{-2} appropriate for metals but not for degenerate semiconductors. Although a better choice for thermoelectric materials is 2×10^{-8} V^2K^{-2}, it is still a somewhat arbitrary value. Moreover, unlike in metals, the Lorenz number in semiconductors is not a temperature-independent constant.

Given the fact that there is a way to experimentally obtain the proper value, as well as the temperature dependence of the Lorenz number, it is unfortunate that the research community has had such a cavalier attitude toward the choice of the Lorenz number.

The experimental technique to establish a reasonably accurate value of the Lorenz number for a given thermoelectric material is based on measurements of the Seebeck and Hall coefficients. The Fermi energy is extracted from the carrier concentration determined from the Hall coefficient and is inserted into the expression for the Seebeck coefficient to obtain the scattering parameter r. Having thus determined the Fermi energy and the scattering parameter, their temperature-dependent values are then used to obtain the Lorenz number. If one is interested only in the values of the Lorenz number at temperatures above the ambient, a reasonable shortcut is to assume that the scattering parameter r is equal to $-.50$, the value corresponding to acoustic phonon scattering that is likely the appropriate scattering mechanism. In this case, one only needs the Seebeck coefficient to determine the Lorenz number.

The problems with the Lorenz number are clearly documented in Figure 5.51, where different approaches of obtaining the lattice thermal conductivity have been used, as described in the legend of the figure. In particular, the exceedingly small lattice thermal conductivity of $Ba_{0.025}Sr_{0.075}Yb_{0.1}Co_4Sb_{12}$ in measurements of Zhang et al. (2009) is purely the consequence of taking a fully degenerate value of the Lorenz number in the Wiedemann-Franz law. On the other hand, measurements by Rogl et al. (2015a) on composite n-type skutterudites consisting of the triple-filled $(Sr_{0.33}Ba_{0.33}Yb_{0.33})_{0.35}Co_4Sb_{12}$ matrix with 9.1 wt% of $In_{0.4}Co_4Sb_{12}$ nano-inclusions revealed that the Lorenz number could actually be significantly lower than even the value of 2×10^{-8} V^2K^{-2}. Assuming the dominance of acoustic phonon scattering and using experimental values of the Seebeck coefficient, the Lorenz number turned out to be in the range 1.7×10^{-8} V^2K^{-2} to 1.62×10^{-8} V^2K^{-2} between 300 K and 700 K. The lesson to the readers is that statements regarding exceptionally low values of the lattice thermal conductivity, often far below estimates of the minimum lattice thermal conductivity, must be critically assessed for the value of the Lorenz number used.

Lattice thermal conductivities of p-type multifilled skutterudites that reached maximum ZT values in excess of unity are displayed in Figure 5.52. While in n-type forms of multifilled skutterudites depicted in Figure 5.51 the trend was more or less uniform, typified by a monotonically decreasing lattice thermal conductivity, p-type multifilled structures reveal a variety of trends. Some samples display a monotonically decreasing behavior, but more samples reverse the trend near 600 K–700 K and their lattice thermal conductivity increases. The onset of intrinsic excitations at elevated temperatures resulting in the bipolar thermal conductivity is likely the reason for the increased lattice thermal conductivity. Again, there is a considerable spread of the values of the lattice thermal conductivity, partly a result of using different values of the Lorenz number. I wish to draw attention to two traces of the lattice thermal conductivity: the one representing a sample of $DD_{0.60}Fe_3CoSb_{12}$ (solid red squares) measured by Rogl et al. (2013), which was not subjected to a high-pressure torsion (HPT) treatment, and the sample with the same composition that was subjected to 4 GPa pressure between anvils while undergoing the HPT treatment. Although the Lorenz number in both cases was taken at its fully degenerate value and thus the lattice thermal conductivities are

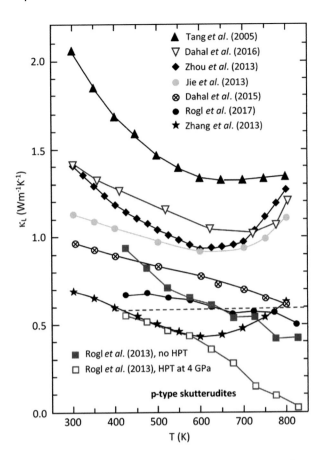

FIGURE 5.52 Lattice thermal conductivity of p-type skutterudites that reached the figure of merit ZT exceeding unity. Solid black up-triangles represent the lattice thermal conductivity of $Ce_{0.28}Fe_{1.5}Co_{2.5}Sb_{12}$ calculated with the constant Lorenz number $L = 2 \times 10^{-8}$ V^2K^{-2}. Circled crosses and open blue down-triangles stand for the lattice thermal conductivity of $Ca_{0.35}Ce_{0.35}Nd_{0.35}Fe_{3.5}Co_{0.5}Sb_{12}$ and $La_{0.68}Ce_{0.22}Fe_{3.5}Co_{0.5}Sb_{12}$ of Dahal et al. (2015) and Dahal et al. (2016), respectively, both using the Lorenz number calculated from the Seebeck coefficient and assuming acoustic phonon scattering ($r = -1/2$). However, in neither case, the plot of L vs. T was provided. Solid black diamonds are for the lattice thermal conductivity of $Yb_{0.25}La_{0.60}Fe_{2.7}Co_{1.3}Sb_{12}$ calculated with the Lorenz number $L = 2 \times 10^{-8}$ V^2K^{-2}. Full orange circles are the data for $Ce_{0.45}Nd_{0.45}Fe_{3.5}Co_{0.5}Sb_{12}$ obtained with the Lorenz number calculated from the Seebeck coefficient and assuming acoustic phonon scattering. Again, no plot of L vs. T was provided. Solid black circles designate the lattice thermal conductivity of a composite p-type sample with the $DD_{0.60}Fe_3CoSb_{12}$ matrix and 0.5 wt% of $Ta_{3.5}Zr_{0.5}B$ calculated with the Lorenz number determined from the Seebeck coefficient and assuming acoustic phonon scattering. A plot of L vs. T was not provided. Black stars stand for the lattice thermal conductivity of $Nd_{0.6}Fe_2Co_2Sb_{11.7}Ge_{0.3}$ determined by taking the Lorenz number as $L = 2 \times 10^{-8}$ V^2K^{-2}. Solid red squares and open red squares are lattice thermal conductivities of $DD_{0.60}Fe_3CoSb_{12}$ before and after applying the HPT treatment, respectively. Both sets of data were calculated with a fully degenerate Lorenz number $L = 2.44 \times 10^{-8}$ V^2K^{-2}, and thus underestimate the lattice thermal conductivity. Nevertheless, the effect of the HPT treatment is evident clearly. The broken blue line represents the minimal lattice thermal conductivity estimated by Rogl et al. (2017) based on the model of Cahill and Pohl (1989).

underestimated, there is nevertheless very clear evidence that HPT has a rather dramatic effect on the lattice thermal conductivity.

A more revealing influence of the HPT treatment on the thermal conductivity of p-type $DD_{0.60}Fe_3CoSb_{12}$ is shown in Figure 5.53. In this case, the total thermal conductivity is plotted as a function of the applied pressure by which a disk-shaped sample is held between two anvils, one of

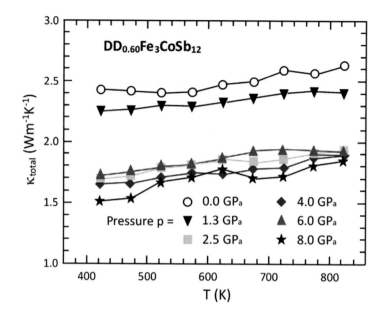

FIGURE 5.53 The effect of severe plastic deformation (high-pressure torsion, HPT) on the total thermal conductivity of $DD_{0.60}Fe_3CoSb_{12}$. The deformation is specified by the applied pressure exerted on a disc-shaped sample held by two anvils, one of which is rotated, applying torsional strain to the sample at a temperature of 675 K. The sample was prepared by a standard reaction process of the stoichiometric constituents, powdered by high-energy ball milling, and consolidated by hot pressing at 873 K under 56 MPa applied for 2 hr. The figure is plotted with the data taken from G. Rogl et al., *Acta Materialia* 61, 6778 (2013). With permission from Elsevier.

which is rotated with respect to the other one (Rogl et al. 2013). The drop in the thermal conductivity is particularly notable between pressures of 1.3 GPa and 2.4 GPa. The effect of HPT has resulted in the highest values of figure of merit. However, it remains to be seen whether repeated cycling of samples between room temperature and the highest operational temperature would eventually not anneal the strain that gives rise to such low values of the thermal conductivity. Readers interested in more detail concerning the HPT processing of skutterudites are referred to a paper by Rogl et al. (2015b).

Thermal conductivity is an important transport parameter for any technological application. In the field of thermoelectricity, it is the key input determining the likelihood of a structure to be an efficient energy conversion material. With their high thermal conductivity, the binary skutterudites are clearly not suitable for the task. However, the worldwide effort of researchers has shown that the thermal conductivity can be dramatically reduced by appropriate filling of the structural voids of skutterudites, fabrication of composite structures having the skutterudite matrix, and inducing large strains in the crystalline lattice. As a result, exceptionally low values of the lattice thermal conductivity, approaching the theoretical limit, have been achieved, enabling such skutterudites to be seriously considered for thermoelectric applications. In the next chapter, I discuss the thermoelectric performance of skutterudites and their prospective applications as power conversion thermoelectric devices.

REFERENCES

Abeles, A., *Phys. Rev.* **131**, 1906 (1963).

Agne, M. T., R. Hanus, and G. J. Snyder, *Energy & Environ. Sci.* **11**, 609 (2018).

Akai, K., H. Kurisu, T. Shimura, and M. Matsuura, *Proc. 16th Int. Conf. on Thermolectrics*, IEEE Catalog Number 97TH8291, Piscataway, NJ, p. 334 (1997).

Allen, P. B. and J. L. Feldman, *Phys. Rev. Lett.* **62**, 645 (1989).

Allen, P. B., J. L. Feldman, J. Fabian, and F. Wooten, *Philos. Mag. B* **79**, 1715 (1999).

Anno, H., K. Matsubara, Y. Notohara, T. Sakakibara, and H. Tashiro, *J. Appl. Phys.* **86**, 3780 (1999a).

Anno, H., H. Tashiro, and K. Matsubara, *Proc. 18th Int. Conf. on Thermoelectrics*, IEEE Catalog Number 99TH8407, Piscataway, NJ, p. 169 (1999b).

Anno, H. and K. Matsubara, *Recent Res. Devel. Appl. Phys.* **3**, 47 (2000).

Anno, H., Y. Nagamoto, K. Ashida, E. Taniguchi, T. Koyanagi, and K. Matsubara, *Proc. 19th Int. Conf. on Thermoelectrics*, Barrow Press, Wales, UK, p. 90 (2000).

Anno, H., K. Ashida, K. Matsubara, G. S. Nolas, K. Akai, M. Matsuura, and J. Nagao, *Mat. Res. Soc. Symp. Proc.* Vol. 691, G2.4.1 (2002).

Arita, Y., T. Ogawa, H. Kobayashi, K. Iwasaki, T. Matsui, and T. Nagasaki, *J. Nuclear Mater.* **344**, 79 (2005).

Ballikaya, S., N. Uzar, S. Yildirim, J. R. Salvador, and C. Uher, *J. Solid State Chem.* **193**, 31 (2012).

Ballikaya, S., N. Uzar, S. Yildirim, and J. R. Salvador, *J. Solid State Chem.* **197**, 440 (2013).

Bang, S., D.-H. Wee, A. Li, M. Fornari, and B. Kozinsky, *J. Appl. Phys.* **119**, 205102 (2016).

Baron, A. Q. R., *J. Spectroscopical Soc. Jpn.* **58**, 205 (2009).

Bashir, M. B. A., S. M. Said, M. F. M. Sabri, Y. Miyazaki, D. A. Shnawah, M. Shimada, M. F. M. Salleh, M. S. Mahmood, E. Y. Salih, F. Fitriani, and M. H. Elsheikh, *J. Electron. Mater.* **47**, 2429 (2018a).

Bashir, M. B. A., S. M. Said, M. F. M. Sabri, Y. Miyazaki, D. A. A. Shnawah, M. Shimada, and M. H. Elsheikh, *Sains Malaysiana* **47**, 181 (2018b).

Bauer, E., A. Galatanu, H. Michor, G. Hilscher, P. Rogl, P. Boulet, and H. Noël, *Eur. Phys. J. B* **14**, 483 (2000).

Bérardan, D., C. Godart, E. Alleno, St. Berger, and E. Bauer, *J. Alloys Compd.* **351**, 18 (2003).

Bérardan, D., E. Alleno, and C. Godart, *J. Appl. Phys.* **98**, 033710 (2005).

Berger, S., C. Paul, E. Bauer, A. Grytsiv, P. Rogl, D. Kaczorowski, A. Saccone, R. Ferro, and C. Godart, *Proc. 20th Int. Conf. On Thermoelectrics*, IEEE Catalog Number 01TH8589, Piscataway, NJ, p. 77 (2001).

Bergman, D. J. and O. Levy, *J. Appl. Phys.* **70**, 6821 (1991).

Bernstein, N., J. L. Feldman, and D. J. Singh, *Phys. Rev. B* **81**, 34301 (2010).

Bertini, L., K. Billquist, M. Christensen, C. Gatti, L. Holmgren, B. Iversen, E. Müller, M. Muhammed, G. Noriega, A. Palmqvist, D. Platzek, D. M. Rowe, A. Saramat, C. Stiewe, M. Toprak, S. G. Williams, and Y. Zhang, *Proc. 22th Inter. Conf. on Thermoelectrics*, IEEE Catalog Number 03TH8726, Piscataway, NJ, p. 93 (2003a).

Bertini, L., C. Stiewe, M. Toprak, S. Williams, D. Platzek, A. Mrotzek, Y. Zhang, C. Gatti, E. Müller, M. Muhammed, and D. M. Rowe, *J. Appl. Phys.* **93**, 438 (2003b).

Bhattacharya, S., M. J. Skove, M. Russell, T. M. Tritt, Y. Xia, V. Ponnambalam, S. J. Poon, and N. Thadhani, *Phys. Rev. B* **77**, 184203 (2008).

Borshchevsky, A., T. Caillat, and J.-P. Fleurial, *Proc. 15th Int. Conf. on Thermoelectrics*, IEEE Catalog Number 96TH8169, Piscataway, NJ, p. 112 (1996).

Braun, J. L., D. H. Olson, J. T. Gaskins, and P. E. Hopkins, *Rev. Sci. Instrum.* **90**, 024905 (2019).

Broido, D. A., M. Malorny, G. Birner, N. Mingo, and D. A. Stewart, *Appl. Phys. Lett.* **91**, 231922 (2007).

Cahill, D. G. and R. O. Pohl, *Solid State Commun.* **70**, 927 (1989).

Cahill, D. G., H. E. Fischer, T. Klitsner, E. Swartz, and R. O. Pohl, *J. Vac. Sci. Technol. A* **7**, 1259 (1989).

Cahill, D. G., *Rev. Sci. Instrum.* **61**, 802 (1990).

Caillat, T., A. Borshchevsky, and J.-P. Fleurial, *Proc. 13th Inter. Conf. on Thermoelectrics*, AIP Conference Proceedings 316, American Institute of Physics, NY, p. 58 (1995a).

Caillat, T., A. Borshchevsky, and J.-P. Fleurial, *Proc. 13th Inter. Conf. on Thermoelectrics*, AIP Conference Proceedings 316, American Institute of Physics, NY, p. 31 (1995b).

Caillat, T., J.-P. Fleurial, and A. Borshchevsky, *Proc. 30th Intersoc. Energy Conversion Engin. Conf.*, American Society of Mechanical Engineering, D. Y. Goswami, L. D. Kannberg, T. R. Mancini, and S. Somasundaram, eds., Vol. 3, p. 83 (1995c).

Caillat, T., A. Borshchevsky, and J.-P. Fleurial, *Proc. 13th Inter. Conf. on Thermoelectrics*, AIP Conference Proceedings 316, American Institute of Physics, NY, p. 209 (1995d).

Caillat, T., A. Borshchevsky, and J.-P. Fleurial, *J. Appl. Phys.* **80**, 4442 (1996a).

Caillat, T., J. Kulleck, A. Borshchevsky, and J.-P. Fleurial, *J. Appl. Phys.* **79**, 8419 (1996b).

Callaway, J., *Phys. Rev.* **113**, 1046 (1959).

Callaway, J. and H. C. von Baeyer, *Phys. Rev.* **120**, 1149 (1960).

Carruthers, P. A., *Rev. Mod. Phys.* **33**, 92 (1961).

Casimir, H. B. G., *Physica* **5**, 495 (1938).

Chen, B., J.-H. Xu, C. Uher, D. T. Morelli, G. P. Meisner, J.-P. Fleurial, T. Caillat, and A. Borshchevsky, *Phys. Rev. B* **55**, 1476 (1997).

Chen, L. D., X. F. Tang, T. Goto, T. Hirai, *J. Mater. Res.* **15**, 2276 (2000a).

Chen, L. D., T. Kawahara, X. F. Tang, T. Goto, T. Hirai, J. S. Dyck, W. Chen, and C. Uher, *Proc. 19th Int. Conf. on Thermoelectrics*, Babrow Press, Wales, UK, p. 348 (2000b).

Chen, L. D., T. Kawahara, X. F. Tang, T. Goto, T. Hirai, J. S. Dyck, W. Chen, and C. Uher, *J. Appl. Phys.* **90**, 1864 (2001a).

Chen, L. D., X. F. Tang, T. Kawahara, J. S. Dyck, W. Chen, C. Uher, T. Goto, and T. Hirai, *Proc. 20th Inter. Conf. on Thermoelectrics*, IEEE Catalog Number 01TH8589, Piscataway, NJ, p. 57 (2001b).

Chen, W., J. S. Dyck, C. Uher, L. D. Chen, X. F. Tang, and T. Hirai, *Proc. 20th Int. Conf. on Thermoelectrics*, IEEE Catalog Number 01TH8589, Piscataway, NJ, p. 69 (2001).

Chen, Y., Y. Kawamura, J. Hayashi, K. Takeda, and C. Sekine, *J. Appl. Phys.* **120**, 235105 (2016).

Cho, J. Y., Z. Ye, M. M. Tessema, R. A. Waldo, J. R. Salvador, J. Yang, W. Cai, and H. Wang, *Acta Mater.* **60**, 2104 (2012).

Cho, J. Y., Z. Ye, M. M. Tessema, J. R. Salvador, R. A. Waldo, Jiong Yang, W. Q. Zhang, Jihui Yang, W. Cai, and H. Wang, *J. Appl. Phys.* **113**, 143708 (2013).

Choi, D.-Y., Y.-E. Cha, and I.-H. Kim, *Korean J. Met. Mater.* **56**, 822 (2018).

Choi, S., K. Kurosaki, Y. Ohishi, H. Muta, and S. Yamanaka, *J. Appl. Phys.* **115**, 023702 (2014).

Christensen, M., A. B. Abrahamsen, N. B. Christensen, F. Juranyi, N. H. Andersen, K. Lefmann, J. Andreasson, C. R. H. Bahl, and B. B. Iversen, *Nature Mater.* **7**, 811 (2008).

Clarke, D. R., *Surf. Coat. Technol.* **163**, 67 (2003).

Cohn, J. L., G. S. Nolas, V. Fessatidis, T. H. Metcalf, and G. A. Slack, *Phys. Rev. Lett.* **82**, 779 (1999).

Dahal, T., Q. Jie, G. Joshi, S. Chen, C. F. Guo, Y. C. Lan, and Z. F. Ren, *Acta Mater.* **75**, 316 (2014).

Dahal, T., Q. Jie, W. Liu, K. Dahal, C. F. Guo, Y. C. Lan, and Z. F. Ren, *J. Alloys Compd.* **623**, 104 (2015).

Dahal, T., H.-S. Kim, S. Gahlawat, K. Dahal, Q. Jie, W. Liu, Y. C. Lan, K. White, and Z. F. Ren, *Acta Mater.* **117**, 13 (2016).

Da Ros, V., B. Lenoir, A. Dauscher, C. Candolfi, C. Bellouard, C. Stiewe, E. Müller, and J. Hejtmánek, *Proc. 25th Inter. Conf. on Thermoelectrics*, IEEE Catalog Number 06TH8931, Piscataway, NJ, p. 155 (2006).

Deng, L., J. Ni, J. M. Qin, and X. P. Jia, *Mater. Letters* **205**, 110 (2017).

Dilley, N. R., E. J. Freeman, E. D. Bauer, and M. B. Maple, *Phys. Rev. B* **58**, 6287 (1998).

Dilley, N. R., E. D. Bauer, M. B. Maple, S. Dordevic, D. N. Basov, F. Freibert, T. W. Darling, A. Migliori, B. C. Chakoumakos, and B. C. Sales, *Phys. Rev. B* **61**, 4608 (2000a).

Dilley, N. R., E. D. Bauer, M. B. Maple, and B. C. Sales, *J. Appl. Phys.* **88**, 1948 (2000b).

Dimitrov, I. K., M. E. Manley, S. M. Shapiro, J. Yang, W. Q. Zhang, L. D. Chen, Q. Jie, G. Ehlers, A. Podlesnyak, J. Camacho, and Q. Li, *Phys. Rev. B* **82**, 174301 (2010).

Ding, J., H. Gu, P. F. Qiu, X. H. Chen, Z. Xiong, Q. Zheng, X. Shi, and L. D. Chen, *J. Electron. Mater.* **42**, 382 (2012).

Dong, Y., P. Puneet, T. M. Tritt, and G. S. Nolas, *Phys. Stat. Solidi RRL* **7**, 418 (2013).

Dong, Y., P. Puneet, T. M. Tritt, and G. S. Nolas, *J. Solid State Chem.* **209**, 1, (2014).

Dudkin, L. D. and N. Kh. Abrikosov, *Zh. Neorg. Khim.* **2**, 212 (1957).

Dudkin, L. D. and N. Kh. Abrikosov, *Sov. Phys.-Solid State* **1**, 126 (1959).

Dyck, J. S., W. Chen, C. Uher, L. D. Chen, X. F. Tang, and T. Hirai, *J. Appl. Phys.* **91**, 3698 (2002).

Einstein, A., *Ann. Phys.* **35**, 679 (1911).

Eucken, A., *Ann. Phys.* **34**, 185 (1911).

Evers, C. B. H., W. Jeitschko, L. Boonk, D. J. Braun, T. Ebel, and U. D. Scholz, *J. Alloys Compd.* **224**, 184 (1995).

Feldman, J. L., M. D. Kluge, P. B. Allen, and F. Wooten, *Phys. Rev. B* **48**, 12589 (1993).

Feldman, J. L. and D. J. Singh, *Phys. Rev. B* **53**, 6273 (1996).

Feldman, J. L., D. J. Singh, I. I. Mazin, D. Mandrus, and B. C. Sales, *Phys. Rev. B* **61**, R9209 (2000).

Feldman, J. L., D. J. Singh, C. Kendziora, D. Mandrus, and B. C. Sales, *Phys. Rev. B* **68**, 094301 (2003).

Feldman, J. L., P. C. Dai, T. Enck, B. C. Sales, D. Mandrus, and D. J. Singh, *Phys. Rev. B* **73**, 014306 (2006).

Feldman, J. L., D. J. Singh, and N. Bernstein, *Phys. Rev. B* **89**, 224304 (2014).

Fleurial, J.-P., T. Caillat, and A. Borshchevsky, *Proc. 13th Int. Conf. on Thermoelectrics*, AIP Conference Proceedings 316, American Institute of Physics, New York, p. 40 (1995).

Fleurial, J.-P., A. Borshchevsky, T. Caillat, D. T. Morelli, and G. P. Meisner, *Proc. 15th Inter. Conf. on Thermoelectrics*, IEEE Catalog Number 96TH8169, Piscataway, NJ, p. 91 (1996).

Fleurial, J.-P., T. Caillat, and A. Borshchevsky, *Proc. 16th Int. Conf. on Thermoelectrics*, IEEE Catalog Number 97TH8291, Piscataway, NJ, p. 1 (1997).

Fu, L. W., J. Y. Yang, J. Y. Peng, Q. H. Jiang, Y. Xiao, Y. Luo, D. Zhang, Z. W. Zhou, M. Y. Zhang, Y. D. Cheng, and F. Q. Cheng, *J. Mater. Chem. A* **3**, 1010 (2015).

Fu, Y., D. J. Singh, W. Li, and L.-J. Zhang, *Phys. Rev. B* **94**, 075122 (2016).

Fu, Y., X. He, L. Zhang, and D. J. Singh, *Phys. Rev. B* **97**, 024301 (2018).

Geng, H. Y., S. Ochi, and J. Q. Cuo, *Appl. Phys. Lett.* **91**, 022106 (2007).

Ghosez, P. and M. Veithen, *J. Phys.: Condens. Matter* **19**, 096002 (2007).

Giri, R., N. Yanase, C. Sekine, I. Shirotani, A. Yamamoto, and C. H. Lee, *Memoirs of the Muraton Inst. of Technol.* **52**, 133 (2002).

Goto, T., Y. Nemoto, K. Sakai, T. Yamaguchi, M. Atatsu, T. Yanagisawa, H. Hazama, K. Onuki, H. Sugawara, and H. Sato, *Phys. Rev. B* **69**, 180511 (2004).

Grytsiv, A., P. Rogl, S. Berger, C. Paul, E. Bauer, C. Godart, B. Ni, M. M. Abd-Elmeguid, A. Saccone, R. Ferro, and D. Kaczorowski, *Phys. Rev. B* **66**, 094411 (2002).

Grytsiv, A., P. Rogl, H. Michor, E. Bauer, and G. Giester, *J. Electron. Mater.* **42**, 2940 (2013).

Guo, J. Q., H. Y. Genk, S. Ochii, H. K. Kim, H. Hyoda, and K. Kimura, *Proc. 26th Int. Conf. Thermpoelectrics*, IEEE, Piscataway, NJ, p. 183 (2007).

Guo, L., X. Xu, J. R. Salvador, and G. P. Meisner, *Appl. Phys. Lett.* **102**, 111905 (2013).

Harnwunggmoung, A., K. Kurosaki, A. Kosuga, M. Ishimaru, T. Plirdpring, R. Yimnirun, J. Jutim, *J. Appl. Phys.* **112**, 043509 (2012).

Hasegawa, T., Y. Takasu, N. Ogita, and M. Udagawa, *J. Phys. Soc. Jpn.* **77**, 248 (2008).

He, Q. Y., S. J. Hu, X. G. Tang, Y. C. Lan, J. Yang, X. W. Wang, Z. F. Ren, Q. Hao, and G. Chen, *Appl. Phys. Lett.* **93**, 042108 (2008).

He, T., J. Z. Chen, H. D. Rosenfeld, and M. A. Subramanian, *Chem. Mater.* **18**, 759 (2006).

He, Z., C. Stiewe, D. Platzek, G. Karpinski, E. Müller, S.-H. Li, M. Toprak, and M. Muhammed, *Nanotechnology* **18**, 235602 (2007).

Hermann, R. P., R. Y. Jin, W. Schweika, F. Grandjean, D. Mandrus, B. C. Sales, and G. J. Long, *Phys. Rev. Lett.* **90**, 135505 (2003).

Herring, C., *Phys. Rev.* **95**, 954 (1954).

Hobbis, D., Y. Liu, K. Wei, T. M. Tritt, and G. S. Nolas, *Crystals* **7**, 256 (2017).

Huang, B. L. and M. Kaviany, *Acta Mater.* **58**, 4516 (2010).

Iwasa, K., M. Kohgi, H. Sugawara, and H. Sato, *Physica B* **378–380**, 194 (2006).

Joo, G.-S., D.-K. Shin, and I.-H. Kim, *J. Electron. Mater.* **44**, 1383 (2015).

Jung, J.-Y., S.-C. Ur, and I.-H. Kim, *Mater. Chem. Phys.* **108**, 431 (2008).

Jung, J.-Y., K.-H. Park, S.-C. Ur, and I.-H. Kim, *Mater. Sci. Forum* **658**, 17 (2010a).

Jung, J.-Y., K.-H. Park, I.-H. Kim, S.-M. Choi, and W.-S. Seo, *J. Korean Phys. Soc.* **57**, 773 (2010b).

Kaltzoglou, A., P. Vaqueiro, K. S. Knight, and A. V. Powell, *J. Solid State Chem.* **193**, 36 (2012).

Kaneko, K., N. Metoki, T. D. Matsuda, and M. Kohgi, *J. Phys. Soc. Jpn.* **75**, 034701 (2006).

Katsuyama, S., Y. Kanayama, M. Ito, K. Majima, and H. Nagai, *Proc. 17th Int. Conf. on Thermoelectrics*, IEEE Catalog Number 98TH8365, Piscataway, NJ, p. 342 (1998a).

Katsuyama, S., Y. Shichijo, M. Ito, K. Majima, and H. Hagai, *J. Appl. Phys.* **84**, 6708 (1998b).

Katsuyama, S., M. Watanabe, M. Kuroki, T. Maehata, and M. Ito, *J. Appl. Phys.* **93**, 2758 (2003).

Keppens, V., D. Mandrus, B. C. Sales, B. C. Chakoumakos, P. Dai, R. Coldea, M. B. Maple, D. A. Gajewski, E. J. Freeman, and S. Bennington, *Nature (London)*, **395**, 876 (1998).

Kim, H., M. Kaviany, J. C. Thomas, A. van der Ven, C. Uher, and B. Huang, *Phys. Rev. Letters* **105**, 265901 (2010).

Kim, J. M., K. Kurosaki, S. Choi, Y. Ohishi, H. Muta, S. Yamanaka, M. Takahashi, and J. Tanaka, *Mater. Transactions* 58, 1207 (2017).

Kim, M.-J., S.-W. You, S.-C. Ur, and I.-H. Kim, *Proc. 25th Int. Conf. on Thermoelecrics*, IEEE Catalog Number 06TH8931, Piscataway, NJ, p. 439 (2006).

Kim, S. I., K. H. Lee, H. A. Mun, H. S. Kim, S. W. Hwang, J. W. Roh, D. J. Yang, W. H. Shin, X. S. Li, and Y. H. Lee, G. J. Snyder, and S. W. Kim, *Science* **348**, 109 (2015).

Kitagawa, H., M. Wakutsuki, H. Nagaoka, H. Noguchi, Y. Isoda, K. Hasezaki, and Y. Noda, *J. Phys. Chem. Solids*, **66**, 1635 (2005).

Kittel, C., *Phys. Rev.* **75**, 972 (1949).

Klemens, P. G., *Proc. Roy. Soc. A* **208**, 108 (1951).

Klemens, P. G., *Phys. Rev. A* **68**, 1113 (1955a).

Klemens, P. G., *Proc. Phys. Soc. A* **68**, 1113 (1955b).

Klemens, P. G., in *Solid State Physics*, eds. F., Seitz and D. Turnbull, Acad. Press, NY, Vol.7 (1958).

Kliche, G. and W. Bauhofer, *J. Phys. Chem. Solids* **49**, 267 (1988).

Koyanagi, T., T. Tsubouchi, M. Ohtani, K. Kishimoto, H. Anno, and K. Matsubara, *Proc. 15th Int. Conf. on Thermoelectrics*, IEEE Catalog Number 96TH8169, Piscataway, NJ, p. 107 (1996).

Koza, M. M., M. R. Johnson, R. Viennois, H. Mutka, L. Girard, and D. Ravot, *Nature Mater.* **7**, 805 (2008).

Koza, M. M., L. Capogna, A. Leithe-Jasper, H. Rosner, W. Schnelle, H. Mutka, M. R. Johnson, C. Ritter, and Y. Grin, *Phys. Rev. B* **81**, 174302 (2010).

Koza, M. M., A. Leithe-Jasper, H. Rosner, W. Schnelle, H. Mutka, M. R. Johnson, M. Krisch, L. Capogna, and Y. Grin, *Phys. Rev. B* **84**, 014306 (2011).

Koza, M. M., D. Adroja, N. Takeda, Z. Henkie, and T. Cichorek, *J. Phys. Soc. Jpn.*, **82**, 114607 (2013).

Koza, M. M., A. Leithe-Jasper, H. Rosner, W. Schnelle, H. Mutka, M. R. Johnson, and Y. Grin, *Phys. Rev. B* **89**, 014302 (2014).

Koza, M. M., M. Boehm, E. Sischka, W. Schnelle, H. Mutka, and A. Leithe-Jasper, *Phys. Rev. B* **91**, 014305 (2015).

Kuznetsov, V. L. and D. M. Rowe, *J. Phys.: Condens. Matter* **12**, 7915 (2000).

Lamberton, G. A. Jr., S. Bhattacharya, R. T. Littleton, M. A. Kaeser, R. H. Tedstrom, T. M. Tritt, J. Yang, and G. S. Nolas, *Appl. Phys. Lett.* **80**, 598 (2002).

Lee, C. H., I. Hase, H. Sugawara, H. Yoshizawa, and H. Sato, *J. Phys. Soc. Jpn.* **75**, 123602 (2006).

Lee, S., K. H. Lee, M.-M. Kim, H. S. Kim, G. J. Snyder, S. Baik, and S. W. Kim, *Acta Mater.* **142**, 8 (2018).

Leithe-Jasper, A., D. Kaczorowski, P. Rogl, J. Bogner, M. Reissner, W. Steiner, G. Wiesinger, and C. Godart, *Solid State Commun.* **109**, 395 (1999).

Leithe-Jasper, A., W. Schnelle, H. Rosner, N. Senthilkumaran, A. Rabis, M. Baenitz, A. Gippius, E. Morozova, J. A. Mydosh, and Y. Grin, *Phys. Rev. Lett.* 91, 037208 (2003).

Leibfried, G. and E. Schlömann, *Nachr. Akad, Wiss. Goettingen, Math.-Physik* Kl. **4**, 71 (1954).

Leszczynski, J., A. Dauscher, P. Masschelein, and B. Lenoir, *J. Electron. Mater.* **39**, 1764 (2010).

Leszczynski, J., V. Da Ros, B. Lenoir, A. Dauscher, C. Candolfi, P. Masschelein, J. Hejtmánek, K. Kutorasinski, J. Tobola, R. I. Smith, C. Stiewe, and E. Müller, *J. Phys. D: Appl. Phys.* **46**, 495106 (2013).

Li, H., X. F. Tang, X. L. Su, W. Q. Cao, and Q. J. Zhang, *Proc. 26th Int. Conf. on Thermoelectrics*, IEEE, Piscataway, NJ, p. 193 (2007).

Li, H., X. F. Tang, X. L. Su, and Q. J. Zhang, *Appl. Phys. Lett.* **92**, 202114 (2008).

Li, H., X. F. Tang, Q. J. Zhang, and C. Uher, *Appl. Phys. Lett.* **93**, 252109 (2008).

Li, H., X. F. Tang, Q. J. Zhang, and C. Uher, *Appl. Phys. Lett.* **94**, 102114 (2009).

Li, H., X. F. Tang, and Q. J. Zhang, *J. Electron. Mater.* **38**, 1224 (2009).

Li, J. L., B. Duan, H. J. Yang, H. T. Wang, G. D. Li, J. Yang, G. Chen, and P. C. Zhai, *J. Mater. Chem. C* **7**, 8079 (2019).

Li, W., L. Lindsey, D. A. Broido, D. A. Stewart, and N. Mingo, *Phys. Rev. B* **86**, 174307 (2012).

Li, W., J. Carrete, N. A. Katcho, and N. Mingo, Comput. Phys. Commun. **185**, 1747 (2014).

Li, W. and N. Mingo, *Phys. Rev. B* **90**, 094302 (2014a).

Li, W. and N. Mingo, *Phys. Rev. B* **89**, 184304 (2014b).

Li, W. and N. Mingo, *Phys. Rev. B* **91**, 144304 (2015).

Li, W. J., J. Wang, Y. T. Xie, J. L. Gray, J. J. Heremans, H. B. Kang, B. Poudel, S. T. Huxtable, and S. Priya, *Chem. Mater.* **31**, 862 (2019).

Li, X. Y., L. D. Chen, J. F. Fan, and W. B. Zhang, *J. Appl. Phys.* **98**, 083702 (2005).

Li, Y. L., P. F. Qiu, Z. Xiong, J. Chen, R. Nunna, X. Shi, and L. D. Chen, *AIP Adv.* **5**, 117239 (2015).

Liu, L. Q., X. B. Zhao, T. J. Zhu, and Y. J. Gu, *J. Rare Earths* **30**, 456 (2012).

Liu, R. H., J. Yang, X. H. Chen, X. Shi, L. D. Chen, and C. Uher, *Intermetallics* **19**, 1747 (2011).

Liu, T. X., X. F. Tang, W. J. Xie, Y. G. Yan, and Q. J. Zhang, *J. Rare Earth* **25**, 739 (2007).

Liu, W.-S., B.-P. Zhang, J.-F. Li, and L.-D. Zhao, *J. Phys. D: Appl. Phys.* **40**, 566 (2007a).

Liu, W.-S., B.-P. Zhang, J.-F. Li, H.-L. Zhang, and L.-D. Zhao, *J. Appl. Phys.* **102**, 103717 (2007b).

Liu, W.-S., B.-P. Zhang, L.-D. Zhao, and J.-F. Li, *Chem. Mater.* **20**, 7526 (2008).

Lorenz, L., *Annln. Phys.* **13**, 422 (1881).

Long, G. J., R. P. Hermann, F. Grandjean, E. E. Alp, W. Sturhahn, C. E. Johnson, D. E. Brown, O. Leupold, and R. Rüffer, *Phys. Rev. B* **71**, 140302(R) (2005).

Lu, Q. M., J. X. Zhang, X. Zhang, Y. Q. Liu, D. M. Liu, and M. L. Zhu, *J. Appl. Phys.* **98**, 196107 (2005).

Lu, Q. M., J. X. Zhang, X. Zhang, and Q. Wei, *Proc. 25th Inter. Conf. on Thermoelectrics*, IEEE Catalog Number 06TH8931, Piscataway, NJ, p. 148 (2006).

Lutz, H. D. and G. Kliche, *J. Solid State Chem.* **40**, 64 (1981).

Lutz, H. D. and G. Kliche, *Phys. Stat. Sol. (b)* **112**, 549 (1982).

Mallik, R. C., J.-Y. Jung, V. D. Das, S.-C. Ur, and I.-H. Kim, *Proc. 25th Inter. Conf. on Thermoelectrics*, IEEE Catalog Number 06TH8931, Piscataway, NJ, p. 431 (2006).

Mallik, R. C., J.-Y. Jung, V. D. Das, S.-C. Ur, and I.-H. Kim, *Solid State Commun.* **141**, 233 (2007).

Mallik, R. C., J.-Y. Jung, S.-C. Ur, and I.-H. Kim, *Metals Mater. Int.* **14**, 223 (2008).

Mallik, R. C., C. Stiewe, G. Karpinski, R. Hassdorf, and E. Müller, *J. Electron. Mater.* **38**, 1337 (2009).

Matsubara, M. and R. Asahi, *J. Electron. Mater.* **45**, 1669 (2016).

McGaughey, A. J. H. and M. Kaviany, *Adv. Heat Transport* **39**, 169 (2006).

Meisner, G. P., D. T. Morelli, S. Hu, J. Yang, and C. Uher, *Phys. Rev. Lett.* **80**, 3551 (1998).

Meng, X. F., Z. H. Liu, B. Cui, D. D. Qin, H. Y. Geng, W. Cai, L. W. Fu, J. Q. He, Z. F. Ren, and J. H. Sui, *Adv. Energy Mater.* **7**, 1602582 (2017).

Mi, J. L., X. B. Zhao, T. J. Zhu, and J. P. Tu, *Appl. Phys. Lett.* **91**, 172116 (2007).

Mi, J. L., M. Christensen, E. Nishibori, and B. B. Iversen, *Phys. Rev. B* **84**, 064114 (2011).

Morelli, D. T. and G. P. Meisner, *J. Appl. Phys.* **77**, 3777 (1995).

Morelli, D. T., T. Caillat, J.-P. Fleurial, A. Borshchevsky, J. Vandersande, B. Chen, and C. Uher, *Phys. Rev. B* **51**, 9622 (1995).

Morelli, D. T., G. P. Meisner, B. Chen, S. Hu, and C. Uher, *Phys. Rev. B* **56**, 7376 (1997).

Morelli, D. T., J. P. Heremans, and G. A. Slack, *Phys. Rev. B* **66**, 195304 (2002).

Mori, H., H. Anno, and K. Matsubara, *Mater. Transact.* **46**, 1476 (2005).

Möchel, A., I. Sergueev, H.-C. Wille, J. Voigt, M. Prager, M. B. Stone, B. C. Sales, Z. Guguchia, A. Shengelaya, V. Keppens, and R. P. Hermann, *Phys. Rev. B* **84**, 184306 (2011a).

Möchel, A., I. Sergueev, N. Nguyen, G. J. Long, F. Grandjean, D. C. Johnson, and R. P. Hermann, *Phys. Rev. B* **84**, 064302 (2011b).

Muta, H., K. Kurosaki, M. Uno, and S. Yamanaka, *J. Alloys Compd.* **359**, 326 (2003).

Nabarro, F. R. N., *Proc. Roy. Soc. A* **209**, 278 (1951).

Nagamoto, Y., K. Tanaka, and T. Koyanagi, *Proc. 17th Int. Conf. on Thermoelectrics*, IEEE Catalog Number 98TH8365, Piscataway, NJ, p. 302 (1998).

Nagao, J., M. Ferhat, H. Anno, K. Matsubara, E. Hatta, and K. Mukasa, *Appl. Phys. Lett.* **76**, 3436 (2000).

Nakagawa, H., H. Tanaka, A. Kasama, K. Miyamura, H. Masumoto, and K. Matsubara, *Proc. 15th Int. Conf. on Thermoelectrics*, IEEE Catalog Number 96TH8169, Piscataway, NJ, p. 117 (1996).

Nakagawa, H., H. Tanaka, A. Kasama, H. Anno, and K. Matsubara, *Proc. 16th Int. Conf. on Thermoelectrics*, IEEE Catalog Number 97TH8291, Piscataway, NJ, p. 351 (1997).

Nan, C.-W., *Prog. Mater. Sci.* **37**, 1 (1993).

Nan, C.-W. and R. Birringer, *Phys. Rev. B* **57**, 8264 (1998).

Narazu, S., Y. Hadano, K. Suekuni, T. Takabatake, K. Suzuki, and H. Anno, *Proc. 26th Int. Conf. on Thermoelectrics*, IEEE Catalog Number CFP07404, Piscataway, NJ, p. 197 (2007).

Navrátil, J., F. Laufek, T. Plecháček, and J. Plášil, *J. Alloys Compd.* **493**, 50 (2010).

Nolas, G. S., G. A. Slack, D. T. Morelli, T. M. Tritt, and A. C. Ehrlich, *J. Appl. Phys.* **79**, 4002 (1996a).

Nolas, G. S., V. G. Harris, T. M. Tritt, and G. A. Slack, *J. Appl. Phys.* **80**, 6304 (1996b).

Nolas, G. S., H. B. Lyons, J. L. Cohn, T. M. Tritt, and G. A. Slack, *Proc. 16th Int. Conf. on Thermoelectrics*, IEEE Catalog Number 97TH8291, Piscataway, NJ, p. 321 (1997).

Nolas, G. S., J. L. Cohn, and G. A. Slack, *Phys. Rev. B* **58**, 164 (1998).

Nolas, G. S. and C. A. Kendziora, *Phys. Rev. B* **59**, 6189 (1999).

Nolas, G. S., M. Kaeser, R. T. Littleton IV, and T. M. Tritt, *Appl. Phys. Lett.* **77**, 1855 (2000a).

Nolas, G. S., H. Takizawa, T. Endo, H. Sellingschegg, and D. C. Johnson, *Appl. Phys. Lett.* **77**, 52 (2000b).

Nolas, G. S., J. Yang, and R. W. Ertenberg, *Phys. Rev. B* **68**, 193206 (2003).

Nolas, G. S. and G. Fowler, *J. Mater. Res.* **20**, 3234 (2005).

Nolas, G. S., M. Beekman, R. W. Ertenberg, and J. Yang, *J. Appl. Phys.* **100**, 036101 (2006).

Paddock, C. A. and G. L. Eesley, *J. Appl. Phys.* **60**, 285 (1986).

Park, K.-H., S.-W. You, S.-C. Ur, and I.-H. Kim, *Proc. 25th Int. Conf. on Thermoelecrics*, IEEE Catalog Number 06TH8931, Piscataway, NJ, p. 435 (2006).

Park, K.-H., J.-Y. Jung, S.-C. Ur, and I.-H. Kim, *J. Electron. Mater.* **39**, 1750 (2010).

Park, K.-H., S.-W. You, S.-C. Ur, I.-H. Kim, S.-M. Choi, and W.-S. Seo, *J. Electron. Mater.* **41**, 1051 (2012).

Park, K.-H., W.-S. Seo, D.-K. Shin, and I.-H. Kim, *J. Korean Phys. Soc.* **65**, 491 (2014).

Parrott, J. E. and A. D. Stuckes, in *Thermal Conductivity of Solids*, Pion Ltd., London (1975).

Pei, Y. Z., J. Yang, L. D. Chen, W. Zhang, J. R. Salvador, and J. Yang, *Appl. Phys. Lett.* **95**, 042101 (2009).

Peierls, R., Ann. Phys., (Leipzig) **3**, 1055 (1929).

Peng, J. Y., P. N. Alboni, J. He, B. Zhang, Z. Su, T. Holgate, N. Gothard, and T. M. Tritt, *J. Appl. Phys.* **104**, 053710 (2008).

Peng, J. Y., J. He, Z. Su, P. N. Alboni, S. Zhu, and T. M. Tritt, *J. Appl. Phys.* **105**, 084907 (2009).

Peng, J. Y., X. Y Liu, J. He, and J. Y. Yang, *Procedia Engin.* **27**, 121 (2012a).

Peng, J. Y., W. Xu, Y. G. Yan, J. Y. Yang, L. W. Fu, H. J. Kang, and J. He, J. Appl. Phys. 112, 024909 (2012b).

Pippard, A. B., *Philos. Mag.* **46**, 1104 (1955).

Pohl, R. O., *Phys. Rev. Lett.* **8**, 418 (1962).

Price, P. I., *Phil. Mag.* **46**, 1252 (1955).

Puyet, M., B. Lenoir, A. Dauscher, M. Dehmas, C. Stiewe, and E. Müller, *J. Appl. Phys.* **95**, 4852 (2004)

Qiu, P. F., J. Yang, R. H. Liu, X. Shi, X. Y. Huang, G. J. Snyder, W. Q. Zhang, and L. D. Chen, *J. Appl. Phys.* **109**, 063713 (2011).

Rogl, G., A. Grytsiv, N. Melnychenko-Koblyuk, E. Bauer, S. Laumann, and P. Rogl, *J. Phys. Condens. Matter* **23**, 275601 (2011).

Rogl, G., A. Grytsiv, P. Rogl, E. Royanian, E. Bauer, J. Horky, D. Setman, E. Schafler, and M. Zehetbauer, *Acta Materialia* **61**, 6778 (2013).

Rogl, G., A. Grytsiv, P. Rogl, E. Bauer, M. Hochenhofer, R. Anbalagan, R. C. Mallik, and E. Schafler, *Acta Mater.* **76**, 434 (2014).

Rogl, G., A. Grytsiv, K. Yubuta, S. Puchegger, E. Bauer, C. Raju, R. C. Mallik, and P. Rogl, *Acta Mater.* **95**, 201 (2015a).

Rogl, G., A. Grytsiv, J. Bursik, J. Horky, R. Anbalagan,. Bauer, R. C. Mallik, P. Rogl, and M. Zehetbauer, *Phys. Chem. Chem. Phys.* **17**, 3715 (2015b).

Rogl, G., A. Grytsiv, P. Rogl, F. Failamani, M. Hochenhofer, E. Bauer, and P. Rogl, *J. Alloys Compd.* **695**, 682 (2017).

Rotter, M., P. Rogl, A. Grytsiv, W. Wolf, M. Krisch, and A. Mirone, *Phys. Rev. B* **77**, 144301 (2008).

Ryll, B., A. Schmitz, J. de Boor, A. Franz, P. S. Whitfield, M. Reehuis, A. Hoser, E. Müller, K. Habicht, and K. Fritsch, *ACS Appl. Energy Mater.* **1**, 113 (2018).

Sales, B. C., D. Mandrus, and R. K. Williams, *Science* **272**, 1325 (1996).

Sales, B. C., D. Mandrus, B. C. Chakoumakos, V. Keppens, and J. R. Thompson, *Phys. Rev. B* **56**, 15081 (1997).

Sales, B. C., B. C. Chakoumakos, R. Jin, J. R. Thompson, and D. Mandrus, *Phys. Rev. B* **63**, 245113 (2001).

Sales, B. C., B. C. Chakoumakos, D. Mandrus, and J. W. Sharp, *J. Solid State Chem.* **146**, 528 (1999).

Sales, B. C., B. C. Chakoumakos, and D. Mandrus, *Phys. Rev. B* **61**, 2475 (2000).

Salvador, J. R., J. Yang, X. Shi, H. Wang, A. A. Wereszczak, H. Kong, and C. Uher, *Phil. Mag.* **89**, 1517 (2009).

Salvador, J. R., J. Yang, H. Wang, and X. Shi, *J. Appl. Phys.* **107**, 043705 (2010).

Salvador, J. R., R. A. Waldo, C. A. Wong, M. M. Tessema, D. N. Brown, D. J. Miller, H. Wang, A. A. Wereszczak, and W. Cai, *Mater. Sci. Engin. B* **178**, 1087 (2013a).

Salvador, J. R., J. Y. Cho, Z. Ye, J. E. Moczygemba, A. J. Thompson, J. W. Sharp, J. D. König, R. Maloney, T. Thompson, J. Sakamoto, H. Wang, A. A. Wereszczak, and G. P. Meisner, *J. Electron. Mater.* **42**, 1389 (2013b).

Sato, H., H. Sugawara, Y. Aoki, and H. Harima, in *Handbook of Magnetic Materials*, Vol. 18, Ch. 1, Elsevier, p. 1 (2009).

Savvides, N. and H. J. Goldsmid, *J. Phys. C* **6**, 1701 (1973).

Schmidt, A. J., R. Cheaito, and M. Chiesa, *Rev. Sci. Instrum.* **80**, 094901 (2009).

Sellinschegg, H., D. C. Johnson, G. S. Nolas, G. A. Slack, S. B. Schujman, F. Mohammed, T. M. Tritt, and E. Nelson, *Proc. 17th Int. Conf. on Thermoelectrics*, IEEE Catalog Number 98TH8365, Piscataway, NJ, p. 338 (1998a).

Sellinschegg, H., S. L. Stuckmeyer, M. D. Hornbostel, and D. C. Johnson, *Chem. Mater.* **10**, 1096 (1998b).

Sharp, J. W., E. C. Jones, R. K. Williams, P. M. Martin, and B. C. Sales, *J. Appl. Phys.* **78**, 1013 (1995).

Shi, X., L. D. Chen, S. Q. Bai, X. Y. Huang, and X. F. Tang, *Proc. 21st Inter. Conf. on Thermoelectrics*, IEEE Catalog Number 02TH8657, Piscataway, NJ, p. 68 (2002).

Shi, X., L. D. Chen, J. Yang, and G. P. Meisner, *Appl. Phys. Lett.* **84**, 2301 (2004).

Shi, X., H. Kong, C.P. Li, C. Uher, J. Yang, J. R. Salvador, H. Wang, L. D. Chen, and W. Q. Zhang, *Appl. Phys. Lett.* **92**, 182101 (2008).

Shi, X., J. R. Salvador, J. Yang, and H. Wang, *J. Electron. Mater.* **38**, 930 (2009).

Shi, X., J. Yang, J. R. Salvador, M. F. Chi, J. Y. Cho, H. Wang, S. Q. Bai, J. H. Yang, W. Q. Zhang, and L. D. Chen, *J. Amer. Chem. Soc.* **133**, 7837 (2011a).

Shi, X., Y. Pei, G. J. Snyder, and L. D. Chen, *Energy & Environ. Sci.* **4**, 4086 (2011b).

Singh, D. J., L. Nordstrom, W. E. Pickett and J. L. Feldman, *Proc. 15th Int. Conf. on Thermoelectrics*, IEEE Catalog Number 96TH8169, Piscataway, NJ, p. 84 (1996).

Singh, D. J., I. I. Mazin, J. L. Feldman, and M. Fornari, *Mat. Res. Soc. Symp. Proc.* **545**, 3 (1999).

Slack, G. A., *Phys. Rev.* **105**, 829 (1957).

Slack, G. A., *Phys. Rev.* **126**, 427 (1962).

Slack, G. A. and S. Galginaitis, *Phys. Rev. B* **133**, A253 (1964).

Slack, G. A., *Proc. Int. Conf. on Phonon Scattering*, ed. H. J. Albany, p. 24. Service de Documentation du CEN Saclay (1972).

Slack, G. A., in *Solid State Physics*, F. Seitz and D. Turnbull eds., Academic Press, N.Y., Vol. 34, p.1, (1979).

Slack, G. A. and V. G. Tsoukala, *J. Appl. Phys.* **76**, 1665 (1994).

Slack, G. A., in *CRC Handbook of Thermoelectrics*, ed. D. M. Rowe, CRC Press, Boca Raton, FL, pp. 407–440 (1995).

Slack, G. A., J.-P. Fleurial, and T. Caillat, *Naval Research News* **18**, 23 (1996).

Sproull, R. L., M. Moss, and H. Weinstock, *J. Appl. Phys.* **30**, 334 (1959).

Stackhouse, S. and L. Stixrude, *Review in Mineralogy & Geochemistry* **71**, 253 (2010).

Stokes, K. L., A. C. Ehrlich, and G. S. Nolas, *Mat. Res. Soc. Symp. Proc.* **545**, 339 (1999).

Takizawa, H., K. Miura, M. Ito, T. Suzuki, and T. Endo, *J. Alloys Compd.* **282**, 79 (1999).

Tan, G. J., W. Liu, H. Chi, X. Su, S. Wang, Y. G. Yan, X. F. Tang, W. Wong-Ng, and C. Uher, *Acta Mater.* **61**, 7693 (2013).

Tang, G. D., Z. H. Wang, X. N. Xu, Y. He, L. Qiu, and Y. W. Du, *J. Electron. Mater.* **40**, 611 (2011).

Tang, X. F., L. D. Chen, T. Goto, and T. Hirai, *J. Mater. Res.* **16**, 837 (2001).

Tang, X. F., H. Li, Q. J. Zhang, M. Goto, and T. Goto, *J. Appl. Phys.* **100**, 123702 (2006).

Tang, X. F., Q. J. Zhang, L. D. Chen, T. Goto, and T. Hirai, *J. Appl. Phys.* **97**, 093712 (2005).

Tang, Y., Y. Qiu, L. Xi, X. Shi, W. Zhang, L. D. Chen, S.-M. Tseng, S.-W. Chen, and G. J. Snyder, *Energy Environ. Sci.* **7**, 812 (2014).

Tang, Y., S.-W. Chen, and G. J. Snyder, *Materiomics* **1**, 75 (2015).

Tashiro, H., Y. Notohara, T. Sakakibara, H. Anno, and K. Matsubara, *Proc. 16th Int. Conf. on Thermoelectrics*, IEEE Catalog Number 97TH8291, Piscataway, NJ, p. 326 (1997).

Thompson, D. R., C. Liu, Jiong Yang, J. R. Salvador, D. B. Haddad, N. D. Ellison, R. A. Waldo, and Jihui Yang, *Acta Mater.* **92**, 152 (2015).

Tobola, J., K. Wojciechowski, J. Cieslak, and J. Leszczynski, *Proc. 18th Inter. Conf. on Thermoelectrics*, IEEE Catalog Number 99TH8407, Piscataway, NJ, p. 169 (1999).

Tobola, J., K. Wojciechowski, J. Cieslak, and J. Leszczynski, *Proc. 22nd Inter. Conf. on Thermoelectrics*, IEEE Catalog Number 03TH8726, Piscataway, NJ, p. 76 (2003).

Tritt, T. M., G. S. Nolas, G. A. Slack, A. C. Ehrlich, D. J. Gillespie, and J. L. Cohn, *J. Appl. Phys.* **79**, 8412 (1996).

Tsubota, M., S. Tsutsui, D. Kikuchi, H. Sugawara, H. Sato, and Y. Murakami, *J. Phys. Soc. Jpn.* **77**, 073601 (2008).

Tsutsui, S., H. Kobayashi, J. Umemura, Y. Yoda, H. Onodera, H. Sugawara, D. Kikuchi, H. Sato, C. Sekine, and I. Shirotani, *Physica B* **383**, 142 (2006a).

Tsutsui, S., J. Umemura, H. Kobayashi, T. Tazaki, S. Nasu, Y. Kobayashi, Y. Yoda, H. Onodera, H. Sugawara, T. D. Matsuda, D. Kikuchi, H. Sato, C. Sekine, and I. Shirotani, *Hyperfine Interact.* **168**, 1073 (2006b).

Tsutsui, S., Y. Yoda, and H. Kobayashi, *J. Phys. Soc. Jpn.*, **76**, 065003 (2007).

Tsutsui, S., H. Kobayashi, D. Ishikawa, J. P. Sutter, A. Q. R. Baron, T. Hasegawa, N. Ogita, M. Udagawa, Y. Yoda, H. Onodera, D. Kikuchi, H. Sugawara, C. Sekine, I. Shirotani, and H. Sato, *J. Phys. Soc. Jpn.*, **77**, 033601 (2008a).

Tsutsui, S., H. Kobayashi, J. P. Sutter, H. Uchiyama, A. Q. R. Baron, Y. Yoda, D. Kikuchi, H. Sugawara, C. Sekine, I. Shirotani, A. Ochiai, and H. Sato, *J. Phys. Soc. Jpn.*, **77**, 257 (2008b).

Tsutsui, S., M. Mizumaki, M. Tsubota, H. Tanida, T. Uruga, Y. Murakami, D. Kikuchi, H. Sugawara, and H. Sato, *J. Phys.: Conf. Series* **150**, 042220 (2009).

Tsutsui, S., H. Uchiyama, J. P. Sutter, A. Q. R. Baron, H. Sugawara, J. Yamaura, Z. Hiroi, A. Ochiai, and H. Sato, *J. Phys.: Conf. Series* **200**, 012213 (2010).

Tsutsui, S., H. Uchiyama, J. P. Sutter, A. Q. R. Baron, M. Mizumaki, N. Kawamura, T. Uruga, H. Sugawara, J.-I Yamaura, A. Ochiai, T. Hasegawa, N. Ogita, M. Udagawa, and H. Sato, *Phys. Rev. B* **86**, 195115 (2012).

Uher, C., B. Chen, S. Hu, D. T. Morelli, and G. P. Meisner, *Mat. Res. Soc. Symp. Proc.* **478**, 315 (1997).

Uher, C., S. Hu, and J. Yang, *Proc. 17th Int. Conf. on Thermoelectrics*, IEEE Catalog Number 98TH8365, Piscataway, NJ, p. 306 (1998).

Uher, C., X. Shi, and H. Kong, *Mater. Res. Soc. Symp. Proc.* Vol. 1044, 191 (2008).

Uher, C., J. S. Dyck, W. Chen, G. P. Meisner, and J. Yang, *Mat. Res. Soc. Symp. Proc.* Vol. 626, Z10.3.1 (2000).

Uher, C., C.P. Li, and S. Ballikaya, *J. Electron. Mater.* **39**, 2122 (2010).

Vaqueiro, P. and G. G. Sobany, *Mat. Res. Soc. Symp. Proc.*, Vol. 1044, 1044-U05-08 (2008).

Viennois, R., L. Girard, M. M. Koza, H. Mutka, D. Ravot, F. Terki, S. Charar, and J.-C. Tedenac, *Phys. Chem. Chem. Phys.* **7**, 1617 (2005).

Visnow, E., C. P. Heinrich, A. Schmitz, J. de Boor, P. Leidich, B. Klobes, R. P. Hermann, E. Müller, and W. Tremel, *Inorg. Chem.* **54**, 7818 (2015).

Wang, L., K. F. Cai, Y. Y. Wang, H. Li, and H. F. Wang, *Appl. Phys. A: Mater. Sci & Proces.* **97**, 841 (2009).

Wang, S., J. R. Salvador, Jiong Yang, P. Wei, B. Duan, and Jihui Yang, *NPG Asia Mater.* **8**, e285 (2016).

Wang, Y., X. Xu, and J. Yang, *Phys. Rev. Lett.* **102**, 175508 (2009).

Wang, Y., J. Mao, Q. Jie, B. H. Ge, and Z. F. Ren, *Appl. Phys. Lett.* **110**, 163901 (2017).

Wang, Y. C., H. L. Yang, W. J. Qiu, Jiong Yang, Jihui Yang, and W. Q. Zhang, *Phys. Rev. B* **98**, 054304 (2018).

Watcharapasorn, A., R. C. DeMattei, R. S. Feigelson, T. Caillat, A. Borshchevsky, G. J. Snyder, and J.-P. Fleurial, *J. Appl. Phys.* **86**, 6213 (1999).

Wee, D. H., B. Kozinsky, N. Marzari, and M. Fornari, *Phys. Rev. B* **81**, 045204 (2010).

Wei, P., W.-Y. Zhao, C.-L. Dong, B. Ma, and Q.-J. Zhang, *J. Electron. Mater.* **39**, 1803 (2010).

Wei, P., W.-Y. Zhao, C.-L. Dong, X. Yang, J. Yu, and Q.-J. Zhang, *Acta Mater.* **59**, 3244 (2011).

Wiedemann, G. and R. Franz, *Annln. Phys.* **89**, 497 (1853).

Wille, H.-C., R. P. Hermann, I. Sergueev, O. Leupold, P. van der Linden, B. C. Sales, F. Grandjean, G. J. Long, R. Rüffer, and Yu. V. Shvyd'ko, *Phys. Rev. B* **76**, 140301(R) (2007).

Wojciechowski, K. T., Mat. Res. Bull. **37**, 2023 (2002).

Wojciechowski, K. T., J. Tobola, and J. Leszczynski, *J. Alloy Compd.* **361**, 19 (2003).

Xia, X. G., P. F. Qiu, X. Shi, X. Y. Li, X. Y. Huang, and L. D. Chen, *J. Electron. Mater.* **41**, 2225 (2012).

Xu, W., J. Y. Peng, J. He, M. H. Zhou, J. Y. Yang, and L. W. Fu, *J. Wuhan Univ. of Technol.-Mater. Sci. Ed.*, p. 677 (Aug. 2013).

Yamakage, A. and Y. Kuramoto, *J. Phys. Soc. Jpn.* **78**, 064602 (2009).

Yamaura, J. and Z. Hiroi, *J. Phys. Soc. Jpn.* **80**, 054601 (2011).

Yang, J., G. P. Meisner, D. T. Morelli, and C. Uher, *Phys. Rev. B* **63**, 014410 (2000).

Yang, J., G. P. Meisner, W. Chen, J. S. Dyck, and C. Uher, *Proc. 20th Int. Conf. on Thermoelectrics*, IEEE Catalog Number 01TH8589, Piscataway, NJ, p.73 (2001).

Yang, J., D. T. Morelli, G. P. Meisner, W. Chen, J. S. Dyck, and C. Uher, *Phys. Rev. B* **65**, 094115 (2002).

Yang, J., D. T. Morelli, G. P. Meisner, W. Chen, J. S. Dyck, and C. Uher, *Phys. Rev. B* **67**, 165207 (2003).

Yang, J., G. P. Meisner, and L. D. Chen, *Appl. Phys. Lett.* **85**, 1140 (2004a).

Yang, J., W. Zhang, S. Q. Bai, Z. Mei, and L. D. Chen, *Appl. Phys. Lett.* **90**, 192111 (2007).

Yang, J., Q. Hao, H. Wang, Y. C. Lan, Q. Y. He, A. Minnich, D. Z. Wang, J. A. Harriman, V. M. Varki, M. S. Dresselhaus, G. Chen, and Z. F. Ren, *Phys. Rev. B* **80**, 115329 (2009).

Yang, J., Y. C. Wang, H. L. Yang, W. Tang, J. Yang, L. D. Chen, and W. Q. Zhang, *J. Phys.: Condens. Matter* **31**, 183002 (2019).

Yang, J., J. S. Wu, and L. T. Zhang, *J. Alloys Compd.* **364**, 83 (2004b).

Yang, L., H. H. Hng, D. Li, Q. Y. Yan, J. Ma, T. J. Zhu, X. B. Zhao, and H. Huang, *J. Appl. Phys.* **106**, 013705 (2009).

Yu, B. L., X. F. Tang, Q. J. Zhang, T. X. Liu, P. F. Luo, and J. Wang, *Proc. 22nd Inter. Conf. on Thermoelectrics*, IEEE Catalog Number 03TH8726, Piscataway, NJ, p. 101 (2003).

Zevalkink, A., K. Star, U. Aydemir, G. J. Snyder, J.-P. Fleurial, S. Bux, T. Vo, and P. von Allmen, *J. Appl. Phys.* **118**, 035107 (2015).

Zhang, L., A. Grytsiv, P. Rogl, E. Bauer, and M. Zehetbauer, *J. Phys. D: Appl. Phys.* **42**, 225405 (2009).

Zhang, L., F. F. Duan, X. D. Li, X. L. Yan, W. T. Hu, L. M. Wang, Z. Y. Liu, Y. J. Tian, and B. Xu, *J. Appl. Phys.* **114**, 083715 (2013).

Zhang, X., Q. M. Lu, J. X. Zhang, Q. Wei, D. M. Liu, and Y. Q. Liu, *J. Alloys Compd.* **457**, 368 (2008).

Zhao, W. Y., C. L. Dong, P. Wei, W. Guan, L. S. Liu, P. C. Zhai, X. F. Tang, and Q. J. Zhang, *J. Appl. Phys.* **102**, 113708 (2007).

Zhao, W. Y., P. Wei, Q. J. Zhang, C. L. Dong, L. S. Liu, and X. F. Tang, *J. Amer. Chem. Soc.* **131**, 3713 (2009).

Zhao, X. Y., X. Shi, L. D. Chen, W. Q. Zhang, S. Q. Bai, Y. Z. Pei, X. Y. Li, and T. Goto, *Appl. Phys. Lett.* **89**, 092121 (2006).

Zhou, C., D. T. Morelli, X. Y. Zhou, G. Wang, and C. Uher, *Intermetallics* **19**, 1390 (2011).

Zhou, L., P. F. Qiu, C. Uher, X. Shi, and L. D. Chen, *Intermetallics* **32**, 209 (2013).

Zhou, X. Y., G. W. Wang, L. J. Guo, H. Chi, G. Y. Wang, Q. F. Zhang, C. Q. Chen, T. Thompson, J. Sakamoto, V. P. Dravid, G. Z. Cao, and C. Uher, *J. Mater. Chem. A* **2**, 20629 (2014).

Ziman, J. M., Philos. Mag. 1, 191 (1956).

Ziman, J. M., Philos. Mag. 2, 292 (1957).

Zink, B., R. Pietri, and F. Hellman, *Phys. Rev. Lett.* **96**, 055902 (2006).

Zobrina, B. N. and L. D. Dudkin, *Sov. Phys.-Solid State* **1**, 1668 (1960).

6 Thermoelectric Properties of Skutterudites

6.1 INTRODUCTION

Chapter 4 and 5 introduced and discussed fundamental electronic properties and lattice dynamics of various forms of skutterudites. With that as a background, we are now in a position to consider in detail the performance of skutterudites as viable thermoelectric (TE) materials.

A vast number of reports has accumulated over the period of some 25 years since skutterudites became the major focus of interest. The promulgation of the concept of phonon-glass-electron-crystal (PGEC) by Slack (1995) kickstarted the TE research in skutterudites. The relevance of the PGEC concept to skutterudites was documented in measurements of the thermal conductivity by Morelli and Meisner (1995). I will not be able to reference all publications, but I will address those that made a major impact on the development of the field.

The chapter will start with the description of the three phenomena underpinning the field of thermoelectricity: the Seebeck, Peltier, and Thomson effects. Next will be presented formulae for the efficiency of thermoelectric conversion, which will lead to the definition of the thermoelectric figure of merit Z (and its dimensionless form ZT, where T is the absolute temperature). Then, we consider optimization of the thermoelectric performance by using the Chalmers-Stratton equation. With the concept of weighted mobility, the discussion will zero in on several phenomena that maximize the carrier mobility yet impede phonon transport to the maximum extent.

Beyond outstanding thermoelectric performance, a practical thermoelectric power generator must also withstand the harsh environmental conditions of cycling over a wide range of temperatures, show excellent resistance to oxidation, and possess robust mechanical properties. The relevance of the above aspects to skutterudites is also covered in this chapter.

The discussion will focus chiefly on filled skutterudites and composite skutterudite structures, as they offer the best prospect for achieving superior thermoelectric performance. Several examples of skutterudites attaining ZT values well in excess of unity (some approaching $ZT \sim 2$) will be presented, and prototype skutterudite-based thermoelectric modules will be illustrated. The chapter will conclude with remarks on future possible improvements of the skutterudite-based thermoelectric materials.

6.2 THERMOELECTRIC PHENOMENA

The discovery of thermoelectricity, or at least its first manifestation, is credited to Thomas Seebeck, who in 1821 observed a voltage developed by heating a junction of a couple made from Bi and Sb 'wires' while the other ends were kept at room temperature. Although Seebeck (1822) completely misunderstood the physical origin of what he observed, he was, nevertheless, an excellent and tenacious observer and documented the effect that carries his name on a large number of different junctions.

A schematic of the Seebeck effect is illustrated in Figure 6.1. A junction is formed of two dissimilar conductors, a and b, and is held at temperature T_1. The free ends of the two conductors are at temperature T_2 (could be room temperature) and are connected by wires (e.g., copper) to a high impedance voltmeter. The imbalance in electron (or hole) concentration created by the junctions being at different temperatures, assume $T_1 > T_2$, leads to diffusion of the carriers from the junction to the free ends of the conductors. This gives rise to the electric field that opposes further diffusion,

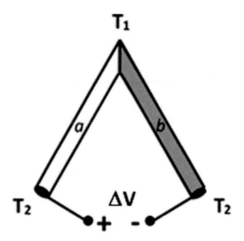

FIGURE 6.1   Schematics of the Seebeck effect. Conductors *a* and *b* form a junction maintained at temperature T_J. The other ends of the conductors are held at temperature T_2 and are connected by wires to a high
impedance voltmeter (essentially an open circuit situation). The voltmeter detects a voltage ΔV proportional
to the temperature difference $\Delta V \sim T_1 - T_2$. For the voltage polarity shown and with $T_1 > T_2$, then $S_a > S_b$.

and a steady state is established with the voltage difference ΔV. The higher the temperature difference between the junction and the free ends of the conductors, the higher the voltage difference.
Under open circuit conditions, the Seebeck coefficient is then

$$S_{ab} = \frac{\Delta V}{\Delta T}, \text{ where } \Delta T = T_1 - T_2. \tag{6.1}$$

The sign of the Seebeck coefficient is determined by the sign of the dominant charge carrier: negative for electrons and positive for holes.

It is essential to understand that the appearance of the Seebeck voltage requires a junction of two
conductors, hence the subscript *ab*. One can, nevertheless, express this differential Seebeck coefficient in terms of the absolute Seebeck coefficients S_a and S_b of the two conductors as $S_{ab} = S_a - S_b$.
Absolute values of the Seebeck coefficients can be conveniently determined by making a couple
between the material of interest and a superconducting material operating at temperatures below
its superconducting transition temperature, i.e., in the regime where every superconductor has zero
Seebeck coefficient, as it must. An example of establishing the absolute Seebeck coefficient of coo-
per wires against a high-Tc superconducting material was described by Uher (1987).

Thirteen years after the discovery by Seebeck, another outstanding amateur experimenter, Jean
Peltier (1834), reported on a complimentary phenomenon, now referred to as the Peltier effect. In
this case, a passage of the current through a circuit formed by two dissimilar conductors, Figure 6.2,
leads to the junction absorbing the heat (junction cools) or evolving the heat (junction heats). Upon
reversal of the current direction, the heating/cooling at the junction is reversed. The reversible heating has nothing to do with the more familiar and dissipative effect of Joule heat that occurs in any
conductor with finite resistance. The amount of heat, *Q*, absorbed or released in the Peltier effect is
proportional to the current *I* passed through the circuit and depends on the materials of the junction.
Again, the observation of the Peltier effect necessitates the presence of a junction of two conductors,
and such differential Peltier coefficient is designated as Π_{ab}. The heat absorbed or evolved is then

$$Q = \Pi_{ab} I = (\Pi_a - \Pi_b) I, \tag{6.2}$$

where Π_a and Π_b are the absolute Peltier coefficients of the two junction materials.

Evolution or absorption of heat Q
at the junction

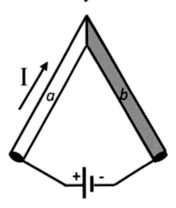

FIGURE 6.2 Schematic representation of the Peltier effect. A direct current source is connected by wires to the free ends of two dissimilar conductors a and b that form a junction. Upon passing a direct current through the circuit, absorption (junction cools) or release of heat (junction heats) takes place at the junction, depending on the direction of the current. With the current direction as shown, the Peltier heat is taken as positive, $\Pi_{ab} = \Pi_a - \Pi_a > 0$, if the heat evolves at the junction. Heating or cooling of the junction is reversible and has nothing to do with the Joule heat that evolves in any conductor having a finite resistance.

William Thomson, one of the fathers of thermodynamics, was much interested in the observations of Seebeck and Peltier and tried to find a thermodynamic connection between the two. Realizing that this is not possible unless there is a third effect tying the Peltier and Seebeck phenomena, he conjectured what we now call the Thomson effect (Thomson 1851). Unlike the Seebeck and Peltier effects, which require the participation of two dissimilar conductors, the Thomson effect is a single conductor effect. It arises when a conductor carries current while subjected to a thermal gradient, a situation depicted in Figure 6.3.

The heat evolved or absorbed along the length of a conductor is proportional to the current the conductor carries and to the imposed thermal gradient along the conductor. The proportionality constant is the Thomson coefficient τ,

$$\frac{dQ}{dx} = \tau I \frac{dT}{dx}.$$ (6.3)

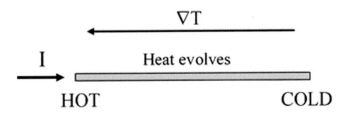

FIGURE 6.3 Schematic description of a single conductor generating Thompson heat when a current I is passed through the conductor while there is also temperature gradient imposed over the conductor. The Thomson coefficient is positive (situation depicted) when the heat evolves upon the current passes from the hot to the cold end of the conductor, and when the heat is absorbed when the current passes from the cold to the hot end.

The sign of the Thomson coefficient is positive when the heat evolves as the current flows from the hot to the cold end (against the thermal gradient) of the wire. With this additional thermoelectric coefficient, Thomson (1851) was able to link all three effects via two important equations known as the Kelvin relations (the same person, Thomson, was elevated to the peerage as Lord Kelvin),

$$\Pi = ST \tag{6.4}$$

and

$$\tau = T \frac{dS}{dT}. \tag{6.5}$$

Because the Thomson effect is directly measurable for a single conductor, while a junction of two dissimilar conductors is required to observe both the Seebeck and Peltier effects, it is the measurement of the Thomson effect that allows, via the second Kelvin relation, the determination of the absolute Seebeck and Peltier coefficients, as done originally by Borelius (1930). For a more recent account of such measurements, see Roberts (1977).

6.3   OPERATION OF A THERMOELECTRIC ENERGY CONVERTER

From the form of the differential Seebeck coefficient, $S_{ab} = S_a - S_b$, it follows that a couple made from materials that have the same sign of charge carriers, e.g., both electron conductors, will not generate a large Seebeck voltage. Thus, to realize a substantial Seebeck voltage, the couple should combine an n-type and a p-type material. The simplest couple is made from rectangular or cylindrical shapes of n-type and p-type semiconductors, illustrated in Figure 6.4. The left-hand panel shows a power generator, while the right-hand panel depicts a thermoelectric cooler. While efficiencies for both thermoelectric energy conversion processes will be given, the emphasis will be on power generation as skutterudites perform best in the temperature interval between 500 K and 900 K and are thus candidates for thermoelectric power generators.

The initial assessment of efficiency of a thermoelectric cooler was made by Altenkirch (1911) more than 100 years ago, who considered metals, and concluded that the conversion efficiency is grossly inadequate. He certainly made the right conclusion for the wrong candidate material and this threw cold water on the progress of thermoelectricity for the next 40 years. Thermoelectricity started to gain attention with the advent of semiconductors during the early 1950s. Abram Ioffe's team paved the way for practical applications of thermoelectricity by rederiving the formulae for

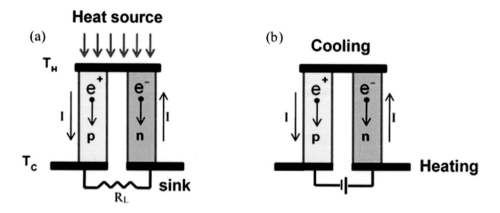

FIGURE 6.4 Operations of a thermoelectric couple as a) a power generator and b) a thermoelectric cooler. The couple combines an n-type thermoelectric element (leg) with a p-type element in the Π-like configuration, i.e., connected electrically in series while the heat passes through them in parallel.

conversion efficiency, introducing the thermoelectric figure of merit in the form $Z = \dfrac{S^2 \sigma}{k_L}$, and

showing that semiconductors are promising thermoelectric materials, particularly if they have large mean atomic weight (Ioffe 1957). Julian Goldsmid's realization that the important parameter identifying a good thermoelectric material is a high ratio of carrier mobility to thermal conductivity, led him to identify and explore Bi_2Te_3 (Goldsmid and Douglas 1954), which has become the mainstay material for thermoelectric applications ever since.

Referring to Figure 6.4(a), an external heat source supplies heat to the hot junction at temperature T_H and creates a temperature difference $\Delta T = T_H - T_C$ across the couple. The cold junction (thermal sink) is maintained at a fixed temperature T_C. The imposed temperature difference gives rise to the Seebeck voltage (electromotive force) of magnitude $(S_p - S_n)(T_H - T_C)$. Given electrical resistance of the couple $R = R_n + R_p$ and a load resistor R_L connected across the thermocouple legs at the cold junction, the electrical current flowing around the circuit is

$$I = \frac{(S_p - S_n)(T_H - T_C)}{R + R_L}. \tag{6.6}$$

The above current flows through the load resistor R_L, generates power P of

$$P = I^2 R_L = \frac{(S_p - S_n)^2 (T_H - T_C)^2}{(R + R_L)^2} R_L, \tag{6.7}$$

and facilitates a direct conversion of heat to electricity. The conversion efficiency η of the process is defined as the power P developed at the load per heat, Q_H, absorbed at the hot junction. In the derivation of the conversion efficiency, we make simplifying assumptions that there are neither thermal nor electrical resistances at the ends of the thermoelements (at the heat source and heat sink), that the heat flows only through the p- and n-legs of the couple (no convective and radiative losses), and no Thompson heat is generated in the legs, i.e., the transport parameters are temperature-independent. The Peltier heat developed at the top junction (heat source), $(S_p - S_n)T_H I$, flows down to the sink in the same direction as the heat withdrawn from the hot junction by thermal conductivity, $K(T_H - T_C)$, where $K = K_n + K_p$ is the thermal conductance of the couple. Both heats are taken as positive. The diffusion of electrons and holes from the heat source to the sink represents the electrical current that passes through the legs of the combined resistance R and generates Joule heat. One-half of this heat deposits at the hot junction, i.e., the amount of heat $-I^2 R / 2$ tends to support the temperature difference across the couple. Thus, accounting for all the contributions, the heat absorbed at the hot junction is

$$Q_H = (S_p - S_n)T_H I + K(T_H - T_C) - I^2 R / 2. \tag{6.8}$$

Consequently, the conversion efficiency of a TE generator is given by

$$\eta = \frac{I^2 R_L}{(S_p - S_n)T_H I + K(T_H - T_C) - I^2 R / 2}$$

$$= \frac{\dfrac{(S_p - S_n)^2 (T_H - T_C)^2}{(R + R_L)^2} R_L}{(S_p - S_n)T_H I + K(T_H - T_C) - I^2 R / 2}. \tag{6.9}$$

In the case of a thermoelectric cooler, it is the coefficient of performance (COP), ϕ, defined as the heat extracted from the source per electrical energy expended to do so, which characterizes the efficiency of cooling. Similar careful accounting for various heats involved, see, e.g., Goldsmid (1986), leads to the COP of

$$\phi = \frac{\left(S_p - S_n\right)T_C I - K\left(T_H - T_C\right) - I^2 R / 2}{\left(S_p - S_n\right)\left(T_H - T_C\right)I + I^2 R}.$$

(6.10)

Back to the generators, it follows from Eq. 6.9 that the efficiency depends on the ratio of the load resistance and the resistance of couple's legs. Alternatively, the efficiency is a function of the current, and two distinct situations are usually considered: the current I_p that delivers the maximum power, and the current I_n with which the couple operates most efficiently.

6.3.1 THERMOELECTRIC GENERATOR OPERATING WITH THE MAXIMUM POWER

A device will deliver maximum power, P_{max}, when the load resistance R_L equals the internal resistance, in our case the resistance of thermocouple legs R. Thus, with $R_L = R$, we obtain from Eq. 6.6,

$$I_P = \frac{\left(S_p - S_n\right)\left(T_H - T_C\right)}{2R},$$

(6.11)

$$P_{max} = \frac{\left(S_p - S_n\right)^2 \left(T_H - T_C\right)^2}{4R},$$

(6.12)

and the efficiency becomes

$$\eta_P = \frac{\dfrac{\left(S_p - S_n\right)^2 \left(T_H - T_C\right)^2}{4R}}{\dfrac{\left(S_p - S_n\right)^2 \left(T_H - T_C\right)T_H}{2R} + K\left(T_H - T_C\right) - \dfrac{\left(S_p - S_n\right)^2 \left(T_H - T_C\right)^2}{2R}}$$

(6.13)

$$= \frac{Z\left(T_H - T_C\right)}{4 + Z\left(T_M + T_H\right)} = \frac{1}{2} \frac{\left(T_H - T_C\right)}{T_H + \dfrac{2}{Z} - \dfrac{\left(T_H - T_C\right)}{4}}.$$

The efficiency of operation under the maximum power output in Eq. 6.13 takes a neater form in the second expression, while the last equality was derived originally by Ioffe (1957). In the two expressions, the parameter Z stands for

$$Z = \frac{\left(S_p - S_n\right)^2}{KR},$$

(6.14)

and is called the *thermoelectric figure of merit*. K and R are the electrical resistance and the thermal conductance of the couple, respectively. The mean temperature T_M in Eq. 6.13 is given by $T_M = \left(T_H + T_C\right)/2$.

Assuming the couple is made of homogeneous p- and n-type legs having cross sections of A_p and A_n, and lengths L_p and L_n, their serial electrical resistance is

$$R = R_p + R_n = \frac{\rho_p L_p}{A_p} + \frac{\rho_n L_n}{A_n},$$

(6.15)

where ρ_p and ρ_n are the respective electrical resistivities of the p- and n-type material. Similarly, because the p- and n-type legs operate thermally in parallel, the overall thermal conductance of the couple is

$$K = K_p + K_n = \frac{\kappa_p A_p}{L_p} + \frac{\kappa_n A_n}{L_n}.$$

(6.16)

where κ_p and κ_n are the thermal conductivities of the p- and n-type material. The product KR in the denominator of Eq. 6.14 is thus

$$KR = \left(\frac{\kappa_p A_p}{L_p} + \frac{\kappa_n A_n}{L_n} \right) \left(\frac{\rho_p L_p}{A_p} + \frac{\rho_n L_n}{A_n} \right)$$

$$= \kappa_p \rho_p + \kappa_n \rho_n + \kappa_p \rho_n \frac{A_p L_n}{A_n L_p} + \kappa_n \rho_p \frac{A_n L_p}{A_p L_n}.$$

(6.17)

To get maximum efficiency, Eq. 6.13 dictates that the thermoelectric figure of merit, Z, should be maximized., i.e., the product KR in the denominator of Eq. 6.14 should be minimized. Differentiating Eq. 6.17 with respect to either the cross-sectional area or the length of the legs, and setting the expression equal to zero, yields

$$\frac{A_p L_n}{A_n L_p} = \left(\frac{\kappa_n \rho_p}{\kappa_p \rho_n} \right)^{1/2}.$$

(6.18)

Substituting this optimal geometrical factor back into Eq. 6.14, the figure of merit becomes

$$Z = \frac{(S_p - S_n)^2}{KR} = \frac{(S_p - S_n)^2}{\left[(\kappa_p \rho_p)^{1/2} + (\kappa_n \rho_n)^{1/2} \right]^2} = \frac{(S_p - S_n)^2}{\left[\left(\frac{\kappa_p}{\sigma_p} \right)^{1/2} + \left(\frac{\kappa_n}{\sigma_n} \right)^{1/2} \right]^2},$$

(6.19)

where, in the last term, electrical resistivities were replaced by electrical conductivities. Strictly speaking, p- and n-type legs should be chosen to maximize Z in Eq. 6.19 rather than simply taking the best p-type and the best n-type materials. However, because often the best p- and n-type materials have similar figures of merit, it is convenient to define a figure of merit for a single element or leg, be it a p-type or an n-type conductor, as

$$Z_{p,n} = \frac{S_{p,n}^2}{\kappa_{p,n} \rho_{p,n}} \equiv \frac{S_{p,n}^2 \sigma_{p,n}}{\kappa_{p,n}}.$$

(6.20)

The true figure of merit Z for the couple is then approximately equal to the average of Z_p and Z_n. The product $S^2\sigma$ in the numerator of the figure of merit contains information about electronic properties of a thermoelectric material and is usually called the material's *power factor*.

6.3.2 Thermoelectric Generator Operating with the Maximum Efficiency

In this case, we need to find the load resistance R_L that will maximize the conversion efficiency. Expressing the conversion efficiency in terms of the ratio R_L/R, differentiating the expression with respect to R_L/R, setting it equal to zero, the form of R_L in terms of R can be found. Proceeding in steps, Eq. 6.9 is written as

$$\eta = \frac{(T_H - T_C) \dfrac{R_L}{R}}{T_M + T_H \dfrac{R_L}{R} + \dfrac{\left(1 + {R_L}/{R} \right)^2}{Z}}.$$

(6.21)

Differentiating with respect to R_L/R, and setting the result to zero, one obtains

$$R_L = R\sqrt{ZT_M + 1}. \qquad (6.22)$$

Substituting Eq. 6.22 back, the electrical current, the power delivered to the load, and the maximum conversion efficiency become

$$I_\eta = \frac{(S_p - S_n)(T_H - T_C)}{R(\sqrt{ZT_M + 1} + 1)^2}, \qquad (6.23)$$

$$P_{I_\eta} = \frac{(S_p - S_n)^2 (T_H - T_C)^2 \sqrt{ZT_M + 1}}{R(\sqrt{ZT_M + 1} + 1)^2}, \qquad (6.24)$$

$$\eta_{max} = \frac{T_H - T_C}{T_H} \frac{\sqrt{ZT_M + 1} - 1}{\sqrt{ZT_M + 1} + \dfrac{T_C}{T_H}}. \qquad (6.25)$$

The first term in Eq. 6.25 is simply the thermodynamic efficiency of a reversible Carnot cycle. The term becomes modified (decreased) due to the irreversible nature of the heat flow along the length of the TE elements and the Joule heat dissipated in them. It thus depends primarily on the material factor Z (figure of merit) and the mean temperature of operation T_M. Again, the figure of merit appearing in Eqs. 6.23–6.25 is the figure of merit for the couple and, to a first approximation, one would average individual figures of merit Z_p and Z_n as described by Eq. 6.20.

All formulae describing the efficiency of a TE power generator derived in this section assumed no generation of the Thomson heat within the body of the thermoelectric elements. In other words, the transport coefficients entering in the figure of merit, and most notably the Seebeck coefficient (see Eq. 6.5), were assumed temperature-independent. The neglect of the Thomson heat in the overall energy balance equation represents a significant simplification of the problem, and it might lead to either an overestimation or underestimation of the actual efficiency of a TE device, depending on the sign of the Thomson coefficient. Such simplification is widely accepted primarily to keep the equations tractable. Several attempts have been made in the past to take the Thomson heat into account, and they span from full modeling of TE generators in the presence of the Thomson heat (Sherman et al. 1960, Kim et al. 2015, Zhang et al. 2017), to including the Thomson effect but taking it as constant at the mean temperature $T_M = (T_c + T_h)/2$ (Sunderland and Burak 1964, Lampinen 1991), to an operation of TE devices under large temperature differences (Min et al. 2004), and to more tractable treatments where either all transport coefficients were assumed linearly dependent on the temperature (Yamashita 2008, Zabrocki et al. 2011), or just the Seebeck coefficient was taken as temperature-dependent while the resistivity and the thermal conductivity were kept constant (Chen et al. 1996). Two complementary scenarios were considered by Mahan (1991): In the first one, the dominant temperature dependence came from the T^{-1}- varying thermal conductivity while the resistivity and the Seebeck coefficient were constant. In the second one, the Seebeck coefficient and the thermal conductivity were related to the electrical conductivity and, thus, varying the latter across the length of the legs, automatically took care of the variation of $S[\sigma(x)]$ and $\kappa[\sigma(x)]$. Whatever corrections in the conversion efficiency are made to account for the Thomson heat, in the case of power generators their effect is considerably smaller than in the case of TE coolers. The reason is that, in the case of TE generators, the Peltier heat and the heat conducted via TE legs (Fourier heat) have the same sign (both flow from the hot junction to the cold junction), and their sum is larger in comparison to the Joule and Thomson heats. In contrast, in the case of a TE cooler, the Peltier

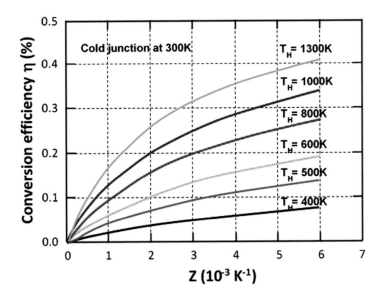

FIGURE 6.5 Conversion efficiency as a function of the figure of merit Z, based on Eq. 6.25, for different temperatures of the hot junction, with the cold junction kept at 300 K.

heat has an opposite sign to that of the Fourier heat, and the net cooling is the difference between the Peltier heat and the combined Fourier plus Joule heats. Against this relatively small net cooling, any contribution of the Thomson heat makes a significantly larger influence than in the case of power generators. Often, having available mathematically easily tractable and physically transparent expressions is more beneficial even at the expense of an error of several percent. However, for precise modeling of the performance of thermoelectric converters, such as commercial modules, Thomson heat should not be neglected.

Figure 6.5 shows the expected conversion efficiency as a function of the figure of merit for a thermoelectric generator operating over different temperature differences with the cold junction (sink) maintained at ambient temperature. It follows that with a temperature difference of 300 K (T_H = 600 K, T_C = 300 K) a thermoelectric couple with both p- and n-type elements having the average figure of merit of 2×10^{-3} K^{-1} across the temperature range (a stringent condition but within the realm of the best laboratory-developed materials), is predicted to convert heat to electricity with the efficiency of about 10%. If the same average figure of merit was maintained across the operational range of 500 K (T_H = 800 K, T_C = 300 K), the conversion efficiency would reach 15%. An alternative plot, providing the conversion efficiency as a function of the temperature of the hot junction T_H (keeping the cold junction at 300 K) for materials with different figures of merit is shown in Figure 6.6. Both figures convey the importance of having available p- as well as n-type thermoelectric materials with high figures of merit that are sustained over a broad temperature range. While not easy to achieve with any single thermoelectric element, one may combine different materials, each peaking its performance in different but partly overlapping temperature range to obtain excellent performance over the desired broad temperature range (see Figure 6.7). Such an approach is the essence of fabricating segmented thermoelectric couples sketched in the same figure.

Employing segmented legs, considerations must be given to the compatibility of the individual segments (Snyder and Ursell 2003). By compatibility, one understands how well the electrical and thermal flux densities of the two thermoelectric segments match. If there is a great disparity in the compatibility factor of the segments (Si–Ge alloys are particularly poor partners in relation to

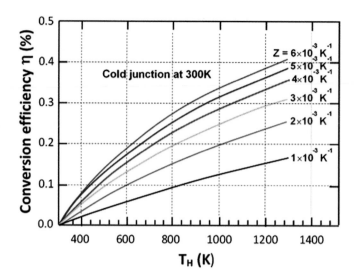

FIGURE 6.6 Conversion efficiency as a function of temperature of the hot junction for various values of the figure of merit, based on Eq. 6.25. The cold junction is kept at 300 K.

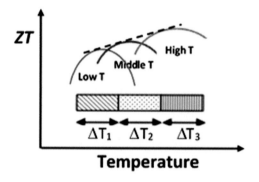

FIGURE 6.7 Averaged *ZT* of a composite thermoelectric element consisting of three material segments, each having its optimal performance in different but closely lying temperature ranges. The dashed line represents the overall *ZT* behavior. Practical realization of a high averaged *ZT* by fabricating a segmented couple from three sections, each operating efficiently over different temperature ranges designated by ΔT_1, ΔT_2, and ΔT_3, is also sketched.

the other thermoelectric materials), they cannot simultaneously operate efficiently, and the actual efficiency of the segmented leg could be even lower than if a single element was used. The compatibility factor s is defined as

$$s = \frac{\sqrt{1+zT}-1}{ST},\tag{6.26}$$

where S in the denominator is the average Seebeck coefficient. Snyder (2006) has computed the compatibility factor for several n-and p-type thermoelectric materials to 1000°C, and the data are plotted in Figures 6.8(a) and 6.8(b). The results vividly document that Si–Ge alloys, in both n- and

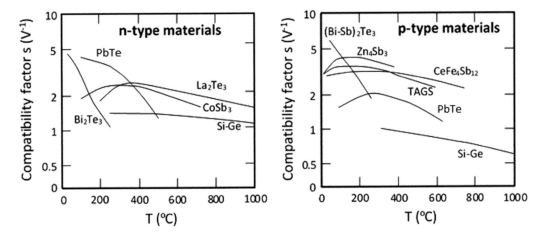

FIGURE 6.8 Compatibility factors of (a) several n-type thermoelectric materials, and (b) some p-type materials. Note a significantly different compatibility factor of Si–Ge alloys, particularly in its p-type form, compare to other thermoelectric materials. On the other hand, skutterudites would make excellent n-type segmented couples with La$_2$Te$_3$ and p-type couples with Zn$_4$Sb$_3$ and, over a limited temperature range, with PbTe and (Bi–Sb)$_2$Te$_3$ alloys, respectively. Redrawn from the data of G. J. Snyder, *Thermoelectric Handbook, Macro to Nano*, ed. D. M. Rowe, CRC Press, Ch. 9 (2006). With permission from Taylor & Francis.

p-type forms, are not great choices for segmented couples with most other thermoelectrics, while skutterudites can couple effectively with Zn$_4$Sb$_3$ and La$_2$Te$_3$ and, over a limited temperature range, with PbTe and (Bi–Sb)$_2$Te$_3$ alloys.

6.4 OPTIMIZATION OF THE THERMOELECTRIC FIGURE OF MERIT

From the form of the power factor $S^2\sigma$ and the dimensionless thermoelectric figure of merit,

$$ZT = \frac{S^2\sigma}{\kappa}T,$$ it may be tempting to seek materials with the largest electrical conductivity, the largest

Seebeck coefficient, and the smallest thermal conductivity. Unfortunately, such a simplistic approach is futile. The reason is that the three transport parameters in question are strongly dependent on the carrier concentration, which plays a different role in each transport coefficient, making them negatively correlated, as shown in Figure 6.9. Thus, while large carrier concentrations characteristic of metals result in large electrical conductivities, such carrier concentrations lead to very small Seebeck coefficients. Moreover, large electrical conductivities inevitably give rise to large electronic thermal conductivities. On the other hand, poorly conducting structures, although they may offer very high Seebeck coefficients, will not perform well either because of grossly inadequate electrical conductivity. The contrasting and uncooperative trends among the electrical conductivity, Seebeck coefficient, and the electronic thermal conductivity make it challenging to design thermoelectric materials with the outstanding performance. One must aim at optimizing the transport parameters rather than maximizing them individually as this inevitably leads to a serious deterioration of the other two parameters.

It follows from Figure 6.9 that the useful domain of carrier concentrations for thermoelectricity lies in the range of 10^{18}–10^{21} cm^{-3}, i.e., the regime of degenerate semiconductors, and possibly also a select group of semimetals that have wide disparities in the mobility or effective mass of electrons

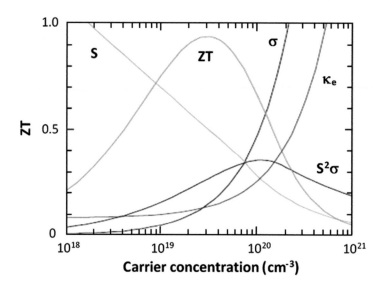

FIGURE 6.9 Schematic of the dependence of the electrical conductivity, Seebeck coefficient, and electronic thermal conductivity on the carrier concentration, which vividly illustrates challenges of designing and fabricating materials with an exceptionally high thermoelectric figure of merit. Reproduced from G. J. Snyder et al., *Nature Materials* 7, 105 (2008). With permission from Springer Nature.

and holes. In this case, the presence of minority carriers is less detrimental to the thermoelectric performance. It is also interesting to note in Figure 6.9 that the power factor typically peaks at a substantially higher carrier concentration than does the figure of merit. The reader should keep that in mind in cases where the premium is placed on maximizing the power factor rather than aiming at maximum values of ZT.

Early in the 1950s, Ioffe (1953) attempted to make a systematic assessment of the thermal conductivity of semiconductors and noted that low values, desirable for good thermoelectric materials, are found in structures having large mean atomic weights. Shortly after, Goldsmid (1954) explored the variation of the ratio of the carrier mobility and thermal conductivity, μ/κ, on the mean atomic weight. Assuming the mobility is under the influence of acoustic deformation potential scattering developed by Bardeen and Shockley (1950), and taking the Peierl's $1/T$ dependence of the thermal conductivity, the mobility to thermal conductivity ratio was expressed in terms of the ratio of the mean free path of electrons, ℓ_e, and the mean free path of phonons, ℓ_p, as

$$\frac{\mu}{\kappa} = \frac{4q\rho_m}{cv_s\left(2\pi\, m_e k_B T\right)^{1/2}}\frac{\ell_e}{\ell_p}. \tag{6.27}$$

Here, ρ_m is the mass density, c is the specific heat, v_s the sound velocity, and q and m_e are the charge and mass of an electron. Plotting the above ratio *versus* the mean atomic weight of several semiconductors, Figure 6.10, indicated a definite trend that inspired Goldsmid to consider and experimentally confirm Bi_2Te as an exceptional thermoelectric material, Goldsmid and Douglas (1954).

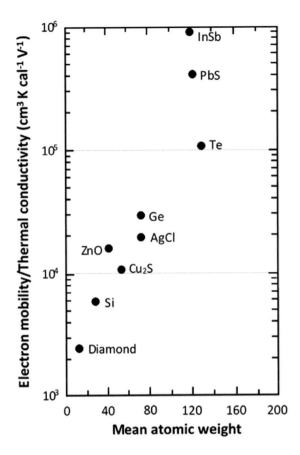

FIGURE 6.10 Ratio of the electron mobility to thermal conductivity plotted against the mean atomic weight for several semiconducting structures. Clearly, the 'heavy' semiconductors are favored over the 'light' semiconductors as potential thermoelectric materials. Redrawn from H. J. Goldsmid, *Proc. Phys. Soc. London B* 67, 360 (1954). With permission from the Institute of Physics.

6.4.1 OPTIMAL VALUE OF THE SEEBECK COEFFICIENT THAT MAXIMIZES ZT

As the Seebeck coefficient plays a pivotal role in the thermoelectric energy conversion, it is interesting to inquire what is the optimal value, S_{opt}, that maximizes the figure of merit. Formally, it means seeking an extreme value of ZT as a function of S, i.e., carrying out the first derivative of the figure of merit as a function of S and setting it equal to zero,

$$\frac{\partial}{\partial S}(ZT) \equiv \frac{\partial}{\partial S}\left[\frac{S^2 \sigma T}{\kappa_e + \kappa_L}\right]$$

$$= \frac{\left(\sigma LT + \kappa_L\right)\left[2S\sigma T + S^2 T \dfrac{\partial \sigma}{\partial S}\right] - S^2 \sigma T\left[LT \dfrac{\partial \sigma}{\partial S} + \sigma T \dfrac{\partial L}{\partial S}\right]}{\left(\sigma LT + \kappa_L\right)^2} = 0. \tag{6.28}$$

After simple algebra, and using $\kappa_e = \sigma TL$, one arrives at a formula for the optimal value of the Seebeck coefficient derived originally by Chasmar and Stratton (1959),

$$S_{opt}\frac{\partial ln\sigma}{\partial S} = -2\left[1+\frac{\kappa_e}{\kappa_L}\left(1-\frac{S}{2}\frac{\partial lnL}{\partial S}\right)\right].$$

(6.29)

It is essential that the derivatives in Eq. 6.29 are taken at the optimal value of the Seebeck coefficient S_{opt}. In the case of an arbitrary degeneracy, the identification of the optimal Seebeck coefficient from Eq. 6.29 is quite messy. In the nondegenerate case, it simplifies dramatically because the Lorenz number is a constant and its derivative is thus zero, and the $\partial ln\sigma / \partial S$ becomes

$$\frac{\partial ln\sigma}{\partial S} = \frac{\partial}{\partial S}\left[ln\sigma_0 + \eta\right] = \frac{\partial \eta}{\partial S} = -\frac{q}{k_B}.$$

(6.30)

Hence, the optimal value of the Seebeck coefficient that maximizes the figure of merit in the case of nondegenerate carriers is

$$S_{opt} = 2\frac{k_B}{q}\left[1+\frac{\kappa_e}{\kappa_L}\right] = 172.4\left[1+\frac{\kappa_e}{\kappa_L}\right]\mu VK^{-1}.$$

(6.31)

Again, the electronic thermal conductivity κ_e in Eq. 6.31 must be taken at the optimal position of the Fermi energy that yields the optimal magnitude of the Seebeck coefficient S_{opt}. Substituting this optimal value of the Seebeck coefficient back into the expression for the figure of merit, the maximum figure of merit becomes

$$Z_{max}T = \frac{S_{opt}^2\sigma}{\kappa}T = \frac{4\left(\frac{k_B}{q}\right)^2\left(1+\frac{\kappa_e}{\kappa_L}\right)^2\frac{\kappa_e}{L}}{\kappa_e+\kappa_L} = \frac{4}{r+\frac{5}{2}}\left(1+\frac{\kappa_e}{\kappa_L}\right)\frac{\kappa_e}{\kappa_L}$$

(6.32)

$$= \frac{4\left(\frac{k_B}{q}\right)^2}{L}\left(1+\frac{S_{opt}}{2\frac{k_B}{q}}\right)\frac{S_{opt}}{2\frac{k_B}{q}}.$$

Equation 6.32 provides a few different forms of the maximum figure of merit obtained by substitution of variables. It is clear that to achieve high values of the Seebeck coefficient in Eq. 6.31, and therefore high values of the figure of merit in Eq. 6.32, the electronic thermal conductivity cannot be too small. This statement should not be misinterpreted. We really do not want large κ_e, but rather very small κ_L, so that the electronic term would be a significant fraction of the lattice thermal conductivity, without being too large in the absolute sense. Taking a reasonable value of κ_e as one-half of the lattice thermal conductivity, the optimized value of the Seebeck coefficient would be 259 μVK^{-1}. Assuming the dominant scattering mechanism is acoustic phonon scattering, $r = -1/2$, the highest $Z_{max}T$ value turns out to be about 1.5. Assuming further that the lattice thermal conductivity is 1 Wm^{-1}K^{-1}, and the Lorenz number for nondegenerate carriers under acoustic phonon scattering is 1.5×10^{-8} V^{2}K^{-2}, we can backtrack the corresponding optimal electrical conductivity $\sigma_{opt} = \kappa_e/LT = \kappa_L/2LT \approx 1.1\times10^5\Omega^{-1}m^{-1} \equiv 1100$ Scm^{-1}. Maintaining acoustic phonon scattering, the corresponding reduced Fermi energy is $\eta \approx -1$. It is an interesting and somewhat surprising result as it places the Fermi level close to the band edge and makes the system partly degenerate, in spite of the fact that all along we assumed a nondegenerate carrier system and used the nondegenerate formulae throughout. However, such an outcome seems to be a general result obtained regardless of an optimization approach used, e.g., Goldsmid (1955), Rossi (1968), and Wood (1988).

The above exercise points out the magnitudes of several transport parameters that a promising thermoelectric material should possess, and the need to achieve as low lattice thermal conductivity as possible. However, what we really need to know is how heavily to dope the material so that the maximum figure of merit can be achieved. There are several strategies outlined in the literature, e.g., Goldsmid (1958), Ioffe and Stilbans (1959), Rossi (1968), each using different simplifying assumptions. The approach that seems most rigorous is due to Chasmar and Stratton (1959), discussed in the next section.

6.4.2 THE CHASMAR-STRATTON MATERIAL PARAMETER β

Let us again look at the form of the dimensionless figure of merit, this time considering an arbitrarily degenerate system. Introducing parameter δ and parameter Δ in the general expressions for the Seebeck coefficient and the Lorenz number of an arbitrarily degenerate carrier system (see, e.g., Appendix A1 in the online version of the book), the dimensionless figure of merit can be written as

$$ZT = \frac{S^2 \sigma}{\kappa_e + \kappa_L} T = \frac{\left(\frac{k_B}{q}\right)^2 (-\eta + \delta)^2 \sigma T}{L\sigma T + \kappa_L}$$

$$= \frac{(-\eta + \delta)^2}{\Delta + \dfrac{\kappa_L}{\left(\dfrac{k_B}{q}\right)^2 \sigma T}} = \frac{(-\eta + \delta)^2}{\Delta + \left[\beta \dfrac{2}{\sqrt{}} F_{1/2}(\eta)\right]^{-1}}. \tag{6.33}$$

For convenience of readers, I also include here the often-quoted nondegenerate expression,

$$ZT = \frac{\left(-\eta + r + \frac{5}{2}\right)^2}{\left(r + \frac{5}{2}\right) + \left(\beta e^\eta\right)^{-1}}. \tag{6.34}$$

In either case, the above expressions feature the parameter β, known as the Chasmar-Stratton *material parameter*.[*] Its form is

$$\beta = \left(\frac{k_B}{q}\right)^2 \frac{\sigma_0 T}{\kappa_L} = \left(\frac{k_B}{q}\right)^2 \frac{2q(k_B T)^{3/2} T}{(2\pi)^{3/2} \hbar^3} \frac{\mu \left(m_{DOS}^*\right)^{3/2}}{\kappa_L}$$

$$= 5.778 \times 10^{-6} \frac{N_v \mu}{\kappa_L} \left(\frac{m_b^*}{m_0}\right)^{3/2} T^{5/2}, \tag{6.35}$$

where σ_0 is the electrical conductivity at the band edge (setting $\eta = 0$), expressed as

$$\sigma_0 = n_0 q \mu = 2q\mu \left(\frac{2\pi m^* k_B T}{h^2}\right)^{3/2} = 778\mu \left(\frac{m^*}{m_0}\right)^{3/2} T^{3/2} \left(\Omega^{-1} m^{-1}\right), \tag{6.36}$$

and for the density of states effective mass was used Eq. 4.20, to cover cases when the material has N_v symmetrically equivalent energy valleys. Other forms of the parameter β are readily obtained by

---

[*] I should mention that a very similar analysis with slightly different variables was carried out the same year by Rittner (1959) but has not attained a track in subsequent studies.

substitutions, one of them highlighting the role of the deformation potential acoustic phonon scattering (low values desired) when the mobility from Eq. 4.46 for arbitrary degeneracy or from Eq. 4.47 for nondegenerate cases is substituted in Eq. 6.35.

The product of μ and $\left(m^*/m_e\right)^{3/2}$ is referred to as the *weighted mobility* and was introduced earlier by Goldsmid and Douglas (1954) as an important criterion by which to judge suitable thermoelectric materials. Values of the Chasmar-Stratton material parameter β at room temperature are typically smaller than 0.25, with the largest value of 0.245 for Bi_2Te_3. Material parameters of skutterudites are smaller, in the range 0.05–0.16 depending on whether binary or filled structures are concerned.

From the form of Eqs. 6.33 and 6.34, it is clear that for any scattering parameter, r, and any value of the reduced Fermi energy, η, the figure of merit increases with the increasing value of β. From Eq. 6.35, it also follows that the parameter β depends on the effective mass of the carriers m^*, the carrier mobility μ, and the lattice thermal conductivity κ_L. From the simple and compact form of Eq. 6.35, it immediately follows that what benefits thermoelectric energy conversion is a large weighted mobility and a low lattice thermal conductivity, the two parameters that are substantially independent of each other. However, the weighted mobility contains the carrier mobility and the effective mass, and it is difficult to maximize both simultaneously as heavy carriers cannot nimbly respond to an applied electric field. The relationship between the mobility and the effective mass is typically manifested in the form of the scattering mechanism discussed in Chapter 4.

Equation 6.33 expresses the functional dependence of the figure of merit on the reduced Fermi energy η for an arbitrary degenerate system. In principle, Eq. 6.33 can be differentiated with respect to η, set equal to zero, and the Fermi energy that maximizes the figure of merit can be obtained in a similar way as done in Eq. 6.28. For an arbitrary degenerate system, the algebra is quite messy, and it is much easier to extract the desired value of the reduced Fermi energy simply by inspecting the trend in a few plots of ZT vs. η, a trivial task these days with laptop computers. For the brave souls who wish to tackle the general problem analytically, they might like to follow the treatment by Wasscher et al. (1963). Again, taking the carrier system as nondegenerate, the task is much simplified, and the optimal value of the reduced Fermi energy, η_{opt}, is given by equation

$$e^{\eta_{pt}} = \frac{-\eta_{opt} + r + \frac{1}{2}}{2\beta\left(r + \frac{5}{2}\right)}. \tag{6.37}$$

Equation 6.37 is of the form $e^x = -Ax + B$ and, once solved numerically, provides the desired optimal position of the Fermi energy. Substituting such optimal value of η_{opt} back into Eq. 6.34, the maximum value of $Z_{max}T$ for the nondegenerate system in terms of the optimal value of the reduced Fermi energy is

$$\left(Z_{max}T\right)_{nd} = \frac{\left(-\eta_{opt} + r + \frac{5}{2}\right)^2}{r + \frac{5}{2} + 2\left(r + \frac{5}{2}\right)\left(-\eta_{opt} + r + \frac{1}{2}\right)^{-1}}. \tag{6.38}$$

As already noted, even in the best cases of Bi_2Te_3 and some chalcogenide semiconductors, the material parameter β does not exceed values of 0.25. In skutterudites, the parameter β is even smaller, in the range 0.05–0.16, Caillat et al. (1996a), Uher et al. (1997), Sales et al. (1997). Since the largest negative value that a scattering parameter r can have is −3/2, Eq. 6.37 implies that the optimal value of the reduced Fermi energy is not smaller than $\eta_{opt} = -1.15$ (in the case of skutterudites not smaller than about $\eta_{opt} = -1.1$). If one takes acoustic phonon scattering as a more realistic scattering mechanism above 300 K, the magnitude of the reduced Fermi energy would be even smaller, i.e., the position of the Fermi level is much closer to the band edge. In either case, we see again that, in spite of assuming nondegenerate charge carriers, the optimization procedure returns a partly degenerate system. Had the arbitrary degenerate formulae with the Fermi integrals been used, the Fermi level would have fallen even closer to the band edge. The lesson learned is that good thermoelectric materials are systems with partly degenerate charge carriers, as already hinted from Figure 6.9.

Equation 6.35 implies that the Chasmar-Stratton parameter β, and therefore the figure of merit, rises rapidly with the temperature. While this is so, the rise is nevertheless limited by the fact that an extrinsic semiconductor sooner or later reaches a temperature where intrinsic (across the band gap) excitations commence. Once this happens, the presence of minority charge carriers rapidly diminishes the magnitude of the Seebeck coefficient. Moreover, the generated electron-hole pairs bring into play an additional heat conducting mechanism, namely the bipolar flow, described in Chapter 4, which dramatically enhances the overall thermal conductivity and compromises the performance of a thermoelectric material (Davydov and Shmushkevich 1940, Price 1955, 1956).

A factor that determines whether a material still operates as an extrinsic semiconductor is the size of its band gap E_g. Various estimates, e.g., Chasmar and Stratton (1959), Goldsmid (1986), Sofo and Mahan (1994), Mahan (1997), Nolas et al. (2001), place this limit in the range of $E_g \sim$ (6–10) $k_B T$, depending on the prevailing mode of scattering but regardless of the band gap being direct or indirect. Clearly, thermoelectric materials with small band gaps cannot be used at too high temperatures. On the other hand, for semiconductors with a very large band gap, $E_g > 10\ k_B T$, the figure of merit no longer rises with the carrier concentration, but rapidly saturates and eventually falls (Mahan 1989). Thus, while the Chasmar-Stratton material parameter has played a guiding role in identifying promising thermoelectric materials, it had to be accompanied by considerations regarding the size of the band gap and the possible onset of bipolar thermal conduction. It would be convenient to have at hand a generalized material parameter, which would combine all of the above important attributes – the weighted mobility, the lattice thermal conductivity, and the band gap value – in a single form. Such a task was carried out recently by Liu et al. (2016), who derived a generalized parameter B, accomplishing just that. We shall return to this issue in the next section, after a two-band conduction is taken into account.

Let us look at how the thermoelectric figure of merit is modified in the presence of two types of carriers, taking specifically the case of majority electrons and minority holes, and considering arbitrary degeneracy. The first attempt at such analysis was made by Simon (1962), who also explored a possible upper limit to $Z_{max}T$ (no limit found). The treatment by Liu et al. (2016), to which I have already alluded, followed closely the analysis of Chasmar and Stratton, but employed an arbitrary degenerate two-carrier system.

Substituting the appropriate two-band forms of the transport parameters from Eqs. 4.11, 4.12, and 4.16 into the dimensionless figure of merit yields

$$ZT = \frac{\left(\dfrac{\sigma_1 S_1 + \sigma_2 S_2}{\sigma_1 + \sigma_2}\right)^2 (\sigma_1 + \sigma_2) T}{\kappa_1 + \kappa_2 + \dfrac{\sigma_1 \sigma_2}{\sigma_1 + \sigma_2}(S_2 - S_1)^2 T + \kappa_L}$$

$$= \frac{\dfrac{\left(\sigma_1 S_1 + \sigma_2 S_2\right)^2}{\sigma_1 + \sigma_2} T}{\sigma_1 L_1 T + \sigma_2 L_2 T + \dfrac{\sigma_1 \sigma_2}{\sigma_1 + \sigma_2}(S_2 - S_1)^2 T + \kappa_L}.$$

(6.39)

Introducing the ratio of electrical conductivities, $\gamma = \dfrac{\sigma_1}{\sigma_2}$, and substituting the appropriate formu-

lae for the electrical conductivity and the Seebeck coefficient under arbitrary degeneracy (see, e.g, Appendix A1 in the online version of the book), the following expression is obtained:

$$ZT = \dfrac{\dfrac{\left(\eta_1 - \delta_1 - \dfrac{\eta_2}{\gamma} + \dfrac{\delta_2}{\gamma}\right)^2}{1 + \dfrac{1}{\gamma}}}{\left(\dfrac{e}{k_B}\right)^2 \left(L_1 + \dfrac{L_2}{\gamma}\right) + \dfrac{1}{\gamma+1}\left(-\eta_1 - \eta_2 + \delta_1 + \delta_2\right)^2 + \dfrac{\kappa_L}{\left(\dfrac{k_B}{e}\right)^2 \sigma_1 T}}. \tag{6.40}$$

Taking the reduced band gap as

$$\eta_G = -\eta_1 - \eta_2, \tag{6.41}$$

see Figure 6.11, and writing the electrical conductivity σ_1 as $\sigma_1 = \sigma_0 \dfrac{F_{r+\frac{1}{2}}(\eta_1)}{\Gamma\left(r + \frac{3}{2}\right)}$, where σ_0 is the

conductivity at the band edge from Eq. 6.36, Eq. 6.41 is modified and becomes

$$ZT = \dfrac{\dfrac{\gamma}{\gamma+1}\left(\eta_1 - \delta_1 + \dfrac{\eta_1 + \eta_G + \delta_2}{\gamma}\right)^2}{\left(\dfrac{e}{k_B}\right)^2 \left(L_1 + \dfrac{L_2}{\gamma}\right) + \dfrac{1}{\gamma+1}\left(\eta_G + \delta_1 + \delta_2\right)^2 + \dfrac{\kappa_L}{\left(\dfrac{k_B}{e}\right)^2 \sigma_0 \dfrac{F_{r+\frac{1}{2}}(\eta_1)}{\Gamma\left(r + \frac{3}{2}\right)} T}}. \tag{6.42}$$

Defining a new generalized dimensionless material parameter B as

$$B = \dfrac{1}{k_B}\left(\dfrac{k_B}{q}\right)^2 \dfrac{\sigma_0}{\kappa_L} E_G, \tag{6.43}$$

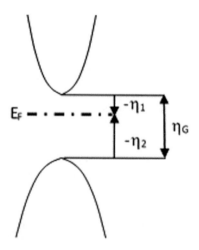

FIGURE 6.11 A sketch of a band of electrons and a band of holes with their respective reduced Fermi energies. The reduced band gap, η_G, can be written as $\eta_G = -\eta_1 - \eta_2$.

the final form of the figure of merit is

$$ZT = \frac{\dfrac{\gamma}{\gamma+1}\left(\eta_1 - \delta_1 + \dfrac{\eta_1 + \eta_G + \delta_2}{\gamma}\right)^2}{\left(\dfrac{e}{k_B}\right)^2\left(L_1 + {L_2}\!\big/\!{\gamma}\right) + \dfrac{1}{\gamma+1}\left(\eta_G + \delta_1 + \delta_2\right)^2 + \left(\dfrac{B}{\eta_G}\dfrac{F_{r+\frac{1}{2}}(\eta_1)}{\Gamma\left(r+\frac{3}{2}\right)}\right)^{-1}}.\tag{6.44}$$

In comparison to the material parameter β of Chasmar and Stratton that did not capture the dependence on the band gap and implied a continuously rising figure of merit with the rising temperature, the new generalized material parameter B explicitly contains the band gap E_G. Substituting numerical values of the constants appearing in Eq. 6.43, one can write the generalized material parameter as

$$B = 6.66 \times 10^{-2}\,\mu_1\left(\frac{m_1^*}{m_e}\right)^{\!3/2} T^{3/2}\,\frac{E_G\,(in\,eV)}{\kappa_L}.\tag{6.45}$$

Plots of ZT as a function of η_1, η_G, and B are shown in Figures 6.12(a) and 6.12(b).

The reader should be aware of a typo in the important paper by Liu et al. (2016) on which the above treatment is based, namely the parameter B (the authors' parameter B^*) is not dimensionless, as it incorrectly contains $\left(\dfrac{e}{k_B}\right)^2$ instead of $\left(\dfrac{k_B}{e}\right)^2$.

From the form of Eqs. 6.33 and 6.44, it follows that the thermoelectric figure of merit depends on three parameters: the scattering factor, r, the material parameter β (or parameter B if the width of the band gap is considered as a limiting factor), and on the reduced Fermi energy, η. Of the three parameters, the reduced Fermi energy is the easiest to adjust by doping (the scattering factor and the parameter β or B are substantially immune to variations of the carrier concentration). Thus, doping, including filling of structural voids in skutterudites, which is an exceptionally effective mode of doping and results in dramatic changes in the carrier concentration, is the primary knob available to an experimenter to alter and fine tune the electronic properties of thermoelectric materials.

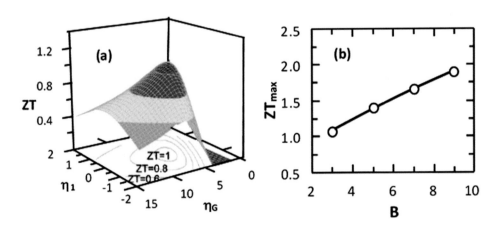

FIGURE 6.12 Dimensionless figure of merit ZT plotted as a function of (a) the reduced Fermi energy of electrons, η_1, and reduced band gap, η_G, with $B = 3$; and (b) the maximum ZT value as function of the material parameter B. Modified from W. Liu et al., *Energy & Environmental Science* 9, 530 (2016). With permission from the Royal Society of Chemistry.

6.4.3 BAND ENGINEERING TO ENHANCE THERMOELECTRIC PERFORMANCE

Effective ways to minimize the thermal conductivity of prospective thermoelectric materials have been discussed in Chapter 5. Here, in the next couple of sections, will be presented interesting and important modifications of the band structure that have proved effective in enhancing the electronic properties, i.e., the power factor, by partly breaking the counterproductive link between the electrical conductivity and the Seebeck coefficient. The focus will be on the multivalley structure of a semiconductor; on attempts to line up the band edges, referred to as band convergence; on a possibility of dopants inducing resonant states within the conduction or valence band of the host semiconductor; and on the effect of energy filtering that enhances the average charge carrier energy and thus enhances the Seebeck coefficient. The above four processes are examples of what is referred to as band engineering. Although the approaches are very effective when successfully implemented, their realization requires to satisfy quite stringent conditions.

6.4.3.1 Multivalley Semiconductors

The expression for the Seebeck coefficient in Eq. 4.52 depends on the effective mass, m^*. The effective mass in question is actually the density-of-states effective mass m_b^* (the band mass) that enters via the density of charge carriers. In general, semiconductors possessing large m_b^* will have large Seebeck coefficients. However, this does not automatically mean that such semiconductors will exhibit large power factors and large ZTs, because such heavy charge carriers would necessarily have low mobilities and thus low electrical conductivity. As discussed in Section 4.4, in order for the charge carriers to have high mobility, it is the inertial (transport) mass, m_i^*, which reflects the curvature of energy bands, that must be low. The problem is that, although somewhat different in cases of anisotropic semiconductors, the band mass and the inertial mass are closely related and, in the case of isotropic and parabolic bands, they are identical. Thus, the Seebeck coefficient and the electrical conductivity impose contrasting requirements on the effective mass.

A promising approach to resolve the dilemma concerning the effective carrier mass is to seek and identify semiconductors with a large number of symmetrically related carrier pockets (valleys) that have a small inertial mass along at least one direction. In this case, the light charge carriers assure high mobilities, while the effective mass entering the Seebeck coefficient is given by $m_{DOS}^* = N_v^{2/3} \left(m_1^* m_2^* m_3^* \right)^{1/3}$, where masses m_i^* are inertial masses in principal directions of an ellipsoid that contains the carriers. The term N_v is the valley degeneracy, which reflects the number of symmetrically related carrier pockets. In the case of an isotropic semiconductor, the density-of-states effective mass is simply $N_v^{2/3} m_t^*$, i.e., enhanced by a factor $N_v^{2/3}$. The enhanced band mass supports a large Seebeck coefficient with no adverse effect on the electrical conductivity. Large numbers of symmetrically related valleys are typically found in high symmetry structures, including skutterudites. Such 'valley engineering' works well provided the intervalley scattering is weak and does not limit the carrier mobility.

6.4.3.2 Band Convergence

Band convergence is a band engineering approach aiming at achieving energy coincidence, or at least closeness within a few $k_B T$ of energy, of band edges of two or more bands by chemical manipulation of the composition, by applied stress, or by changes in the temperature. Originally envisaged and applied to low-dimensional superlattice structures (Koga et al. 1998), the idea was to rely on carrier confinement and tune the superlattice parameters (layer thickness, superlattice period, composition) to achieve energy coincidence of the band edges and thus increase the valley degeneracy. In the context of bulk thermoelectrics, the band convergence was demonstrated by Pei et al. (2011) with p-type $PbTe_{1-x}Se_x$, where doping and the formation of solid solutions led to the convergence of L-point and Σ-point valence bands (Figure 6.13(a)). Since the normally inaccessible lower lying Σ band has valley degeneracy of 12, while the higher lying L-point valence band only 4, at some particular composition and temperature the descending band edge of the L-point band will align with

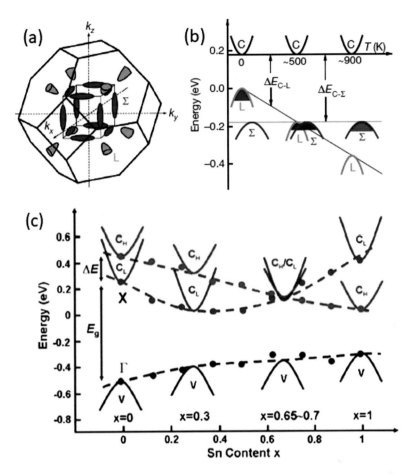

FIGURE 6.13 (a) Brillouin zone displaying the low degeneracy hole pockets (orange) centered at the L point and the high degeneracy hole pockets (blue) along the Σ line. The figure shows eight half-pockets at the L point, so the number of valleys is $N_v = 4$. The valley degeneracy of the Σ band is $N_v = 12$. (b) Relative energy in valence bands in PbTe$_{1-x}$Se$_x$. At $T \sim 500$ K, the two valence bands converge, resulting in transport contributions from both L and Σ bands. C stands for the conduction band, L for the low degeneracy valence band, and Σ for the high degeneracy hole band. Reproduced from Y. Z. Pei et al., *Nature* 473, 66 (2011). (c) Band convergence of the conduction bands in Mg$_2$Si$_{1-x}$Sn$_x$ having the origin in the inverted order of the two bands in the end-member compounds. At $x = 0.7$, the two-band cross, giving rise to increased valley degeneracy and very high ZT. Reproduced from W. Liu et al., *Physical Review Letters* 108, 166601 (2012). With permission from Springer Nature and the American Physical Society, respectively.

the band edge of the stationary Σ band (Figure 6.13(b)). The consequence of the band convergence is the dramatically enhanced valley degeneracy to 16, which in turn results in a much-enhanced density-of-states effective mass that supports high Seebeck coefficients.

The band convergence can also be achieved with the conduction bands. An example is Mg$_2$Si$_{1-x}$Sn$_x$ solid solutions, where the end members of the solid solution, namely Mg$_2$Si and Mg$_2$Sn, have the band edges of their two lowest lying conduction bands, C$_H$ and C$_L$, reversed (Zaitsev et al. 2006). Taking advantage of the situation, Liu et al. (2012) determined the composition $x = 0.7$ at which the two conduction band edges cross and achieved the desired band convergence (Figure 6.13(c)). A high performance with $ZT = 1.3$ at 700 K was obtained.

As discussed in Section 6.5.1, the band convergence is relevant and perhaps even the key to the understanding of exceptionally high Seebeck coefficients observed in n-type skutterudites with electron concentrations in the 10^{20}–10^{21} cm^{-3} range. To take full advantage of band convergence, it is essential to have a good understanding of the band structure as well as of the phase diagram that guides which dopants have high enough solubility in order to achieve carrier concentrations necessary to induce significant band movements.

6.4.3.3 Resonance States

Electronic transport is enabled by high density of electronic states and the presence of nearly free charge carriers. The transport is limited by scattering processes the charge carriers are subjected to. The Seebeck coefficient, the pivotal parameter as far as thermoelectricity is concerned, is particularly sensitive to the density of electronic states, and its magnitude is enhanced anytime the density of states develops sharp features in a narrow energy range in the vicinity of the Fermi energy. The desire for sharp features in the density of states follows from the form of the Seebeck effect, which in the Mott's interpretation is given by

$$S = \frac{\pi^2}{3}\frac{k_B}{q}k_B T \left\{ \frac{1}{n}\left(\frac{dn(E)}{dE} \right)_{E=\xi} + \frac{1}{\mu}\left(\frac{d\mu(E)}{dE} \right)_{E=\xi} \right\}. \qquad (6.46)$$

Remembering that the carrier concentration is given by $n(E) = D(E)f(E)$, where $D(E)$ is the density of states and $f(E)$ is the Fermi-Dirac distribution function, the desire for sharp features in the density of states is apparent as it will lead to a large derivative and, hence, a large Seebeck coefficient.

Unfortunately, sharp features in the density of states are not naturally available for bulk thermoelectrics because the three-dimensional density of states is a smoothly varying function of energy. Realizing advantages of lower dimensional structures described by densities of states that have inherently sharp discontinuities in their energy dependence, Hicks and Dresselhaus (1993) focused the attention to thin films and narrow wires as great prospects to realize exceptionally large ZT values and superior thermoelectric performance. While promising results have been obtained, e.g., Venkatasubramanian et al. (2001) with Bi$_2$Te$_3$/Sb$_2$T$_3$ superlattices, it would be impractical and exceedingly expensive to use lower-dimensional structures in large-scale power generation thermoelectric applications. On the other hand, it makes every sense to try to mimic and import the features that are at the heart of excellent performance of lower-dimensional structures into the realm of bulk thermoelectric materials. The most notable is this respect is the way to engineer distinct local changes (bumps) in the otherwise smoothly varying density of states.

As shown by Heremans et al. (2008), it is possible, with appropriate dopants, to arrange for a resonance of either valence or conduction bands of a semiconducting matrix with an energy level of a localized impurity. The resonance gives rise to a small bump on the smoothly changing density of states with energy, schematically depicted in Figure 6.14. Provided the dopant's atomic orbitals hybridize with the orbitals of the host semiconductor, the position of the impurity states is inside the band but close to the band edge. If now the Fermi energy can be tuned by doping into the region of the enhanced density of states, a large Seebeck coefficient is expected.

Although there is inevitably some degradation in the carrier mobility associated with the presence of impurity, the gain from the enhanced density of states should weigh over in a successful implementation of the resonance level. Whether this, indeed, happens can be verified by looking for significant departures from the Pisarenko plot, i.e., the dependence of the Seebeck coefficient on the carrier concentration. Values of the Seebeck coefficient that fall significantly above the Pisarenko curve would be an indication of significantly enhanced density of states, likely arising from the presence of a resonant level.

Unfortunately, resonant states do not form with just any arbitrarily picked dopant introduced into the semiconductor, and there are no simple theoretical guidelines helping to identify prospective candidates. It is more or less a trial-and-error exercise. Nevertheless, finding the lucky dopants

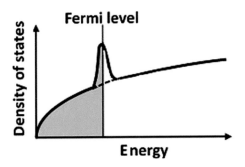

FIGURE 6.14 Schematic illustration of a resonant state created in a semiconducting matrix by an impurity atom that gives rise to a bump in the density of states. For the Seebeck coefficient to benefit from the existence of the resonant state, it is critical that the structure can be doped such as to bring the Fermi level in the range of the disturbed density of states.

that induce the resonance makes it worthwhile as a means of improving the Seebeck coefficient, the power factor, and ultimately the figure of merit. Band resonant states have been observed in many semiconductors, but meaningful enhancements in the Seebeck coefficient have been noted in only few cases, mostly with the PbTe matrix and in the case of Sn in Bi_2Te_3, Jaworski et al. (2009). Readers interested to learn more about the physical basis of band resonant states and their effectiveness are referred to an excellent review article by Heremans et al. (2012).

Studies to explore the presence of resonant states in skutterudites have been very few. The first attempt was made by Nieroda et al. (2013), who studied theoretically and experimentally the influence of Ag on the band structure of $CoSb_3$. Silver was considered to enter the skutterudite at three different lattice sites: substituting for Co, replacing a small fraction of Sb, and filling the structural voids. The RKK–CPA calculations indicated that the first two entry sites into the lattice, i.e., substitutions for Co and Sb, had no influence on the band structure, apart from shifting the position of the Fermi level into the conduction and valence band, respectively. In contrast, placing Ag in the voids, the calculations revealed a dramatic change with a pronounced Ag-dominated peak in the density of states appearing at the valence band edge of $CoSb_3$. Although it was difficult to resolve the presence of Ag in the matrix grains (very low solubility), a marked enhancement in the Seebeck coefficient was measured in all Ag-filled structures, implying that Ag acted as a resonant state in the $[Co_4Sb_{12}]$ framework. The second, and so far the last, attempt was made by Hui et al. (2014), who computed the band structure of $CoSb_3$ containing Sn at the sites of Sb and noted the progressively greater distortion in the density of states in the valence band with the increasing content of Sn. Moreover, since Sn acts as a p-type dopant at the Sb site, the Fermi level shifted into the range of the perturbed density of states when the Sn content in $CoSb_{3-x}Sn_x$ reached x = 0, a clear sign that conditions for a resonant state existed, as shown in Figures 6.15(a)–6.15(c). With such an encouraging scenario for the existence of a band resonant state based on computations, the authors synthesized a series of $CoSb_{3-x}Sn_x$ samples with x up to 1.5 and carried out detailed structural and transport studies, hoping to detect the influence of the band resonant state. Unfortunately, except for some minor features in the Seebeck coefficient near 150 K, no convincing evidence has emerged for the presence of the resonant state. Very simply, the level of doping, $x \approx 0.1$, expected theoretically to bring the Fermi level into the region of the Sn-derived states in the valence band was well beyond the solubility limit of Sn in $CoSb_3$, indicated from the structural and morphological examination to be not much higher than $x \sim 0.05$, consistent with the early estimates of Zobrina and Dudkin (1960). I am not aware of other studies aiming to generate band resonant states in the skutterudite matrix. Apparently, low solubility limits of the prospective dopants that might give rise to band resonant states in skutterudites are the major stumbling block. However, it would be interesting to try to promote a greater shift of the Fermi level by enhancing the level of p-type doping by a small fraction

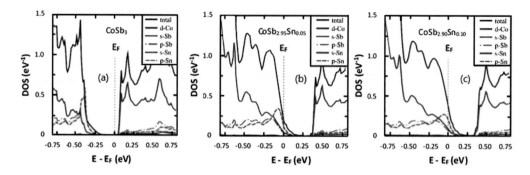

FIGURE 6.15 Progressive formation of the Sn-derived states in the valence band of $CoSb_3$ upon doping with Sn at the site of Sb. As the concentration of Sn in $CoSb_{3-x}Sn_x$ increases, the Fermi level shifts to the valence band, and at $x = 0.1$, enters the Sn-derived states, signaling a prospect for the formation of a band resonant state. (a) The band structure of $CoSb_{3-x}Sn_x$ with $x = 0$; (b) the band structure of $CoSb_{3-x}Sn_x$ with $x = 0.05$; and (c) the band structure of $CoSb_{3-x}Sn_x$ with $x = 0.10$. Adapted from S. Hui et al., *Journal of Applied Physics* 115, 103704 (2014). In particular, the original energy and DOS value in units of Rydbergs were converted to units of eV. With permission from the American Institute of Physics.

of Fe substituting for Co, hoping that this would not inhibit the formation of the Sn-derived states in the valence band of $CoSb_3$.

6.4.3.4 Energy Filtering

The charge carriers with energies above and below the Fermi energy give opposite contributions to the Seebeck coefficient. With a flat density of states near the Fermi energy, the case of metals, the two contributions nearly cancel, and that is the reason why metals have small Seebeck coefficients. In semiconductors, the density of states has a high asymmetry, particularly when the Fermi energy lies near the band edge, and thus the Seebeck coefficients of semiconductors are generally much larger. In the general expression for the Seebeck coefficient (see, Eq. A1.27 in the Appendix A1 in the online version of the book), the density of states $D(E)$ is accompanied by the relaxation time $\tau(E)$. The concept of energy filtering is an attempt to increase the asymmetry of the relaxation time near the Fermi energy, and hence further enhance the Seebeck coefficient. In essence, one tries to create conditions for strongly energy-dependent scattering mechanism, which effectively prevents the passage of low energy carriers while higher energy carriers pass unaffected. In practice, the filtering may be achieved by introducing barriers in the structure that block low energy electrons. The barriers typically form at the grain boundaries and nanopores. How to optimize the barrier height with respect to the Fermi energy was discussed by Narducci et al. (2012). While the electrical conductivity is slightly reduced because of the presence of the barriers, the increase in the magnitude of the Seebeck coefficient is often more significant, resulting in an overall gain in the power factor. In nanocomposites, particularly those formed by dispersing metallic nanoinclusions in the semiconducting matrix, band bending at the metal-semiconductor interface is often sufficient to scatter low energy charge carriers (Faleev and Léonard 2008). While the intensity of energy filtering depends on the grain size (Kishimoto and Koyanagi 2002), not all structures respond the same way and show the same benefit. Energy filtering is generally more effective with wider band-gap semiconductors, and its effectiveness has been well documented with chalcogenides, e.g., Hsu et al. (2004) and Heremans et al. (2005), and with half-Heusler matrices containing full-Heusler nanoinclusions (Liu et al. 2013). For the recent survey of energy filtering in thermoelectric materials, the reader is advised to consult Gayner and Amouyal (2020). The only vaguely related energy filtering in skutterudites is the work by Zhao et al. (2017) with superparamagnetic metallic nanoparticles $BaFe_{12}O_{19}$ in $Ba_{0.3}In_{0.3}Co_4Sb_{12}$, in which the authors reported a notable increase in the thermoelectric properties compared to pure $Ba_{0.3}In_{0.3}Co_4Sb_{12}$. Although the electrical conductivity decreased somewhat upon the presence of the nanoparticles, the gain in the Seebeck coefficient, likely due to energy filtering at the nanoparticle-matrix interfaces, was large enough, which, together with the suppressed thermal conductivity, resulted in the *ZT* value enhanced from 1.3 to 1.65 at 850 K.

While energy filtering is often claimed to be the origin of the enhanced Seebeck coefficient, judged primarily based on deviations from the Pisarenko plot of the Seebeck coefficient for a single parabolic band model vs. the carrier concentration, thorough structural and spectroscopic verification is rarely provided to back up the claim.

6.5   THERMOELECTRIC SKUTTERUDITES

Outstanding electronic properties of binary skutterudites were known since the measurements of Dudkin and colleagues during the first half of 1950s. Unfortunately, the structures also possessed very high thermal conductivity on the order of 10 $Wm^{-1}K^{-1}$, and this precluded any considerations of using binary skutterudites as possible thermoelectrics. The discovery of filled skutterudites by Jeitschko and Braun (1977) and the idea of Slack (1995) that the rattling motion of fillers might dramatically lower the thermal conductivity, promptly confirmed in experiments by Morelli and Meisner (1995), opened the door for exploration of thermoelectricity in this family of compounds. Among the early papers are Mandrus et al. (1995, 1997), Morelli et al. (1995), Sharp et al. (1995), Borshchevsky et al. (1996), Sales et al. (1996), Nolas et al. (1996), Matsubara et al. (1996), Koyanagi et al. (1996), Caillat et al. (1996a, 1996b), Fleurial et al. (1995, 1996), Kloc et al. (1996), Arushanov et al. (1997), Chen et al. (1997), and Sales et al. (1997). In fact, the researchers at the Jet Propulsion Laboratory started to work on thermoelectric properties of skutterudites even before Slack proposed his PGEC concept in 1995, with the publications reported in the Proceedings of the 11th–14th International Conferences on Thermoelectrics.

As discussed in Chapter 5, various attempts to disrupt lattice vibrations and impede the flow of heat in the skutterudite matrix have been exceptionally successful, and it was possible to reduce the thermal conductivity close to the limit imposed by wave mechanics, referred to as the minimum thermal conductivity. It is thus unlikely that further significant reduction in the thermal conductivity could be obtained, and one has to turn attention to the electronic properties as the primary source of any future gains in the figure of merit of skutterudites.

In Chapter 1, I noted that the natural tendency of stoichiometric binary skutterudites, and $CoSb_3$ in particular, is to form p-type semiconductors. The reason is the native defects that form during the solidification of the structure. According to DFT calculations by Park and Kim (2010) discussed in detail in Section 1.1.5, the defects with the lowest formation energy are an isolated interstitial Co_i and a much smaller density of pairs of interstitials $Co_i–Co_i$. Isolated Co_i are deficient of one electron, while pairs of cobalt interstitials alter their charge state according to the position of the Fermi level and have marginally smaller formation energy. In stoichiometric $CoSb_3$, the large density of Co_i pins the Fermi level close to the valence band edge, yielding the p-type nature of transport. Even when some small nonstoichiometry favoring electrons over holes (often a trace amount of Ni in Co used in the synthesis is enough) is present and the structure shows an n-type behavior at low temperatures, the mode of transport invariably switches to a p-type semiconductor at elevated temperatures. The switchover takes place because the pairs of cobalt interstitials, which pinned the Fermi energy in such Co-rich structures close to the conduction band edge at low temperatures, rapidly dissociate as the temperature increases and single cobalt interstitials take over as the dominant intrinsic charged defect.

In Chapters 1, 3, and 4, we also considered the effect of doping, filling, and formation of composite structures on electronic energy bands and the electronic transport. I reiterate that there is a notable difference in the energy bands, and thus electronic transport properties, between n-type and p-type skutterudites, regardless of whether they are binary or filled structures. In contrast, the thermal conductivity does not differentiate between n-type and p-type forms of skutterudites and responds in a similar fashion to attempts to degrade it regardless of the dominant type of charge carriers. In other words, the density of phonon states and phonon dispersion are substantially similar in n-type and p-type skutterudites, comparing separately the binary and filled skutterudites. In view of the very different structure of conduction and valence bands, the thermoelectric properties of n-type and p-type forms of skutterudites will be discussed separately. Again, most of the attention will be given to skutterudites having either $[Co_4Sb_{12}]$ or compensating $[Co_xFe_{4-x}Sb_{12}]$ frameworks. Because of their inferior thermoelectric properties and, above all, the prohibitive cost, skutterudites with the $[Pt_4Ge_{12}]$ framework have not been considered seriously as thermoelectrics.

6.5.1 Thermoelectric Performance of n-Type Skutterudites

The existence of triply degenerate conduction bands at the Γ-point, coupled with the proximity of a higher lying heavy conduction band along the Γ-N direction, revealed in recent DFT calculations (see Section 3.1), called for reevaluation of the opinion why n-type skutterudites are able to maintain high Seebeck coefficients in spite of possessing high electron densities on the order of 10^{21} cm^{-3}. Prior to having solid evidence of the existence of the heavy secondary conduction band along the Γ-N direction, the triply degenerate conduction band edge at the Γ-point (one band being a mirror image of the light, linearly dispersing Γ-point valence band, and the other two bands harboring not much heavier electrons) could not provide for sufficiently heavy band masses to maintain high Seebeck coefficients. In the absence of a reasonably heavy band of electrons, it was conjectured that the rising effective masses of electrons with their concentration (see Figure 4.20), are associated with the non-parabolic nature of the conduction bands, invoking specifically the Kane model discussed in Section 4.5 (Caillat et al. 1996a, Anno et al. 1999, Kuznetsov et al. 2003, and Salvador et al. 2010), among others. In retrospect, nonparabolicity has nothing to do with the heavy masses of electrons, as shown by detailed modeling by Tang et al. (2015). The results indicate that for the same value of the effective mass at the conduction band edge, the Kane model actually yields lower Seebeck coefficients (implying smaller effective mass) than does the parabolic band. The new detailed assessment of the band structure by Tang et al. (2015), combining DFT computations with infrared absorption measurements, revealed a fascinating scenario whereby the higher lying secondary conduction band, located about 80 meV above the Γ-point band edge along the Γ–N direction (see Figure 3.10), moves rapidly down as the temperature increases, and its band edge approaches the essentially stationary Γ-point conduction band manifold, as shown in Figure 3.11. The actual convergence of the bands takes place at about 800 K. The secondary band position along the Γ–N direction brings into play 12 other symmetry related carrier pockets accessible by heavy doping. Upon the convergence, the conduction bands act as parallel conductors with now 12 + 3 = 15-fold valley degeneracy that dramatically enhances the density of states effective mass by about a factor of 6. Such enhancement is more than sufficient to maintain high Seebeck coefficient at high carrier concentrations. It is instructive to see in Figure 6.16 not only how much the secondary conduction band contributes to the figure of merit at 800 K but also the gain attained when its band edge lines up with the band edges of the Γ-point conduction bands. I should note that a prospect for the convergence of the conduction bands in skutterudites was also seen in the study by Shi et al. (2015) exploring dual site occupancy by Ga in Ga-doped $Yb_xCo_4Sb_{12}$ skutterudites. The interaction between Ga at the filler site and Ga on the framework not only enhanced the filling fraction occupancy of Yb in the structure, but it also sharply lowered the upper conduction bands and brought them closer to the Γ-point bands. Analysis of X-ray absorption fine structure (EXAFS) and X-ray absorption near-edge structure (XANES) by Wei et al. (2016) indicated that the Sb_4 rings become more square as the double-filling content of Ba and In in $Ba_xIn_yCo_4Sb_{12}$ increases. The authors argued that such a transition in the shape of the pnicogen rings increases the density of the Sb–Sb bonding and antibonding states, and the observed enhancement in the thermoelectric performance was attributed to the band convergence.

While fine-tuning the Fermi level to further optimize the performance is possible via doping on either cation or anion sides of the framework, it is rarely done for that purpose in n-type skutterudites

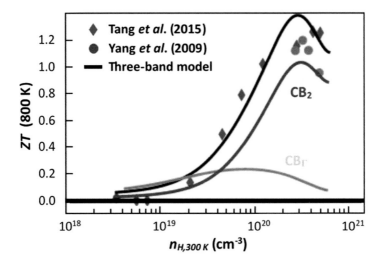

FIGURE 6.16 *ZT* at 800 K plotted as function of the electron concentration for $Yb_xCo_4Sb_{12}$. Contributions of the secondary CB_2 conduction band and the Γ-point conduction band CB_Γ to the figure of merit are shown separately. The fit of the experimental data using a three-band model is indicated by a black curve. Adapted from Y. Tang et al., *Nature Materials* 14, 1223 (2015). With permission from Springer Nature.

because the doping efficiency is generally low (not much better than 30%). Moreover, filling the voids is a far more efficient form of doping as the valence electrons of the filler are stripped off and donated to the framework (see Figure 1.24). As already noted, the sweet spot for the optimal carrier concentration is in the range 0.4–0.6 electrons per primitive cell (Yang et al. 2009), corresponding to high electron concentrations of around 10^{21} cm^{-3} compared to the classical thermoelectrics, such as Bi_2Te_3, PbTe, and Si-Ge alloys, where optimal concentrations are in the $10^{19} - 10^{20}$ cm^{-3} range.

An example of power factors achieved with n-type filled skutterudites is given in Figure 6.17, where I have selected n-type structures that attained $(ZT)_{max}$ values in excess of an arbitrarily set value of 1.3.

The power factors of n-type skutterudites attain high values, some of them exceed 6 mWm^{-1}K^{-2} at temperatures above 700 K. In comparison to other thermoelectrics, power factors of skutterudites are among the highest, certainly on par with the benchmark values set by Bi_2Te_3-based alloys with their room temperature power factors near 5 mWm^{-1}K^{-2}. Among the skutterudites, the highest value was set by the lightest filler, lithium, in the $[Co_4Sb_{12}]$ framework. Because of its very small ionic radius, Li cannot be trapped in skutterudite voids under ordinary ambient pressure synthesis. However, it can fill the $[Co_4Sb_{12}]$ framework using the high-pressure synthesis (Zhang et al. 2011). With the Li content $y = 0.4$, resulting in the carrier density of 2×10^{20} cm^{-3}, electrons maintain high mobility of 61 cm^{2}V^{-1}s^{-1}. Even though, as expected, the Seebeck coefficient decreases with the increasing level of filling, the high electrical conductivity results in a record high power factor of 5.8 mWm^{-1}K^{-2} at 300 K, which increases well above 6 mWm^{-1}K^{-2} at high temperatures. Another light filler, sodium, also shows quite high power factor at room temperature (black, six-point star in Figure 6.17) (Pei et al. 2009). Such light fillers, although not reaching the most impressive figures of merit because of their inability to scatter effectively heat-conducting phonons, might be of interest in applications where the premium is on the magnitude of the power factor. In the case of lithium-filled skutterudites, however, one should be aware of their metastable nature, which leads to the degradation above 700 K. Other high-power factors among n-type skutterudites are associated with double- and triple-filled structures, typically containing either Yb or In (or both) in the voids. Interestingly, even an unfilled skutterudite, albeit heavily doped on the pnicogen rings, shown with black up-triangles in Figure 6.17, exhibits a reasonably large power factor.

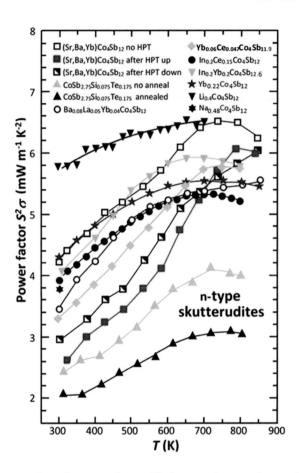

FIGURE 6.17 Temperature-dependent power factors $S^2\sigma$ for several n-type skutterudites that attained maximum ZT values in excess of 1.3. Among the samples are single-filled, multifilled, and unfilled skutterudites. Black down-triangles are the data for $Li_{0.4}Co_4Sb_{12}$ measured by Zhang et al. (2011). Square symbols stand for the data of Rogl et al. (2014) obtained on triple-filled $(Sr_{0.25}Ba_{0.25}Yb_{0.50})_{0.5}Co_4Sb_{12.5}$ with an obvious excess of the fillers and extra Sb to drive the formation of secondary phases, primarily Yb_2O_3 and $YbSb_2$. The samples are labeled as $(Sr,Ba,Yb)Co_4Sb_{12}$, and were measured before applying high-pressure torsion (HPT) processing (open squares), on the first run up in temperature following HPT (red squares), and on the run down after HPT (half while/half black squares). Light orange down-triangles are power factors of $In_{0.2}Yb_{0.2}Co_4Sb_{12}$ measured by Lee et al. (2018). Open circles represent measurements by Shi et al. (2011) on triple-filled $Ba_{0.10}La_{0.05}Yb_{0.07}Co_4Sb_{12}$. Blue stars are power factors for $Yb_{0.22}Co_4Sb_{12}$ measured by Wang et al. (2016). Solid black circles are the data measured by Li et al. (2009) for $In_{0.2}Ce_{0.15}Co_4Sb_{12}$. Orange diamonds is the power factor of Ballikaya et al. (2012) for the nominal sample of $Yb_{0.20}Ce_{0.15}Co_4Sb_{12}$ that, after processing, ended up with the composition $Yb_{0.06}Ce_{0.047}Co_4Sb_{11.9}$. Gray and black up-triangles are measurements by Khan et al. (2017) on an unfilled sample of $CoSb_{2.75}Si_{0.075}Te_{0.175}$ prior and after annealing, respectively. The black star stands for the room temperature power factor of $Na_{0.48}Co_4Sb_{12}$ measured by Pei et al. (2009).

Much interest was generated by reports of the beneficial influence of severe plastic deformation on the thermoelectric properties introduced via High-Pressure Torsion (HPT) (Rogl et al. 2014). Indeed, such processing resulted in record-high ZT values for both n- and p-type skutterudites. The benefit comes exclusively from a lower thermal conductivity. The HPT process is not friendly to the power factor because the carrier mobility is adversely affected. For instance, the mobility drops from 26.7 $cm^2V^{-1}s^{-1}$ before HPT down to 12–15 $cm^2V^{-1}s^{-1}$ after the HPT processing. Clear evidence of the effect of HPT on the power factor is apparent in Figure 6.17, where the power factors of the triple-filled skutterudites $(Sr_{0.25}Ba_{0.25}Yb_{0.50})_{0.5}Co_4Sb_{12.5}$ before and after HPT processing are presented and can be compared. Since the Seebeck coefficient is immune to the introduced severe plastic deformation, the degradation of the power factor arises from roughly a doubling of the electrical resistivity as additional defects are introduced during the HPT process.

The dimensionless figure of merit achieved with n-type skutterudites is presented in Figure 6.18. To aid the reader, many of the entries are plotted with the same symbols as their corresponding power factors in Figure 6.17. Included are also some surprising compositions, such as a partially-filled $Sm_{0.15}Co_4Sb_{12}$. The best performing structures are clearly multiple-filled skutterudites. However, even some single-filled skutterudites, the already noted $Sm_{0.15}Co_4Sb_{12}$ and particularly $Yb_{0.22}Co_4Sb_{12}$, are strong competitors with the $ZT \approx 1.35$ and 1.5, respectively. Another exceptional single filled skutterudite with the ZT ~1.5 at 725 K is $In_xCo_4Sb_{12}$ synthesized by induction melting with an excess of In, as reported by Khovaylo et al. (2017).

The dimensionless figure of merit for most of skutterudites rises substantially linearly with temperature, the trend being broken eventually by the onset of intrinsic excitations. The triple-filled skutterudites, designated in Figures 6.17 and 6.18 as $(Sr,Ba,Yb)Co_4Sb_{12}$, were synthesized by extensive ball

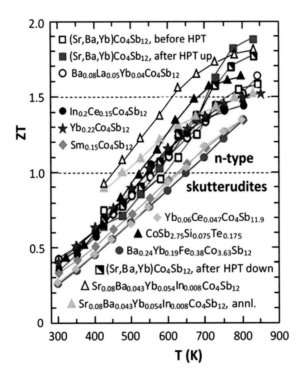

FIGURE 6.18 Temperature dependence of the dimensionless figure of merit for several exceptionally performing n-type skutterudites. Eight samples are identical with those in Figure 6.17 and are plotted with the same symbols. New entries here are $Sm_{0.15}Co_4Sb_{12}$ (light blue diamonds) measured by Li et al. (2016), $Ba_{0.24}Yb_{0.19}Fe_{0.38}Co_{3.62}Sb_{12}$ (solid red circles) investigated by Ballikaya and Uher (2014), and two quaternary skutterudites, $Sr_{0.08}Ba_{0.043}Yb_{0.054}In_{0.008}Co_4Sb_{12}$ before and after annealing reported on by Rogl et al. (2015b). The last two structures were prepared by ball milling of nominally triple-filled $(Sr_{0.33}Ba_{0.33}Yb_{0.33})_{0.35}Co_4Sb_{12}$ and 9.1 wt.% of $In_{0.4}Co_4Sb_{12}$ followed by hot pressing. In the preparation of the quaternary skutterudite labeled with open up-triangles was used high-energy ball milling, and the structure had uniform small size grains of 100 nm and no InSb secondary phase was detected. The other quaternary skutterudite with the same nominal composition, labeled with orange up-triangles, was made by regular ball milling, had somewhat larger grains, and InSb nanoinclusions were detected. The sample was subsequently annealed at 873 K for three days. Although the power factor was slightly affected by the annealing, the thermal conductivity was lower and a competitive $ZT = 1.5$ was obtained. Further annealing resulted in no structural or transport changes, implying that that the quaternary structure was stable.

milling from the nominal starting composition $(Sr_{0.25}Ba_{0.25}Yb_{0.50})_{0.5}Co_4Sb_{12.5}$. Thus, they contained a significant excess of fillers and of Sb to promote the formation of minute amounts of Yb_2O_3 and $YbSb_2$ secondary phases, shown previously to enhance the thermoelectric performance by providing additional scattering for heat-conducting phonons. The reader may note a rather sharp increase in the figure of merit of $(Sr_{0.25}Ba_{0.25}Yb_{0.50})_{0.5}Co_4Sb_{12.5}$ above about 650 K following the HPT processing (the sample is shown with red squares), which makes this material the highest performing skutterudite with $ZT = 1.9$. Surprisingly, the rise in ZT is not due to further enhanced scattering of phonons (the lattice thermal conductivity is the same before and after HPT), but due to the halving of the electronic thermal conductivity contribution associated with the doubling of the electrical resistivity, as already noted. Since the structural defects responsible for the higher electrical resistivity are partly 'annealed', as seen on the cool down path of the measurements, the figure of merit settles at $ZT = 1.8$. The same ZT value was obtained a year later with quadruple-filled $(Sr_{0.08}Ba_{0.043}Yb_{0.054}In_{0.008})Co_4Sb_{12}$ structures (actual composition) synthesized using high energy ball milling by mixing triple-filled $(Sr_{0.33}Ba_{0.33}Yb_{0.33})_{0.35}Co_4Sb_{12}$ skutterudite with 9.1 wt% of $In_{0.4}Co_4Sb_{12}$ (Rogl et al. 2015b).

The formation of composite skutterudite structures has attracted considerable attention and numerous attempts were made to use this approach to enhance the thermoelectric performance. Although not all such attempts succeeded, in some cases significant enhancements in the figure of merit, gained primarily by lowering the thermal conductivity, were achieved. More success seems to have been achieved with the intrinsically formed nanocomposites whereby finely distributed nanophases form due to the overstoichiometry of filler species or pnicogens, or upon rapid cooling, or long-term annealing. A canonical example is the formation of an InSb nanostructure when one attempts to force a large amount of In into the voids of $CoSb_3$. The resulting InSb nanophase scatters effectively mid- to high frequency phonons, reduces the thermal conductivity, and results in high ZT values. The case in point is double-filled $In_{0.2}Ce_{0.15}Co_4Sb_{12}$ measured by Li et al. (2009) and quaternary annealed $Sr_{0.08}Ba_{0.043}Yb_{0.054}In_{0.008}Co_4Sb_{12}$ mentioned above, both displayed in Figure 6.18. A similar situation is encountered with Ga fillers. Attempting to double-fill $CoSb_3$ with Yb and Ga, Xiong et al. (2011) observed that upon annealing the structure, Ga is expelled from the voids and reacts with Sb forming a distributed nanodispersion of GaSb (average size 11 nm) in the $Yb_yCo_4Sb_{12}$ matrix. One such nanocomposite, labeled as $Yb_{0.26}Co_4Sb_{12}/0.2$ GaSb, reached $ZT = 1.45$ at 850 K (not shown in Figure 6.18 due to an already overcrowded figure).

Extrinsic composite skutterudites have attracted comparable interest. Here, various powders were admixed to the powder of the skutterudite matrix (see Section 1.4.2). Recent efforts by Rogl et al. (2017) to admix several different kinds of additives are summarized in Figure 1.42[*]. Yes, it is sometimes possible to improve the figure of merit of skutterudites by forming extrinsic composites, but it is rather unpredictable which additive will achieve the desired outcome. Moreover, the gains are typically not more than 10–15% and, given an uncertainty in the figure of merit (accounting for errors from three independent measurements), the gain is often within the error bars.

Before turning to p-type skutterudites, I wish to mention two more studies reporting rather high ZT values in n-type skutterudites. The first one describes a valiant effort to maximize phonon scattering by designing a seven element filled compensated skutterudites with composition $Ba_{0.1}Yb_{0.2}Al_{0.1}La_{0.05}Eu_{0.05}Ga_{0.1}In_{0.1}Co_{3.75}Fe_{0.25}Sb_{12}$, where the overall high filling fraction was enabled by partly replacing Co with Fe (Matsubara and Asahi 2016). The structure achieved a high $ZT = 1.5$ at 800 K. The other study, Khan et al. (2017), describes an *unfilled* skutterudite heavily co-doped on the pnicogen ring with Si and Te. The samples were prepared by a more-or-less standard solid-state synthesis with twice heating the powders to 1323 K in quartz tubes, and finally using SPS to consolidate the material into a dense pellet. A section of the pellet was then annealed for 15 hr at 873 K. Remarkably, the thermal conductivity of the annealed pellet exhibited exceptionally low values,

---

[*] The composite skutterudites reported by Rogl et al. (2017) were prepared from commercially available n-type $(Mm,Sm)_yCo_4Sb_{12}$ and p-type $(DD)_yFe_3CoSb_{12}$ skutterudite powders supplied by Treibacher Industry, AG. I mention this to inform the readers that high-quality skutterudites are now available on the market.

much lower than a sample from the other section of the pellet that was not annealed, and even lower than the typical thermal conductivity of competitive triple-filled skutterudites. Upon examination by mercury intrusion porosimetry, the annealed sample revealed a porous structure with a spectrum of pores spanning from the nanometer to micrometer scale and having irregular shapes and varied orientation. With the large excess of Te in the starting material that had the nominal composition (normalized to the Co constituency of one) of $CoSb_{2.75}Si_{0.075}Te_{0.175}$, the annealing process evaporated a large fraction of Te, leaving behind a myriad of misoriented pores. Apparently, such a hierarchical system of pores was exceptionally effective in scattering heat-conducting phonons and the lattice thermal conductivity at 323 K fell below 0.9 $Wm^{-1}K^{-1}$. As is usually the case, the Seebeck coefficient was unaffected by the presence of the pores, and the electrical conductivity dropped by only about 16% compared to the sample not annealed. Thus, the power factor remained quite high (see gray up-triangles in Figure 6.17), and with the much degraded thermal conductivity, the figure of merit reached $ZT = 1.6$ at 773 K, the value unprecedented for an unfilled skutterudite, black up-triangles in Figure 6.18. Previously, a general guidance concerning the effectiveness of filling was broadly understood to mean that unfilled (but doped) skutterudites are limited to values of $ZT < 0.7-0.9$, good single-filled skutterudites reach $ZT \sim 1$, judiciously chosen double-filling may enhance the figure of merit to perhaps 1.3, and triple-filled skutterudites can achieve $ZT \geq 1.5$. Obviously, the results of Khan et al. break this categorization and show that, via the presence of a medium that scatters phonons exceptionally strongly, even an unfilled skutterudite can attain a highly competitive performance.

6.5.2 THERMOELECTRIC PERFORMANCE OF P-TYPE SKUTTERUDITES

In general, p-type filled skutterudites have lagged behind the efficiency of their n-type forms, with the lower power factors and ZTs usually not much more than the unity. The problem rests with a somewhat schizophrenic way p-type filled skutterudites are made, and with their less favorable valence band structure. To start with, to make a p-type filled skutterudite with electropositive fillers, one must compensate for the electrons introduced by filling, which is needed to lower the thermal conductivity. The charge compensation is typically done by a partial or complete replacement of Co by Fe, a process that is highly deleterious to the mobility of holes. An alternative charge compensation, *via* doping the elements of Column 14 at the sites of Sb, is unlikely to overcome a very efficient electron generation when the fillers enter the structure, and the damage to the mobility of holes would not be much less significant. Regarding the valence bands structure, in binary skutterudites there is only a single light, linearly dispersing valence band at the Γ point, and its valley degeneracy is one. Although this band is responsible for exceptionally large hole mobilities, exceeding those of n-type skutterudites (see Figure 4.8), the band offers no prospect for a reasonably large band effective mass essential to support high Seebeck coefficients. The lower lying valence bands are far away, having band edges some 0.5 eV below the Γ-point valence band. It is not obvious how to make a good use of them to enhance the band effective mass. The energy separation is likely too large to breech by doping, and it is difficult to imagine a sufficiently large temperature dependence of the valence band edges that would bring the bands closer together within the range of temperatures that do not exceed sample decomposition. The only hope rests with altering band positions by forming ternary skutterudites or by promoting the formation of additional bands by filling the voids. In the former scenario, DFT calculations by Zevalkink et al. (2015) indicated a significant upward shift of the secondary valence bands in ternary $CoSn_{1.5}Te_{1.5}$ as a consequence of isoelectronic replacement of Sb by equal amounts of Sn and Te. However, because of a rather poor electrical conductivity of most ternary skutterudites, the approach is unlikely a viable solution. In the case of filling, DFT calculations by Zhou et al. (2011) indicated that in AFe_4Sb_{12} skutterudites the interaction of certain fillers with the framework atoms might elevate the secondary valence band manifold and bring it close to the Γ-point valence band. In fact, in the case of Sn and In fillers, a new valence band has emerged at the H point, which in $InFe_4Sb_{12}$ is at nearly the same energy as the Γ-point valence band edge. If this could be realized in experiments, a significantly elevated band effective mass of holes might be at hand, benefiting from an additional valley degeneracy associated with the symmetry position of the H-point. On the other hand, it is well known that partial or complete replacement of Co by Fe has a deleterious effect on the carrier mobility, and the penalty to pay by working with the $[Fe_4Sb_{12}]$ framework may be more than the benefit gained from the enhanced band mass. It is fair to say that the valence band manifold of filled skutterudites has not yet been sufficiently studied to identify and explore various possible band engineering approaches that might lead to the optimal configuration. As far as the experimental work is concerned, it is certainly true.

One might raise a question about the use of skutterudites with electronegative fillers. After all, they are naturally p-type semiconductors, require no charge compensation, and no undue damage is done to the hole mobility. Unfortunately, the synthesis of skutterudites with electronegative fillers, such as iodine and halogens, requires the use of high pressure (see Section 3.2.6). Moreover, and more important, such structures are notoriously unstable and decompose at elevated temperatures. Thus, there is currently little prospect of using electronegative fillers in practical thermoelectric applications.

Figure 6.19 collects the power factors of several more efficient p-type skutterudites that achieved ZT values of at least unity. Compared to n-type skutterudites, the best of p-type materials attain no more than two-thirds of the power factor values, and the majority barely reaches half the power factors of n-type skutterudites. Apart from what has already been said above regarding low power

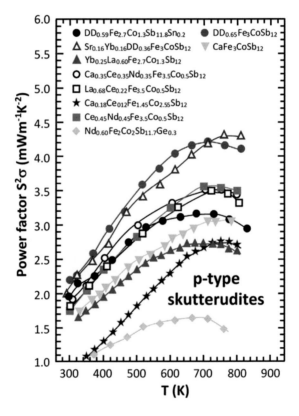

FIGURE 6.19 Power factors of a select group of p-type skutterudites that attained ZT values of at least the unity. Solid red circles, solid black circles, and empty blue up-triangles are measurements on p-type skutterudites containing didymium in combination with other fillers or a single filler (Rogl et al. 2011, Rogl et al. 2015a, and Rogl et al. 2010a, respectively). Open black circles and open black squares are measurements of Dahal et al. (2015) and Dahal et al. (2016), respectively. Black stars are the data by Tang et al. (2006). Solid gray squares are power factors measured by Jie et al. (2013). Orange diamonds are the data of Zhang et al. (2013). Light blue solid down-triangles are from Thompson et al. (2015). Solid red up-triangles are measurements of Zhou et al. (2013).

factors of p-type filled skutterudites, a contributing factor is also their generally high carrier (hole) densities needed to mitigate the early onset of intrinsic excitations. Such high hole concentrations drive the Fermi level deep into the valence band where the carriers acquire the heavy band character (see, e.g., Cho et al. 2013), and their mobility plummets. As the data in Figure 6.19 indicates, multiple filling is, again, highly beneficial and one of the best p-type fillers is didymium (DD), either alone or in combination with other fillers (Rogl et al. 2010a, 2011). Didymium is a commercially available mixture of 95.24 wt% Nd and 4.76 wt% Pr. The reader may recall that Nd is even more effective than Yb in degrading the heat transport, as noted in Section 5.13.3.2, and the dominant presence of Nd in didymium makes the filler an obvious choice to minimize the thermal conductivity. With rather large filling fractions used to assure low thermal conductivity, charge compensation (Fe for Co) is essential to maintain the p-type nature of transport. As the composition of the structures depicted in Figure 6.19 indicates, the content of Fe is very high, which, as already noted, affects the mobility of holes.

Thermoelectric figures of merit achieved with the above p-type skutterudites are plotted in Figure 6.20. A glance at Figures 6.18 and 6.20 immediately points out the difference between the performances of n- and p-type skutterudites. While n-type skutterudites still show a robustly rising figure of merit at the 800 K–850 K range, p-type structures are well past their peak performance already by 750 K. The highest values of the figure of merit were obtained with the didymium-filled structures, closely followed by other compositions. The absolute maximum ZT value was measured on filled and doped $DD_{0.59}Fe_{2.7}Co_{1.3}Sb_{11.8}Sn_{0.2}$, reaching slightly above 1.3 near 750 K (Rogl et al. 2015a). The sample also had, by far, the best average ZT value over the temperature range from 300 K to 800 K.

Heavy reliance on rare-earth fillers prompted several attempts to identify possible replacement fillers that would be less expensive and of less strategic importance, yet support good thermoelectric

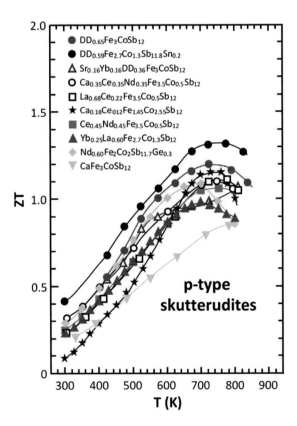

FIGURE 6.20 Thermoelectric figure of merit of several high performing p-type skutterudites. The symbols and samples are identical to those used in Figure 6.19.

performance of p-type skutterudites. In this regard, alkaline earth elements were considered viable candidates and were frequently tried as fillers. Three entries in Figures 6.19 and 6.20 demonstrate the effectiveness of Ca, the best option among alkaline earth fillers. In combination with the least expensive rare earth element Ce, the double-filled $Ca_{0.18}Ce_{0.12}Fe_{1.45}Co_{2.55}Sb_{12}$ proved very competitive, although its optimal performance was restricted to a rather narrow range of temperatures (Tang et al. 2006). Figures 6.19 and 6.20 also include the data for a single filled $CaFe_3CoSb_{12}$. On its own, this inexpensive filler supports a competitive power factor (Thompson et al. 2015). However, even

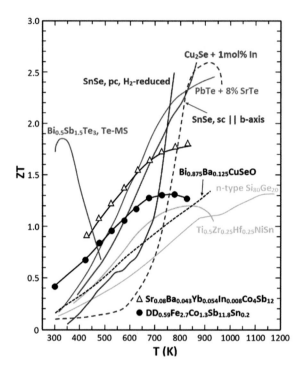

FIGURE 6.21 Comparison of the best n- and p-type skutterudites (the same symbols used as in Figures 6.18 and 6.20) with other high performing thermoelectric materials. In the mid-temperature range 500 K–800 K, skutterudites are an excellent choice for modules. $Bi_{0.5}Sb_{1.5}Te_3$, Te-MS is a high performing p-type Bi_2Te_3-based structure synthesized by Kim et al. (2015) with a significant excess of Te that, upon sintering, excaped a left behind dense dislocation arrays that very effectively reduced the thermal conductivity. $Ti_{0.5}Zr_{0.25}Hf_{0.25}NiSn$ was fabricated by Schwall and Balke (2013) with a small fraction (0.002) of Sb substituting for Sn and is one of the best half-Heusler thermoelectric alloys. Because of their good thermoelectric properties, excellent thermal stability, and mechanical robustness, half-Heusler alloys are the strongest competitor for skutterudites. $Si_{80}Ge_{20}$ was prepared by Wang et al. (2008) and is an example of Si–Ge alloys used for the high temperature range of thermoelectric applications. $Bi_{0.875}Ba_{0.125}CuSeO$ was synthesized by Sui et al. (2013) and is an example of a recently discovered family of promising thermoelectric materials. PbTe+8%SrTe is a composite nanostructured p-type chalcogenides explored by Tan et al. (2016) that hold promise for future applications. Cu_2Se + 1 mol% In prepared by Olvera et al. (2017) shows excellent thermoelectric properties. However, Cu_2Se are superionic semiconductors and, as such, they typically possess structural and chemical instabilities. Major efforts are needed to stabilize the structure and make Cu_2Se-based materials practical thermoelectrics. The dashed blue curve is the data for a single crystal of SnSe measured parallel to the b-axis by L.-D. Zhao et al. (2014). The exceptionally small thermal conductivity of this crystal was a major puzzle because the value was much smaller than the thermal conductivity of SnSe polycrystals. The puzzle was solved when Lee et al. (2019) realized that the ball-milled polycrystalline samples very easy oxidize and contain large quantities of SnO that have a high thermal conductivity. Upon reducing the oxide in H_2, the thermal conductivity of the polycrystalline SnSe dropped dramatically, and its figure of merit (solid blue curve) became comparable to that of the single crystal.

though Ca filling leads to a surprisingly low thermal conductivity measured with respect to skutterudites filled with Ba or Sr, it is not able to degrade the heat transport to the same degree as the small and heavier rare-earth fillers. Consequently, the ZT value of $CaFe_3CoSb_{12}$ is markedly lower. On the other hand, the ZT has not yet reached its maximum by 800 K, where all other p-type skutterudites are already past their peak performance, and the figure of merit of the compound does not fall as rapidly with the decreasing temperature as in the other p-type structures. At the other extreme of power factors, $Nd_{0.60}Fe_2Co_2Sb_{11.7}Ge_{0.3}$ is grossly inadequate, yet its conversion efficiency, measured by the figure of merit, is among the best of the p-type skutterudites (Zhang et al. 2013). The structure clearly benefits from the exceptionally low lattice thermal conductivity due to the presence of Nd, the filler able to scatter heat-conducting phonons most effectively, as noted in Section 5.13.3.2 and Figure 5.42.

A comparison of the best n- and p-types skutterudites with other efficient thermoelectric materials is given in Figure 6.21. Indeed, in the mid-temperature range of 500 K–800 K, skutterudites are an excellent choice for thermoelectric modules. Although above 600 K to 700 K, there are thermoelectrics with higher figures of merit, they are often unstable or available only with one type of charge carriers. Recently, SnSe has generated considerable interest following exceptional thermoelectric properties ($ZT \sim 2.5$) reported by L.-D. Zhao et al. (2014) on single crystals measured along the crystallographic b-axis. The puzzling result of the measurements was a very low value of the thermal conductivity, much lower than in several measurements exploring properties of polycrystalline forms of SnSe. The puzzle was solved lately by Lee et al. (2019), who realized that the ball-milled polycrystalline samples of SnSe readily oxidize and contain large quantities of SnO and SnO_2 that have a high thermal conductivity. Upon reducing the oxide in H_2, the thermal conductivity of the polycrystalline SnSe dropped dramatically, and its figure of merit (solid blue curve) became comparable to that of the single crystal. Further explorations of SnSe might, indeed, result in a superior thermoelectric material.

6.6 STABILITY OF SKUTTERUDITES

While skutterudites show exceptional thermoelectric properties in a laboratory environment, transferring their attributes to a practical operational setting of thermoelectric modules is a different matter. The following sections will consider several important topics that affect reliable long-term operation of modules based on thermoelectric skutterudites.

6.6.1 COMPOSITIONAL STABILITY

According to the phase diagram in Figure 2.1, skutterudites are peritectic systems with the MX_3 phase decomposing at the peritectic temperature given in Table 2.1. In the case of $CoSb_3$, the skutterudite of the greatest interest, this is a rather low temperature of 1146 K (873°C). In itself, this would not be a major impediment because the temperature is well above the peak performance of both n- and p-type skutterudites. However, because of elevated sublimation rates of Sb and high rates of oxidation in air, significant structural changes start to take place in skutterudites at much lower temperatures, often already at 600 K.

Problems with the decomposition and oxidation of skutterudites were recognized early on in the works of Wilson and Mikhail (1989), Wojciechowski et al. (2002), Hara et al. (2003), and Snyder and Caillat (2004). As the research progressed and the optimistic prognoses about the thermoelectric performance of skutterudites were further validated, the need to look closely into their chemical stability became more urgent. Possible decomposition routes and oxide reactions taking place in different atmospheres became the primary target of studies (Wen et al. 2009, Sklad et al. 2010, Godlewska et al. 2010a, and Leszczynski et al. 2011). The activation energy for oxidation in $CoSb_3$ was reported by Zhao et al. (2010a) to be 37.4 kJ/mol, while the activation energy for Sb sublimation was found by Zhao et al. (2011) as 44.5 kJ/mol. Problems with sublimation and oxidation were not limited to $CoSb_3$

but encompassed also filled skutterudites, as documented in reports by Wei et al. (2011), Biswas et al. (2011), Park et al. (2012), Xia et al. (2012), Tafti et al. (2015), and Wen et al. (2016). Sometimes, one might not have noticed that structural changes are taking place in the sample because the transport parameters have not changed abruptly. There are even occasions when the initial structural degradation actually improved the figure of merit, as reported by Wen et al. (2013) during their measurements of composite structures consisting of 1 vol% TiN nanoparticles dispersed in $Co_4Sb_{11.5}Te_{0.5}$. In this case, upon annealing at 773 K in vacuum for 100 hr, the porosity increased substantially as Sb evaporated, yet, due to a lower thermal conductivity degraded by enhanced phonon scattering on the pores, the figure of merit increased by 20%. However, setting aside such rare occasions, the structural degradation and the eventual electronic failure of skutterudites was well documented and called for action to mitigate the influence of Sb sublimation and sample oxidation.

Clearly, some kind of protective coating applied to the surface of skutterudite elements (legs) was highly desirable. But this had to be done judiciously in order not to shunt too much heat along the coated legs and degrade the performance. Moreover, considerations had to be given to the adherence of the coatings to the surface of skutterudites, while also trying to minimize differences between the thermal expansion of the coating and the skutterudite to avoid cracks in the coverage.

The early attempts to limit the oxidation and evaporation of Sb focused on magnetron sputter-deposited thin metallic coatings of Mo, V, Ta, and Ti applied to the skutterudite legs (El-Genk et al. 2006, 2007). Subsequent aging tests documented that the technique works and is able to mitigate the loss of Sb to temperatures of at least 900 K (D.-G. Zhao et al. 2014). Metal coatings were also applied by electrochemical deposition, an example being a double-layer of Al-Ni on $Yb_{0.3}Co_4Sb_{12}$ that, according to accelerated aging tests (873 K for 30 days) carried out by Bao et al. (2019), provided an excellent protection from Sb sublimation and oxidation. Not all metal coatings, though, were equally effective, and some failed due to cracking of the film at elevated temperatures (Godlewska et al. 2010b). Overall, the above studies served as an excellent proof of the principle that oxidation and Sb sublimation can be prevented. On the other hand, such metal coating are, of course, not a viable solution in the form they were used in the tests by applying them directly to the surface of the skutterudite legs because they electrically short the thermoelectric elements. Therefore, the interest has shifted to the exploration of electrically insulating coatings, among them aerogels (Sakamoto et al. 2011), enamels (Zawadzka et al. 2012, Park et al. 2015), and inorganic/organic silica layers that often contained dispersed glass or particles of granular alumina supposed to inhibit cracking during solidification of the gel (Dong et al. 2012, 2013). In general, such insulating coatings performed well in aging tests to temperatures of 873 K in vacuum. Perhaps the most convincing demonstration of the efficacy of enamel coatings to suppress oxidation is the work by Park et al. (2015). The authors used a silica-based enamel powder (0.1 to 1 μm size), consisting of 66 at% oxygen, 11 at% silicon, 11 at% sodium, 6 at% fluorine, 2 at% aluminum, 2 at% nitrogen, 1 at% barium, and 1 at% calcium, which was applied to the surface of both n- and p-type skutterudite samples in a graphite crucible. The samples, enamel powder, and crucible were heated in a furnace located in an argon-filled glove box at 700°C for 1 hr. The enamel-coated skutterudite samples were subjected to extended (8 days) isothermal air-stability tests at 500 and 600°C, respectively, as well as to thermal cycling consisting of 20 cycles between room temperature and 600°C at a rate of 100°C/hour, with the holding time of 1 hr at 600°C. The best way to show the effectiveness of the enamel coating is to compare the effect of heating on the topography and composition of samples without and with the coating. As Figures 6.22a and 6.22b indicate, unprotected skutterudite samples start to undergo structural changes at 300°C (573 K), in the case of n-type forms of the material, and even earlier at 200°C (473 K) in the case of p-type samples, after 1 hr in air at the respective temperature. In general, because of a larger content of air sensitive rare earth elements in p-type skutterudites than in n-type skutterudites, the former is more prone to oxidation.

With the enamel coating covering the skutterudite samples, the outcome was very different in both isothermal heating tests and thermal cycling. The enamel remained intact, no surface oxide was detected, and the element composition across the enamel-skutterudite interface was uniform in both n- and p-type skutterudites (Figure 6.23).

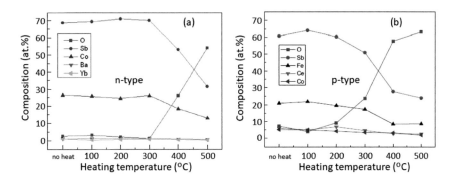

FIGURE 6.22 Changes in the composition of unprotected (a) n-type $Ba_{0.05}Yb_{0.025}Co_4Sb_{12}$ and (b) p-type $Ce_{0.9}Fe_{3.5}Co_{0.5}Sb_{12}$ skutterudites after heating to temperatures of 100°C, 200°C, 300°C, 400°C and 500°C and parking there for 1 hr in air. Reprinted from Y.-S. Park et al., *Journal of Materials Science* 50, 1500 (2015). With permission from Springer.

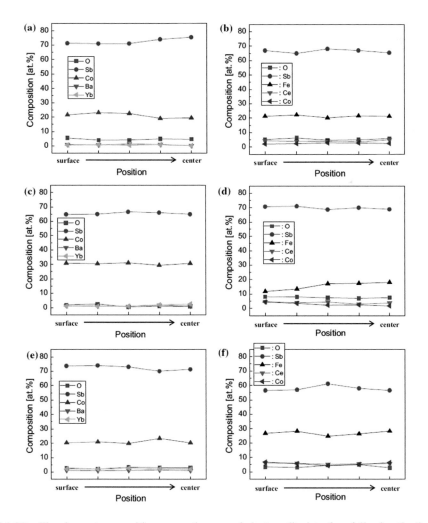

FIGURE 6.23 The element composition across the enamel-skutterudite interface following the thermal stability test at 600°C for 8 days for (a) n-type skutterudite and (b) p-type skutterudite. (c) and (d) are element composition across the enamel-skutterudite interface after 20 cycles between room temperature and 600°C for n-type skutterudite and p-type skutterudite, respectively. Reproduced from Y.-S. Park et al., *Journal of Materials Science* 50, 1500 (2015). With permission from Springer.

The results indicate that enamels are able to combat effectively oxidation problems associated with skutterudites being exposed to high temperatures in air. Moreover, their low thermal conductivity assures a minimal thermal shunt, provided the coating is sufficiently thin. As an example, for typical leg dimensions of 3.5 mm × 3.5 mm × 6 mm, an enamel coating of 0.1 mm is estimated by the authors to represent no more than 1–2% parasitic heat loss.

6.6.2 MECHANICAL INTEGRITY OF SKUTTERUDITES

Apart from excellent thermoelectric performance and thermal stability, a practically viable thermoelectric material must also possess robust mechanical properties. This is particularly important for power generator applications where a thermoelectric module might be subjected to large external and internal stresses and could be exposed to intense vibrations. Elastic moduli, which determine mechanical properties of any structure, depend critically on processing and microstructure of a material, as documented by the fracture strength being inversely proportional to the square of the grain size (Rice 2000). The implication is that samples prepared from sub-micron size powders, typically obtained by ball milling, should have significantly higher fracture strength than the cast ingots having the grain size of typically tens of microns.

It is remarkable that in spite of possessing large structural voids, skutterudites are quite stable compounds. For instance, $IrSb_3$ can withstand the pressure of at least 42 GPa generated in diamond anvil cells (Snider et al. 2000). The existing studies of skutterudites have attested to the fact that they are among the best thermoelectrics as far as mechanical properties are concerned. Moreover, they are easily machineable.

The cubic structure of skutterudites is characterized by three elastic constants C_{11}, C_{12}, and C_{44}. Molecular dynamics simulations (Yang et al. 2011) give values C_{11} = 185.13 GPa, C_{12} = 46.95 GPa, and C_{44} = 43. 02 GPa at 300 K, in excellent agreement with the first-principles calculations. The values are expected to decrease monotonically with increasing temperature because the lattice becomes softer. The three important parameters, the Young's modulus, the Poisson ratio, and the ultimate strength under tension at 300 K are estimated to have values 165.9 GPa, 0.193 GPa, and 22.18 GPa, respectively. Under compression, skutterudites are notably more robust, with the compressive strength of 39.44 GPa for $CoSb_3$, the value supported by the fact that $CoSb_3$ remains intact in diamond anvil cells at pressures well above 20 GPa (Kraemer et al. 2007).

One problematic area for skutterudites is the failure under a relatively small applied shear deformation. An atomistic explanation for such catastrophic failure of $CoSb_3$ was given recently by Li et al. (2016), based on their large-scale molecular dynamics simulations that focused on the shear response along the most likely (010)/<100> slip system. The fracture apparently starts with the failure of the shorter Sb–Sb bond at the shear stress of 14.04 GPa, resulting in the destruction of the Sb_4 ring. In turn, the collapse of the Sb_4 rings results in a progressively greater slippage of Co-octahedrons and the severance of the Co–Sb bond. Once this happens, cavitation develops, which leads to an opening of a crack and the ensuing brittle fracture. The critical shear stress decreases with the increasing temperature and is 13.46 GPa at 800 K.

A collection of mechanical properties of various skutterudites is presented in Table 6.1. The data are reproduced from Zhang et al. (2010) and, unless otherwise noted, were obtained by the time-of-flight pulse technique (TOF) (Ledbetter et al. 1981), and by resonant ultrasound spectroscopy (RUS) (Migliori and Maynard 2005). The TOF technique measures the sound velocity from which the eleastic constants are determined via the following equations:

$$C_{11} = \rho_d v_L^2, \tag{6.47}$$

$$C_{44} = \rho_d v_T^2, \tag{6.48}$$

TABLE 6.1

Elastic Properties of Skutterudites at Room Temperature. The Material Density is Given as a Percentage of the X-Ray Density

Skutterudite	ρ_d	v_L	v_T	v_m	C_{11}	C_{44}	E	ν	B	θ_D	VH
CoSb$_3$[a]	99	4.76	2.83	3.14	-	61	148	0.23	90	327	-
CoSb$_3$[b]	95	4.65	2.80	-	158	57	-	-	82	319	-
CoSb$_3$[c]	>95	4.62	2.79	-	-	-	-	-	-	321	-
CoSb$_3$[d]	99	-	-	-	-	60	136	0.20	-	-	-
CoSb$_3$[e]	99.5	4.59	2.64	2.93	-	-	-	-	-	307	-
La$_{0.75}$Fe$_3$CoSb$_{12}$[b]	98	4.53	2.68	-	157	55	-	-	84	310	-
La$_{0.75}$Fe$_3$CoSb$_{12}$[c]	>95	4.48	2.66	-	-	-	-	-	-	308	-
Ce$_{0.75}$Fe$_3$CoSb$_{12}$[c]	>95	4.56	2.71	-	-	-	-	-	-	312	-
CeFe$_3$RuSb$_{12}$[d]	99	-	-	-	-	44	133	0.25	-	-	-
(Ba,Yb)$_{0.03}$Co$_4$Sb$_{12}$	99.2	4.59	2.75	3.04	159	57	140	0.22	84	323	542
Ca$_{0.07}$Ba$_{0.23}$Co$_{3.95}$Ni$_{0.05}$Sb$_{12}$	96	4.52	2.71	3.00	153	55	134	0.22	80	311	495
Ba$_{0.075}$Sr$_{0.025}$Yb$_{0.1}$Co$_4$Sb$_{12}$	99.4	4.65	2.79	3.08	165	59	144	0.22	86	327	562
LaFe$_4$Sb$_{12}$[a]	99	4.60	2.71	3.00	-	57	141	0.23	89	315	-
Mm$_{0.7}$Fe$_4$Sb$_{12}$	96.7	4.10	2.70	2.96	126	55	122	0.12	53	306	-
Mm$_{0.86}$Fe$_4$Sb$_{12}$[f]	99.7	4.34	2.55	2.83	147	51	126	0.24	80	298	470
DD$_{0.86}$Fe$_4$Sb$_{12}$	99.4	4.41	2.55	2.83	147	49	123	0.25	82	294	390
Ca$_{0.75}$Fe$_4$Sb$_{12}$[f]	99	4.35	2.59	2.87	132	49	121	0.25	73	300	411
DD$_{0.25}$Ca$_{0.75}$Fe$_4$Sb$_{12}$[f]	98.6	4.42	2.58	2.86	146	50	123	0.24	79	300	-
DD$_{0.75}$Ba$_{0.25}$Fe$_4$Sb$_{12}$	98.7	4.37	2.57	2.85	149	51	127	0.24	80	299	-
Mm$_{0.7}$Fe$_3$CoSb$_{12}$	98.2	4.43	2.72	3.00	151	57	136	0.20	75	313	528
DD$_{0.68}$Fe$_3$CoSb$_{12}$[f]	98.5	4.34	2.55	2.83	149	51	127	0.24	80	298	399
Ca$_{0.75}$Fe$_4$Sb$_{12}$	94.9	3.50	-	-	91	-	-	-	-	-	467
DD$_{0.68}$Fe$_4$Sb$_{12}$	88.8	4.05	2.43	2.68	119	43	105	0.22	62	276	253
Mm$_{0.68}$Fe$_{3.5}$Ni$_{0.5}$Sb$_{12}$	96.2	3.84	2.33	2.58	109	41	99	0.22	58	267	-
DD$_{0.25}$Fe$_{2.5}$Ni$_{1.5}$Sb$_{12}$	88.7	4.05	2.41	2.67	115	41	100	0.23	61	271	206

Longitudinal (v_L), transverse (v_T) and mean (v_m) sound velocities are in units of 10^5 cm/s, elastic constants C_{11} and C_{44} are in units of GPa, the Young's modulus E and the bulk modulus B are in GPa, and the Debye temperature θ_D is in K. VH stands for the Vickers hardness. Adapted from L. Zhang et al., *Mater. Sci. Technol. B* **170**, 26 (2010). With permission from Elsevier.

[a] Data from Recknagel et al. (2007).
[b] Data from Keppens et al. (1998).
[c] Data from Sales et al. (1997).
[d] Data from Ravi et al. (2008).
[e] Data from Caillat et al. (1996a).
[f] Sample was hand milled and hot pressed. Other samples were ball milled and hot pressed.

$$E = \rho_d \frac{3v_L^2 v_T^2 - 4v_T^4}{v_L^2 - v_T^2}, \tag{6.49}$$

$$\nu = \frac{E}{2C_{44}} - 1, \tag{6.50}$$

$$B = \frac{E}{3(1 - 2\nu)}. \tag{6.51}$$

Here, ρ_d is the density of the material, v_L and v_T are the longitudinal and transverse sound velocities, respectively, E is the Young's modulus, ν is the Poisson's ratio, and B is the bulk modulus. In
the RUS technique, the elastic constants are extracted from the frequency spectrum of the material.
The Vickers hardness entered in Table 6.1 was measured under a load of 1 N applied for 10 s in a
microhardness tester. Table 6.1 also contains values of the Debye temperature based on the velocity
measurements via an equation

$$\theta_D = \frac{h}{k_B}\left(\frac{3nN_A\rho_d}{4\pi M}\right)^{1/3} v_m, \tag{6.52}$$

where h is the Planck's constant, k_B is the Boltzmann constant, N_A is the Avogadro's number, n is
the number of atoms in the molecule, and M is the molecular weight. The mean velocity v_m follows
from the averaging of the longitudinal and transverse sound velocities assuming an isotropic polycrystalline material,

$$\frac{3}{v_m^3} = \frac{1}{v_L^3} + \frac{2}{v_T^3}. \tag{6.53}$$

Another important mechanical aspect of a material is its response to cycling loading referred to as
mechanical fatigue. Fatigue tests were conducted by Ruan et al. (2010) on well-compacted $CoSb_3$
and $Ce_{0.5}Fe_{1.5}Co_{2.5}Sb_{12}$ samples having 99% theoretical density. The samples were subjected to
repeated loading with 60% and 80% maximum compressive strength (determined as 384 MPa, an
average value obtained on numerous identical samples) and the residual compressive strength was
measured after a predetermined number of cycles. In addition, the number of cycles leading to a
catastrophic failure was also recorded. After 5,000 loading cycles with 60% compressive strength,
the test revealed a 19% decrease in the residual compressive strength to 312 MPa. As the number
of loading cycles increased, the residual compressive strength decreased and, after 18,000 cycles
became only 222 MPa, a decrease of some 42%. The total number of loading cycles with 60%
compressive strength prior to a catastrophic failure was in the range of 35–45 thousands. The number was roughly halved when the loading cycles were performed with 80% compressive strength.
The experiments revealed a rather strong fatigue effect on the residual compressive strength of
skutterudites.

As pointed out several times, pores, particularly when having a range of sizes from nanometers to microns, are very effective in scattering phonons leading to a significantly reduced thermal conductivity. In the case of skutterudites, an exemplary case is an unfilled double-doped
$CoSb_{2.95}Si_{0.064}Te_{0.256}$, which, upon annealing, lost a large fraction of its overstoichiometric amount
of Te by evaporation. The pores left behind ranged in size from several hundred nanometers to
several micrometers (Khan et al. 2017), and were hailed as being responsible for the exceptionally high value of $ZT = 1.6$ at 800 K. This is obviously an exciting discovery, especially given an
unfilled nature of skutterudites, but a question looms to what extent the presence of pores affects
the mechanical stability of the structure. A preliminary estimate, based on molecular dynamics
simulations carried out by Li et al. (2015), indicates that both the tensile and compressive elastic
moduli decrease with the porosity, leading to about halving the strain a single crystal of $CoSb_3$ can
withstand prior to a catastrophic failure. Not surprisingly, the tensile strength is reduced much more
than the compressive strength. Interestingly, the effect does not seem to depend strongly on the pore
radius in the range 10–30 Å. If the presence of pores becomes an important aspect how to improve
the thermoelectric performance of skutterudites, more studies of their effect on the mechanical
properties will be called for to make sure that the structural integrity of the material is not compromised and skutterudites remain viable thermoelectrics.

6.6.3 Thermal Expansion of Skutterudites

Power-generating thermoelectric modules operate over a wide temperature range of several hundred degrees. Since n- and p-type elements (legs) are made of different materials, the issue of thermal expansion is of major relevance regarding the integrity of the module. Even when both n- and p-type legs are made of the same or very similar skutterudite framework, different fillers, dopants, and substitutions of Fe for Co all may alter the linear thermal expansion coefficient. Any significant asymmetrical strain developed between n- and p-type legs may lead to a fracture and failure of the module. Consequently, thermal expansion of various forms of skutterudites has been of interest over the years. Of course, because the thermoelectric elements must be electrically connected, compatible thermal expansion of solders and welds is of equal importance. This topic will be considered in the following sections dedicated to skutterudite modules. Here, the discussion will be limited to thermal expansion of the skutterudite legs themselves.

The first report on the thermal expansion of skutterudites, $IrAs_3$ and $IrSb_3$, was made by Kjekshus and Rakke (1974), giving the average linear expansion coefficients of 6.7×10^{-6} K^{-1} and 8.0×10^{-6} K^{-1}, respectively, for the temperature range 300 K–740 K. The next batch of data, again mostly on the binary forms of the structure, were provided by Slack and Tsoukala (1994) and by Caillat et al. (1994, 1996). Since then, several other entries were made in the literature regarding thermal expansion of skutterudites. The most comprehensive collection of the data can be found in Table II of Rogl et al. (2010b). A subset of the entries, indicating thermal expansion of different forms of skutterudites, is presented in Table 6.2.

In general, phosphide-based skutterudites have a small thermal expansion coefficient, followed by arsenides and skutterudites with the $[Pt_4Ge_{12}]$ framework. Antimonide skutterudites show the largest thermal expansion. However, there is considerable variation within each pnicogen family, depending on the filling fraction and the degree of substitution of Co by Fe. High filling levels and large contents of Fe tend to increase the thermal expansion coefficient. Consequently, n-type skutterudites with the substantially $[Co_4Sb_{12}]$ framework expand less than p-type skutterudites with high filling fractions and the essentially $[Fe_4Sb_{12}]$ framework.

6.7 SKUTTERUDITE THERMOELECTRIC MODULES

There is no question that skutterudites are excellent thermoelectric materials for power conversion applications in the temperature range 500 K–800 K. Following the worldwide intensive effort, both n- and p-type forms of the structure have attained figures of merit well in excess of the unity, the benchmark considered important for a reasonably efficient energy conversion. Moreover, skutterudites are mechanically robust, their fabrication is straightforward, and the raw materials they are made from are plentiful and among the least expensive of all thermoelectric materials. The most problematic issue concerning skutterudites, namely the volatilization of Sb and oxidation in air environment, seems to be successfully solved by applying protective coatings for the skutterudite legs.

In spite of these clear advantages of skutterudites, it is surprising that there are no commercially available modules on the market. I see a number of reasons why this may be so. Perhaps it is the reluctance of investors and module manufacturers to commit to the development of a new type of the thermoelectric module, given that Bi_2Te_3-based, PbTe-based, and Si-Ge technology has served well for the past 50 years. It is also possible that frequent reports in the literature regarding problems with the loss of Sb and oxidation in air operations that require special attention, and may even necessitate hermetically sealing the modules, are sufficient disincentives to go ahead and develop skutterudite modules.

Perhaps the biggest blow to the development of skutterudite modules came from the area where they could have played the most prominent role and, by far, would represent the largest scale applications of thermoelectricity. A few years ago, many leading car companies had a vibrant research

TABLE 6.2

Lattice Parameter and Linear Coefficient of Thermal Expansion of Various Forms of Skutterudites

Compound	a (Å)	α (10^{-6} K^{-1})	Range (K)	Reference
$CoSb_3$	9.03484	9.1	120–220	Rogl et al. (2010)
$IrSb_3$	9.2503	6.6	300–673	Slack & Tsoukala (1994)
$RhSb_3$	9.2255	12.7		Caillat et al. (1994)
$IrAs_3$	8.4691	8	293–746	Kjekshus (1961)
$Fe_{0.2}Co_{3.8}Sb_{11.5}Te_{0.5}$	9.0435	8.8	300–700	Mallik et al. (2013)
$Co_{0.95}Pd_{0.05}Te_{0.05}Sb_3$	9.0547	10.7	303–773	Schmidt et al. (2012)
$LaRu_4P_{12}$	8.0605	4.7	10–300	Rogl et al. (2010)
$PrRu_4P_{12}$	8.0493	5.6	10–300	Rogl et al. (2010)
$GdRu_4P_{12}$	8.0375	5.4	150–300	Rogl et al. (2010)
$LaOs_4P_{12}$	8.0932	5.1	100–300	Rogl et al. (2010)
$CeOs_4P_{12}$	8.0751	5.08	100–300	Rogl et al. (2010)
$PrOs_4P_{12}$	8.0813	4.77	100–300	Rogl et al. (2010)
$NdOs_4P_{12}$	8.0790	5.8	100–300	Rogl et al. (2010)
$SmOs_4P_{12}$	8.0731	5.39	100–300	Rogl et al. (2010)
$LaFe_4As_{12}$	8.3273	9.25	160–250	Rogl et al. (2010)
$PrFe_4As_{12}$	8.3100	10.3	160–250	Rogl et al. (2010)
$NaFe_4Sb_{12}$	9.1759	11	100–300	Leithe-Jasper et al. (2004)
$CaFe_4Sb_{12}$	9.171	10.9	150–300	Rogl et al. (2010)
$Tl_{0.22}Co_4Sb_{12}$	9.056	7.4	180–300	Sales et al. (2000)
$Tl_{0.8}Co_3FeSb_{12}$	9.1120	9.5	180–300	Sales et al. (2000)
$La_{0.9}Fe_4Sb_{12}$	9.1503	11.7	100–300	Rogl et al. (2010)
$Ce_{0.9}Fe_4Sb_{12}$	9.1406	12.7	100–300	Rogl et al. (2010)
$Ce_{0.9}Fe_{3.5}Co_{0.5}Sb_{12}$	9.1260	13.3	303–773	Schmidt et al. (2012)
$PrFe_4Sb_{12}$	9.1290	11.21	125–300	Rogl et al. (2010)
$Nd_{0.85}Fe_4Sb_{12}$	9.1412	12.5	100–300	Rogl et al. (2010)
$Eu_{0.93}Fe_4Sb_{12}$	9.1725	10.6	100–296	Rogl et al. (2010)
$Yb_{0.95}Fe_4Sb_{12}$	9.1586	12	110–295	Rogl et al. (2010)
$Yb_xCo_4Sb_{12}$	9.048	8.17	100–300	Da Ros et al. (2008)
$DD_{0.86}Fe_4Sb_{12}$	9.1357	11.26	160–245	Rogl et al. (2010)
$DD_{0.44}Fe_{2.1}Co_{1.9}Sb_{12}$	9.0878	9.45	160–245	Rogl et al. (2010)
$DD_{0.68}Fe_3NiSb_{12}$	9.1208	12.1	160–245	Rogl et al. (2010)
$DD_{0.08}Fe_2Ni_2Sb_{12}$	9.0927	9.8	160–600	Rogl et al. (2010)
$Mm_{0.76}Fe_4Sb_{12}$, BM	9.1370	13.43	160–250	Rogl et al. (2010)
$Mm_{0.70}Fe_3CoSb_{12}$, BM	9.1165	11.33	160–245	Rogl et al. (2010)
$Mm_{0.68}Fe_3CoSb_{12}$	9.1167	11.4	300–500	Rogl et al. (2010)
$Mm_{0.05}FeCo_3Sb_{12}$	9.0624	9.97	300–500	Rogl et al. (2010)
$La_{0.743}Fe_{2.74}Co_{1.26}Sb_{12}$	9.0971	9	300	Chakoumakos et al. (1999)
$Ce_{0.3}Co_{2.57}Fe_{1.43}Sb_{12}$	-	11.5	300–723	Salvador et al. (2011)
$Ca_{0.07}Ba_{0.23}Co_{3.5}Ni_{0.5}Sb_{12}$	9.0665	9.19	160–280	Rogl et al. (2010)
$Sr_{0.02}Ba_{0.075}Yb_{0.1}Co_4Sb_{12}$	9.0617	8.85	160–245	Rogl et al. (2010)
$Pt_4Sn_{4.4}Sb_{7.6}$	9.3304	6.94	130–230	Rogl et al. (2010)
$BaPt_4Ge_{12}$	8.6928	10.2	160–245	Rogl et al. (2010)

The bulk of the entries are measurements of Rogl et al. (2010). With permission from Elsevier.

program to develop skutterudites modules for harvesting waste heat along the tail pipe and converting it to electricity. There were two primary motivations: to improve the gas mileage by as much as 5% by eliminating the alternator, and to replace the electricity it provided by the dc power generated by thermoelectric modules. Although the initial tests were promising and saving of typically 2–3% in gas mileage were achieved, over the past five years, the interest of car companies has unfortunately veined, as the long-term strategy has shifted to battery-operated vehicles rather than gasoline and diesel engines. In essentially all tests of thermoelectric modules to recover heat generated by car engines, the thermoelectric stack of modules was placed far down the tail pipe, behind the catalytic converter, where the streaming gas temperature has dropped dramatically and the necessary high ΔT was no longer available to assure efficient operation of thermoelectric modules. Placing the modules closer to the engine block, where high streaming gas temperatures are available, was conflicting with the requirement to heat the catalytic converter as soon as possible to guarantee its operation in order to fulfill the stringent emission standards. In my opinion, the dilemma could have been solved by designing a simple bypass that would feed the streaming hot gas to the modules after the operation of the catalytic converter was established. However, making such changes, and taking into account the added cost of the thermoelectric modules, was deemed too expensive. In spite of the above setback, several large research groups have tried to develop skutterudite-based thermoelectric modules, and I am pleased to describe several of them here.

As part of the research program of NASA to develop a new generation of thermoelectric power generators for space applications, where they convert heat from radioactively decaying $^{238}PuO_2$ pellets and convert it for electricity to operate all on-board instrumentation during the space flights, Caillat et al. (1999, 2001) at the Jet Propulsion Laboratory, reported on the first attempt to assemble a couple made from the Te-doped $CoSb_3$ n-type leg in combination with the p-type leg of $Ce_{0.8}Fe_{3.5}Co_{0.5}Sb_{12}$. The team developed all essential interconnect technology by brazing a Nb bridge across the n- and p-type elements at the hot junctions and used Bi–Sn solder to couple the cold end of the legs to a Cu-plated Al_2O_3. The unicouple operated over the temperature difference of 510 K, and with the hot side at 873 K delivered 1.3 W of electrical power at a current of 15 A, operating with approximately a 10% efficiency. The group also developed the technology for segmented unicouples that combined short sections of n- and p-type Bi_2Te_3-based materials (for the low temperature end of the operation) with the skutterudite legs effective above 500 K. By pointing out the critical issues a successful module must satisfy, the above pioneering work opened the door for several future initiatives to develop skutterudite-based module. Among the critical issues are preventing diffusion between the TE legs and electrodes, keeping the internal resistance to a minimum, minimizing interconnect resistivity, and assuring robustness of modules during thermal cycling to high temperatures, especially in an air environment. Of course, as with every new effort, the experience with the development of skutterudite modules was gained through successive experimental work.

Diffusion barriers, applied in the form of plasma sprayed Mo to both ends of the skutterudite legs by Zhao et al. (2010b), proved very satisfactory during the operation and after five repeated cycles to high temperatures. As n-type legs was used $Yb_{0.25}Co_4Sb_{12}/Yb_2O_3$ (excess Yb oxidized as the ingot was powdered in air prior to SPS compaction), while p-type legs were made from $Ce_{0.45}Co_{2.5}Fe_{1.5}Sb_{12}$. The hot side of the legs was attached to a 1-mm-thick Mo–Cu plate, found to match the thermal expansion of $CoSb_3$. The actual joint was realized by a short SPS processing at 1023 K under 10–20 MPa using Ag–Cu solder. The low temperature ends of the elements were joined to Cu electrodes using ordinary Sn-based solder. Operating over a temperature difference of 490 K, the two connected couples generated 210 mV open voltage and delivered 140 mW when the load resistance matched the internal resistance of 73 mΩ. The design called for the thermoelectric efficiency of 8.4%, while the device operated with 6.4% efficiency, the performance degraded by the contact resistance and heat losses. More details concerning the development of skutterudite modules, their testing, and the use of aerogel-based thermal insulations were described by Sakamoto et al. (2011). Suitable brazing materials, taking into account thermal expansion in relation to $CoSb_3$, were explored by Lee et al. (2017). The most thorough examination of the reaction zone formed at the interface of skutterudites with metals was made recently by Grytsiv et al. (2020). The authors made numerous diffusion couples between n-and p-type skutterudites with a variety of metals likely to be used as diffusion barriers, and inspected the interfacial region following heating to 873 K for 1100 hr. Based on the study, to assure chemically and mechanically stable diffusion barriers for skutterudite-based modules, the diffusion barriers should satisfy two main criteria: (1) the barrier should be in equilibrium with the skutterudite element (leg) and (2) the constituting elements of the barrier should have low solubility in the skutterudite material. The findings provide a good guidance for manufacturing skutterudite-based thermoelectric modules.

Multifilled skutterudites, n-type legs made from $Yb_{0.3}Ca_{0.1}Al_{0.1}Ga_{0.1}In_{0.1}Co_{3.75}Fe_{0.25}Sb_{12}$ with the $ZT = 1$ and p-type legs from $La_{0.7}Ba_{0.1}Ga_{0.1}Ti_{0.1}Fe_3CoSb_{12}$ with the $ZT = 0.65$, were used to assemble a 32-unicouple module by Guo et al. (2012). The team used Co–Fe–Ni-based alloys that served a dual purpose as both diffusion barriers and electrodes. The composition of the alloy was readily adjusted to match the thermal expansion of n- and p-type elements. The module was tested in operations over the temperature difference of 550 K, and developed the maximum power output of 32 W, with the conversion efficiency of about 8%. The modules proved very stable, with the performance unchanged after 100 heating/cooling cycles between 873 K and 473 K. Even after additional 240 hr at 873 K in vacuum, the operational parameters have not changed. Moreover, no interdiffusion was observed at the skutterudite-diffusion barrier interface even after one-month-long annealing at 873 K. A year later, replacing expensive Yb with much cheaper La in their n-type skutterudite legs, the team reproduced the power output of 32 W with a similar 32-unicouple module (Geng et al. 2013). Of course, operating in vacuum as they did, the authors avoided the major problem affecting skutterudites, which is severe oxidation in air environment. This was not the case of 9-unicouple modules fabricated from skutterudites of similar composition by García-Cañadas et al. (2013) but operated in air. Even during relatively low temperature operations with the hot junction not exceeding 573 K, a dramatic increase in the internal resistance of the module was noted just after four thermal cycles. A similar outcome was reported by Katsuyama et al. (2019) when operating in air their 14-couple module fabricated from n-type $Yb_{0.15}Co_4Sb_{12}$ and p-type $CeFe_3CoSb_{12}$ with Ag electrodes bonded with a Ag-23 mass% Pd paste sintered at 723 K for 7 hr in vacuum under the uniaxial pressure of 14.7 MPa. While the performance was satisfactory when operating below 573 K, increasing the temperature of the hot side above 573 K led to an increase in the resistance of the couple and the reduced power output. The inspection of the module revealed the presence of Sb_2O_3 oxide at the hot end of the p-type legs. The performance of n-type skutterudite in this module was very disappointing, and was actually inferior to the performance of p-type skutterudite.

Among the car companies actively engaged in the development of thermoelectric modules for the recovery of waste heat in cars and trucks, the team led by GM Global Research & Development with additional support provided by funding from the US Department of Energy, has played an important role. The GM-led consortium supported not only the development of skutterudite materials but also the design, assembly, and performance evaluation of the modules under operational conditions in the company's test vehicles. The key achievements are described in several publications, among them Salvador et al. (2013, 2014). The actual skutterudite modules were assembled at Marlow Industries, and are shown in Figure 6.24. Each module consisted of 32 p–n couples of nominal composition $Yb_{0.09}Ba_{0.05}La_{0.05}Co_4Sb_{12}$ (n-type) and $Mm_{0.30}Fe_{1.46}Co_{2.54}Sb_{12.05}$ (p-type, with Mm standing for the mischmetal). The thermoelectric materials were prepared by arc-melting followed by long-term annealing. The ingots were powdered and consolidated by SPS. The dimensionless figure of merit of n-type legs of around 1.2 at 750 K was quite competitive, while p-type legs were distinctly inferior with the $ZT \sim 0.5$ at 750 K. A diffusion barrier of Mo was arc-sprayed between the skutterudite legs and electrodes. The legs were brazed to Al pads that were directly bonded to thin plates of alumina. To minimize Sb sublimation, oxidation, as well as to limit convective and radiative heat losses between the legs of the couples, aerogel thermal insulation was applied, as indicated in the lower left part of Figure 6.24. The performance of modules was tested at the Fraunhofer Institute for

FIGURE 6.24 A photograph of nine skutterudite modules assembles at Marlow Industries as part of the GM Global Research & Development efforts to develop skutterudite-based thermoelectric modules for the recovery of waste heat in cars and trucks and its conversion to electricity. Each 5-cm-by-5-cm module consists of 32 p-n skutterudite couples with a Mo diffusion barrier between the elements and the electrodes. At the lower left-hand corner, there is a module encapsulated with aerogel that protects against Sb sublimation and oxidation, and also provides thermal insulation to minimize convective and radiative heat losses through the space between the legs. Reproduced from J. R. Salvador et al., *J. Electron. Mater.* 42, 1389 (2013). With permission from Springer.

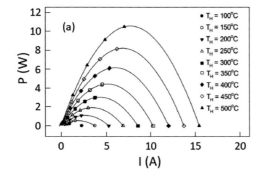
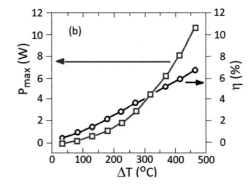

FIGURE 6.25 (a) The power generated by the skutterudite module as a function of the current at different temperatures of the hot junction, with the cold junction maintained at room temperature. (b) The maximum output power (left-hand scale) and the conversion efficiency (right-hand scale) achieved when the skutterudite module was tested by being clamped between the heater and the sink with the pressure of 0.9 MPa. Adapted from Salvador et al., *Physical Chemistry and Chemical Physics* 16, 12510 (2014). With permission from the Royal Society of Chemistry.

Physical Measurement Techniques and the data obtained are shown in Figures 6.25a and 6.25b. The maximum conversion efficiency under the temperature difference of 450 degrees across the module was about 7%, the performance undoubtedly affected by rather poorly performing p-type legs, parasitic heat losses, and the effect of contact resistances. Nevertheless, the performance was comparable to the results reported on skutterudite modules by Zhao et al. (2010b), Guo et al. (2012), Ochi et al. (2014), and Nie et al. (2017), and a bit lower than 9.1%, obtained on a skutterudite unicouple by Muto et al. (2013). The last mentioned authors made an interesting variant of a unicouple in an

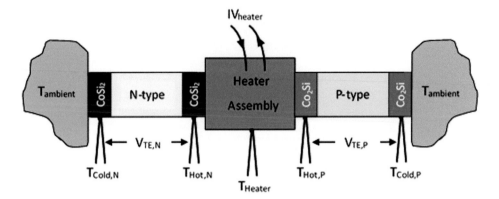

FIGURE 6.26   A sketch of an in-line unicouple consisting of n-type $Yb_{0.35}Co_4Sb_{12}$ and p-type $NdFe_{3.5}Co_{0.5}Sb_{12}$ with $CoSi_2$ electrodes at both ends of the n-type leg and Co_2Si electrodes at both ends of the p-type leg. The unicouple operated between 70°C and 550°C, i.e., across the temperature difference of 480 degrees with the efficiency of 9.1%. Adapted from A. Muto et al., *Advanced Energy Materials* 3, 245 (2013). With permission from Wiley-VCH Verlag.

in-line configuration, sketched in Figure 6.26, with $CoSi_2$ electrodes at both ends of the n-type legs and Co_2Si electrodes at both ends of the p-type legs, all applied during hot pressing of the ingots.

Efficiency of 9.15% was also reported by Nie et al. (2019) from the Tsukuba Development Center for their 32-couple module made with multiple-filled n-type $(Yb,Ca,Ga,In)_{0.7}Fe_{0.25}Co_{3.75}Sb_{12}$ and p-type $(La,Ba,Ga,Ti)_{0.9}Fe_3CoSb_{12}$ legs operated across the temperature difference of 600 degrees. The module was fabricated using n-type legs of a bulk heterojunction material formed by mixing 1:1 two skutterudite compositions with different Yb content. The enhanced electrical conductivity of n-legs was believed to arise due to the carrier transfer from the heavily Yb-doped regions to the lightly Yb-doped regions of the sample, while the presence of two phases lowered the lattice thermal conductivity. As p-type legs was used a multi-filled $Pr_{0.8}Ti_{0.1}Ga_{0.1}Ba_{0.1}Fe_3CoSb_{12}$ skutterudite. Regarding the performance, however, it should be noted that the $\Delta T \approx 600°C$ is some 100 degrees higher than the usual temperature difference of around 500°C applied in most other studies. Based on the authors' data for $\Delta T \approx 500°C$, the efficiency of about 7.3% would be on par with other reports. I should also mention a module developed by Park et al. (2018) that used $(Mm,Sm)_yCo_4Sb_{12}$ (n-type) and $DD_yFe_3CoSb_{12}$ (p-type, with DD standing for didymium) legs. An interesting point about this module is a diffusion barrier of Fe–Ni (alloy of 65 at% Fe + 35 at% Ni) that was applied at the hot junction between the skutterudite elements and Cu electrodes. The measured specific resistance of such contacts was (2.2–2.5) $\mu\Omega$ cm^2, some of the lowest values reported in the literature. The authors did not provide the efficiency of their module but noted that it delivers the maximum power of 8 W under the current of 14 A and the hot junction at 873 K.

Recently, the Shanghai Institute of Ceramics fabricated a series of skutterudite-based modules, each consisting of 36 p-n couples and developing power density of 1 Wcm^{-2} with the outstanding conversion efficiency of 10%. The modules used $Yb_{0.3}Co_4Sb_{12}$ as n-type legs and $Ce_{0.85}Fe_3CoSb_{12}$ for p-type legs. The diffusion barrier layer was Ti–Al alloys, and brazing to electrodes was done with $Mo_{50}Cu_{50}$. A collection of the modules is shown in Figure 6.27.

In Section 1.4, discussing properties of composites based on the skutterudites matrix, we have seen that various forms of graphite tended to enhance the figure of merit. In particular, 0.72 vol% of reduced graphene oxide, rGO, added to $Yb_{0.27}Co_4Sb_{12}$ and 1.4 vol% of the same added to $Ce_{0.85}Fe_3CoSb_{12}$ seemed very effective, and Zong et al. (2017) assembled a 16-couple module using the above n- and p-type skutterudites and compared its performance to the identical 16-couple module that had no rGO additives. The modules were operated across 577 K, with the hot side at 873 K. The composite module delivered the highest power output of 3.8 W with the efficiency of

FIGURE 6.27 A collection of skutterudite modules fabricated by the Shanghai Institute of Ceramics, Chinese Academy of Science. Each module consists of 18 p-n couples with the leg dimensions of $4 \times 4 \times 8$ mm^3, and the overall module size of $30 \times 30 \times 10$ mm^3. The n-type legs are $Yb_{0.3}Co_4Sb_{12}$ and p-type elements are $Ce_{0.85}Fe_3CoSb_{12}$. The diffusion barrier layer of Ti-Al is coupled with a welding layer of Ni, and the electrodes are attached by brazing with $Mo_{50}Cu_{50}$. The modules develop maximum conversion efficiency of 8% and the power density of 1 Wcm^{-2}. Courtesy of Prof. S. Q. Bai of the Shanghai Institute of Ceramics.

8.4%, while the module without rGO additives attained the highest power of 3.1 W, operating with the efficiency of 6.8%.

The common feature of many of the above skutterudite modules was the distinctly inferior performance in comparison to the expected outcome based on the properties of the thermoelectric elements used in the construction of the modules (even though on several occasions, better choices could have been made for the p-type legs). The results vividly documented the important role of interconnects and the serious effect of the parasitic losses in the overall performance of the modules. Moreover, even the time-honored design based on a 1D heat flow may not be adequate enough to capture the complexity of parameter optimization in the operation of modules at high temperatures. Clearly, to make a high-performing thermoelectric module, one has to use a design strategy that fully integrates optimal approaches that minimize energy losses. An example of such a rational strategy was a fabrication of thermoelectric modules comprising segmented legs of skutterudites with Bi_2Te_3, which achieved the conversion efficiency of 12% (Zhang et al. 2017), the highest efficiency of any family of thermoelectric materials when the temperature difference over the module is not to exceed 600 K. The design started with 3D numerical modeling of a segmented module that included the temperature-dependent material properties, determination of optimal geometrical factors, and the effect of all relevant interconnects based on the developed bonding schemes. An 8-couple module used n-type $Bi_2Te_{2.5}Se_{0.5}$ and p-type $Bi_{0.4}Sb_{1.6}Te_3$ for the low temperature range, segmented with n-type $Yb_{0.3}Co_4Sb_{12}$ and p-type $CeFe_{3.85}Mn_{0.15}Sb_{12}$, respectively, that generated the power at the higher temperature range. The space between the legs was filled with low thermal conductivity glass fibers to minimize the heat loss. The fabrication flow is depicted in Figure 6.28. The efficiency of the module at various temperatures of the hot and cold sides, as well as the 3D model of the segmented unicouple with all essential fabrication steps, are shown in Figure 6.29. Testing the stability of the operation under the imposed 520°C temperature difference across the module for 570 hr in the argon atmosphere resulted in the 1.8% increased

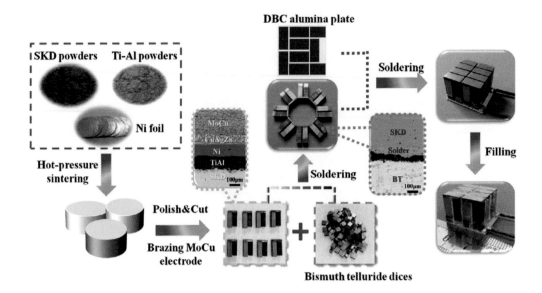

FIGURE 6.28 Fabrication flow diagram indicating manufacturing steps in the preparation and assembly of modules using segmented skutterudite/Bi$_2$Te$_3$ elements, which achieved conversion efficiency of 12% operating under the temperature difference of 541 K, with the hot end at 849 K. Reproduced from Q. Zhang et al., *Energy & Environmental Science* 10, 956 (2017). With permission from the Royal Society of Chemistry.

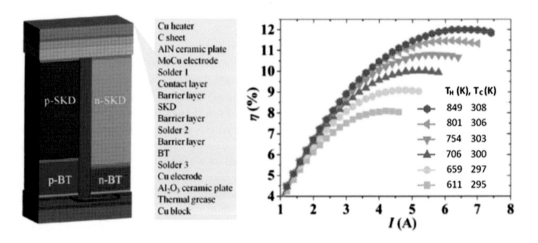

FIGURE 6.29 The panel on the left shows the 3D finite element model of the segmented Bi$_2$Te$_3$/skutterudite unicouple with the legend indicating details of all fabrication steps from the top to the bottom. The panel on the right depicts the measured conversion efficiency of an eight-couple module assembled from the unicouples shown on the left. Adapted from Q. H. Zhang et al., *Energy & Environmental Science* 10, 956 (2017). With permission from the Royal Society of Chemistry.

internal resistance and only 1.8% degradation in the power generated. Of course, some protective coating on the skutterudite segments would be necessary for operations in air to prevent oxidation and Sb evaporation.

It is interesting to compare the performance of various skutterudite modules described in the literature with the best reported efficiencies obtained with modules built using other high-performing thermoelectric materials. The comparative performance is shown in Figure 6.30. The skutterudite

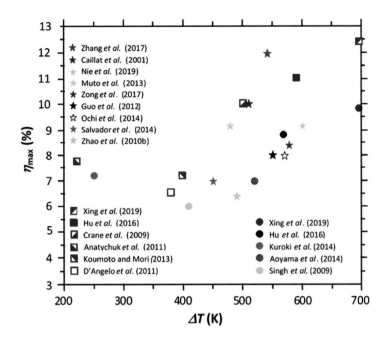

FIGURE 6.30 Comparison of the conversion efficiency of skutterudite modules designated by different color stars. Also included are several reported conversion efficiencies of single (circles) and segmented (squares) modules fabricated with different thermoelectric materials. Modules with segmented legs are indicated by various squares, and typically use Bi_2Te_3-based materials for the low temperature stage. Modules with legs of various single thermoelectric materials are designated by circles of different color. The highest conversion efficiency of 12.4% was measured by Xing et al. (2019) on a module with segmented legs of Bi_2Te_3 and half-Heusler alloys, designated by squares with the filled upper left half. It was operated with the temperature difference across the module of 698 K. At temperature differences comparable to those used across skutterudite modules (typically around 550 K), the efficiency dropped to just above 10%, making skutterudite modules the best-performing thermoelectric energy converters in the temperature range 500 K–800 K.

modules are shown by stars of various color. Modules with segmented legs of other thermoelectrics, typically using Bi_2Te_3-based materials for the lower temperature range of operation, are shown with different squares. Modules fabricated with legs of various single thermoelectric materials are depicted as circles with different colors.

The highest documented conversion efficiency of 12.4% was attained recently with a carefully designed module consisting of segmented legs of Bi_2Te_3 and half-Heusler (HH) alloys (Xing et al. 2019). Without segmentation with Bi_2Te_3, the otherwise identical HH module operated with the efficiency of 9.6%. However, both efficiencies quoted above were achieved with the temperature difference across the modules of nearly 700 K, i.e., the hot side of the module was maintained at 1013 K. Such a temperature is too high for any $CoSb_3$-based skutterudite material. When the hot side temperature was reduced to a comparable temperature used in tests of skutterudite modules, namely about 850 K, the efficiency of the Bi_2Te_3/HH module was reduced to just above 10%. Thus, in the temperature range of 500 K–800 K, the best-performing modules are those based on skutterudites, and carefully designed, assembled, and well protected ones should be able to operate reliably with the conversion efficiency of 10%, and perhaps even somewhat better.

Of course, high ZT thermoelectric materials are expected to operate with higher conversion efficiency than materials with low ZTs. However, fabrication of a thermoelectric module is a complex process that involves applying diffusion barriers, low resistance interconnects, and above all, efficient heat exchanges that couple heat to the module at the hot end and enable withdrawal of heat at the cold end. Thus, optimistic projections based on the transport performance of the thermoelectric material can be easily lost during the assembly of modules if insufficient care is paid to the engineering steps one must take to make a high-performing module, such as noted in Figures 6.28 and 6.29.

Finally, a few comments regarding evaluation of the conversion efficiency of thermoelectric modules. As noted in Section 6.3.2, and specifically Eqs. 6.13 and 6.25, the conversion efficiency depends on the temperature of the hot and cold sides of the module (the Carnot efficiency) that is modified (lowered) by the properties of the thermoelectric material expressed through the figure of merit Z. Thus, obtained conversion efficiency reflects the transport properties specific to the thermoelectric material, and is the highest efficiency a given thermoelectric can attain. Fabrication of modules introduces contact resistances, and the operation itself suffers from parasitic heat losses, both inevitably reducing the conversion efficiency. How much is the performance degraded depends on the design, choices of interconnects, effective mitigation of heat losses, and the care used during the assembly. To estimate the efficiency of the module thus calls for a direct measurement of its thermoelectric performance.

In principle, there are two basic approaches how to determine the module's conversion efficiency: (a) using a heat flow meter technique, and (b) relying on the Harman method. I outline both in turn.

(a) From the definition of the efficiency of a power generator, i.e., the ratio of the power generated to the heat drawn from the source, it follows that the critical parameter is the heat flux entering the module at the hot side. Heat is always more difficult to measure and control than the electrical current because it cannot be easily constrained to pass through a specific cross-sectional area on account of convective and radiative losses. Hence, great care must be used in the design of the heat flow meter to assure that convective and radiative losses are minimized and bulk of the measured heat, indeed, passes through the module. This is accomplished by sandwiching the module between metal plates of known thermal conductivity provided with thermocouples to determine the amount of heat passing through the plate from the dimensions of the plate, the placement of the thermocouples, and the known thermal conductivity of the plate. On top of the metal plate is placed a thin heater plate, isolated from above by a layer of thermally insulating material. To enhance the heat transfer between the heater plate and the metal plate and between the metal plate and the module, a thin sheet of graphite is often used. On the cold side of the assembly is used a high conducting grease that serves the same purpose. The assembly is clamped under pressure of typically 100 psi (about 0.7 MPa) and a cylindrical radiation shield if often placed around the assembly in order to minimize radiation losses from the stack by trying to match the thermal gradient along the stack. Convective losses are avoided by enclosing the entire assembly under a bell jar that is evacuated to about 10^{-3} Pa. A sketch of the setup is provided in Figure 6.31. One can improve the accuracy of the system by placing the identical metal plate also below the module and compare the amount of heat that actually passed through the module with the heat that entered it. One could even fancy a completely symmetrical setup where a heater is placed centrally and heats identical modules on both sides of it as well as identical heat flow meters on the cold side of the modules. Representative heat flow meter setups used in several laboratories worldwide are described by Wang et al. (2014). The power output of the module is obtained from the V–I curves collected at each predetermined temperature. Alternatively, the power P generated by the module is

$$P = I^2 R_L = \frac{V_0^2}{\left(R_m + R_L\right)^2} R_L,$$ where R_L and R_m are the load and module resistances, and V_0 is

the open-circuit voltage. The maximum power output is obtained when the load and module resistances are the same. Hence, $P_{max} = \frac{V_0^2}{\left(2R_L\right)^2} R_L = \left(\frac{V_0}{2}\right)^2 \frac{1}{R_L}.$

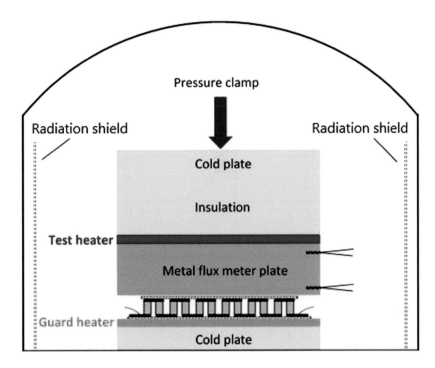

FIGURE 6.31 A sketch of the test system to determine conversion efficiency of a thermoelectric module by measuring the heat flux passing through the module.

(b) The essence of the Harman method is a difference between isothermal and adiabatic voltages developed across a module when a small current (small for reasons of keeping the Joule heat negligible) passes through the module. In practice, the conditions are finessed by suspending the module by its current wires or thermocouple leads in air or in an oven, making sure that no thermal disturbances are present. Upon supplying a small current to the module, the resistive voltage across the module, V_R, immediately sets in, and it takes some time before Peltier heating at the junctions results in a noticeably rising temperature difference across the module, which, in turn, gives rise to the Seebeck voltage, V_S, that adds to the resistive voltage. After more time has elapsed, the total voltage across the module reaches a steady state and no longer changes. On switching the current off, the voltage immediately drops by the amount of the resistive voltage and then slowly decreases further until no temperature difference, i.e., no Seebeck voltage is present. From the components of the resistive and Seebeck voltages, the figure of merit of the module is obtained as $ZT = \dfrac{V_{total}}{V_R} - 1 \equiv \dfrac{V_S + V_R}{V_R} - 1$. Figure 6.32 shows a typical voltage response of the module when the small dc current is switched on and off. With thus obtained ZT values at different temperatures, the conversion efficiency is calculated from Eqs. 6.13 or 6.23.

More details concerning the design, construction, operation, and testing of thermoelectric modules can be found in several publications, including among others monographs by Ioffe (1957), Goldsmid (1960, 1986, 2016), Burshteyn (1964), Harman and Honig (1967), Rowe and Bhandari (1983), and Nolas, Sharp and Goldsmid (2001). An excellent source is also the *Thermoelectric Handbook, Macro to Nano*, edited by D. M. Rowe (2006), and proceedings of the International Conference on Thermoelectrics published annually, in the last several years appearing as special volumes of the *Journal of Electronic Materials*.

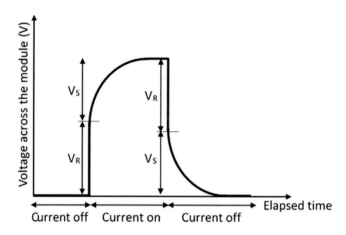

FIGURE 6.32 Resistive and Seebeck components of the voltage developed across a module in response to passing a small DC current during the so-called Harman method of measurement of the module's figure of merit.

6.8   APPLICATIONS OF SKUTTERUDITE MODULES

Many of the skutterudite modules were originally developed with an eye on applications in the area of waste heat recovery from the exhaust gas stream of cars and trucks. Special design considerations for this hostile environment of constantly changing temperatures and strong vibrations are amply described in the literature. However, because of the placement of the thermoelectric generators behind the catalytic converter, the heat content of the gas stream is low and most devices switched to Bi_2Te_3-based technology that is better suited to such lower temperature environment. Skutterudite-based module were used in several stationary applications, such as collecting heat from cement plants and steel mills. Skutterudites are also being considered as viable thermoelectrics for the NASA's new generation of radionuclei thermoelectric generators (RTGs) or possibly small portable nuclear reactors, replacing the existing technology based on Si–Ge alloys. The idea is to increase the operational temperature and thus generate more on-board power efficiently. Since Si–Ge alloys have particularly poor compatibility with other thermoelectrics, they cannot be readily segmented with other materials to optimize the performance of modules. Skutterudites are far more compatible and would likely be used in segmented couples generating power in the mid-temperature range of modules' operation. It is also likely that skutterudites will be used on Mars to supply power for the planned missions. In either application, on board deep-space crafts or on the surface of Mars, the operation of skutterudite modules will obviously not be compromised by oxidation.

Beyond waste heat recovery and power generation for deep space missions, an interesting application put in place is the solar photovoltaic-thermoelectric (PV-TE) hybrid power generation system, installed in Qinghai, China in 2016. The system was projected to supply 100 kW of electricity with the overall efficiency of 30%. Originally envisioned by Niino (2002) and by Zhang et al. (2005), using a wavelength splitter, the concentrated solar light is separated into the ultraviolet/visible range of wavelengths (200–800 nm) that impinge on the solar cell, while the long-wavelength light (800–3000 nm) heats the sun-absorbing membrane attached to a heat collector situated on top of a $CoSb_3/Bi_2Te_3$-segmented thermoelectric module. While the majority of power is generated by the PV cells, the thermoelectric generator contributes about 4–5% of the power. Beyond its power contribution, by splitting the light and avoiding the long-wavelength spectrum of light (heat) to interact with PV cells, their operation is protected from undesirable heating, detrimental to their operation. Integration of the system, starting with the beam splitter, individual PV and TE cells arranged in 6 × 4 panels, to the panels interconnected in the system is shown in Figure 6.33.

FIGURE 6.33 Elements of integration of the PV-TE hybrid power system installed by the Wuhan University of Technology at Qinghai, China. Highly concentrated solar light is split by a wavelength separator marked (1) into a short wavelength beam that passes vertically down toward a PV cell marked (6). Long-wavelength light (800–3000 nm) is collimated (2) and impinges on a thermoelectric module fabricated from $CoSb_3/Bi_2Te_3$ segmented legs at the Shanghai Institute of Ceramics, CAS. The heat passing through the module is dispersed with the aid of fins, marked (4). Each PV-TE unit is assembled into 6 × 4 blocks (middle panel), which, in turn, are interconnected to form a large power-generating system (right-hand panel). Courtesy of Prof. X. F. Tang.

I have no doubts that well designed, carefully fabricated and air-protected skutterudite-based module can develop reliable, unrivaled performance as thermoelectric power generators in the temperature interval between about 500 K and 800 K, converting waste heat into electricity with the efficiency of 10% and perhaps even higher. I should also note that the main chemical elements Co, Fe, and Sb used in the synthesis of skutterudites are, in general, plentiful and environmentally not objectionable. The more expensive rare earth elements are used only as fillers at quite limited quantities, and one can often use much cheaper Ce and La instead of Yb with only a modest effect on the performance. Compared to Bi_2Te_3- and PbTe-based thermoelectric modules that rely heavily on expensive and low-abundant Te, skutterudites modules hold a great advantage in the cost of materials. In this respect, skutterudites are comparable to half-Heusler alloys, their main competitor in the mid-to-high temperature range of operation. Readers interested in the cost-performance matrix of various thermoelectric materials may like to read articles by Yee et al. (2013) and LeBlanc et al. (2014).

REFERENCES

Altenkirch, E., *Physicalische Zeitschrift* **12**, 920 (1911).

Anatychuk, L. L., L. N. Vikhor, L. T. Strutynska, and I. S. Termena, *J. Electron. Mater.* **40**, 957 (2011).

Anno, H., K. Matsubara, Y. Notohara, T. Sakakibara, and H. Tashiro, *J. Appl. Phys.* **86**, 3780 (1999).

Aoyama, I., H. Kaibe, and L. Rauscher, *Jpn. J. Appl. Phys.* **44**, 4275 (2005).

Arushanov, E., K. Fess, W. Kaefer, C. Kloc, and E. Bucher, *Phys. Rev. B* **56**, 1911 (1997).

Ballikaya, S., N. Uzar, S. Yilderim, J. R. Salvador, and C. Uher, *J. Solid State Chem.* **193**, 31 (2012).

Ballikaya, S. and C. Uher, *J. Alloys Compd.* **585**, 168 (2014).

Bao, X., M. Gu, Q. H. Zhang, Z. H. Wu, S. Q. Bai, L. D. Chen, and H. Q. Xie, *J. Electron. Mater.* **48**, 5523 (2019).

Bardeen, J and W. Shockley, *Phys. Rev.* **80**, 72 (1950).

Biswas, K., M. A. Subramanian, M. S. Good, K. C. Roberts, and T. J. Hendricks, *J. Electron. Mater.* **41**, 1615 (2011).

Borelius, G, *Proc. Roy. Acad. Amsterdam* **33**, 17 (1930).

Borshchevsky, A., T. Caillat, and J.-P. Fleurial, *Proc. 15th Int. Conf. on Thermoelectrics*, IEEE Catalog Number 96TH8169, p. 112, Piscataway, NJ. (1996).

Burshteyn, A. J., in *Semiconductor Thermoelectric Devices*, Temple Press, London (1964).

Caillat, T., A. Borshchevsky, and J.-P. Fleurial, *AIP Conf. Proc.* **301**, 517 (1994).

Caillat, T., A. Borshchevsky, and J.-P. Fleurial, *J. Appl. Phys.* **80**, 4442 (1996a).

Caillat, T., J. Kulleck, A. Borshchevsky, and J.-P. Fleurial, *J. Appl. Phys.* **79**, 8419 (1996b).

Caillat, T., J.-P. Fleurial, and A. Borshchevsky, *J. Cryst. Growth* **166**, 722 (1996c).

Caillat, T., J.-P. Fleurial, G. J. Snyder, A. Zoltan, D. Zoltan, and A. Borshchevsky, *Proc. 18th Inter. Conf. on Thermoelectrics*, IEEE Catalog Number 99TH8407, p. 473, Piscataway, NJ (1999).

Caillat, T., J.-P. Fleurial, G. J. Snyder, and A. Borshchevsky, *Proc. 20th Inter. Conf. on Thermoelectrics*, IEEE Catalog Number 01TH8589, 282 (2001).

Chakoumakos, B. C., B. C. Sales, D. Mandrus, and V. Keppens, *Acta Crystallog. Section B: Struct. Sci.* **55**, 341 (1999).

Chasmar, R. P. and R. Stratton, *J. Electron. Control* **7**, 52 (1959)

Chen, J., Z. Yan, and L. Wu, *J. Appl. Phys.* **79**, 8823 (1996).

Chen, B., J. H. Xu, C. Uher, D. T. Morelli, G. P. Meisner, J.-P. Fleurial, T. Caillat, and A. Borshchevsky, Phys. Rev. B **55**, 1476 (1997).

Cho, J. Y., Z. X. Ye, M. M. Tessema, J. R. Salvador, R. A. Waldo, Jiong Yang, W. Q. Zhang, Jihui Yang, W. Cai, and H. Wang, *J. Appl. Phys.* **113**, 143708 (2013).

Crane, D. T., D. Kossakovski, and L. E. Bell, *J. Electron. Mater.* **38**, 1382 (2009).

Dahal, T., Q. Jie, W. S. Liu, K. Dahal, C. F. Guo, Y. C. Lan, and Z. F. Ren, *J. Alloys Compd.* **623**, 104 (2015).

Dahal, T., H. S. Kim, S. Gahlawat, K. Dahal, Q. Jie, W. S. Liu, Y. C. Lan, K. White, and Z. F. Ren, *Acta Mater.* **117**, 13 (2016).

D'Angelo, J., E. D. Case, N. Matchanov, C. Wu, T. P. Hogan, J. Barnard, C. Cauchy, T. Hendricks, and M. G. Kanatzidis, *J. Electron. Mater.* **40**, 2051 (2011).

Da Ros, V., C. Candolfi, J. Leszczynski, A. Daucher, B. Lenoir, S. J. Clarke, and R. Smith, *Proc. 6th Europ. Conf. on Thermoelectrics*, Paris, France, (2008), unpublished.

Davydov, B. J. and J. M. Shmushkevich, *Uspechy Phys.* Nauk (USSR) **24**, 21 (1940).

Dong, H. L., X. Y. Li, Y. S. Tang, J. Zou, X. Y. Huang, Y. F. Zhou, W. Jiang, G.-J. Zheng, and L. D. Chen, *J. Alloys Compd.* **527**, 247 (2012).

Dong, H. L., X. Y. Li, Y. Y. Zhou, W. Jiang, and L. D. Chen, *Ceram. Int.* **39**, 4551 (2013).

El-Genk, M. S., H. H. Saber, T. Caillat, and J. S. Sakamoto, *Energy Convers. Manage.* **47**, 174 (2006).

El-Genk, M. S., H. H. Saber, T. Caillat, and J. S. Sakamoto, *Energy Convers. Manage.* **48**, 1383 (2007).

Faleev, S. V. and F. Léonard, *Phys. Rev. B* **77**, 214304 (2008).

Fleurial, J.-P., T. Caillat, and A. Borshchevsky, *Proc. 13th Int. Conf. on Thermoelectrics*, *AIP Conf. Proc.* **316**, p. 40 (1995).

Fleurial, J.-P., A. Borshchevsky, T. Caillat, D. T. Morelli, and G. P. Meisner, *Proc. 15th Int. Conf. on Thermoelectrics*, IEEE Catalog Number 96TH8169, p. 91, Piscataway, NJ (1996).

García-Cañadas, J., A. V. Powell, A. Kaltzoglou, P. Vaqueiro, and G. Min, *J. Electron. Mater.* **42**, 1369 (2013).

Gayner, C. and Y. Amouyal, *Adv. Funct. Mater.* **30**, 1901789 (2020).

Geng, H. Y., T. Ochi, S. Suzuki, M. Kikuchi, S. Ito, and J. Q. Guo, *J. Electron. Mater.* **42**, 1999 (2013).

Godlewska, E, K. Zawadzka, A. Adamczyk, M. Mitoraj, and M. Mars, *Oxid. Met.* **74**, 113 (2010a).

Godlewska, E, K. Zawadzka, M. Mars, R. Mania, K. T. Wojciechowski, and A. Opoka, *Oxid. Met.* **74**, 205 (2010b).

Goldsmid, H. J., *Proc. Phys. Soc. London* **67**, 360 (1954).

Goldsmid, H. J. and R. W. Douglas, *Brit. J. Appl. Phys.* **5**, 386 (1954).

Goldsmid, H. J., *J. Electronics* **1**, 218 (1955).

Goldsmid, H. J., *Proc. Phys. Soc. London* **71**, 633 (1958).

Goldsmid, H. J., in *Applications of Thermoelectricity*, Methuen Monograph, London (1960).

Goldsmid, H. J., in *Electronic Refrigeration*, Pion, London (1986).

Goldsmid, H. J., in *Introduction to Thermoelectricity*, Springer, Berlin (2016).

Grytsiv, A., G. Rogl, E. Bauer, and P. Rogl, *J. Phase Equilib. Diffus.* (2020). doi:10.1007/s11669-020-00799-0

Guo, J. Q., H. Y. Geng, T. Ochi, S. Suzuki, M. Kikuchi, Y. Yamaguchi, and S. Ito, *J. Electron. Mater.* **41**, 1036 (2012).

Hara, R. S. Inoue, H. T. Kaibe, and S. Sano, *J. Alloys Compd.* **349**, 297 (2003).

Harman, T. C. and J. M. Honig, in *Thermoelectric and Thermomagnetic Effects and Applications*, McGraw-Hill, NY (1967).

Heremans, J. P., C. M. Thrush, and D. T. Morelli, *J. Appl. Phys.* **98**, 063703 (2005).

Heremans, J. P., V. Jovovic, E. S. Toberer, A. Saramat, K. Kurosaki, A, Charoenphakdee, S. Yamanaka, and G. J. Snyder, *Science* **321**, 554 (2008).

Heremans, J. P., B. Wiendlocha, and A. M. Chamoire, *Energy Environ. Sci.* **5**, 5510 (2012).

Hicks, L. D., and M. S. Dresselhaus, *Phys. Rev. B* **47**, 12727 (1993).

Hsu, K. F., S. Loo, F. Guo, W. Chen, J. S. Dyck, C. Uher, T. Hogan, E. K. Polychroniadis, and M. G. Kanatzidis, *Science* **303**, 818 (2004).

Hu, X. K., P. Jood, M. Ohta, M. Kunii, K. Nagase, H. Nishiate, M. G. Kanatzidis, and A. Yamamotou, *Energy & Environ. Sci.* **9**, 517 (2016).

Hui, S., M. D. Nielsen, M. R. Homer, D. L. Medlin, J. Tobola, J. R. Salvador, J. P. Heremans, K. P. Pipe, and C. Uher, *J. Appl. Phys.* **115**, 103704 (2014).

Ioffe, A. F, *Zhurnal Tekhnicheskoi Fiziki* **23**, 1452 (1953).

Ioffe, A. F., in *Semiconductor Thermoelements and Thermoelectric Cooling*, Infosearch, London (1957).

Ioffe, A. F. and L. S. Stilbans, *Rep. Prog. Phys.* **22**, 167–203 (1959).

Jaworski, C. M., V. A. Kulbachinski, and J. P. Heremans, *Phys. Rev. B* **80**, 233201 (2009).

Jeitschko, W. and D. J. Braun, *Acta Crystallog.* B 33, 3401 (1977).

Jie, Q., H. Z. Wang, W. S. Liu, H. Wang, G. Chen, and Z. F. Ren, *Phys. Chem. Chem. Phys.* **15**, 6809 (2013).

Katsuyama, S., W. Yamakawa, Y. Matsumura, and R. Funahashi, *J. Electron. Mater.* **48**, 5257 (2019).

Keppens, V., D. Mandrus, B. C. Sales, B. C. Chakoumakos, P. Dai, R. Coldea, M. B. Maple, D. A. Gajewski, E. J. Freeman, and S. Bennington, *Nature* **395** 876 (1998).

Khan, A. U., K. Kobayashi, D.-M. Tang, Y. Yamauchi, K. Hasegawa, M. Mitime, Y. M. Xue, B. Z. Jiang, K. Tsuchiya, D. Goldberg, Y. Bando, and T. Mori, *Nano Energy* **31**, 152 (2017).

Khovaylo, V. V., T. A. Korolkov, A. I. Voronin, M. V. Gorshenkov, and A. T. Burkov, *J. Mater. Chem. A* **5**, 3541 (2017).

Kim, H. S., W. S. Liu, G. Chen, C.-W. Chu, and Z. F. Ren, *Proc. Nat. Acad. Sci.* **112**, 8205 (2015).

Kim, S. I., K. H. Lee, H. A. Mun, H. S. Kim, S. W. Hwang, J. W. Roh, and D. J. Yang, *Science* **348**, 109 (2015).

Kishimoto, K. and T. Koyanagi, *J. Appl. Phys.* **92**, 2544 (2002).

Kjekshus, A., *Acta Chem. Scand.* **15**, 678 (1961).

Kjekshus, A. and T. Rakke, *Acta Chemica Scandinavica A* **28**, 99 (1974).

Kloc, C., K. Fess, W. Kaefer, K. Friemelt, H. Riazi-Najad, M. Wendl, and E. Bucher, *Proc. 15th Int. Conf. on Thermoelectrics*, IEEE Catalog Number 96TH8169, p. 155, Piscataway, NJ (1996).

Koga, T., X. Sun, S. B. Cronin, and M. S. Dresselhaus, *Appl. Phys. Lett.* **73**, 2950 (1998).

Koumoto, K. and T. Mori, in *Thermoelectric Nanomaterials: Materials Design and Application*, Springer (2013).

Koyanagi, T., T. Tsubouchi, M. Ohtani, K. Kishimoto, H. Anno, and K. Matsubara, *Proc. 15th Int. Conf. on Thermoelectrics*, IEEE Catalog Number 96TH8169, p. 107, Piscataway, NJ (1996).

Kraemer, A. C., M. R. Gallas, J. A. H. da Jornada, and C. A. Perottoni, *Phys. Rev. B* **75**, 024105 (2007).

Kuroki, T., K. Kabeya, K. Makino, T. Kajihara, H. Kaibe, H. Hachiuma, H. Matsuno, and A. Fujibayashi, *J. Electron. Mater.* **43**, 2405 (2014).

Kuznetsov, V. L., L. A. Kuznetsova, and D. M. Rowe, *J. Phys.: Condens. Matter* **15**, 5035 (2003).

Lampinen, M. J., *J. Appl. Phys.* **69**, 4318 (1991).

LeBlanc, S., S. K. Yee, M. L. Scullin, C. Dames, and K. E. Goodson, *Renew. Sustain. Energy Rev.* **32**, 313 (2014).

Ledbetter, H. M., *J. Appl. Phys.* **52**, 1587 (1981).

Lee, S., K. H. Lee, Y.-M. Kim, H. S. Kim, G. J. Snyder, S. Baik, and S. W. Kim, Acta Mater. **142**, 8 (2018).

Lee, Y. K., Z. Z. Luo, S. P. Cho, M. G. Kanatzidis, and I. Chung, *Joule* **3**, 719 (2019).

Lee, Y. S., S. J. Kim, B. G. Kim, S. Lee, W.-S. Seo, I.-H. Kim, and S.-M. Choi, *J. Electron. Mater.* **46**, 3083 (2017).

Leithe-Jasper, A., W. Schnelle, H. Rosner, M. Baenitz, A. Rabis, A. A. Gippius, E. N. Morozova, H. Borrmann, U. Burkhardt, R. Ramlau, U. Schwarz, J. A. Mydosh, and Y. Grin, *Phys. Rev. B* **70**, 214418 (2004).

Leszczynski, J., K. T. Wojciechowski, and A. L. Malecki, *J. Therm. Anal. Calorim.* **105**, 211 (2011).

Li, G. D., Q. An, W. A. Goddard III, R. Hanus, P. C. Zhai, Q. J. Zhang, and G. J. Snyder, *Acta Mater.* **103**, 775 (2016).

Li, H., X. F. Tang, Q. J. Zhang, and C. Uher, *Appl. Phys. Lett.* **94**, 102114 (2009).

Li, W. J., G. D. Li, X. Q. Yang, L. S. Liu, and P. C. Zhai, *J. Electron. Mater.* **44**, 1477 (2015).

Liu, W., X. Tan, K. Yin, H. Liu, X. F. Tang, J. Shi, Q, Zhang, and C. Uher, *Phys. Rev. Lett.* **108**, 166601 (2012).

Liu, W., J. Zhou, Q. Jie, Y. Li, H.-S. Kim, J. Bao, G. Chen, and Z. F. Ren, *Energy Environ. Sci.* **9**, 530 (2016).

Liu, Y., P. Sahoo, J. Makongo, X. Zhou, S.-J. Kim, H. Chi, C. Uher, X. Pan, and P. F. P. Poudeu, *J. Amer. Chem. Soc.* **135**, 7486 (2013).

Mahan, G. D., *J. Appl. Phys.* **65**, 1578 (1989).

Mahan, G. D., *J. Appl. Phys.* **70**, 4551 (1991).

Mahan, G. D., in *Solid State Physics*, Vol. 51, p. 82, Academic Press (1997).

Mallik, R. C., R. Anbalagan, G. Rogl, E. Royanian, P. Heinrich, E. Bauer, P. Rogl, and S. Suwas, *Acta Mater.* **61**, 6698 (2013).

Mandrus, D., A. Magliori, T. W. Darling, M. F. Hundley, E. J. Peterson, and J. D. Thompson, *Phys. Rev. B* **52**, 4926 (1995).

Mandrus, D., B. C. Sales, V. Keppens, B. C. Chakoumakos, P. Dai, L. A. Boatner, R. K. Williams, J. R. Thompson, T. W. Darling, A. Migliori, M. B. Maple, D. A. Gajewski, and E. J. Freeman, *Mater. Res. Soc. Symp. Proc.* **478**, 199 (1997).

Matsubara, K., T. Sakakibara, Y. Notohara, H. Anno, H. Shimizu, and T. Koyanagi, *Proc. 15th Int. Conf. on Thermoelectrics*, IEEE Catalog Number 96TH8169, p. 107, Piscataway, NJ, (1996).

Matsubara, M. and R. Asahi, *J. Electron. Mater.* **45**, 1669 (2016).

Migliori, A. and J. D. Maynard, *Rev. Sci. Instrum.* **76**, 121301 (2005).

Min, G., D. M. Rowe, and K. Kontostavlakis, *J. Appl. Phys. D-Appl. Phys.* **37**, 1301 (2004).

Morelli, D. T. and G. P. Meisner, *J. Appl. Phys.* **77**, 3777 (1995).

Morelli, D. T., T. Caillat, J.-P. Fleurial, J. Vandersande, B. Chen, and C. Uher, *Phys. Rev. B* **51**, 9622 (1995).

Muto, A., J. Yang, B. Poudel, Z. F. Ren, and G. Chen, *Adv. Energy Mater.* **3**, 245 (2013).

Narducci, D., E. Selezneva, G. Cerofolini, S. Frabboni, and G. Ottaviani, *J. Solid State Chem.* **193**, 19 (2012).

Nie, G., S. Suzuki, T. Tomida, A. Sumiyoshi, T. Ochi, K. Mukaiyama, M. Kikuchi, J. Q. Guo, A. Yamamoto, and H. Obara, *J. Electron. Mater.* **46**, 2640 (2017).

Nie, G., W. J. Li, J. Q. Guo, A. Yamamoto, K. Kimura, X. Zhang, E. B. Isaacs, V. Dravid, C. Wolverton, M. G. Kanatzidis, and S. Priya, *Nano Energy* **66**, 104193 (2019).

Nieroda, P., K. Kutorasinski, J. Tobola, and K. T. Wojciechowski, *J. Electron. Mater.* **43**, 1 (2013).

Niino, M., in *Research Report on the Hybrid System of Solar-Thermo Power Generator System*, National Aerospace Laboratory, Japan, (2002) (in Japanese).

Nolas, G. S., G. A. Slack, D. T. Morelli, T. M. Tritt, and A. C. Ehrlich, *J. Appl. Phys.* **79**, 4002 (1996).

Nolas, G. S., J. Sharp, and H. J. Goldsmid, in *Thermoelectrics: Basic Principles and New Materials Developments*, Springer, p. 292 (2001).

Ochi, T., G. Nie, S. Suzuki, M. Kikuchi, S. Ito, and J. Q. Guo, *J. Electron. Mater.* **43**, 2344 (2014).

Olvera, A. A., N. A. Moroz, P. Sahoo, P. Ren, T. P. Bailey, A. A. Page, C. Uher, and P. F. P. Poudeu, *Energy Environ. Sci.* 10, 1668 (2017).

Park, C.-H., and Y.-S. Kim, *Phys. Rev. B* **81**, 085206 (2010).

Park, K.-H., S.-W. You, S.-C. Ur, I.-H. Kim, S.-M. Choi, and W.-S. Seo, *J. Electron. Mater.* **41**, 1051 (2012).

Park, Y.-S., T. Thompson, Y. Kim, J. R. Salvador, and J. S. Sakamoto, *J. Mater. Sci.* **50**, 1500 (2015).

Park, S. H., Y. H. Jin, and J. Cha, K. Hong, Y. S. Kim, H. Yoon, C.-Y. Yoo, and I. Chung, *ASC Appl. Energy Mater.* **1**, 1603 (2018).

Park, Y.-S., T. Thompson, Y. Kim, J. R. Salvador, and J. S. Sakamoto, *J. Mater. Sci.* **50**, 1500 (2015).

Pei, Y. Z., Jeong Yang, L. D. Chen, W. Q. Zhang, J. R. Salvador, and Jihui Yang, *Appl. Phys. Lett.* **95**, 042101 (2009).

Pei, Y. Z., X. Shi, A. LaLonde, H. Wang, L. D. Chen, and G. J. Snyder, *Nature* **473**, 66 (2011).

Peltier, J. C., *Ann. Chem.* LVI, 371 (1834).

Price, P. I., *Phil. Mag.* **46**, 1252 (1955).

Price, P. I., *Phys. Rev.* **102**, 1245 (1956).

Ravi, V., S. Firdosy, T. Caillat, B. Lerch, A. Calamino, R. Pawlik, M. Nathal, A. Sechrist, J. Buchhalter, and S. Nutt, *American Institute of Physics (AIP) Conf. Proc.* **969**, 656 (2008)

Recknagel, C., N. Reinfried, P. Höhn, W. Schnelle, H. Rosner, Yu. Grin, A. Leithe-Jasper, *Sci. Technol. Adv. Mater.* **8** 357 (2007).

Rice, R. W. in *Mechanical Properties of Ceramics and Composites: Grains and Particle Effects*, Marcel-Dekker, N.Y. (2000).

Rittner, E. S., *J. Appl. Phys.* **30**, 702 (1959).

Roberts, R. B., *Nature* **265**, 226 (1977).

Rogl, G., A. Grytsiv, P. Rogl, E. Bauer, M. B. Kerber, M. Zehetbauer, and S. Puchegger, *Intermetallics* **18**, 2435 (2010).

Rogl, G., L. Zhang, P. Rogl, A. Grytsiv, M. Falmbigl, D. Rajs, M. Kriegisch, H. Müller, E. Bauer, J. Koppensteiner, W. Schranz, M. Zehetbauer, Z. Henkie, and M. B. Maple, *J. Appl. Phys.* **107**, 043507 (2010b).

Rogl, G., A. Grytsiv, P. Rogl, E. Bauer, and M. Zehetbauer, *Intermetallics* **19**, 546 (2011).

Rogl, G., A. Grytsiv, P. Rogl, N. Peranio, E. Bauer, M. Zehetbauer, and O. Eibl, *Acta Mater.* **63**, 30 (2014).

Rogl, G., A. Grytsiv, P. Heinrich, E. Bauer, P. Kumar, N. Peranio, O. Eibl, J. Horky, M. Zehetbauer, and P. Rogl, *Acta Mater.* **91**, 227 (2015a).

Rogl, G., A. Grytsiv, K. Yubuta, S. Puchegger, E. Bauer, C. Raja, R. C. Malik, and P. Rogl, *Acta Mater.* **95**, 201 (2015b).

Rogl, G., A. Grytsiv, F. Failamani, M. Hochenhofer, E. Bauer, and P. Rogl, *J. Alloys Compd.* **695**, 682 (2017).

Rossi, F. D., *Solid State Electron.*, **11**, 833–868 (1968).

Rowe, D. M. and C. M. Bhandari, in *Modern Thermoelectrics*, Holt Technology (1983).

Rowe, D. M., ed., *Thermoelectrics Handbook, Macro to Nano*, CRC press, Taylor & Francis, Boca Raton, FL (2006).

Ruan, Z.-W., L.-S. Liu, P.-C. Zhai, P.-F. Wen, and Q.-J. Zhang, *J. Electron. Mater.* **39**, 2029 (2010).

Sakamoto, J. S., H. Schock, T. Caillat, J.-P. Fleurial, R. Maloney, M. Lyle, T. Ruckle, E. Timm, and L. Zhang, *Sci. Adv. Mater.* **3**, 621 (2011).

Sales, B. C., D. Mandrus, and R. K. Williams, *Science* **272**, 1325 (1996).

Sales, B. C., D. Mandrus, B. C. Chakoumakos, V. Keppens, and J. R. Thompson, *Phys. Rev. B* **56**, 15081 (1997).

Sales, B. C., B. C. Chakoumakos, and D. Mandrus, *Phys. Rev. B* **61**, 2475 (2000).

Salvador, J. R., J. Yang, H. Wang, and X. Shi, *J. Appl. Phys.* **107**, 043705 (2010).

Salvador, J. R., J. Y. Cho, Z. X. Ye, J. E. Moczygemba, A. J. Thompson, J. W. Sharp, J. D. König, R. Maloney, T. Thompson, J. Sakamoto, H. Wang, A. A. Wereszczak, and G. P. Meisner, *J. Electron. Mater.* **42**, 1389 (2013).

Salvador, J. R., J. Y. Cho, Z. X. Ye, J. E. Moczygemba, A. J. Thompson, J. W. Sharp, J. D. König, R. Maloney, T. Thompson, J. Sakamoto, H. Wang, A. A. Wereszczak, *Phys. Chem. Chem. Phys.* **16**, 12510 (2014).

Schmidt, R. D., E. D. Case, J. E. Ni, J. S. Sakamoto, R. M. Trejo, E. Lara-Curzio, E. A. Payzant, M. J. Kirkham, and R. A. Peascoe-Meisner, *Phyl. Mag.* **92**, 1261 (2012).

Seebeck, T, J., Abhand Deut. Akad. Wiss. Berlin, 265 (1822).

Schwall, M. and B. Balke, *Phys. Chem. Chem. Phys.* **15**, 1868 (2013).

Sharp, J. W., E. C. Jones, R. K. Williams, P. M. Martin, and B. C. Sales, *J. Appl. Phys.* **78**, 1013 (1995).

Sherman, B., R. R. Heikes, and R. W. Ure, Jr., *J. Appl. Phys.* **31**, 1 (1960).

Shi, X., J. Yang, J. R. Salvador, M. F. Chi, J. Y. Cho, H. Wang, S. Q. Bai, J. H. Yang, W. Q. Zhang, and L. D. Chen, *J. Amer. Chem. Soc.* **133**, 7837 (2011).

Shi, X., J. Yang, L. J. Wu, J. R. Salvador, C. Zhang, W. L. Villaire, D. Haddad, J. Yang, Y. M. Zhu, and Q. Li, *Sci. Rep.* **5**, 14641 (2015).

Simon, R., *J. Appl. Phys.* **33**, 1830 (1962).

Singh, A., S. Bhattacharya, C. Thinaharan, D. K. Aswal, S. K. Gupta, J. V. Yakhmi, and K. Bhanumurthy, *J. Phys. D: Appl. Phys.* **42**, 15502 (2009).

Sklad, A. C., M. W. Gaultois, and A. P. Grosvenor, *J. Alloys Compd.* **505**, L6 (2010).

Slack, G. A. and V. G. Tsoukala, *J. Appl. Phys.* **76**, 1665 (1994).

Slack, G. A., in *CRC Handbook of Thermoelectrics*, ed. D. M. Rowe, CRC Press, Boca Raton, FL (1995).

Snider, T. S., J. V. Badding, S. B. Schujman, and G. A. Slack, *Chem. Mater.* **12**, 697 (2000).

Snyder, G. J. and T. Caillat, *TE Workshop*, San Diego, Feb. 17–20 (2004).

Snyder, G. J., in *Thermoelectrics Handbook, Macro to Nano*, ed. by D. M. Rowe, Ch. 9, Taylor & Francis, Boca Raton, FL, (2006).

Snyder, G. J. and T. Ursell, *Phys. Rev. Lett.* **91**, 148301 (2003).

Sofo, J. O. and G. D. Mahan, *Phys. Rev. B* **49**, 4565 (1994).

Sui, J. H., J. Li, J. Q. He, Y.-L. Pei, D. Berardan, H. J. Wu, N. Dragoe, W. Caia, and L.-D. Zhao, *Energy Environ. Sci.* **6**, 2916 (2013).

Sunderland, J. E. and N. T. Burak, *Solid State Electron.* **7**, 465 (1964).

Tafti, M. Y., M. Saleemi, M. Johnsson, A. Jacquot, and M. S. Toprak, *Mater. Res. Soc. Symp. Proc.* Vol. 1735, 14 (2015).

Tan, G. J., F. Y. Shi, S. Q. Hao, L.-D. Zhao, H. Chi, X. M. Zhang, C. Uher, C. Wolverton, V. P. Dravid, and M. G. Kanatzidis, *Nature Commun.* **7**, 12167 (2016).

Tang, X. F., H. Li, Q. J. Zhang, M. Niino, and T. Goto, *J. Appl. Phys.* **100**, 123702 (2006).

Tang, Y., Z. M. Gibbs, L. A. Agapito, G. D. Li, H.-S. Kim, M. B. Nardelli, S. Curtarolo, and G. J. Snyder, *Nature Materials* **14**, 1223 (2015).

Thompson, D. R., C. Liu, Jiong Yang, J. R. Salvador, D. B. Haddad, N. D. Ellison, R. A. Waldo, and Jihui Yang, *Acta Mater.* **92**, 152 (2015).

Thomson, W., *Proc. Roy. Soc. Edinburgh* 91 (1851).

Uher, C., *J. Appl. Phys.* **62**, 4636 (1987).

Uher, C., B. Chen, S. Hu, D. T. Morelli, and G. P. Meisner, *Mater. Res. Soc. Symp. Proc.* Vol. 478, 315 (1997).

Venkatasubramanian, R., R. E. Siivola, T. Colpitts, and B. O'Quinn, Nature **413**, 597 (2001).

Wang, H., R. McCarty, J. R. Salvador, A. Yamamoto, and J. König, *J. Electron. Mater.* **43**, 2274 (2014).

Wang, S., J. R. Salvador, Jiong Yang, P. Wei, B. Duan, and Jihui Yang, *NPG Asia Mater.* **8**, e285 (2016).

Wang, X. W., H. Lee, Y. C. Lan, G. H. Zhu, G. Joshi, D. Z. Wang, J. Yang, A. J. Muto, M. Y. Tang, J. Klatsky, S. Song, M. S. Dresselhaus, G. Chen, and Z. F. Ren, *Appl. Phys. Lett.* **93**, 193121 (2008).

Wasscher, J. D., W. Albers, and C. Haas, *Solid State Electron.* **6**, 261 (1963).

Wei, P., W. Y. Zhao, C. L. Dong, X. Yang, J. Yu, and Q. J. Zhang, *Acta Mater.* **59**, 3244 (2011).

Wei, P., W. Y. Zhao, D. G. Tang, W. T. Zhu, X. L. Nie, and Q. J. Zhang, *J. Materiomics* **2**, 280 (2016).

Wen, P. F., P. Li, Q. J. Zhang, F. J. Yi, L. S. Liu, and P. C. Zhai, *J. Electron. Mater.* **38**, 1200 (2009).

Wen, P. F., B. Duan, P. C. Zhai, P. Li, and Q. J. Zhang, *J. Mater. Sci.: Mater. Electron.* **24**, 5155 (2013).

Wen, P. F., Y. Zhu, J. Chen, H. J. Yang, and P. C. Zhai, *J. Mater. Engin. Perf.* **25**, 4764 (2016).

Wilson, L. J. and S. A. Mikhail, *Thermochimica Acta* **156**, 107 (1989).

Wojciechowski, K. T., J. Leszczynski, and R. Gajerski, *Proc. 7th Europ. Workshop on Thermoelectrics*, Pamplona, Spain (2002).

Wood, C., Rep. Prog. Phys., **51**, 459 (1988).

Xia, X. G., P. F. Qiu, X. Shi, X. Y. Li, X. Y. Huang, and L. D. Chen, *J. Electron. Mater.* **41**, 2225 (2012).

Xing, Y. F., R. H. Liu, J. C. Liao, Q. H. Zhang, X. G. Xia, C. Wang, H. Huang, J. Chu, M. Gu, T. J. Zhu, C. X. Zhu, F. F. Xu, D. X. Yao, Y. P. Zeng, S. Q. Bai, C. Uher, and L. D. Chen, *Energy Environ. Sci.* **12**, 3390 (2019).

Xiong, Z., L. Xi, J. Ding, X. H. Chen, X. Y. Huang, H. Gu, L. D. Chen, and W. Q. Zhang, *J. Mater. Res.* **26**, 1848 (2011).

Yamashita, O., *Energy Convers. Manag.* **49**, 3163 (2008).

Yang, Jiong, L. Xi, W. Zhang, L. D. Chen, and J. Yang, *J. Electron. Mater.* **38**, 1397 (2009).

Yang, X. Q., P. C. Zhai, L. S. Liu, and Q. J. Zhang, *J. Appl. Phys.* **109**, 123517 (2011).

Yee, S. K., S. LeBlanck, K. E. Goodson, and C. Dames, *Energy Environ. Sci.* **6**, 2561 (2013).

Zabrocki, K., E. Müller, W. Seifert, and S. Trimper, *J. Mater. Res.* **26**, 1963 (2011).

Zaitsev, V. K., M. I. Fedorov, E. A. Gurieva, I. S. Eremin, P. P. Konstantinov, A. Yu. Samunin, and M. V. Vedernikov, *Phys. Rev. B* **74**, 045207 (2006).

Zawadzka, K., E. Godlewska, K. Mars, and M. Nocun, *AIP Conf. Proc.* **1449**, 231 (2012).

Zevalkink, A., K. Star, U. Aydemir, G. J. Snyder, J.-P. Fleurial, S. Bux, T. Vo, and P. von Allmen, *J. Appl. Phys.* **118**, 035107 (2015).

Zhang, L., G. Rogl, A. Grytsiv, S. Puchegger, J. Koppensteiner, F. Spieckermann, H. Kabelka, M. Reinecker, P. Rogl, W. Schranz, M. Zehetbauer, and M. A. Carpenter, Mater. Sci. Engin. B **170**, 26 (2010).

Zhang, J. J., B. Xu, J.-M. Wang, D. L. Yu, Z. Y. Liu, J. L. He, and Y. J. Tian, *Appl. Phys. Lett.* **98**, 072109 (2011).

Zhang, L., F. F. Duan, Z. D. Li, X. L. Yan, W. T. Hu, L. M. Wang, Z. Y. Liu, Y. J. Tian, and B. Xu, *J. Appl. Phys.* **114**, 083715 (2013).

Zhang, Q. H., J. C. Liao, Y. S. Tang, M. Gu, C. Ming, P. F. Qiu, S. Q. Bai, X. Shi, C. Uher, and L. D. Chen, *Energy Environ. Sci.* **10**, 956 (2017).

Zhang, Q. J., X. F. Tang, P. C. Zhai, M. Niino, and C. Endo, *Mater. Sci. Forum* **492**, 135 (2005).

Zhao, D. G., C. W. Tian, S. Q. Tang, Y. T. Liu, and L. D. Chen, *J. Alloys Compd.* **504**, 552 (2010a).

Zhao, D. G., C. W. Tian, S. Q. Tang, Y. T. Liu, L. K. Jiang, and L. D. Chen, *Mater. Sci. in Semicond. Process.* **13**, 221 (2010b).

Zhao, D. G., C. W. Tian, Y. T. Liu, C. W. Zhan, and L. D. Chen, *J. Alloys Compd.* **509**, 3166 (2011).

Zhao, D. G., M. Zuo, Z. Q. Wang, X. Y. Teng, and H. R. Geng, *Appl. Surf. Sci.* **305**, 86 (2014).

Zhao, L.-D., S.-H. Lo, Y. S. Zhang, H. Sun, G. J. Tan, C. Uher, C. Wolverton, V. P. Dravid, and M. G. Kanatzidis, *Nature* **508**, 373 (2014).

Zhao, W. Y., Z. Y. Liu, P. Wei, Q. J. Zhang, W. T. Zhu, X. L. Su, X. F. Tang, J. Yang, Y. Lin, J. Shi, Y. M. Chao, S. Q. Lin, and Y. Z. Pei, *Nature Nanotechnology* **12**, 55 (2017).

Zhou, L., P. F. Qui, C. Uher, X. Shi, and L. D. Chen, *Intermetallics* **32**, 209 (2013).

Zobrina, B. N. and L. D. Dudkin, *Sov. Phys. Solid State* **1**, 1668 (1960).

Zong, P. A., R. Hanus, M. Dylla, Y. S. Tang, J. C. Liao, Q. H. Zhang, G. J. Snyder, and L. D. Chen, *Energy Environ. Sci.* **10**, 183 (2017).

Index